Living in a Seasonal World

Thomas Ruf · Claudia Bieber
Walter Arnold · Eva Millesi
Editors

Living in a Seasonal World

Thermoregulatory and Metabolic Adaptations

Springer

Thomas Ruf
Department of Integrative Biology and Evolution
University of Veterinary Medicine
Savoyenstr. 1
1160 Vienna, Austria

Walter Arnold
Department of Integrative Biology and Evolution
University of Veterinary Medicine
Savoyenstr. 1
1160 Vienna, Austria

Claudia Bieber
Department of Integrative Biology and Evolution
University of Veterinary Medicine
Savoyenstr. 1
1160 Vienna, Austria

Eva Millesi
Department of Behavioural Biology
University of Vienna
Althanstr. 14
1090 Vienna, Austria

ISBN 978-3-642-28677-3 e-ISBN 978-3-642-28678-0
DOI 10.1007/978-3-642-28678-0
Springer Heidelberg New York Dordrecht London

Library of Congress Control Number: 2012935676

Printed on acid-free paper

Springer is part of Springer Science+Business Media (www.springer.com)

Preface

"Climate plays an important part in determining the average numbers of a species, and periodical seasons of extreme cold or drought, I believe to be the most effective of all checks." With this remark in his book "On the Origin of Species", Charles Darwin (1859) identified the seasonality of environments as one of the most important forces in natural selection. Darwin was well aware that "characters appearing periodically at different seasons" are widespread among animals (ibid.), and that many species, including those in the tropics, respond to unfavourable conditions by using torpor and "hybernation". Darwin was, however, largely ignorant about the physiological mechanisms underlying these phenomena, as physiology was still in its infancy during his time. Only a few years later though, in 1865, Claude Bernard described several fundamental characteristics of hibernation, and in the 1870s Alexis Horvath wrote a series of influential papers on hibernation in European rodents. In 1896 Raphael Dubois then published a seminal study on the physiology of the hibernating marmot. Still, work on this topic remained sporadic in the late nineteenth and early twentieth century. In the 1930s and 1940s research on hibernation and torpor notably picked up, however, largely due to the highly productive and insightful work of Martin Eisentraut, Paavo Suomalainen, and Charles Kayser. Research in the field gained even more momentum in the late 1940s and 1950s when these early pioneers were joined by other physiologist, such as Charles P. Lyman, Peter Morrison, Edgar Folk, or Felix Strumwasser, to name but a few. It was only logical then, that all leading researchers in the field as well as "newcomers" (such as George A. Bartholomew who had started to study dormancy only a few years earlier) and graduate students (e.g., Michael Menaker who received his Ph.D. later in the same year) would present and discuss their work during the "First International Symposium on Natural Mammalian Hibernation" which was held at Massachusetts Institute of Technology's Endicott House in Dedham, Mass., from May 13–15, 1959.

Since 1959, International Hibernation Symposia were held at 3–5 year intervals at changing locations throughout the world. This book contains the proceedings of the 14th International Hibernation Symposium, held from August 9–13, 2012 in Semmering, Austria. The peer-reviewed chapters of this book provide a comprehensive overview over the current state of research on not only torpor and hibernation, but seasonal adaptations of animals in general. This broadening of topics reflects a general trend in this conference series. Whereas the first symposium in 1959 was largely dedicated to hibernation, subsequent meetings have continuously increased the scope, and included other aspects of energetics and thermoregulation, biochemical mechanisms, endogenous clocks, signal processing, life-history tactics and further ecological topics. As in the present book, the range of animal models studied was also extended to include not just mammalian species, but also birds, reptiles, and invertebrates.

Due to this steady extension of topics, the title "Mammalian Hibernation" eventually seemed too narrow. Hence, starting with the 7th symposium in 1985, the name of the meeting and the book series was changed to "Living (or Life) in the Cold". As pointed out by the editors of the previous book in this series (the proceedings of the 13th International Hibernation Symposium), this title was however somewhat unfortunate, as it implies a restriction to adaptations of animals exposed to cold temperatures in temperate or arctic climates. This would be misleading, as even the nineteenth century naturalists, such as Darwin, were already well aware that there are numerous tropical animals that display seasonal adjustments of morphology, behaviour and physiology, including states of dormancy. Therefore, the name "Life in the Cold" was given up in 2008, at the first meeting of this series held on the African continent.

The title of this book is "Living in a seasonal world: thermoregulatory and metabolic adaptations". The term "seasonal" is by no means meant to refer only to the "four" seasons in temperate or polar regions, which are governed by changes in the amount of sunlight that reaches the Earth's surface, and hence in temperature and photoperiod. It also includes the alternation between wet and dry periods in tropical and subtropical regions, temporal changes in food and water availability due to various reasons, and even irregular, unpredictable fluctuations of the environment. This book, as well as the previous books in the series, describes and discusses the fascinating adaptations of animals to all of the above facets of seasonality. It is this continuous physiological response to changing environmental conditions—a phenomenon called "rheostasis" by Nicholas Mrosovsky, another regular contributor to this symposium series—which is at the core of the proceedings summarized here.

This book was compiled within a few months only, with the help of many contributors. First, we wish to thank all authors for their work, and for their fast responses to editorial requests. We also thank all reviewers, from both within and outside the participants group, for ensuring a high standard of scientific quality of the contents of this book. Further, we thank Ursula Gramm, Brigitte Jandl, Maria Leitgeb, Thomas Paumann, Christian Schwarz, and Barbara Wyatt for their help in organising the symposium and producing this book. Finally, we thank the Austrian Federal Ministry of Science and Research for financial support.

This book is dedicated to:
E. F. Adolph, E.G. Ball, G. A. Bartholomew, G. H. Bishop, B. Brattstrom, M. A. Brock, R.W. Bullard, A. R. Dawe, W. Dawson, H. A. Denyes, M. Eisentraut, K. C. Fisher, E. Flamboe, G. E. Folk Jr., D. R. Griffin, R. J. Hock, B. Johansson, N. I. Kalabukhov, F. G. Kallen, C. Kayser, T. H. Kehl, B. R. Landau, B. J. Luyet, C. P. Lyman, W. V. Mayer, M. Menaker, M. P. Meyer, P. Morrison, X. J. Musacchia, O. P. Pearson, E. T. Pengelley, V. Popovic, C. L. Prosser, M. L. Riedesel, T. K. Ruebush, H. S. S. Sarajas, E. Schönbaum, D. E. Smith, F. E. South, J. Steen, F. Strumwasser, P. Suomalainen, J. S. Willis, W. A. Wimsatt, and M. L. Zimny; the participants of the "First International Symposium on Natural Mammalian Hibernation", Dedham, Massachusetts, May 13–15, 1959.

Thomas Ruf
Claudia Bieber
Walter Arnold
Eva Millesi

Contents

Part I
Evolution and Ecophysiology of Torpor

Chapter 1
A Single Origin of Heterothermy in Mammals

Barry G. Lovegrove

Abstract Some mammal lineages survived the global fires that occurred in the hours following the asteroid impact at Chicxulub, Yucatan, at the Cretaceous/Tertiary Boundary (K/T Boundary) 65 mya. Several studies have proposed that it was the capacity for torpor and refuge underground, in tree holes, caves, and underwater, that ensured the short- and long-term survival of the post-impact conditions. Here I test the hypothesis that heterothermy was a pleisiomorphic condition in ancestral mammals which allowed certain mammal lineages to survive the K/T Boundary. I employed a maximum likelihood approach to reconstruct the likely heterothermic status of the last mammalian ancestor. With our current knowledge, the probability of heterothermy (58%) slightly exceeds that of no heterothermy. However, if some mammals that have yet to be studied, but which have been identified as highly likely heterotherms, are scored as heterotherms, the proportional likelihood of heterothermy in ancestral mammals exceeds the 96% probability. At the least, these data confirm that there was single origin of heterothermy in mammals, but further research is required to determine how extensive heterothermy was in Mesozoic mammals.

1.1 Introduction

The thermal radiation produced by the ballistic re-entry of ejecta after the 10 km wide comet or asteroid that slammed into Chicxulub on the Yucatan Peninsula in Mexico at the K/T Boundary has been estimated to have increased the global

B. G. Lovegrove (✉)
School of Life Sciences, University of KwaZulu-Natal,
Private Bag X01, Scottsville 3209, South Africa
e-mail: lovegrove@ukzn.ac.za

T. Ruf et al. (eds.), *Living in a Seasonal World*,
DOI: 10.1007/978-3-642-28678-0_1, © Springer-Verlag Berlin Heidelberg 2012

infrared radiation by 50–150 times the solar input for several hours (Huber 2008; Melosh et al. 1990). The intensity of the radiation ignited global wildfires that produced the charcoal and soot found across the globe in K/T Boundary deposits (Melosh et al. 1990). Ozone (O_3), and nitric and sulfuric acid rain could also have reached lethal physiological levels for several hours following the impact (Kikuchi and Vanneste 2010; Kring 2007). Equivalent to being exposed to radiation from an oven set at "broil" (Melosh et al. 1990), the immediate post-impact conditions literally cooked all animals and most non-aquatic plants that were exposed above the ground (Robertson et al. 2004). Moreover, in the months that followed the impact, soot and dust prevented sunlight from reaching the Earth's surface leading to surface cooling by several to a few tens of degrees, and the curtailment of photosynthesis. The cooling period was probably followed by greenhouse warming following the addition of huge amounts of CO_2, CH_4, and H_2O that were added to the atmosphere (Kring 2007). These secondary conditions were very challenging to a sustained terrestrial existence.

The differential survival/extinction of Maastrichtian vertebrates at the K/T boundary has been a source of intense debate, especially among palaeontologists (MacLeod 1996; Robertson et al. 2004). The central question is why some animal groups died out, whereas others did not. All large animals that could not avoid the aboveground conditions went extinct; such as the non-avian dinosaurs, pterosaurs, some turtles and lizards, archaic birds, and some eutherian and placental mammals (Robertson et al. 2004). However, many mammals survived the impact, such as the monotreme, egg-laying mammals, marsupials, placentals, and several groups that later went extinct during the Cenozoic (Gondwanatherians, Multituberculates, and Dryolestoids) (Robertson et al. 2004).

The majority of the Maastrichtian mammal survivors were smaller than 1 kg (Luo 2007; Robertson et al. 2004). It has been argued that these small mammals were able to avoid the lethal aboveground conditions immediately following the impact by gaining refuge in burrows, caves, trees holes, or by being submerged under water (Kikuchi and Vanneste 2010; Robertson et al. 2004). Although these mammals survived the immediate impact, they also survived the longer term effects, such as persistent cold and lack of photosynthesis-dependent resources. Theoretically, it is much more difficult for endothermic mammals to endure long periods of limited energy intake compared with ectotherms, such as frogs, lizards, crocodiles, and snakes. Mammalian basal metabolic rate (BMR), the lowest level of metabolism that can sustain life, is considerably higher than it is in ectotherms (Schmidt-Nielsen 1983). Moreover, at low ambient temperatures, mammals need to produce endogenous heat by increasing metabolic rate above the basal rate to offset heat losses in order to defend a constant body temperature.

Several authors have proposed the hypothesis that mammals could have survived the long-term effects by employing torpor (Kikuchi and Vanneste 2010; Robertson et al. 2004). Daily torpor and hibernation profoundly reduce a mammal's energy demands because body temperatures are reduced well below normothermic levels (Geiser and Ruf 1995), in the most extreme case, to −2°C in the Arctic ground squirrel (Barnes 1989). From a palaeontological perspective, we have no way of verifying whether Maastrichtian mammals were heterothermic because fossils provide very limited information on ancestral physiological states (although see Eagle et al. 2010).

Based upon the ontogeny of endothermy, Geiser (2008) suggested that heterothermy evolved independently at least three times in birds and mammals. He argues that heterothermy was pleisiomorphic in birds, evolved independently in marsupials, and is a derived characteristic in placental mammals. Geiser (2008) did not provide any geological time frames for these events, so I presume that, in the case of placental mammals, he is suggesting a Cenozoic evolution of heterothermy in placental mammals.

Here I test the hypothesis that the immediate ancestors of the marsupials and eutherians were heterothermic by utilizing macrophysiological data from modern, crown small mammals (<1 kg) to reconstruct the ancestral state using a maximum likelihood (ML) approach. I constructed a phylogeny for all mammalian families in which the species are <1 kg. In modern mammals the vast majority of modern heterotherms are <1 kg, like their Maastrichtian ancestors. Generally, once mammals exceed 1 kg, they lose the capacity for heterothermy (Geiser 1998; Lovegrove 2012).

1.2 Methods

All data were obtained from Lovegrove (2012). The phylogeny of mammal families were obtained from; Orders (Bininda-Emonds et al. 2007), marsupials (Meredith et al. 2008; Meredith et al. 2009), Xenarthra (Delsuc et al. 2002; Delsuc et al. 2003), Afrotheria (Asher and Helgen 2010; Murphy et al. 2007; Tabuce et al. 2008), Rodentia (Montgelard et al. 2008; Steppan et al. 2004), Primates (Purvis 1995), Eulipotyphla (Douady and Douzery 2003; Dubey et al. 2007), and Chiroptera (Teeling et al. 2005). Binomial character states (1 = observed heterothermy, 0 = no observed heterothermy) were assigned to 72 extant mammalian families. The character history of the binomial variables was traced to the root of the phylogeny using the method of maximum likelihood and a Markov k-state 1 parameter (Mk1) probability model (Maddison and Maddison 2009). The maximum likelihoods are reported as proportional likelihoods (0–1). Two families that were considered homeotherms in the Lovegrove (2011) database, Rhonopomatidae and Sorininae, were coded as heterotherms following Kulzer (1965) and Newman and Rudd (1978).

1.3 Results and Discussion

The proportional ML (0–1) of heterothermy at the root of the phylogeny of 72 mammal families, representing nine of the 11 mammalian orders (Rodentia, Eulipotyphla, Chiroptera, Primates, Scandentia, Dermoptera, Marsupials,

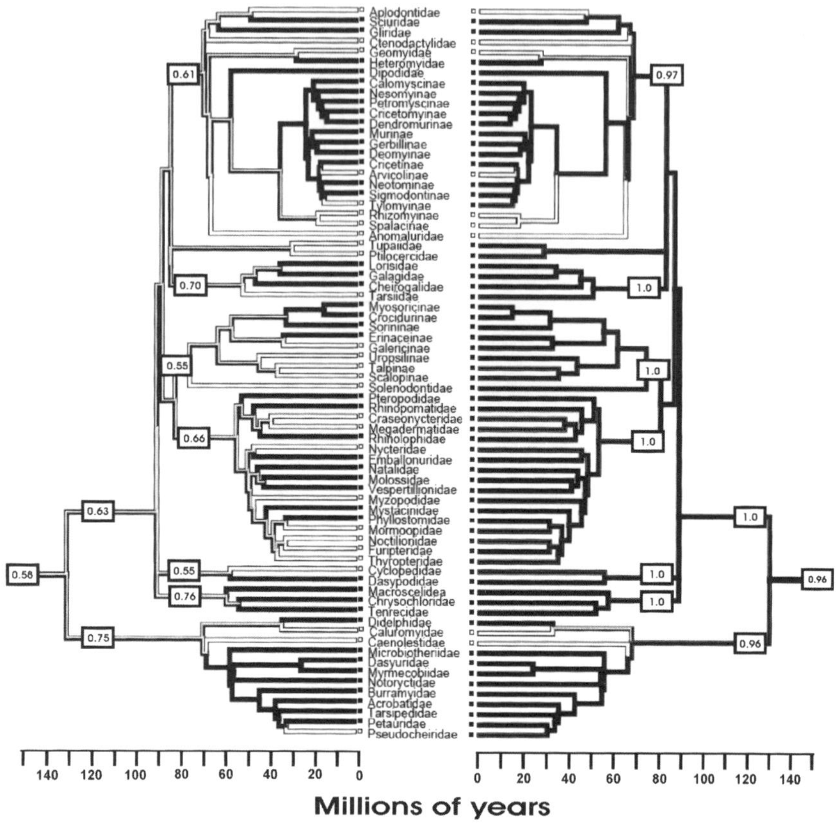

Fig. 1.1 The phylogeny of 72 families of mammals in which at least one species within each family is heterothermic (employs daily torpor and/or hibernation) and is <1 kg. Species on the left phylogeny are coded as either heterothermic (filled branch lengths) or non-heterothermic (clear branch lengths) based on current knowledge. In the right phylogeny, certain non-heterothermic families have been recoded as heterothermic on the basis of insularity, distribution (tropical and semitropical), body size, level of endothermy (based on body temperatures), the amplitude of circadian body temperature rhythms, and phylogenetic affinity to closely related heterotherms (Lovegrove 2012). The numbers at the root of certain clades indicate the proportional (0–1) maximum likelihood of heterothermy

Afrotheria, Xenarthra), was 0.58, compared with a ML of 0.42 of no heterothermy (Fig. 1.1). This node represents the split between the marsupials and placental mammals 144 mya (Fig. 1.1). The highest ML of heterothermy for a particular clade was for the marsupials (0.75). Thus with our current knowledge of heterothermy in crown mammals, at best we can conclude that (a) there was a single origin of heterothermy in mammals, and (b) there is a slightly higher probability that the Maastrichtian ancestors were heterotherms rather than obligatory homeotherms. Note that, in the former instance, a low likelihood of heterothermy would not have traced heterothermy to the root of the phylogeny, but would instead have created multiple cases of apomorphic heterothermy in those clades with high proportions of heterothermic families. This eventuality would have suggested that heterothermy evolved more than once independently.

1.3.1 Educated Recoding

A review of mammalian heterothermy and endothermy identified several unstudied mammal families that are highly likely to be heterotherms on the basis of insularity, distribution (tropical and semitropical), body size, level of endothermy (based on body temperatures), the amplitude of circadian body temperature rhythms, and phylogenetic affinity to closely related heterotherms (Lovegrove 2012). The families discussed below were recoded as heterotherms.

Current data on heterothermy in mammals are woefully biased by disproportional geographic sampling. The data for the Chiroptera, in particular, are highly biased toward Holarctic sampling. For example, a cursory count of the number of publications obtained from Thompson ISI Web of Science for the widespread vesper bats (Vespertilionidae) was 1,108, compared with <20 each for tropical families such as Myzopodidae, Thyropteridae, Nycteridae and Rhinopomatidae. To date, based upon the phylogeny of Teeling et al. (2005), daily torpor or hibernation has been recorded in nine of the 17 bat families (Geiser and Stawski 2011; Lovegrove 2012). Tropical and semi-tropical bats from Southeast Asia, Africa and South America have received very little attention from physiologists. The families include Myzopodidae (endemic to Madagascar), Thyropteridae and Furipteridae (Central and South America), Nycteridae (Southeast Asia and Africa), and Megadermatidae (Southeast Asia, Australia, and Africa). I would argue that there is a high probability that most chiropteran families are heterothermic, and that the poorly studied families can realistically be recoded as heterothermic (see also Geiser and Stawski 2011).

Pseudocheiridae is a marsupial family that comprises the possums. Although most possums weigh between 1 and 2 kg (Strahan 1991), some species, such as the pygmy ringtail possum (*Pseudochirulus mayerii*) from the highland forests of New Guinea, weigh around 150 g and are thought to enter torpor (Flannery 1995).

The pygmy anteater (*Cyclopes didactylus*) is the only xenarthran species in the genus *Cyclopes* in the family Cyclopedidae. It inhabits tropical forests in South America. It is the smallest of the xenarthran anteaters (<400 g), and is a potential heterotherm.

Tylomyinae is a rodent family that includes the vesper and climbing mice with a Central American tropical distribution. Nothing is known about their physiology. However, given their tropical distribution and close phylogenetic relationship to other heterothermic rodent families (e.g. Neotominae and Cricetinae), there is a possibility that some species may be heterothermic.

Dermoptera (Cynocephalidae, culagos) and Scandentia (Ptilocercidae and Tupaiidae, tree shrews), are basal to the primates, and today inhabit tropical forests in southeast Asia (Lovegrove 2012). Although they have yet to be studied in the wild, in the laboratory tree shrews display low BMRs (Bradley and Hudson 1974; Nelson and Asling 1962; Whittow and Gould 1976) and unusually large circadian amplitudes of T_b (Bradley and Hudson 1974; Refinetti and Menaker 1992). Body temperatures below 29°C have been measured in *Tupaia glis* at $T_a = 10$°C (Bradley and Hudson 1974). These conservative endothermic characteristics, as well as their basal phylogenetic relationship with heterothermic primates (e.g. mouse and dwarf lemurs), their ancient Cretaceous origin, and their current rainforest distribution, render tree shrews potentially heterothermic (Lovegrove 2012). Similar arguments can be applied to Dermoptera, although the flying lemurs may be too large to have retained heterothermy.

The phylogentic position of the Tarsiidae (tarsiers) remains uncertain (Matsui et al. 2009). Based upon molecular phylogenetics they may be basal to all primates, that is, basal to both the Strepsirrhini and Anthropoidea, or basal to Anthropoidea within Haplorrhini (Matsui et al. 2009). However, Rosa et al. (1996) suggest that the unusual pattern of their reinogeniculate neural projections that are not found in any other primate, probably render the Tarsiidae basal to all primates. These small (80–160 g) prosimians are strictly nocturnal and, in the laboratory, displayed the lowest body temperature ($T_b = 33.8$°C) and basal metabolic rate (BMR = 65% of expected) of all primates (McNab and Wright 1987). Thus given their tropical rainforest distribution, conservative endotherm characteristics, and their close phylogenetic relationship to other heterothermic primates, it is highly like that this primate family is heterothermic.

There are four families within Eulipotyphla which are probably heterothermic but which have been poorly studied. Solenodons (Solenodontidae) are very rare primitive 'insectivores' that inhabit the tropical islands of Haiti and Cuba. Their ancient origins, insular, tropical distribution, and close phylogenetic relationship to other heterothermic eulipotyphlans (e.g. hedgehogs and shrews), suggest they may be heterotherms. The Scalopinae and Talpinae are moles with a mostly Holarctic distribution. Being subterranean, they are very difficult to study in the wild. On the basis that the American shrew-mole, *Neurotrichus gibsii*, may employ torpor (Campbell and Hochachka 2000), this family can probably be recoded as heterothermic. The moonrats or gymnures (Galericinae) are closely related to hedgehogs (Erinaceidae), which are heterothermic. They inhabit moist jungles in Southeast Asia, including Vietnam, Sumatra, China and the Malaysian Peninsula. Given their tropical distribution and their close relationship to the heterothermic hedgehogs, they are likely to be heterothermic.

When the families discussed above were recoded as heterotherms, the proportional ML of heterothermy at the root of the recoded phylogeny increased to 0.96. The maximum likelihoods of the root of each ordinal clade were 0.97 (Rodentia), 1.0 (Chiroptera), 1.0 (Primates), 1.0 (Eulipotyphla), 0.95 (Marsupials), 1.0 (Afrotheria), and 1.0 (Xenarthra). Therefore, if we assume that the recoded families are indeed heterothermic, then we can conclude that (a) there is a single, pleisiomorphic origin of heterothermy in mammals, and (b) ancestral mammals were heterotherms. Indeed, there is a very high likelihood that Late Cretaceous small mammals survived the long-term effects of the Chicxulub impact by employing heterothermy. A focus on future studies of the Chiroptera and Eulipotyphla in particular, should greatly increase our knowledge on how ubiquitous heterothermy was in Maastrichtian mammals.

1.4 Conclusions

The reconstruction of the ancestral mammalian heterothermic status using a maximum likelihood approach suggests that the mammaliaform ancestors of the mammals were probably heterothermic and that this capacity was passed on to descendent mammalian lineages. Thus controlled hypothermia may be at least 200 million years old, coincident with the fossil presence of *Morganocodon* and *Hadrocodium*. These two mammaliaform ancestors display the first encephalization pulses compared with their cynodont ancestors (Rowe et al. 2011). Increased encephalization was associated with the evolution of more advanced chemosensory, auditary, visual, and tactile ability, as mammaliaforms switched to the nocturnal niche (Rowe et al. 2011). I would also argue that the encephalization pulses provide the earliest putative evidence of true endothermy, that is, the capacity for endogenous heat production. Increased encephalization would have generated increased metabolic demands which could only have been sustained through nocturnal heat production and normothermy in small-sized, nocturnal mammaliaforms (Malan 1996).

However, this does not mean that mammaliaformes and early monotreme, marsupial and placental mammals did not retain the capacity for torpor or hibernation during the daytime rest phase or during unfavorable seasons. This capacity would have been inherited from cynodont ancestors, whose endothermic status remains unclear (Hillenius 1994; Hillenius and Ruben 2004; Kemp 2006). Geiser's (2008) suggestion that "…endothermic thermoregulation evolved during, not prior to, the evolution of mammals…" is inconsistent with my conclusions. What is not clear in Geiser's thesis is when exactly these derived heterothermic events occurred. If they did not occur prior to the K/T boundary, it would be very hard to explain how the mammals survived the global apocalypse at the end of the Cretaceous.

Acknowledgments The research was financed by incentive grants from the University of KwaZulu-Natal and the National Research Foundation, South Africa.

References

Asher RJ, Helgen KM (2010) Nomenclature and placental mammal phylogeny. BMC Evol Biol 10:102

Barnes B (1989) Freeze avoidance in a mammal: body temperatures below 0°C in an arctic hibernator. Science 244:1593–1595

Bininda-Emonds ORP, Cardillo M, Jones KE, Macphee RDE, Beck RMD, Grenyer R, Price SA, Vos RA, Gittleman JL, Purvis A (2007) The delayed rise of present-day mammals. Nature 446:507–512

Bradley SR, Hudson JW (1974) Temperature regulation in the tree shrew, *Tupaia glis*. Comp Biochem Physiol 48A:55–60

Campbell KL, Hochachka PW (2000) Thermal biology and metabolism of the American shrew-mole, *Neurotrichus gibbsii*. J Mammal 81:578–585

Delsuc F, Scally M, Madsen O, Stanhope MJ, de Jong WW, Catzeflis FM, Springer MS, Douzery EJP (2002) Molecular phylogeny of living xenarthrans and the impact of character and taxon sampling on the placental tree rooting. Mol Biol Evol 19:1656–1671

Delsuc F, Stanhope MJ, Douzery EJP (2003) Molecular systematics of armadillos (Xenarthra, Dasypodidae): contribution of maximum liklihood and Bayesian analyses of mitochondrial and nuclear genes. Mol Phyl Evol 28:261–275

Douady CJ, Douzery EJP (2003) Molecular estimation of eulipotyphlan divergence times and the evolution of the "Insectivora". Mol Phyl Evol 28:285–296

Dubey S, Salamin N, Ohdachi SD, Barriere P, Vogel P (2007) Molecular phylogenetics of shrews (Mammalia: Soricidae) reveal timing of transcontinental colonizations. Mol Phyl Evol 44:126–137

Eagle RA, Schauble EA, Tripati AK, Tüken T, Hulbert RC, Eiler JM (2010) Body temperatures of modern and extinct vertebrates from ^{13}C–^{18}O bond abundances in bioapatite. Proc Natl Acad Sci U S A 107:10377–11382

Flannery T (1995) Mammals of new guinea. Cornell University Press, New York

Geiser F (1998) Evolution of daily torpor and hibernation in birds and mammals: importance of body size. Clin Exp Pharm Physiol 25:736–740

Geiser F (2008) Ontogeny and phylogeny of endothermy and torpor in mammals and birds. Comp Biochem Physiol A 150:176–180

Geiser F, Ruf T (1995) Hibernation versus daily torpor in mammals and birds: physiological variables and classification of torpor patterns. Physiol Zool 68:935–966

Geiser F, Stawski C (2011) Hibernation and torpor in tropical and subtropical bats in relation to energetics, extinctions, and the evolution of endothermy. Int Comp Biol 51:337–348

Hillenius WJ (1994) Turbinates in therepasids: evidence for Late Permian origins of mammalian endothermy. Evolution 48:207–229

Hillenius WJ, Ruben JA (2004) The evolution of endothermy in terrestrial vertebrates: who? when? why? Physiol Biochem Zool 77:1019–1042

Huber M (2008) A hotter greenhouse? Science 321:353–354

Kemp TS (2006) The origin of mammalian endothermy: a paradigm for the evolution of complex biological structure. Zool J Linn Soc 147:473–488

Kikuchi R, Vanneste M (2010) A theoretical exercise in the modeling of ground-level ozone resulting from the K–T asteroid impact: its possible link with the extinction selectivity of terrestrial vertebrates. Palaeogeogr Palaeoclimatol Palaeoecol 288:14–23

Kring DA (2007) The Chicxulub impact event and its environmental consequences at the Cretaceous-Tertiary boundary. Palaeogeogr Palaeoclimatol Palaeoecol 255:4–21

Kulzer E (1965) Temperaturregulation bei Fledermausen (Chiroptera) aus verschiedenen Klimatozonen. Z Vergl Physiol 50:1–34

Lovegrove BG (2012) The evolution of endothermy in Cenozoic mammals: a plesiomorphic–apomorphic continuum. Biol Rev 87:128–162

Luo ZX (2007) Transformation and diversification in early mammal evolution. Nature 450:1011–1019

MacLeod N (1996) K/T redux. Paleobiology 22:311–317

Maddison WP, Maddison DR (2009) Mesquite: a modular system for evolutionary analysis. Version 1.12

Malan A (1996) The origins of hibernation: a reappraisal. In: Geiser F, Hulbert AJ, Nicol SC (eds) Adaptations to the cold: tenth international hibernation symposium. University of New England Press, Armidale, pp 1–6

Matsui A, Rakotondraparany F, Munechika I, Hasegawa M, Horai S (2009) Molecular phylogeny and evolution of prosimians based on complete sequences of mitochondrial DNAs. Gene 441:53–66
McNab BK, Wright PC (1987) Temperature regulation and oxygen consumption in the Philippine tarsier, Tarsier syrichta. Physiol Zool 60:596–600
Melosh HJ, Schneider NM, Zahnle KJ, Latham D (1990) Ignition of global wildfires at the Cretaceous/Tertiary boundary. Nature 343:251–254
Meredith RW, Westerman M, Case JA, Springer MS (2008) A phylogeny and timescale for marsupial evolution based on sequences for five nuclear genes. J Mammal Evol 15:1–36
Meredith RW, Westerman M, Springer MS (2009) A phylogeny of Diprotodontia (Marsupialia) based on sequences for five nuclear genes. Mol Phyl Evol 51:554–571
Montgelard C, Forty E, Arnal V, Matthee CA (2008) Suprafamilial relationships among Rodentia and the phylogenetic effect of removing fast-evolving nucleotides in mitochondrial, exon and intron fragments. BMC Evol Biol 8:231
Murphy WJ, Pringle TH, Crider TA, Springer MS (2007) Using genomic data to unravel the root of the placental tree. Genome Res 17:413–421
Nelson LE, Asling CW (1962) Metabolic rate of tree shrews, *Urogale evertii*. Proc Soc Exp Biol Med 46:180–185
Newman JR, Rudd RL (1978) Observations of torpor-like behavior in shrew, *Sorex sinuosus*. Acta Theriol 23:446–448
Purvis A (1995) A composite estimate of primate phylogeny. Philos Trans R Soc Lond B 348:405–421
Refinetti R, Menaker M (1992) Body temperature rhythm of the tree shrew, *Tupaia belangeri*. J Exp Zool 263:453–457
Robertson DS, McKenna MC, Toon OB, Hope S, Lillegraven JA (2004) Survival in the first hours of the Cenozoic. Geol Soc Am Bull 116:760–768
Rosa MGP, Pettigrew JD, Cooper HM (1996) Unusual pattern of retinogeniculate projections in the controversial Primate *Tarsius*. Brain Behav Evol 48:121–129
Rowe TB, Macrini TE, Luo ZX (2011) Fossil evidence on origin of the mammalian brain. Science 332:958–960
Schmidt-Nielsen K (1983) Animal physiology: adaptation and environment. Cambridge University Press, Cambridge
Steppan SJ, Storz BL, Hoffmann RS (2004) Nuclear DNA phylogeny of the squirrels (Mammalia: Rodentia) and the evolution of arboreality from c-myc and RAG1. Mol Phyl Evol 30:703–719
Strahan R (1991) Complete book of Australian mammals. Cornstalk, Sydney
Tabuce R, Asher RJ, Lehmann T (2008) Afrotherian mammals: a review of current data. Mammalia 72:2–14
Teeling EC, Springer MS, Madsen O, Bates P, O'Brien SJ, Murphy WJ (2005) A molecular phylogeny for bats illuminates biogeography and the fossil record. Science 307:580–584
Whittow GC, Gould E (1976) Body temperature and oxygen consumption of the pentail tree shrew (*Ptilocerus lowii*). J Mammal 57:754–756

Chapter 2
Afrotropical Heterothermy: A Continuum of Possibilities

Kathrin H. Dausmann, Julia Nowack, Susanne Kobbe and Nomakwezi Mzilikazi

Abstract The three closely related primate species *Cheirogaleus medius, Microcebus griseorufus,* and *Galago moholi* employ a spectrum of thermoregulatory responses to environmental bottlenecks. *C. medius* is an obligate hibernator, *M. griseorufus* shows extreme flexibility in patterns of heterothermy, ranging from daily torpor to prolonged torpor and hibernation, whereas *G. moholi* becomes heterothermic under extremely adverse conditions only. To gain further insights into the proximate and ultimate factors that favour and constrain torpor use in small primates, we compared the characteristics of *M. griseorufus, C. medius* and *G. moholi* as well as the prevailing ambient conditions where each of the species are found. Our analyses did not reveal any fundamental differences in the physiological parameters of heterothermy between the three species that would indicate different underlying physiological mechanisms. Instead we propose that the different modes of reproduction, connected to climatic differences, are the key ultimate causes of the differences in thermoregulatory strategies between the three species. We suggest that the common ancestor of the strepsirrhine primates was a heterothermic endotherm and that the ability to express daily torpor/hibernation is closely linked to ecological and evolutionary forces that favour and prioritise activities such as reproduction.

K. H. Dausmann (✉) · J. Nowack · S. Kobbe
Department of Animal Ecology and Conservation, Biocentre Grindel, University of Hamburg, Martin-Luther-King-Platz 3, 20146 Hamburg, Germany
e-mail: kathrin.dausmann@uni-hamburg.de

N. Mzilikazi
Centre for African Conservation Ecology, Department of Zoology, Nelson Mandela Metropolitan University, PO Box 77000, Port Elizabeth 6031, South Africa

T. Ruf et al. (eds.), *Living in a Seasonal World*,
DOI: 10.1007/978-3-642-28678-0_2, © Springer-Verlag Berlin Heidelberg 2012

2.1 The Expression of Daily Torpor and Hibernation

The presence of heterothermy in unrelated mammalian groups has intrigued biologists, who have proposed two scenarios for the evolution of heterothermy: (1) the common ancestor of mammals was a heterothermic endotherm and subsequently many lineages lost the ability or propensity to express daily torpor/hibernation, or (2) heterothermy may have evolved repeatedly during the evolution of these lineages and are the result of convergent evolution. These opposing hypotheses are still under contention. For example, monotremes, the so-called primitive mammals, show evidence of 'proto-endothermy'—i.e., heterothermy is present and endothermy is facultative (e.g., Nicol and Andersen 1996), a pattern also observed in tenrecs (Lovegrove and Génin 2008; Wein 2010). This could be evidence that heterothermy is a plesiomorphic trait. More support for this theory comes from a model proposing that heterothermy was the ancestral condition from which homeothermy derived (Lovegrove 2011) and a maximum likelihood approach reconstructing the probability of heterothermy of late Mesozoic mammals, indicating a single origin of heterothermy in mammals (Lovegrove 2012). Furthermore, expression of heterothermy is flexible: within a single species, some individuals can display daily torpor, whereas others display hibernation (Kobbe et al. 2011). Studies on the ontogeny of torpor support the idea that heterothermy evolved more than once; Geiser (2008) suggested that heterothermy may be a derived trait in placental mammals because in some organisms, such as shrews and hamsters, poikilothermy (i.e., lack of effective endogenous temperature regulation) at birth is followed first by a homeothermic phase and the ability to employ torpor develops later. This ontogenetic sequence may signal the order of evolutionary change provided that 'ontogeny recapitulates phylogeny'. This argument, in fact, does not deny the possibility that heterothermy in endotherms has indeed evolved multiple times in several lineages (i.e., highly polyphyletic). Dietary habits and body mass correlate with the expression of torpor/hibernation in mammals (e.g., Geiser 1988). Although these factors cannot be viewed independently from phylogeny, these variables may provide a better explanation for the presence of heterothermy. Within the primates, most of what is known about heterothermic responses has been gained from studies on virtually a single family of Malagasy lemurs, the Cheirogaleidae (e.g., Dausmann et al. 2004; Kobbe and Dausmann 2009; Ortmann et al. 1997; Schmid et al. 2000) whose closest relatives (apart from Lorises) are the galagos from mainland Africa, to which they are comparable in terms of body size and life histories.

2.2 So Near Yet So Far

All habitats of the cheirogaleids in Madagascar, as well as of galagos in South Africa (Limpopo and Mpumalanga provinces), are characterised by seasonality and environmental unpredictability. Unpredictability results in periods of resource scarcity during the austral winter that are highly variable within and from year to year. Additionally, Madagascar forests have comparatively lower fruit production than other tropical forests (Wright 1999). These circumstances pose great energetic constraints, especially to animals relying on highly seasonal high-energy food sources, e.g., small-bodied species with relatively high metabolic rates (MR). Given the relative geographical proximity of Madagascar and South Africa, as well the similarities relating to environmental unpredictability, it seems reasonable to expect comparable adaptations, such as the use of heterothermy, to similar environmental challenges. Admittedly, the geological history and biogeography of Madagascar, especially the long separation from the African mainland (~ 120 mya) may have led to a unique evolution of its mammals, creating an excellent natural experiment.

Currently, within the cheirogaleids heterothermy has been found in mouse lemurs (*Microcebus*), giant mouse lemurs (*Mirza*) and dwarf lemurs (*Cheirogaleus*) (Dausmann 2008; McKechnie and Mzilikazi 2011). A recent study has found torpor in a mainland primate, the African Lesser Bushbaby, *Galago moholi* (Nowack et al. 2010). Among these three closely related genera a spectrum of thermoregulatory responses is evident. *Cheirogaleus medius* is an obligate hibernator (Dausmann et al. 2004). *Microcebus griseorufus* shows extreme flexibility in patterns of heterothermy (Kobbe et al. 2011), whereas *G. moholi* utilises heterothermy under extreme conditions only (Nowack et al. 2010). Together these species provide an excellent opportunity for interrogating the proximate influences on the energy expenditure of small primates. It also raises the question of what the ultimate factors contributing to the reluctant use of torpor by *G. moholi* are. To gain further insights into the proximate and ultimate factors that favour and constrain torpor use in small primates, we compared the characteristics of *M. griseorufus, C. medius* and *G. moholi* as well as the prevailing ambient conditions where each of the species are found. The ultimate aim of this summary is to highlight similarities and the differences that have possibly led to the evolution of different thermoregulatory strategies in primates.

Throughout this summary, data were obtained from the literature and from three studies carried out by us over a time span between 3 and 6 years each. *C. medius* was studied in the forest of Kirindy C.F.P.F./Madagascar, *M. griseorufus* within Tsimanampetsotsa National Park/Madagascar and *G. moholi* in the Nylsvley Nature Reserve/South Africa. In the course of these studies demographic data were obtained by capture–recapture and individual marking. Behavioural data and data on food intake were recorded during nightly tracking of radio-collared individuals. Physiological measurements were performed using temperature-sensitive radio collars and loggers (skin temperature [T_{skin}], as a proxy of body temperature [T_b]), and portable gas analysers (measurement of MR as rate of oxygen consumption). For more detail see Dausmann et al. (2009), Kobbe et al. (2011), and Nowack et al. (2010).

2.3 Habitat and Life Histories

2.3.1 Taxonomy and General Species Description

All three species [*Galago* (Galagidae), *Microcebus* (Cheirogaleidae) and *Cheirogaleus* (Cheirogaleidae)] are strepsirrhine primates and very similar in their morphological appearance, but belong to two different taxa: the lemuriformes, endemic to Madagascar, and the lorisiformes, found in Africa and Asia.

M. griseorufus is the smallest of the three species and it occurs in Southern Madagascar (Table 2.1). *G. moholi* occurs in Southern Africa and is slightly heavier than *C. medius* (Western Madagascar). All three species are nocturnal; *M. griseorufus* and *G. moholi* are vertical clingers and leapers, while *C. medius* is an arboreal quadruped.

2.3.2 Similar Habitats, Similar Diets

The habitats of all three species are distributed at comparable latitudes (around the Tropic of Capricorn) and share general habitat characteristics (Table 2.1). All three habitats are characterised by a marked dry season during the austral winter. The habitat of *M. griseorufus* is the most unpredictable of the three areas with an annual average rainfall of less than 500 mm (November–March) and years with almost no rainfall (Ratovonamana et al. submitted). The habitats of *C. medius* and *G. moholi* are less arid (Fig. 2.1); however, annual rainfall also varies highly in amount (from 300/400 mm to over 1,000 mm per year) and timing. Annual average rainfall is 700 mm in the habitat of *G. moholi* [Limpopo parks data (Nylsvley)]. The dry period is often only restricted to 3 or 4 months during mid-winter (Scholes and Walker 1993). Annual rainfall is higher (about 800 mm) for the habitat of *C. medius* (Fietz and Dausmann 2006), but the dry period is extended to 6 or 7 months per year (April–October).

Although the pattern of annual precipitation is most comparable for the habitats of *G. moholi* and *C. medius,* the biomes of *G. moholi* and *M. griseorufus* are more similar. While *C. medius* occurs in tropical dry forests regions, *G. moholi* and *M. griseorufus* inhabit savannah areas with thorny bushes and trees. Ambient temperature (T_a) also varies between the three habitats (Fig. 2.1) and the lowest temperatures are found in the habitat of *G. moholi*. During the hot-wet season T_a is fairly high during the day and seldom decreases below 15°C during the night, whereas night temperature during winter can reach −5°C (Nowack et al. 2010). Temperatures in South and Western Madagascar are about 10°C warmer during summer nights, but never decrease below 5°C during winter nights (Fietz and Dausmann 2006). Daily amplitudes between day and night are high in all three habitat types (*G. moholi* and *C. medius:* 2–35°C, *M. griseorufus*: 6–35°C).

The pronounced seasonality of all three habitats is not only reflected in a decrease in T_a and precipitation, but also in a decline in food and water availability during the dry season. The diet of *G. moholi* and *M. griseorufus* consists mainly of gum and small arthropods. *M. griseorufus* extends its food repertoire to fruits and nectar. Both species increase gum feeding during the winter months: insect and fruit abundances decrease noticeably in winter, whereas gum is available throughout the year (Bearder and Martin 1980; Génin 2008; Nowack et al. submitted). The diet of *C. medius* does not contain any gum, but consists of fruits, flowers, nectar, and a varying proportion of insects. *C. medius* feeds on sugary fruits before hibernation to build a sufficient fat reservoir to fuel energy demands during hibernation (Fietz and Dausmann 2006).

Table 2.1 Life history parameters for *G. moholi*, *M. griseorufus* and *C. medius*

	G. moholi	*M. griseorufus*	*C. medius*
Distribution	Southern Africa	South-western Madagascar	Western Madagascar
Habitat	Thornveld savannah	Spiny forest	Deciduous dry forest
Weight	~180 g	~50 g[a]	~150 g[b]
Sexual dimorphism in weight	Males heavier	Females heavier (before winter)	None[b]
Seasonal fattening	No	+ (Hibernation: 49–73 g, prolonged torpor: 47–53 g, daily torpor: 43–48 g, short torpor: 44–46 g)	++ (130–250 g)[b]
Body length	19 cm	15.5 cm[c]	20 cm
Diet	Gum, small arthropods[d,e]	Fruits, nectar, insects, small vertebrates, gum[g]	Insects, fruits, flowers, nectar[b]
Seasonal change in diet	Increases gum feeding during winter[e]	Increases gum feeding in winter by 33% of total foraging activity[f]	Eat sugary fruits before hibernation, no food during winter[b]
Gestation time	123 days[g]	52–60 days[f]	60–65 days[b]
Litter size	Mostly twins[g]	1–3, mostly twins[f]	Mostly twins[b]
Birth periods	January/February and September–November[g]	December and February/March[f]	December–February (every second year)[h]
Predation	Mongooses, genets, birds of prey (owls, eagles), snakes[i]	Birds of prey (mainly owls), snakes, civets	Birds of prey (owls, hawks), snakes, civets[b]
Estimated average field life span	3–5 years	3–5 years	5–7 years (up to over 12)

[a] Kobbe et al. (2011)
[b] Fietz and Dausmann (2006)
[c] Rakotondranary et al. (2011)
[d] Bearder and Martin (1980)
[e] Nowack et al. submitted
[f] Génin (2008)
[g] Pullen et al. (2000)
[h] Fietz and Dausmann (2003)
[i] Mzilikazi et al. (2006)
Data not cited is unpublished data by the authors

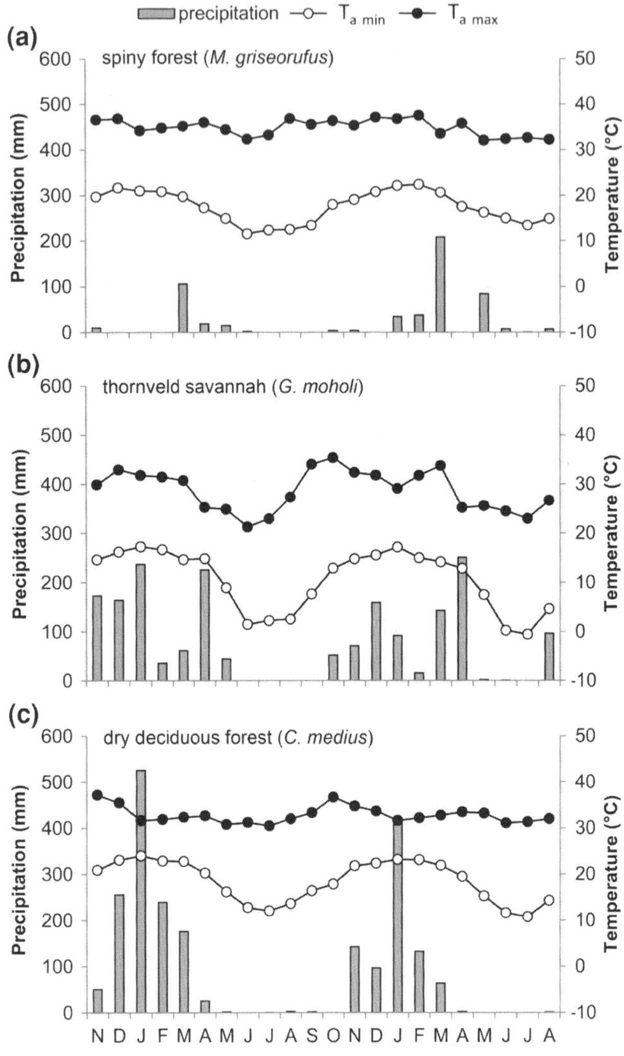

Fig. 2.1 Seasonal variation of ambient temperature (T_a) and precipitation of **a** the spiny forest in South-Western Madagascar 2007–2009 (habitat of *M. griseorufus*), **b** the Thornveld Savannah in South Africa 2009–2011 (habitat of *G. moholi*) and **c** the deciduous dry forest in Western Madagascar 1999–2001 (habitat of *C. medius*) (modified from Fietz and Dausmann 2006)

2.3.3 Social System and Territoriality

G. moholi and *M. griseorufus* have a similar social organisation. They forage solitarily during the night, but females are seldom found nesting solitarily during daytime. They rest in family groups (one or more female and their current off-spring, in *G. moholi* sometimes accompanied by the assumptive father) or pairs

(Génin 2008; Nowack et al. submitted). However, in *G. moholi* some (probably subdominant) males are frequently found resting solitarily. In contrast, *C. medius* forms lifelong pair bonds (social monogamy). The pair defends a lifelong territory and usually shares a resting site during daytime; however, they are also solitary foragers and often hibernate separately (Dausmann et al. 2005).

All three species are territorial, but whereas *G. moholi* has been observed to maintain its territorial boundaries throughout the year (Nowack et al. 2010), territoriality of *C. medius* is restricted to the wet season and abandoned during the cold-dry season, when the animals become heterothermic. In *M. griseorufus* we can observe solitary males and females with overlapping home ranges as well as individuals sharing sleeping sites and home ranges (socio-territoriality; Génin 2010).

Reproductive patterns vary markedly between the three species (Table 2.1). *C. medius* usually reproduce only every second year (Fietz and Dausmann 2003) and give birth to twins after a gestation time of 60–65 days, which is only slightly longer than that of the almost three times smaller *M. griseorufus* (52–60 days) (Fietz and Dausmann 2006; Génin 2008). In the first 2 weeks after birth, both parents share babysitting duties; later, the parents guide the young during foraging (Fietz and Dausmann 2003). *M. griseorufus* and *G. moholi* have two reproductive periods per year, also usually giving birth to twins. Due to the relatively short gestation time, both birth periods of *M. griseorufus* occur during the hot-wet season. Parental care rests solely with the females, and young are independent after a few weeks. Within socio-territorial female pairs, alloparental care is a common phenomenon due to overlapping lactation periods (Génin 2008). Gestation time of *G. moholi* is about twice as long (123 days) and females give birth once shortly before the dry and cold winter time and once shortly after winter (Pullen et al. 2000). The young are weaned after few weeks and infants seldom have any contact with adult individuals other than the mother (Bearder 1969).

2.3.4 Predation and Mortality Rates

Various predators are described for each species (Table 2.1). Mortality rates of *C. medius* differ with habitat, but lie between 16 and 30% within the dry deciduous forest (Lahann and Dausmann 2011). Mortality rates are not known for *M. griseorufus* and *G. moholi*, but life spans in the wild are presumably considerably shorter than for *C. medius*.

2.4 Energy-Saving Responses: Three Species—Three Strategies

Heterothermy is utilised in all three species in order to cope with unfavourable conditions [during the southern winter or cold-dry season, respectively (i.e. between April and October)], however, in very different extent and variation (Table 2.2).

2.4.1 Differences in Energy-Saving Strategies

Most strikingly, there are great differences in the percentage of the population that becomes heterothermic. While 100% of *C. medius* and *M. griseorufus* employ heterothermy throughout (*C. medius*) or at some point (*M. griseorufus*) during winter, barely a fourth of the population of *G. moholi* becomes heterothermic at all (Nowack et al. 2010, unpublished data). Thus, between these three species we observe a continuum of different heterothermic responses: while *C. medius* is an obligate hibernator with all individuals hibernating throughout the dry season between mid-April and mid-October (Dausmann et al. 2004), *M. griseorufus* is physiologically more flexible (Kobbe et al. 2011). A part of the population of *M. griseorufus* stays normothermic most of the winter time and shows only spontaneous short torpor bouts whenever it seems necessary. However, most individuals use daily torpor and some individuals additionally enter prolonged torpor bouts for up to several days. The most extensive heterothermic behaviour we observe in this species is long-term hibernation. Similar to *C. medius,* hibernating individuals of *M. griseorufus* remain completely inactive up to 6 months during the dry season. *G. moholi* is also capable of heterothermy, however, in contrast to *M. griseorufus* and *C. medius* individuals exclusively use short torpor bouts on especially adverse conditions and not as a routine behaviour (Nowack et al. 2010). *M. griseorufus* and *C. medius* use tree holes for torpor and hibernation and the same is assumed, but yet needs to be ascertained, for *G. moholi.*

2.4.2 Influence of Body Mass, Sex and Age on Energy-Saving Strategies

In *C. medius* all individuals hibernate throughout the dry season regardless of their age, body mass or sex. Even juveniles display hibernation although they seem to be active longer than adults (Dausmann et al. 2005). Prior to hibernation *C. medius* accumulates large quantities of fat in the tail and on the body in order to survive many months without food intake (Fietz and Dausmann 2006). This seasonal body fattening also occurs in *M. griseorufus*, but in varying extent between individuals. Thermal behaviour utilised by *M. griseorufus* strongly depends on the individual's body condition before winter. Only the heaviest individuals hibernate up to several months. Individuals without sufficient fat stores occasionally use short torpor bouts and animals with intermediate body masses display daily torpor or prolonged torpor (Kobbe and Dausmann, unpublished data). In contrast to another mouse lemur species (*M. murinus*; Schmid 1999), *M. griseorufus* does not show any sex-specific differences in heterothermy (Kobbe et al. 2011). Compared to the cheirogaleids the thermal behaviour of *G. moholi* differs in so far as the exhibition of heterothermy does not seem to depend on body mass primarily but rather on the individuals' age and/or sex. Only subadults and juveniles, and mainly males use torpor during winter (Nowack et al. 2010, unpublished data).

Table 2.2 Heterothermic parameters for *G. moholi*, *M. griseorufus* and *C. medius* (N gives number of individuals, n number of days)

	G moholi	*M. griseorufus*	*C. medius*
Heterothermy	Torpor under especially adverse conditions[a]	Short, daily or prolonged torpor hibernation[b]	Obligate hibernation[c]
% of heterothermic individuals during winter	26 ($N = 8$)	100 ($N = 16$)[b]	100 ($N = 66$)[d]
Sex-specific heterothermy?	Mostly males	No[b]	No[d]
Differences in age of heterothermic animals?	Subadult/juvenile[a]	Unknown, probably yes due to lower body mass in juveniles	Juveniles shortened hibernation period[e]
Occurrence of heterothermy	April–end of August	April–November[b]	Adults: Mid-April to Mid-October; juveniles might be active until end of May[e]
Absolute $T_{skin\ min}$ during torpor/hibernation (°C)	13.7	Short torpor: 10.5, daily torpor: 7.5, prolonged torpor: 7, hibernation: 6.5[b]	9.3[d]
Mean $T_{skin\ min}$ during torpor/hibernation (°C)	21.2 ± 5 ($N = 8$)	16.42 ± 1.63 ($N = 16$)[b]	18.2 ± 2.8 ($N = 30$)[e]
Mean TMR_{min} (ml O_2 $h^{-1}g^{-1}$)	0.3 ± 0.2 ($N = 4$)	0.13 ± 0.05 ($N = 4$)	0.14 ± 0.058 ($N = 11$)[c]
Mean TMR ($T_{skin} < 30$°C)	0.5 ± 0.2 ($N = 4$)	0.28 ± 0.16 ($N = 4$)	0.29 ± 23.2 ($N = 11$)[c]

(continued)

Table 2.2 (continued)

	G. moholi	*M. griseorufus*	*C. medius*
Mean duration of torpor bouts/hibernation bout[g]	5.5 ± 3 h ($N = 6$, $n = 6$)	Short torpor: 5.6 ± 2.3 h ($N = 4$, $n = 147$)[f] Daily torpor: 8.7 ± 1.5 h ($N = 5$, $n = 504$)[f] Prolonged torpor: ~43.8 ± 19.9 h ($N = 4$, $n = 6$)[f] Hibernation: 12.3 ± 1.3 days ($N = 3$, $n = 17$)[f]	6.7 ± 3.9 days ($N = 5$, $n = 14$)[e]
Frequency of torpor during winter (%)	Seldom[a]	Torpor: 29–80 (between May and October) Hibernation: 100[b]	Hibernation: 100[d]
Daily energy savings through heterothermy ($T_{skin} < 30°C$) (%)	~66	~80	~72[c]

[a] Nowack et al. (2010)
[b] Kobbe et al. (2011)
[c] Dausmann et al. (2009)
[d] Dausmann et al. (2004)
[e] Dausmann et al. (2005)
[f] Kobbe and Dausmann (2009)
[g] Between arousal
Data not cited in unpublished data by the authors

2.4.3 Physiological Characteristics of Heterothermy

2.4.3.1 Hibernation Bout Duration

The duration of torpor bouts are most extensive in those species and individuals that undergo hibernation. The longest hibernation bout (>70 days) without any active heat production was observed in *C. medius* (after that the logger stopped recording; Dausmann et al. 2005). The second longest bouts (45 days) were observed in hibernating individuals of *M. griseorufus* (Kobbe and Dausmann 2009). However, the average hibernation bout length of *M. griseorufus* was 12 days, and was thus twice as long as in *C. medius* with 6 days (Table 2.2). We attribute these differences to variations in insulation capacities of the tree holes used as hibernacula. While *M. griseorufus* uses only comparatively thin-walled poorly insulated tree holes, resulting in fluctuating T_a and T_b, and reducing the need of arousals, *C. medius* additionally uses thick-walled, well-insulated tree holes, necessitating regular arousals, thus decreasing the average value.

2.4.3.2 Torpor Bout Duration

The average duration of daily torpor bouts was longer in *M. griseorufus* than in *G. moholi*. However, in both species torpor bouts can range between 2 and 11 h (Table 2.2). As mentioned above the duration of torpor in *M. griseorufus* is strongly correlated to the body condition of the individual and T_a. *G. moholi* also undergoes torpid states only during the cold and dry season (from April to end of August) when food abundance and T_a are low, but heterothermy appears to be neither physiologically necessary nor possible for all individuals. Food-deprived juvenile and subadult individuals always entered torpor during the cold-dry season, but never during the hot-wet season. The length of torpor bouts seems to depend mainly on T_a (Nowack et al. 2010, unpublished data). *C. medius* only show daily torpor at the beginning of the hibernation season, so-called test drops (Dausmann et al. 2005). Prolonged torpor occurs only in *M. griseorufus* and only on occasions, where T_a decreases below a certain threshold (<9°C, Kobbe et al. 2011). Only individuals that have enough fat reserves to sustain the additional energetic demands before resuming delayed food intake can utilise this option.

2.4.3.3 General Patterns of Body Temperature

All three species show the regular pattern of T_b of small mammals during normothermia: T_b fluctuates around 37°C, with slightly lower values during the day, when the animals are resting (Dausmann et al. 2005; Kobbe et al. 2011; Mzilikazi et al. 2006).

General patterns of T_b in heterothermic individuals are also similar among the three species: T_b passively follows T_a for most of the heterothermic phase, and thus

T_b mainly depends on T_a and the insulation capacities of the resting place. Due to the differences in energy-saving strategies and environmental conditions (i.e. T_a) minimal T_{skin} ($T_{skin\ min}$) during heterothermy shows a significant variation among the species. In *M. griseorufus*, $T_{skin\ min}$ values seem to be strongly associated with the length of the heterothermic period. Both mean and absolute $T_{skin\ min}$ are lowest in hibernating *M. griseorufus* (6°C) and highest in individuals that enter short torpor bouts (absolute $T_{skin\ min}$ 10.5°C) (Kobbe and Dausmann 2009; Kobbe et al. 2011). $T_{skin\ min}$ of hibernating *C. medius* has been found to decrease down to 9.3°C (Dausmann et al. 2005; Table 2.2). The highest mean and absolute $T_{skin\ min}$ during heterothermy occur in *G. moholi*, where T_{skin} rarely decreases below 18°C (Nowack unpublished data). In cases where it did decrease below this value (two juveniles under laboratory conditions), the individuals had substantial difficulties rewarming from torpor.

2.4.3.4 Metabolic Rate

We were able to measure the MR of heterothermic individuals (torpid MR: TMR): of hibernating *C. medius* (Dausmann et al. 2009), torpid (but not hibernating) *M. griseorufus* and torpid *G. moholi*. The energy consumption of hibernating *C. medius* and torpid *M. griseorufus* are comparable (Table 2.2). This is due to the relatively low T_b during torpor bouts that occur in *M. griseorufus*. In contrast, *G. moholi* expends more than twice as much energy as the two cheirogaleid species during heterothermy, as anticipated from the higher minimal T_b during torpor.

In order to compare the potential of energy savings by heterothermy between all three species we compared average TMR to average MR during normothermia. In all three species heterothermic individuals utilise about 70% (between 66 and 80%) less energy per hour than normothermic animals (Table 2.2). Total energy savings depend not only on TMR but decisively also on the duration of the heterothermic period, i.e. on the particular thermal strategy. Thus we find a great variation of total energy savings by heterothermy among species and/or individuals with hibernating *C. medius* and *M. griseorufus* saving most and *M. griseorufus* and *G. moholi* with short torpor bout saving least.

2.5 Résumé

Our analyses did not reveal any fundamental differences in the physiological parameters of heterothermy between the three species that would indicate different basic physiological mechanism. Rather, patterns of T_b and MR during heterothermic phases, as well as length and depth of these, seem to follow general rules in all three species, depending on environmental variables, such as T_a, and individual variables, such as body condition, sex and age.

In our opinion, the finding that galagos and lemurs employ the same basic physiological mechanisms when using heterothermy suggests that the common ancestor of these two groups was also heterothermic. Particularly since galagos only use this energy-saving mechanism sporadically it seems unlikely that they also evolved this capability convergently. Rather: they 'still' possess the ability, but only resort to it under exceptional circumstances. A plesiomorphic origin of heterothermy in mammals has also been suggested by Lovegrove (2011, 2012).

Environmental variables of the different habitats also did not show any substantial differences, the most important one being marked differences in the degree of unpredictability of rainfall. While the habitat of *C. medius* is highly seasonal, this seasonality, including food and water availability, is very predictable. The climate of the habitat of *M. griseorufus*, though showing the same seasonality, has much less precipitation, which additionally occurs very unpredictably within and between years, and can be totally skipped in some years. The habitat of *G. moholi* is less seasonal, rainfall being distributed more evenly over the year in amount and timing.

The most striking difference between *G. moholi*, *M. griseorufus* and *C. medius* is their differing reproductive strategies. Reproduction is a large component of energy budgets. In evolutionary terms, the ultimate goal of any animal should be the maximisation of its proportion of genes in the next generation. This can either be achieved by producing more offspring in a short period, or by reducing fecundity, but reproducing over a longer period. We propose that the different modes of reproduction, connected to climatic differences, are the key to the ultimate causes of the differences in thermoregulatory strategies between the three species. In the dry forest, rainfall is seasonal, but very predictable. Thus, *C. medius* can prepare for a defined period of scarcity when food is reliably abundant and hibernate during winter, reliably encountering food and water during the summer months, enabling itself and its offspring seasonal fattening before, and replenishment after hibernation. Frequent reproduction is not an option for *C. medius*, possibly because of the extensive and time-intensive parental care of both parents that is needed for the successful upbringing of offspring (Fietz and Dausmann 2003). Hibernating animals generally have a higher annual survival (Turbill et al. 2011) and *C. medius* may therefore use this energy-saving strategy during the lean dry season that cannot be used for reproduction anyway, to compensate less frequent reproduction events by a longer life. Other cheirogaleids in the dry forest are either smaller, and thus may not be able to store sufficient amounts of fat to make 7 months long hibernation an option (Dausmann 2008); or they are able to exploit a more exclusive or less seasonal spectrum of food: *Mirza coquereli* includes animal prey in its diet, and *Phaner furcifer* mainly feeds on tree exudates, both of which are available throughout the year.

G. moholi, on the other hand, lives in a less seasonal habitat and can successfully raise young throughout the year and thus reproduce more often, an option not available to *C. medius*. They, therefore, largely forgo the advantages of heterothermy in favour of reproduction possibilities, except in rare events. We assume that the physiological potential of entering energy-saving heterothermic states is only tapped by individuals that are unable to participate in reproduction at that point, e.g. young males without a territory. Females can already become pregnant at the age of 6 months, and due to their long gestation period, they are either pregnant or lactating most of the year. Since both activities are seriously impaired by heterothermy (e.g., Farmer 2000; Wilde et al. 1999), this would explain why females are rarely encountered torpid. Adult males, on the other hand, presumably cannot afford to spend time in a 'suspended' state at any time of the year, neglecting territory defence (Nowack et al. 2010) and possibly mate guarding.

M. griseorufus inhabits the most unpredictable habitat and is the most flexible of the three species. Hibernation seems to be the preferred strategy for these animals, but can only be adopted by individuals whose body condition (fat stores) is sufficient to fuel long-term hibernation over several months with uncertain conditions when resuming activity. When hibernation is not possible, *M. griseorufus* can employ daily torpor or prolonged torpor, 'deciding' on a day-to-day basis, depending on their own current body condition and current ambient conditions. With this very flexible opportunistic strategy, they can brave the unpredictably of their habitat and are best equipped to adjust quickly to changing conditions.

In conclusion, the thermoregulatory physiology of *C. medius*, *M. griseorufus* and *G. moholi* is tightly adapted the environmental conditions of the habitats, where the species evolved, in combination with reproductive possibilities, highlighting how flexibly physiological traits can be expressed.

References

Bearder SK (1969) Territorial and intergroup behaviour of the lesser bushbaby, *Galago senegalensis moholi*, in semi-natural conditions and in the field. Dissertation, University of the Witwatersrand, Johannesburg

Bearder SK, Martin RD (1980) *Acacia* gum and its use by bushbabyies, *Galago senegalensis* (Primates: Lorisidae). Int J Primatol 1(2):103–128

Dausmann KH (2008) Hypometabolism in primates: torpor and hibernation. In: Lovegrove BG, McKechnie AE (eds) Hypometabolism in animals: torpor, hibernation and cryobiology. University of KwaZulu-Natal, Pietermaritzburg, pp 327–336

Dausmann KH, Glos J, Ganzhorn JU, Heldmaier G (2004) Hibernation in a tropical primate. Nature 429(6994):825–882

Dausmann KH, Glos J, Ganzhorn JU, Heldmaier G (2005) Hibernation in the tropics: lessons from a primate. J Comp Physiol B 175(3):147–155

Dausmann KH, Glos J, Heldmaier G (2009) Energetics of tropical hibernation. J Comp Physiol B 179(3):345–357

Farmer CG (2000) Parental care: the key to understanding endothermy and other convergent features in birds and mammals. Am Nat 155(3):326–334

Fietz J, Dausmann KH (2003) Costs and potential benefits of parental care in the nocturnal fat-tailed dwarf lemur (*Cheirogaleus medius*). Folia Primatol 74(5–6):246–258

Fietz J, Dausmann KH (2006) Big is beautiful: fat storage and hibernation as a strategy to cope with marked seasonality in the fat-tailed dwarf lemus (*Cheirogaleus medius*). In: Gould L, Sauther ML (eds) Lemurs: ecology and adaptation. Springer, Berlin, pp 97–111

Geiser F (1988) Daily torpor and thermoregulation in antechinus (Marsupialia): influence of body mass, season, development, reproduction, and sex. Oecologia 77(3):395–399

Geiser F (2008) Ontogeny and phylogeny of endothermy and torpor in mammals and birds. Comp Biochem Physiol A 150(2):176–180

Génin F (2008) Life in unpredictable environments: first investigation of the natural history of *M. griseorufus*. Int J Primatol 29(2):303–321

Génin F (2010) Who sleeps with whom? Sleeping association and socio-territoriality in *Microcebus griseorufus*. J Mammal 91(4):942–951

Kobbe S, Dausmann KH (2009) Hibernation in Malagasy mouse lemurs as a strategy to counter environmental challenge. Naturwissenschaften 96(10):1221–1227

Kobbe S, Dausmann KH, Ganzhorn JU (2011) Extreme individual flexibility of heterothermy in free-ranging Malagasy mouse lemurs (*Microcebus griseorufus*). J Comp Physiol B 181(1):165–173

Lahann P, Dausmann KH (2011) Live fast, die young: flexibility of life-history traits in the fat-tailed dwarf lemur (*Cheirogaleus medius*). Behav Ecol Sociobiol 65(2):381–390

Lovegrove B (2011) The evolution of endothermy in Cenozoic mammals: a plesiomorphic-apomorphic continuum. Biol Rev. doi:10.1111/j.1469-185X.2011.00188.x

Lovegrove B (2012) A single origin of heterothermy in mammals. In: Ruf T, Bieber C, Arnold W, Millesi E (eds) Living in a seasonal world: thermoregulatory and metabolic adaptations. Springer, Berlin

Lovegrove BG, Génin F (2008) Torpor and hibernation in a basal placental mammal, the lesser hedgehog tenrec (*Echinops telfairi*). J Comp Physiol B 178(6):691–698

McKechnie AE, Mzilikazi N (2011). Heterothermy in Afrotropical mammals and birds: a review. Integr Comp Biol:1–15. doi:10.1093/icb/icr035

Mzilikazi N, Masters JC, Lovegrove B (2006) Lack of torpor in free-ranging southern lesser galagos, *Galago moholi*: ecological and physiological considerations. Folia Primatol 77(6):465–476

Nicol SC, Andersen NA (1996) Hibernation in the echidna: not an adaptation to cold? In: Geiser F, Hulbert AJ, Nicol SC (eds) Adaptations to the cold. University of New England Press, Armidale, pp 7–12

Nowack J, Mzilikazi N, Dausmann KH (2010) Torpor on demand: heterothermy in the non-lemur primate *Galago moholi*. PLoS One 5(5):e10797. doi:10.1371/journal.pone.0010797

Nowack J, Mzilikazi N, Dausmann KH (submitted) Surviving the cold and dry period in Africa: behavioural thermoregulation as an alternative to heterothermy in *Galago moholi*

Ortmann S, Heldmaier G, Schmid J, Ganzhorn JU (1997) Spontaneous daily torpor in Malagasy mouse lemurs. Naturwissenschaften 84(1):28–32

Pullen SL, Bearder SK, Dixson AF (2000) Preliminary observations on sexual behavior and the mating system in free-ranging lesser galagos (*Galago moholi*). Am J Primatol 51(1):79–88

Rakotondranary SJ, Hapke A, Ganzhorn JU (2011) Distribution and morphological variation of Microcebus spp. along an environmental gradient in southeastern Madagascar. Int J Primatol 32(5):1037–1057

Ratovonamana RY, Rajeriarison C, Roger E, Kiefer I, Ganzhorn JU (submitted). Impact of livestock grazing on vegetation characteristics in southwest Madagascar. Plant Ecol Evol

Schmid J (1999) Sex-specific differences in activity patterns and fattening in the gray mouse lemur (*Microcebus murinus*) in Madagascar. J Mammal 80(3):749–757

Schmid J, Ruf J, Heldmaier G (2000) Metabolism and temperature regulation during daily torpor in the smallest primate, the pygmy mouse lemur (*Microcebus myoxinus*) in Madagascar. J Comp Physiol B 170(1):59–68

Scholes RJ, Walker BH (1993) An African savannah: synthesis of the Nylsvlei study. Cambridge University Press, Cambridge

Turbill C, Bieber C, Ruf T (2011) Hibernation is associated with increased survival and the evolution of slow life histories among mammals. Proc R Soc B 278(1723):3355–3363

Wein J (2010) Effects of ambient temperature on tropical hibernation in the lesser hedgehog tenrec, *Echinops telfairi*. Dissertation, University Hamburg, Germany

Wilde CJ, Knight CH, Racey PA (1999) Influence of torpor on milk protein composition and secretion in lactating bats. J Exp Zool 284(1):35–41

Wright PC (1999) Lemur traits and Madagascar ecology: coping with an island environment. AJPA 110(29):31–72

Chapter 3
Tropical Heterothermy: Does the Exception Prove the Rule or Force a Re-Definition?

Cindy I. Canale, Danielle L. Levesque and Barry G. Lovegrove

Abstract Recent interest in heterothermy in the tropics and the subtropics has raised issues with the existing definitions of torpor. The current methods used to distinguish and define patterns of heterothermy are insufficient in face of the numerous forms of torpor expression and high daily variation in normothermic body temperature (T_b) observed in species inhabiting the tropics. Tropical heterothermy often occurs at highly variable ambient temperatures that may lead to a continuum between hibernation, daily torpor and normothermia with no clear distinction between states. While we do not seek to redefine torpor in this review, by listing torpor patterns that fall outside the usual categories (the exceptions to the rule), we discuss these thermoregulatory behaviours in terms of the energetics and evolution of heterothermy under warm climates.

3.1 Defining Heterothermy in Endotherms

Strict homeothermy bears a high energetic cost and, therefore heterothermy has been selected in numerous species in response to harsh environmental constraints such as low temperatures or reduced energy supply (Lyman et al. 1982). Torpor, commonly defined as 'a physiological state associated with controlled reductions of metabolism and T_b resulting in energy savings compared to defence of

Cindy I. Canale and Danielle L. Levesque contributed equally to the study.
Cindy I. Canale is Claude Leon Postdoctoral Fellow at University of KwaZulu-Natal

C. I. Canale · D. L. Levesque · B. G. Lovegrove (✉)
School of Biological and Conservation Sciences,
University of KwaZulu-Natal, Private Bag X01,
Scottsville 3209, South Africa
e-mail: lovegrove@ukzn.ac.za

T. Ruf et al. (eds.), *Living in a Seasonal World*,
DOI: 10.1007/978-3-642-28678-0_3,

normothermic T_b' (Geiser 2004), was traditionally thought to be restricted to temperate, cold climate zones, where the energy saved by lowering T_b was high. However, recent studies revealed an increasing number of heterothermic species living in the tropics: bats, birds, echidnas, primates, tenrecs, marsupials (Geiser et al. 2008; Heldmaier et al. 2004; Lovegrove and Genin 2008; McKechnie and Lovegrove 2002; McKechnie and Mzilikazi 2011; Stawski and Geiser 2011). The increasing interest in the study of torpor in tropical and subtropical climates has led to the discovery of more varied patterns of heterothermy than was originally thought possible (Cossins and Barnes 1996; Geiser and Ruf 1995; Lovegrove 2000, 2012; McKechnie and Mzilikazi 2011). Classically, daily torpor was defined as short shallow torpor bouts, generally lasting less than 24 h, with body temperatures remaining slightly elevated above ambient temperatures (T_a). Hibernation involves multiday torpor bouts that may last as long as several days with T_b decreasing to a degree or two above T_a (Geiser and Ruf 1995).

Recent reviews have questioned the accuracy of the current definitions of torpor arguing that they are often arbitrary in application and that the lines between daily variation in T_b and torpor are often blurred (Angilletta et al. 2010; Boyles et al. 2011a; Brigham et al. 2011). Similarly, not only can the definitions of torpor be unclear, but also the degree to which it is expressed. Thus, whether an animal is classed as a daily heterotherm or as a hibernator remains vague, all the more so when torpor patterns appear to be governed largely by environmental temperatures (McKechnie and Mzilikazi 2011). Torpor in the tropics, often uncoupled from low ambient temperatures, provides some of the most clear-cut examples of the failings of the current terminology.

Here, we argue that the current definitions for torpor, daily heterothermy and hibernation are not appropriate to all heterotherms, especially those in the tropical and subtropical habitats. By listing the current methods used to define torpor and reviewing the species that fall outside the usual categories, we hope to illustrate the problems that can occur using current definitions. We wish to call for a systematic review of current terminology and seek to caution the inference of various states (active, resting, torpid) from T_b patterns alone.

3.2 Current Methods for Determining Torpor Expression

The most accurate method for determining torpor expression would be to measure metabolic rate (MR) in free-ranging animals over long time periods. Unfortunately, the complicated logistics of measuring MR in the field has led the vast majority of studies to use T_b as a proxy. Furthermore, as measuring core T_b involves the use of invasive methods (surgery), skin temperature (T_{sk}) is often used instead. Whereas T_{sk} has been shown to approximate T_b over a wide range of activity states and body temperatures (Barclay et al. 1996; Dausmann 2005), it is most accurate when an animal is resting at stable T_a and can be inaccurate when observing active animals in changing thermal environments (Willis and Brigham 2003).

3.2.1 *Threshold* T_b

When using T_b or T_{sk}, the most common method for determining torpor expression is via the use of a threshold value differentiating torpid T_b from normothermic T_b (Barclay et al. 2001; Boyles et al. 2011a; Willis 2007). While there are numerous theoretical and logistical limitations of using this method (see Boyles et al. 2011a) it remains the simplest way to determine the number, depth and duration of torpor bouts. Whereas some studies have provided clear reasons for selecting a particular T_b threshold a large number simply select an arbitrary T_b, usually 30°C, and provide little or no justification for the choice (Barclay et al. 2001). The most common, reliable but underused methods for empirically determining cut-off T_bs are:

(1) *The bi-, multi-modal frequency distribution of T_b measurements* (McKechnie et al. 2007). Ideally torpid and normothermic T_b distributions form clear more or less normally distributed groupings, and cut-off temperature can be determined using the lesser used transitional temperatures (see Geiser and Mzilikazi 2011 for a good example of this). This method works best when T_a is stable and thus has no effect on T_b distribution.
(2) *Recording MR during entry into torpor and calculating the T_b at which MR decreases below basal metabolic rate (BMR)*. Willis (2007) described this method in detail and calculated a general equation taking into account the effects of T_a and body mass on threshold T_b. Unfortunately, there has been little success in establishing the use of either the method or the equation.
(3) *Active T_b*. Barclay et al. proposed that the threshold T_b could be calculated using the lowest T_b recorded immediately prior to an animal exiting its rest site. This method proved controversial as bats and birds leave roosts while still in a semi-torpid state and rely on heat generated by flight muscle activity to complete the final stages of rewarming (Willis and Brigham 2003) and many daily heterotherms rewarm with using T_a well before the start of their activity periods (Geiser et al. 2004). A more accurate application of this concept could be achieved using daily profiles of locomotor activity recorded either by a receiver in the laboratory or movement sensors in the field and to identify the threshold T_b under which locomotor activity rapidly declines (Canale et al. 2011).

3.2.2 *Metrics*

Recently, the concept of using metrics to quantify the level of heterothermy has been proposed. The primary advantage being the inclusion of homeotherms and the potential benefits they gain from circadian changes in T_b (Boyles et al. 2011b; Gordon 2009). The first proposed that the absolute differential in daily body temperatures can be used to calculate the instability of circadian rhythms, in species with regular variation in T_b (Gordon 2009). However, this method highly underestimated the level of heterothermy in boreal and arctic hibernators (Boyles et al. 2011b) as they maintain a low and constant T_b during these periods (Gordon 2009). Moreover, it is only accurate for species that maintain precise and regulated circadian T_b rhythms. A second metric, the heterothermy index (HI), has been proposed for comparison among individuals and species (Boyles et al. 2011b) by estimating the magnitude of the heterothermic response, i.e. the time spent away from normothermic T_b over any time frame. The limitations of this method have been presented in greater detail elsewhere (see review by Brigham 2011), but perhaps the most important of those listed is that the HI places species with shallow, long-duration torpor bouts and species with deep, short-duration bouts on the same scale, despite the fact that these responses are obviously different in terms both of energy savings and the adaptive value of heterothermy.

3.2.3 Thermoregulating Versus Thermoconforming

While not explicitly stated as such, the majority of studies on animals hibernating at high and variable T_a have used $T_b - T_a$ differentials to determine torpor expression (Dausmann et al. 2005; Kobbe et al. 2011; Lovegrove and Genin 2008). Thus when an animal's T_b is only slightly above T_a, and tracks T_a as it changes, it is in torpor, and it is normothermic when T_b is independent of T_a (Arlettaz et al. 2000). However, such methods underestimate the potential occurrence of torpor at T_bs below normothermic levels but higher than T_a and are only applicable when the animal is thermoconforming and not in cases where they defend a set torpid T_b above T_a (Barclay et al. 2001).

3.3 The Exceptions that (Dis)prove the Rule

Whereas the above methods have their advantages, their application can be complicated in certain situations. A growing number of novel patterns of torpor expression, many at high ambient temperatures have been described in the last few decades that question classical definitions.

3.3.1 Continuum Between Daily Torpor and Hibernation?

The most clear-cut of these exceptions are species with levels of torpor expression intermediate to daily torpor and hibernation. In these species, torpor expression appears to be highly flexible in response to current environmental conditions. Whereas many of these species appear capable of long-term hibernation, thus allowing them to be classified as hibernators sensu Geiser and Ruf (1995), torpor expression can vary within and between individuals to such an extent that a strict delineation between daily torpor and hibernation can be difficult. This has been demonstrated in a wide range of species such as echidnas (*Tachyglossus aculeatus*, Brice 2009), bats (*Nyctophilus bifax*, Stawski et al. 2009) and many Afrotropical species including elephant shrews (*Elephantulus edwardii, Elephantulus myurus*, Geiser and Mzilikazi 2011; Mzilikazi and Lovegrove 2004), Egyptian free-tailed bats (*Tadarida aegyptiaca*, Cory Toussaint et al. 2010) and mouse lemurs (*Microcebus murinus, Microcebus griseorufus* (Kobbe et al. 2011; Schmid and Ganzhorn 2009; see McKechnie and Mzilikazi 2011 for a full review). One of the clearest examples of individual flexibility in thermoregulation has been observed in *Microcebus griseorufus* exhibiting different energy saving strategies influenced by ambient temperatures. *M. griseorufus* showed daily torpor bouts and also long-term hibernation. Individuals have been observed to enter daily torpor from the early morning until they warmed up with the rising T_a, with an absolute minimal T_{sk} of 7°C. The longest hibernation phase without arousals lasted 45 days and the lowest T_{sk} observed was 6.5°C (Kobbe et al. 2011). It is unclear whether these patterns are truly distinct intermediates between daily heterothermy and hibernation or simply indicative of the high degree of physiological flexibility in torpor expression selected in response to unpredictable anomalies in rainfall associated with the El Niño Southern Oscillation, widespread in Afrotropical, Australasian and Indomalayan zones (Canale and Henry 2010; Kobbe et al. 2011; Lovegrove 2003).

3.3.2 Continuum Between Normothermy and Torpor?

In addition to increasingly blurred lines between daily torpor and hibernation, the lines between circadian changes in T_b and torpor expression are continually under contention (Angilletta et al. 2010; Boyles et al. 2011b). This debate has been more predominant in birds, where the delineation between nocturnal hypothermia (controlled yet non-torpid nocturnal reductions in T_b) and torpor has been highly contested (Lovegrove and Smith 2003; McKechnie and Lovegrove 2002; Schleucher and Prinzinger 2006). As with the use of a threshold T_b, the distinction between the two states is often arbitrary with studies considering reductions of up to 50% of BMR as circadian changes in MR and not torpor (Lovegrove and Smith 2003).

It is becoming increasingly apparent that T_b alone is not indicative of the physiological state of the animal. Many species show a rather large overlap between active and resting body temperatures (see Boyles et al. 2011a for a review). Multiple species have been shown to be active, and perform optimally, at T_b close to or <30°C, the most commonly used cut-off T_b (Brice et al. 2002). For example, a recent paper shows that torpid planigales (*Planigale gilesi*) move to basking sites with T_b as low as 13.8°C (Warnecke and Geiser 2009). The distinction between the two states is especially difficult in baso- and meso-endotherms (sensu Lovegrove

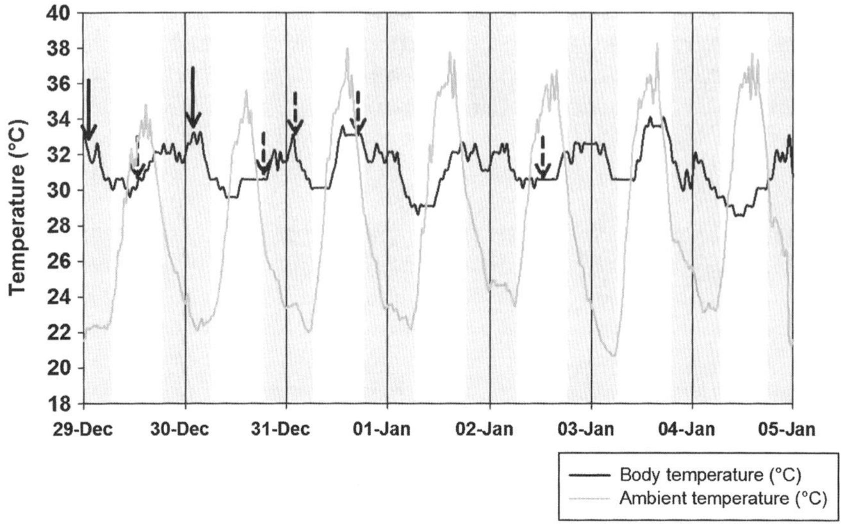

Fig. 3.1 Body temperature of a nocturnal free-ranging male *Setifer setosus* over the course of 7 days during the rainy (breeding season). *Arrows* indicate times which the animal was located and found to be active (solid) or *dashed* (resting). On two occasions, the 31-December and the 4-January, T_b was highest when the animal was in the nest site resting and decreased when the animal became active. *Grey horizontal bars* indicate the dark phase and the *horizontal lines* midnight

2012). These species, most commonly found in the tropics and semi-tropics, have high variations in normothermic T_b and have highly variable torpor patterns. Data on free-ranging echidnas in the hot dry climate of south-west Queensland showed that resting animals had higher T_bs than active animals (Brice et al. 2002). The authors in this study took a novel and cautious view of torpor expression and defined it as occurring when an individual had remained in a rest site for greater than 24 h and had lower T_b than other individuals in similar situations.

Similar difficulties in differentiating torpor from highly variable normothermic T_b patterns have been found in free-ranging greater hedgehog tenrecs (*Setifer setosus*—Levesque and Lovegrove unpublished data). Free-ranging T_bs from the active season (Fig. 3.1) illustrate the caution necessary when interpreting states of activity, rest and torpor from T_b alone. During a 7-day period of T_b measurement from an adult male during the breeding season, higher resting T_bs than active T_bs was observed on two occasions and the lowest recorded T_b (28.6°C) occurred during what may have been a torpor bout (thermoconforming) that was cut short by a rise in T_a. A lack of knowledge on the activity patterns of this species could have led to the false belief that the initiation of activity occurred during the high T_bs shown in late afternoon. However, this animal is strictly nocturnal and, in one case, was in the same rest location from 02:00 to 18:03. The animal became active sometime after T_b had reached its maximum and T_b decreased during activity.

Such species possessing a high range in active and normothermic T_b can also complicate the use of an MR threshold for torpor expression, since it is nearly impossible to measure BMR or calculate a thermal neutral zone. Echidnas (Brice, Levesque and Grigg unpublished data), streaked tenrecs (*Hemicentetes semispinosus*, Stephenson and Racey 1994) and hedgehog tenrecs (*Setifer setosus*, Levesque and Lovegrove unpublished data) all show a high degree of heterothermy and the T_b of resting, but not torpid, animal changes dramatically with ambient temperature making it impossible to determine MR from T_b alone.

3.3.3 Thermoconforming Versus Change in T_b Set Point

One of the clearest exceptions to the classical patterns of torpor expression is the fat-tailed dwarf lemur (*Cheirogaleus medius*, Dausmann et al. 2005). In this species, hibernation occurs in often poorly insulated arboreal nests which are subject to high variation in T_a (from 9.3 to 35.9°C). Thus even if the animal lowers its T_b set point it cannot maintain a continuous low T_b, but rather thermo-conforms throughout the hibernation period. This pattern has subsequently been observed in several species of mouse lemurs from the same family (Kobbe and Dausmann 2009; Kobbe et al. 2011; Schmid and Ganzhorn 2009), as well as another Malagasy heterotherm, the lesser hedgehog tenrec (*Echinops telfairi*, Lovegrove and Genin 2008). Interestingly, whereas individuals in poorly insulated nests did not show the spontaneous arousals common to most hibernators, those in well-insulated nests, in which T_a did not rise above 30°C, actively rewarmed. Similar patterns have also been observed in captive *E. telfairi* held under various T_a regimes (Wein 2010).

3.4 (Re)-Defining Torpor?

Tropical heterothermy is widespread enough to be more than just a single exception to the previously defined patterns of heterothermy. In fact, Lovegrove (2012) argues that heterothermy evolved in the tropics prior to the Cenozoic. However, quantifying it can be extremely difficult, especially with the methods currently available. The concept of a cut-off T_b under which the animal is considered torpid cannot be used in a species where T_b passively follows T_a. Difficulties therefore arise both in the calculation of a threshold T_b and in its application. Frequency distributions of T_b from tropical heterotherms do not show the bi- or multi-modal distributions usually seen in mammalian heterotherms making clear distinctions between thermoregulatory states nearly impossible (Lovegrove and Génin 2008). Identical T_b can be observed in animals that are resting, active and torpid (Brice et al. 2002). Similarly, the concept of active T_b as proposed by Barclay et al. (2001) is not applicable in these cases. High rest-phase ambient temperatures can lead to a decrease in T_b at the onset of activity, in stark contrast to the increase usually observed in temperate climates. Metrics used to evaluate heterothermy usually assume that changes in T_b are controlled by the animal, and whereas tropical heterotherms might lower the T_b set point, highly variable T_a makes low T_b impossible to obtain for long periods of time. Thus a metric might underestimate the use of torpor at temperatures that would be considered normothermic in an active animal.

The method that has so far had the most success in the tropics, thermoconformity, requires accurate measurements of the T_a experienced by the animal and is not applicable in defining most torpor that occurs in temperate zones. Thus we are faced with the dilemma of whether to let the exceptions support the traditional views or to force us to reconsider the original definitions.

3.5 Tropical Heterothermy: Benefits Without Cost?

Is tropical hibernation better at overcoming energetic constraints than temperate hibernation? Torpor is a well-known mechanism that reduces energy expenditure during seasonal energetic bottlenecks in both high and low latitudes. However, the physiological flexibility of torpor in tropical species allows them to rapidly adjust their energy expenditure, and thus overcome sudden energetic limitations in shorter time frames. This flexible, aseasonal use of daily torpor in response to rapid environmental changes confers high energetic benefits even during the reproductive season (Canale et al. 2012; Daniel et al. 2010; Geiser 1996; Stawski and Geiser 2010).

Torpor and hibernation at high and variable T_a provide an excellent opportunity to study the benefits conferred by warmth on torpor expression. Tropical and subtropical climates with high daily variation in T_a could confer to species greater energy savings than similar time spent torpid in arctic and temperate species. Frequent passive, often daily, exposures to high ambient temperatures are more energy efficient than costly arousals, obligatory to hibernation in cold environments (Carey et al. 2003; Humphries et al. 2003). Stripe-faced dunnarts (*Sminthopsis macroura*) can reduce the cost of arousals threefold through the use of exogenous passive heating (Lovegrove et al. 1999a). Periodic arousals in hibernating fat-tailed dwarf lemurs are much less costly with MR increasing by 40% instead of by 1000% as seen in temperate hibernators (Dausmann et al. 2009). In addition, the ability to completely forgo periodic arousals would limit the high level of oxidative stress resulting from the increased production of reactive oxygen species during active rewarming (Giroud et al. 2009). Furthermore, the negative effects of extreme metabolic suppression and long-term hypothermia could be limited at variable T_a. Short bouts of hypometabolism associated with T_b close to high T_a might allow sleep during a semi-torpid state, ensuring brain function and memory consolidation, in contrast to arctic species that suffer a sleep debt (Roth et al. 2010). Moreover, rapid rewarming tracking T_a might allow reactivation of the immune system faster than that of pathogens. Passive thermal responses leading to high T_b would confer an immune advantage for heterothermic hosts (Canale and Henry 2011). Immune activity would regularly return to maximal efficiency and the short arousal periods would give pathogens too short a time to proliferate (Prendergast et al. 2002). With, potentially, such low costs associated, with lower foraging requirements leading to lower risk of predation, it is not surprising that heterothermy is so widespread in the tropics (Geiser and Stawski 2011; Geiser et al. 2011; McKechnie and Mzilikazi 2011). The flexibility in its expression and the relative reduction in the costs are perhaps indicative of a more pleisiomorphic form of torpor (Lovegrove 2012).

3.6 Conclusions

It has become apparent that more time and caution is needed when inferring the expression of hypometabolic states from T_b patterns alone, especially those from individuals living in environments with high and variable T_a. T_b recordings should, as much as possible, be accompanied by accurate measurements of T_a, observations on the state of activity and basic knowledge on the species' MR response to constant ambient temperatures (can BMR and the thermal neutral zone be quantified?). The diversity of patterns of heterothermy in endotherms is much larger than was initially believed. There is likely a continuum between daily torpor and hibernation as well as between torpor and normothermy (Heldmaier et al. 2004). During daily torpor the extent of hypometabolism and hypothermia is usually less pronounced as compared to hypometabolism in hibernation (Geiser and Ruf 1995). However, in some species like dormice or elephant shrews, torpid MRs can be as low as minimum MR in hibernation (Lovegrove et al. 1999b; Wilz and Heldmaier 2000). It has been discussed whether daily torpor and hibernation are based on different physiological adaptations. At present no clear qualitative differentiation is known (Heldmaier et al. 2004). The physiological properties of daily torpor and hibernation seem to be very similar. It remains to be determined if differences observed in torpor patterns and thermal physiology are ecological or physiological. The exceptions and mid-points presented above provide excellent case studies for both the adaptive value of heterothermy and the evolution of homeothermy via heterothermy (Angilletta et al. 2010; Grigg 2004; Lovegrove 2012).

As with the reviews cited throughout, this chapter raises more questions than it answers. More work, in both field and in laboratory settings, is needed to fully characterise both the physiology and the energetics of torpor at high and variable ambient temperatures (>30°C). What is happening during the high and variable torpid T_b that allows the animals to forgo the periodic arousals deemed obligatory in cold temperature hibernators? Should the definitions of daily torpor, hibernation and even torpor itself be broadened to include these different forms or do they require their own category?

Acknowledgments The authors thank Prof. Fritz Geiser for suggested improvements to the manuscript. The research was financed by incentive grants from the University of KwaZulu-Natal, the National Research Foundation, South Africa and a Claude Leon Foundation Postdoctoral Fellowship to CIC and a post-graduate scolarship from National Science and Engineering Research Council (Canada) to DLL.

References

Angilletta MJJ, Cooper BS, Schuler MS, Boyles JG (2010) The evolution of thermal physiology in endotherms. Front Biosci E 2:861–881

Arlettaz R, Ruchet RC, Aeschimann J, Brun E, Genoud M, Vogel P (2000) Physiological traits affecting the distribution and wintering strategy of the bat *Tadarida teniotis*. Ecology 81:1004–1014

Barclay RMR, Kalcounis MC, Crampton LH, Stefan C, Vonhof MJ, Wilkinson L, Brigham RM (1996) Can external radiotransmitters be used to assess body temperature and torpor in bats? J Mammal 77:1102–1106

Barclay RMR, Lausen CL, Hollis L (2001) What's hot and what's not: defining torpor in free-ranging birds and mammals. Can J Zool 79:1885–1890

Boyles JG, Smit B, McKechnie AE (2011a) Does use of the torpor cut-off method to analyze variation in body temperature cause more problems than it solves? J Therm Biol 36:373–375

Boyles JG, Smit B, McKechnie AE (2011b) A new comparative metric for estimating heterothermy in endotherms. Phys Biochem Zool 84:115–123

Brice PH (2009) Thermoregulation in monotremes: riddles in a mosaic. Aust J Zool 57(3–4):255–263

Brice PH, Grigg GC, Beard LA, Donovan JA (2002) Patterns of activity and inactivity in echidnas (*Tachyglossus aculeatus*) free-ranging in a hot dry climate: correlates with ambient temperature, time of day and season. Aust J Zool 50:461–475

Brigham R, Willis C, Geiser F, Mzilikazi N (2011) Baby in the bathwater: should we abandon the use of body temperature thresholds to quantify expression of torpor? J Therm Biol 36:376–379

Canale CI, Henry PY (2010) Adaptive phenotypic plasticity and resilience of vertebrates to increasing climatic unpredictability. Clim Res 43:135–147

Canale CI, Henry PY (2011) Energetic costs of the immune response and torpor use in a primate. Funct Ecol 25:557–565

Canale CI, Perret M, Thery M, Henry PY (2011) Physiological flexibility and acclimation to food shortage in a heterothermic primate. J Exp Biol 214:551–560

Canale CI, Perret M, Henry P-Y (2012) Torpor use during gestation and lactation in a heterothermic primate. Naturwissenschaften 99:159–163

Carey HV, Andrews MT, Martin SL (2003) Mammalian hibernation: cellular and molecular responses to depressed metabolism and low temperature. Physiol Rev 83:1153–1181

Cory Toussaint D, McKechnie AE, van der Merwe M (2010) Heterothermy in free-ranging male Egyptian free-tailed bats (*Tadarida aegyptiaca*) in a subtropical climate. Mammal Biol 75:466–470

Cossins AR, Barnes BM (1996) Southern discomfort. Nature 382:582–583

Daniel S, Korine C, Pinshow B (2010) The use of torpor in reproductive female hemprich's long-eared bats (*Otonycteris hemprichii*). Physiol Biochem Zool 83:142–148

Dausmann KH (2005) Measuring body temperature in the field-evaluation of external vs. implanted transmitters in a small mammal. J Therm Biol 30:195–202

Dausmann KH, Glos J, Ganzhorn JU, Heldmaier G (2005) Hibernation in the tropics: lessons from a primate. J Comp Physiol B 175:147–155

Dausmann KH, Glos J, Heldmaier G (2009) Energetics of tropical hibernation. J Comp Physiol B 179:345–357

Geiser F (1996) Torpor in reproductive endotherms. In: Geiser F, Hulbert AJ, Nicol SC (eds) Adaptations to the cold, 10th international hibernation symposium, University of New England Press, pp 81–86

Geiser F (2004) Metabolic rate and body temperature reduction during hibernation and daily torpor. Annu Rev Physiol 66:239–274

Geiser F, Mzilikazi N (2011) Does torpor of elephant shrews differ from that of other heterothermic mammals? J Mammal 92:452–459

Geiser F, Ruf T (1995) Hibernation versus daily torpor in mammals and birds—physiological variables and classification of torpor patterns. Physiol Zool 68:935–966

Geiser F, Stawski C (2011) Hibernation and torpor in tropical and subtropical bats in relation to energetics, extinctions, and the evolution of endothermy. Integr Comp Biol 51:337–348

Geiser F, Drury RL, Körtner G, Turbill C, Pavey CR, Brigham RM (2004) Passive rewarming from torpor in mammals and birds: energetic, ecological and evolutionary implications. In: Barnes BM, Carey HV (eds) Life in the cold. Evolution, mechanisms, adaptation, and application, vol 27. Biological Papers of the University of Alaska, University of Alaska, Fairbanks, pp 199–208

Geiser F, Christian N, Cooper CE, Körtner G, McAllan BM, Pavey CR, Turner JM, Warnecke L, Willis CKR, Brigham M (2008) Torpor in marsupials: recent advances. In: Lovegrove BG, McKechnie AE (eds) Hypometabolism in animals: hibernation, torpor and cryobiology. Interpack Books, Pietermaritzburg, pp 297–306

Geiser F, Stawski C, Bondarenco A, Pavey CR (2011) Torpor and activity in a free-ranging tropical bat: implications for the distribution and conservation of mammals? Naturwissenschaften 98:447–452

Giroud S, Perret M, Gilbert C, Zahariev A, Goudable J, Le Maho Y, Oudart H, Momken I, Aujard F, Blanc S (2009) Dietary palmitate and linoleate oxidations, oxidative stress, and DNA damage differ according to season in mouse lemurs exposed to a chronic food deprivation. Am J Physiol Reg I 297:R950–R959

Gordon CJ (2009) Quantifying the instability of core temperature in rodents. J Therm Biol 34:213–219

Grigg GC (2004) An evolutionary framework for studies of hibernation and short term torpor. In: Barnes BM, Carey HV (eds) Life in the cold. Evolution, mechanisms, adaptation, and application, vol 27. Biological Papers of the University of Alaska, University of Alaska, Fairbanks, pp 1–11

Heldmaier G, Ortmann S, Elvert R (2004) Natural hypometabolism during hibernation and daily torpor in mammals. Resp Physiol Neurobiol 141:317–329

Humphries MM, Thomas DW, Kramer DL (2003) The role of energy availability in mammalian hibernation: a cost-benefit approach. Physiol Biochem Zool 76:165–179

Kobbe S, Dausmann KH (2009) Hibernation in Malagasy mouse lemurs as a strategy to counter environmental challenge. Naturwissenschaften 96:1221–1227

Kobbe S, Ganzhorn JU, Dausmann KH (2011) Extreme individual flexibility of heterothermy in free-ranging Malagasy mouse lemurs (*Microcebus griseorufus*). J Comp Physiol B 181:165–173

Lovegrove BG (2000) The zoogeography of mammalian basal metabolic rate. Am Nat 156:201–219

Lovegrove BG (2003) The influence of climate on the basal metabolic rate of small mammals: a slow–fast metabolic continuum. J Comp Physiol B 173:87–112

Lovegrove BG (2012) The evolution of endothermy in Cenozoic mammals: a plesiomorphic-apomorphic continuum. Biol Rev 87:128–162

Lovegrove BG, Genin F (2008) Torpor and hibernation in a basal placental mammal, the Lesser Hedgehog Tenrec *Echinops telfairi*. J Comp Physiol B 178:691–698

Lovegrove BG, Smith GA (2003) Is "nocturnal hypothermia" a valid physiological concept in small birds? A study on Bronze Mannikins (*Spermestes cucullatus*). Ibis 145:547–557

Lovegrove BG, Kortner G, Geiser F (1999a) The energetic cost of arousal from torpor in the marsupial *Sminthopsis macroura*: benefits of summer ambient temperature cycles. J Comp Physiol B 169:11–18

Lovegrove BG, Lawes MJ, Roxburgh L (1999b) Confirmation of pleisiomorphic daily torpor in mammals: the round-eared elephant shrew *Macroscelides proboscideus* (Macroscelidea). J Comp Physiol B 169:453–460

Lyman CP, Willis JS, Malan A, Wang LCH (1982) Hibernation and torpor in mammals and birds. Academic Press, New York

McKechnie AE, Lovegrove BG (2002) Avian facultative hypothermic responses: a review. Condor 104:705–724
McKechnie AE, Mzilikazi N (2011) Heterothermy in Afrotropical mammals and birds: a review. Integr Comp Biol 51:1–15
McKechnie AE, Chetty K, Lovegrove BG (2007) Phenotypic flexibility in the basal metabolic rate of laughing doves: responses to short-term thermal acclimation. J Exp Biol 210:97–106
Mzilikazi N, Lovegrove BG (2004) Daily torpor in free-ranging rock elephant shrews, *Elephantulus myurus*: a year-long study. Physiol Biochem Zool 77:285–296
Prendergast BJ, Freeman DA, Zucker I, Nelson RJ (2002) Periodic arousal from hibernation is necessary for initiation of immune responses in ground squirrels. Am J Physiol Reg I 282:R1054–R1062
Roth TC, Rattenborg NC, Pravosudov VV (2010) The ecological relevance of sleep: the trade-off between sleep, memory and energy conservation. Philos Trans Royal Soc B Biol Sci 365:945–959
Schleucher E, Prinzinger R (2006) Heterothermia and torpor in birds: highly specialized physiological ability or just deep "nocturnal hypothermia"? The limitations of terminology. Acta Zool Sinica 52:393–396
Schmid J, Ganzhorn JU (2009) Optional strategies for reduced metabolism in gray mouse lemurs. Naturwissenschaften 96:737–741
Stawski C, Geiser F (2010) Seasonality of torpor patterns and physiological variables of a free-ranging subtropical bat. J Exp Biol 213:393–399
Stawski C, Geiser F (2011) Do season and distribution affect thermal energetics of a hibernating bat endemic to the tropics and subtropics? Am J Physiol Reg I 301:R542–R547
Stawski C, Turbill C, Geiser F (2009) Hibernation by a free-ranging subtropical bat (*Nyctophilus bifax*). J Comp Physiol B 179:433–441
Stephenson PJ, Racey PA (1994) Seasonal variation in resting metabolic rate and body temperature of streaked tenrecs, *Hemicentetes nigriceps* and *H. semispinosus* (Insectivora: Tenrecidae). J Zool 232:285–294
Warnecke L, Geiser F (2009) Basking behaviour and torpor use in free-ranging *Planigale gilesi*. Aust J Zool 57:373–375
Wein J (2010) Effects of ambient temperature on tropical hibernation in the lesser hedgehog tenrec, *Echinops telfairi*. Ph.D. dissertation, Universität Hamburg, Hamburg
Willis CKR (2007) An energy-based body temperature threshold between torpor and normothermia for small mammals. Physiol Biochem Zool 80:643–651
Willis CKR, Brigham RM (2003) Defining torpor in free-ranging bats: experimental evaluation of external temperature-sensitive radiotransmitters and the concept of active temperature. J Comp Physiol B 173:379–389
Wilz M, Heldmaier G (2000) Comparison of hibernation, estivation and daily torpor in the edible dormouse, *Glis glis*. J Comp Physiol B 170:511–521

Chapter 4
Hibernation in Free-Ranging African Woodland Dormice, *Graphiurus murinus*

Nomakwezi Mzilikazi, Zimkitha Madikiza, Rebecca Oelkrug and Roderick M. Baxter

Abstract Although daily torpor is common in African animals, hibernation seems to be uncommon. In this study we investigated the use of hibernation in free-ranging African woodland dormice, *Graphiurus murinus*, during winter. We also investigated if this species made any seasonal adjustments to basal metabolic rates. *G. murinus* were heterothermic on a 100% of the measurement days. The minimum body temperature recorded was 1.5°C and the longest torpor bout without arousal was 8 days. There were no significant differences in basal metabolic rates between seasons and the measured values were similar to those previously reported in laboratory studies. We conclude that hibernation is the main adjustment that *G. murinus* utilise to deal with challenging winter conditions.

N. Mzilikazi (✉)
Centre for African Conservation Ecology, Department of Zoology,
Nelson Mandela Metropolitan University, PO Box 77000,
Port Elizabeth 6031, South Africa
e-mail: nomakwezi.mzilikazi@nmmu.ac.za

Z. Madikiza
Department of Zoology and Entomology, University of Fort Hare,
P/Bag X1314, Alice 5700, South Africa

R. Oelkrug
Department of Animal Physiology, Phillips Universitaet Marburg,
Marburg, Germany

R. M. Baxter
Department of Geology and Resource Management,
School of Environmental Sciences, University of Venda,
P/Bag X5050, Thohoyandou 0950, South Africa

T. Ruf et al. (eds.), *Living in a Seasonal World*,
DOI: 10.1007/978-3-642-28678-0_4,

4.1 Introduction

Heterothermy has been observed in many African mainland small mammal species, with representatives from bats (Cory Toussaint et al. 2010), elephant shrews (Lovegrove et al. 1999; Mzilikazi et al. 2002), golden moles (Jackson et al. 2009), hedgehogs (Hallam and Mzilikazi 2011), rodents (Webb and Skinner 1996), shrews (Baxter 1996) and primates (Nowack et al. 2010). With the exception of the elephant shrews and the golden moles whose heterothermic profiles straddle the line between daily torpor and hibernation, the majority of species tend to display daily torpor and true hibernation has not been widely documented in the mainland (McKechnie and Mzilikazi 2011). One laboratory study found that the spectacled dormouse, *Graphiurus ocularis,* enters hibernation of up to 13 days (Ridgard and Perrin 1999), and another recent study found that semi-captive hedgehogs were also capable of hibernation (Hallam and Mzilikazi 2011). As yet, no study has found hibernation in unrestrained, African mainland animals.

Although laboratory studies provide insights into potentially important physiological mechanisms that enable organisms to live in their environments, the results obtained achieve ecological and evolutionary meaning only if patterns correlate with attributes of organisms in their natural environments (Williams and Tieleman 2002). Furthermore, in order to understand the association between patterns of time allocation to various behaviours, their cost in terms of energy expenditure, their influence on life history parameters, information about variation of allotments of time and energy over the annual cycle is crucial (Stearns 1992). In this regard, data on animals' energy expenditure during different seasons and preferably different years should be obtained.

Several aspects of the woodland dormouse (*Graphiurus murinus*) ecology, behaviour and physiology make this an ideal species in which to investigate energy balance and use of hibernation by small mammals in their natural habitats. *G. murinus* has a wide distribution across South Africa excluding the arid west. They occur in a range of habitats ranging from grassland and rocky areas to woodland and forest (Skinner and Chimimba 2005). Within the Eastern Cape of South Africa, it is found only in forested areas. It is often found in areas where winter temperatures regularly fall below 0°C. Its small size, ca. 30 g (Skinner and Chimimba 2005) means it has high thermoregulatory costs. Furthermore, its diet consists largely of arthropods whose availability is greatly reduced during winter months. Although lone foragers, woodland dormice are known to aggregate during winter, with as many as eleven individuals known to huddle together. In addition, food caches are often observed in artificial nest boxes that they use (Madikiza 2010). Data obtained from a mark-capture-recapture study over a 4-year period reveal that *G. murinus* also show seasonal increases of up to 20% in body mass before the onset of winter in the Eastern Cape (Baxter, unpublished data).

Given the thermoregulatory and energetic challenges faced by this species, especially during the winter time, the aim of this study was to investigate the use of hibernation in free-ranging *G. murinus* as well as to establish whether any seasonal adjustments are made in other parameters associated with expenditure, such as body mass and basal metabolic rate (BMR).

4.2 Materials and Methods

The study was carried out at the Great Fish River Reserve (33°04′–33°09′S and 26°37′–26°49′E) about 40 km northeast of Grahamstown in the Eastern Cape, South Africa from April 2008 to June 2009. The study site is described in detail by Madikiza et al. (2010). The site is dominated by African bushwillows, *Combretum caffrum*, which are prone to rotting and provide numerous nest sites for *G. murinus*. In addition, as part of a long-term project, 70 nest boxes were deployed in the study site and these were regularly used by the dormice.

4.3 Body Temperature Measurements

Nine female *G. murinus* were captured using Sherman traps baited with oats and sunflower oil in April 2008. They were implanted with pre-calibrated, temperature sensitive, data loggers (iBBat—Alpha Mach Inc—Canada, mass ca. 1.3 g, resolution = 0.5°C) for body temperature (T_b) measurement. The mass of the data loggers did not exceed 5% of the animals' body mass. The data loggers were covered in biologically inert wax and programmed to record the body temperature in 15 min intervals, starting on 15 May–27 June 2008, resulting in data obtained over a 42-day period. For the intraperitoneal implantation of the data loggers the animals were anesthetized by inhalation anaesthesia using isoflurane (3% for induction and maintenance) in medical oxygen. Animals usually recovered within 24 h of implantation and were released at their point of capture. After 9–10 weeks a second trapping session was conducted and recaptured animals were explanted using the same procedure detailed above. Each animal was tagged with a subcutaneous transponder (Trovan, Euro I.D.) ensuring that no single animal would be subjected to repeated surgery other than that necessary for implantation and removal of data loggers.

Ambient temperature (T_a) was recorded using three iButtons placed at different locations in the field: one iButton was placed inside a nest box and another was placed in the shade. The third iButton was placed in a black tin, which was exposed to the sun so as to get an idea of passive heating, if any occurred.

4.4 Metabolic Rate Measurements

Metabolic rate measurements were based on the principle of the indirect calorimetry using open flow-through respirometry. Measurements were conducted during February ($N = 9$, summer) 2009 and July ($N = 7$, winter) 2009. None of the animals used in winter had been captured previously. In winter, no animals were captured in traps, and all the individuals were found torpid in nest boxes and hand captured. All animals were allowed to become normothermic and fully capable of coordinated movements before metabolic rate measurements were made. All measurements were carried out at 29 ± 1°C, the thermoneutral zone of this species (Whittington-Jones and Brown 1999). Measurements were conducted at a sampling interval of one sample per second over a 4-h period. BMR was calculated from the lowest mean oxygen consumption over a period of 5 min in a resting animal.

All animal experiments were approved by the Nelson Mandela Metropolitan University (NMMU) animal ethics committee (animal ethics clearance no. A08-SCI-ZOO-002) and complied with the American Physiology Society's "Guiding principles for research involving animals and human being" and the "Code of ethics for animal experimentation" manual adopted by the NMMU.

4.5 Data Analyses

Seasonal changes in mass were assessed using a paired-*t*-test for recaptured individuals equipped with a data logger. Animals were considered torpid if their body temperature decreased below 32°C. This is 3.4°C below published normothermic T_b (Whittington-Jones and Brown 1999). The torpor bout length was therefore taken as the total time during which T_b was maintained below 32°C. The minimum T_b (T_b min) was taken as the lowest T_b measured during a 24 h period. Torpor frequency was calculated for individual animals as the proportion of days during which the animals displayed torpor. The relationship between various torpor variables (T_b min, bout length) with T_a was described using regression equations. Unless otherwise stated, T_a refers to that measured in the nest boxes. All values are represented as mean ± SE.

4.6 Results

4.6.1 Seasonal Changes in Body Mass and Basal Metabolic Rate

Of the nine originally implanted animals only three were recaptured (33% recapture rate). The average body mass in April 2008 (implanted animals) was 33.1 ± 1.7 g ($N = 9$) and had decreased to 22.3 ± 1.5 g ($N = 3$), upon recapture (September) representing a significant mass loss of 32.6% in body mass over winter period ($t = 2.92$, $P < 0.05$). For the non-implanted animals (used in BMR measurements) the mean body mass in summer was 27.6 ± 0.8 g ($N = 9$) and 28.8 ± 1.5 g in winter ($N = 7$). There were no significant differences between these values ($P > 0.05$, $F_{1,16} = 0.365$). The mean BMR during summer was

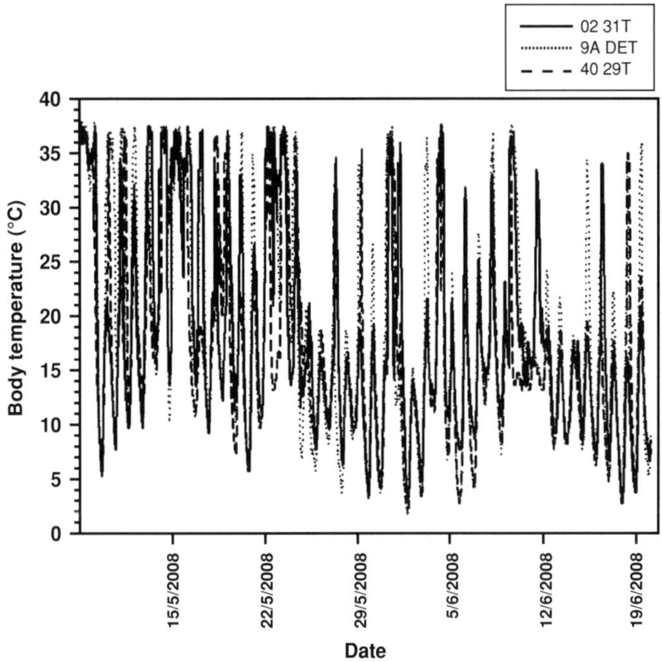

Fig. 4.1 The body temperature profiles of three free-ranging female *G. murinus* in winter 2008

1.160 ± 0.08 ml O_2 g^{-1} h^{-1} and that during the winter was 1.376 ± 0.112 ml O_2 g^{-1} h^{-1}.

4.6.2 Prevalence of Heterothermy and Torpor Parameters

The lowest T_a measured in a nest box during the study period was 2.1°C, and the mean minimum T_a was 11.1 ± 0.6°C and 7.3 ± 0.6°C in May and June, respectively. The mean maximum T_a in the nest boxes was 21.8 ± 0.7°C and 17.7° $\pm$ 0.4°C in May and June, respectively.

All animals entered torpor on all measurement days (Fig. 4.1). The mean torpor T_b min was 9.5 ± 0.5°C and the range of torpor T_b minima was 1.5–29.5°C (Fig. 4.2). The mean daily T_b maximum was 30.9 ± 0.7°C and daily T_b maxima ranged between 13.5 and 38.0°C (Fig. 4.2 inset). The T_b maxima below 32°C corresponds with data where the animals did not arouse from torpor within a 24 h period (00h00–00h00).

The lowest T_b measured was 1.5°C at $T_a = 2.1$°C. In fact, there were a number of data points that appeared below the $T_b = T_a$ line, implying that the T_b was lower than the measured T_a. Body temperatures lower than T_a are only possible if the

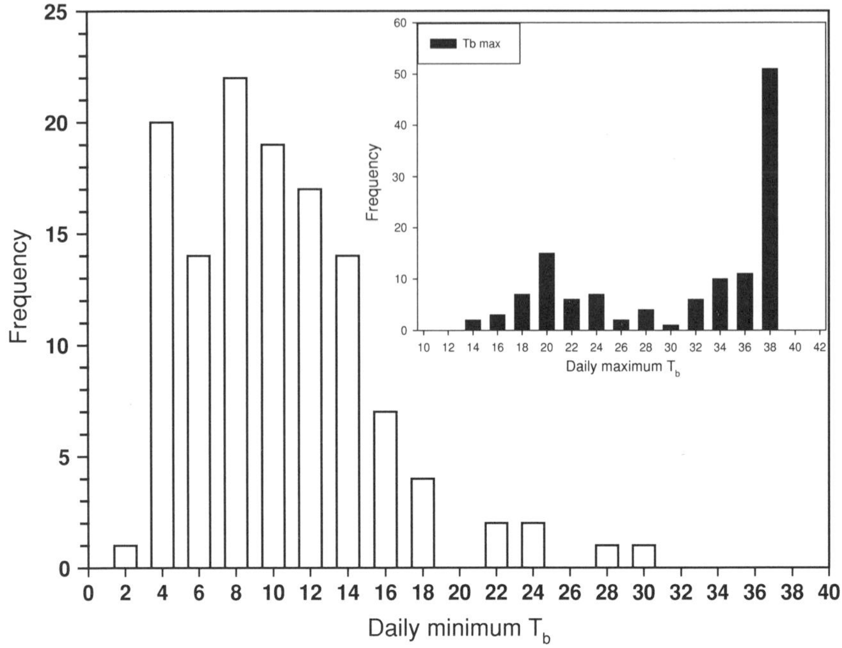

Fig. 4.2 The frequency distribution of minimum T_b measured in free-ranging *G. murinus* during winter 2008. The *inset* graph shows the frequency distribution of maximum T_b

animals were utilising evaporative water loss, a situation that is highly unlikely during winter time. Since we measured T_a in a single nest box, it is possible that at the time that the T_a was measured, the animals may have been in a different nest box or even a tree hole. The largest discrepancy between T_b and T_a (that is, T_b lower than T_a) was 3.1°C. However, the mean T_b–T_a (ΔT) was 0.2°C showing that T_b min was closely coupled to T_a min on any day. The correlation between T_b min and T_a was significant and was described by the equation: T_b min = 1.25 (T_a)–2.05; $R^2 = 0.77$, $P < 0.05$). There was a significant negative correlation between Julian day and T_b min ($R^2 = 0.106$; $P < 0.05$), showing that as winter progressed (and nights became colder) the animals displayed significantly lower T_b minima.

Mean torpor bout length was 32.5 ± 4.7 h. The longest continuous torpor bout was 8 days (Fig. 4.3) and the shortest torpor bout lasted 2.5 h. Even when the animal did not arouse during the longest bout, T_b fluctuated with T_a, although it never increased above 20°C. Although this species is nocturnal, the animals entered torpor during the night on the majority of the days. There was evidence of passive heating in so far as T_b fluctuated with T_a, but a return to normothermic T_bs was presumably achieved through heat production via non-shivering thermogenesis or shivering because most arousals were completed during the night, and full arousal seldom coincided with the increase in T_a.

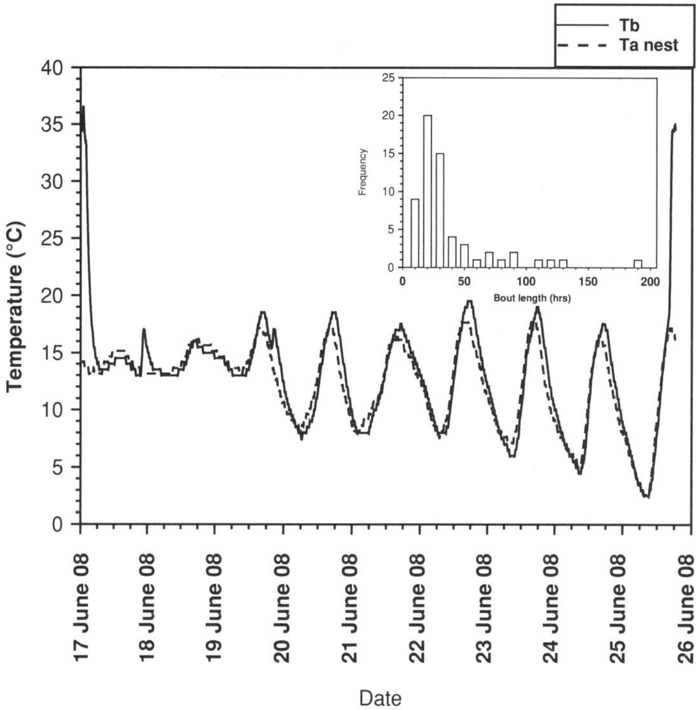

Fig. 4.3 The body temperature trace of the longest torpor bout measured in hibernating *G. murinus*. The *inset* graph shows the frequency distribution of torpor bout lengths

4.7 Discussion

Among non-volant Afrotropical mammals, true hibernation has been described in only in six species, four of them are Malagasy lemur species (McKechnie and Mzilikazi 2011). On the mainland the prevalence of hibernation seems to be uncommon, being so far recorded in *Graphiurus ocularis* (Ridgard and Perrin 1999) and *Atelerix frontalis* (Hallam and Mzilikazi 2011). The heterothermic patterns observed in golden moles, *Amblysomus hottentotus longiceps* (Scantlebury et al. 2008) and tenrecs, *Echinops telfairi* (Lovegrove and Genin 2008) are difficult to characterise as either daily torpor or hibernation and at present are probably best described as prolonged torpor. The T_b data presented here add to the body of literature documenting hibernation in mainland Africa and with a T_b min of 1.5°C represent some of the lowest body temperatures measured in the Afrotropics. This is very similar to that measured in *A. frontalis* recently.

Excluding the bats (Cory Toussaint et al. 2010), the woodland dormouse is currently the smallest known non-volant African mainland hibernator. The body mass of hibernators ranges from 10 to 80,000 g although most hibernators are in the body mass range of 10–1,000 g, with a median of 85 g (Geiser and Ruf 1995).

Consistent with other hibernators which increase body mass by 10–30% at the onset of the hibernation season (Geiser 2001), *G. murinus* increased their body mass from an average of 24.9 g (Webb and Skinner 1996) to 33.1 g (this study) representing an increase of 33% at the end of autumn. This mass gain is enough to sustain the animals through the hibernation season because BMR reduction in small hibernators may be down to 1% of normothermic values (Geiser 2001), allowing them to survive on modest fat stores for extended periods of time.

The longest torpor bout in this study was 8 days. It is unfortunate that we obtained data for only 6 weeks, as longer bouts may have been recorded in July and August, the coldest winter months. Indeed we did observe that the torpor bouts became significantly longer with the progression of winter. Although in this study we did not capture *G. murinus* in traps during winter, Madikiza et al. (2010) have trapped this species throughout the year in our study site, providing evidence of some foraging activity during winter time. It has been noted that there tends to be higher capture success of the late-breeding season offspring during winter. This observation suggests that late season offspring do not gain enough mass to sustain long torpor bouts. Furthermore, it is well known that in some hibernators forage during the hibernation season so as to supplement internal energy stores (Geiser 2001). We suggest that maximum torpor bout length in this species is a trade-off between the need to balance energy savings through extended bout lengths with maintenance of body condition through foraging.

G. murinus has been shown to utilise summer torpor in the laboratory (Webb and Skinner 1996). It would be interesting to investigate whether this species utilises summer torpor in the wild especially in light of the observation that for most rodents, reproductive activity and use of heterothermy are mutually exclusive (Goldman et al. 1986; Lovegrove and Raman 1998). Incidental observations reveal that *G. murinus* do indeed utilise daily torpor in the field during summer (Madikiza 2010) with torpid individuals being found in nest boxes throughout the year. It therefore appears that *G. murinus* is fairly flexible and opportunistic in its use of heretothermy throughout the year, exhibiting daily torpor, hibernation and summer torpor. This is similar to observations made in the edible dormouse, *Glis glis* (Wilz and Heldmaier 2000). Another factor that may potentially influence patterns of heterothermy in *G. murinus* is huddling with conspecifics. Unfortunately we did not obtain behavioural data and cannot comment on whether any of our individuals did huddle with conspecifics.

G. murinus captured in the Drakensburg reached a body mass of 70 g and hibernated continuously without arousal for 3 months (unpublished data). However, it has long been "common" knowledge the *G. murinus* is a complex of medium-sized glirid species awaiting revision. More recently, Kryštifek et al. (2004) have found significant morphometric differences between two Eastern Cape populations occurring 80 km apart. Subsequent genetic analyses have indicated that the Eastern Cape has at least two species of the *G. murinus* complex and that there are considerably more species in the rest of South Africa. Until this problem is resolved, data obtained from *G. murinus* from different localities should be treated cautiously, at least with regards to conclusions on differences in heterothermic responses in different populations.

Seasonal adjustments other than heterothermy are utilised by many small mammals and these include changes in body mass, basal metabolic rates and conductance (Heldmaier 1989; Lovegrove 2005). In this study, the winter and summer BMR were not significantly different and were quite similar to values previously published from laboratory studies (Webb and Skinner 1996; Whittington-Jones and Brown 1999). In this species, the greatest changes in winter energy requirements were presumably met through use of heterothermy (Heldmaier 1989). The animals used in this study were never in the laboratory for more than 12 h and we take our values to be a fair representation of BMR in free-ranging woodland dormice.

Acknowledgments This study was supported by a Thuthuka grant from the National Research Foundation (NRF) of South Africa. Any opinions, findings and conclusions expressed in this material are those of the authors and therefore the NRF does not accept any liability in regard thereto. Mr Brad Fike graciously hosted us and assisted with logistics at the Great Fish River Reserve (GFRR). The Eastern Cape Parks Board granted permission to work at GFRR.

References

Baxter RM (1996) Evidence for spontaneous torpor in *Crocidura flavescens*. Acta Theriol 41:327–330

Cory Toussaint D, McKechnie AE, Van der Merwe M (2010) Heterothermy in free-ranging male Egyptian free-tailed bats (*Tadarida aegyptiaca*) in a subtropical climate. Mamm Biol 75: 466–470

Geiser F (2001) Hibernation: endotherms. In: Encyclopedia of life sciences. Macmillan Publishers, New York

Geiser F, Ruf T (1995) Hibernation versus daily torpor in mammals and birds: physiological variables and classification of torpor patterns. Phys Zoo 68:935–966

Goldman BD, Darrow JM, Duncan MJ, Yogev L (1986) Photoperiod, reproductive hormones, and winter torpor in three hamster species. In: Heller HC, Musacchia XJ, Wang LCH (eds) Living in the cold: physiological and biochemical adaptations. Elsevier, New York, pp 341–351

Hallam SL, Mzilikazi N (2011) Heterothermy in the southern African hedgehog, *Atelerix frontalis*. J Comp Physiol B 181:437–445

Heldmaier G (1989) Seasonal acclimatization of energy requirements in mammals: functional significance of body weight control, hypothermia, torpor and hibernation. In: Wieser W, Gnaiger E (eds) Energy transformations in cells and organisms. Georg Thieme Verlag, Stuttgart, pp 130–139

Jackson C, Setsaas T, Robertson M, Scantlebury M, Bennett N (2009) Insights into torpor and behavioural thermoreguation of the endangered Juliana's golden mole. J Zool 278:299–307

Kryštufek B, Haberl W, Baxter RM, Zima J (2004) Morphology and karyology of two populations of the woodland dormouse, *Graphiurus murinus* in the Eastern Cape, South Africa. Folia Zool 53:339–350

Lovegrove BG (2005) Seasonal thermoregulatory responses in mammals. J Comp Physiol B 175:231–247

Lovegrove BG, Génin F (2008) Torpor and hibernation in a basal placental mammal, the lesser hedgehog tenrec, *Echinops telfairi*. J Comp Physiol B 178:691–698

Lovegrove BG, Lawes MJ, Roxburgh L (1999) Confirmation of pleisiomorphic daily torpor in mammals: the round-eared elephant shrew *Maroscelides proboscideus* (Macroscelidea). J Comp Physiol B 169:453–460

Lovegrove BG, Raman J (1998) Torpor patterns in the pouched mouse (*Saccostomus camprestris*; Rodentia): a model animal for unpredictable environments. J Comp Physiol B 168:303–312

Madikiza ZKJ (2010) Population Biology and aspects of socio-spatial organisation of the woodland dormouse, *Graphiurus murinus* (Desmaret 1822) in the Great Fish River Reserve, South Africa. MSc dissertation, University of Fort Hare

Madikiza ZKJ, Bertolino S, Baxter RM, Do Linh San E (2010) Seasonal, sexual and age related variations in the live trapping success of woodland dormice, *Graphiurus murinus*. Zool Stud 49:797–805

McKechnie AE, Mzilikazi N (2011) Heterothermy in Afrotropical mammals and birds: a review. Integr Comp Biol. doi:10.1093/icb/icr035:1-15

Mzilikazi N, Lovegrove BG, Ribble DO (2002) Exogenous passive heating during torportorpor arousal in free-ranging rock elephant shrews, *Elephantulus myurus*. Oecologia 133:307–314

Nowack J, Mzilikazi N, Dausmann K (2010) Torpor on demand: heterothermy in the non-lemur primate *Galago moholi*. PLoS One 5:e10797

Ridgard BW, Perrin MR (1999) Thermoregulation and patterns of torpor in the spectacled dormouse *Graphiurus ocularis* (Smith 1829):Gliridae. Trop Zool 12:253–266

Scantlebury M, Lovegrove BG, Jackson C, Bennett N, Lutermann H (2008) Hibernation and non-shivering thermogenesis in the Hottentot golden mole *(Amblysomus hottentotus longiceps)*. J Comp Physiol B 178:887–897

Skinner J, Chimimba C (2005) The mammals of the southern African subregion, 3rd edn. Cambridge University Press, Cambridge

Stearns SC (1992) The evolution of life histories. Oxford University Press, Oxford

Webb PI, Skinner JD (1996) Summer torpor in African woodland dormice *Graphiurus murinus* (Myoxidae: Graphiurinae). J Comp Physiol B 166:325–330

Whittington-Jones CA, Brown CR (1999) Thermoregulatory capabilities of the woodland dormouse, *Graphiurus murinus*. S Afr J Zool 34:34–38

Williams JB, Tieleman BI (2002) Ecological and evolutionary physiology of desert birds: a progress report. Intergr Comp Biol 42:68–75

Wilz M, Heldmaier G (2000) Comparison of hibernation, estivation and daily torpor in the edible dormouse, *Glis glis*. J Comp Physiol B 170:511–521

Chapter 5
Evolutionary Ecology of Mammalian Hibernation Phenology

Jeffrey E. Lane

Abstract Hibernation is assumed to have evolved in response to environmental energy and/or water shortages, yet the environment in which it has most often been studied is the laboratory. Our understanding of the ecological and evolutionary significance of natural hibernation expression thus lags behind the impressive body of work that has been done on its physiological and biochemical mechanisms. In this chapter, I review studies that have been done on phenological variation in wild populations and argue for a tightened focus on individual variation. Climate change is altering temporal resource distributions worldwide and the impact that this may have on populations will depend on their ability to adjust their phenologies through phenotypic plasticity and/or microevolution. Making predictions regarding these two phenomena requires detailed information on the environmental and genetic contributions to, and the fitness consequences of, phenological variation. I describe each of these components, in turn, and briefly explain the analytical procedures used to calculate them. Although, to date, empirical information of this sort is relatively sparse for wild hibernators, recent studies have begun to provide it and the theoretical and analytical tools with which to undertake further study are becoming increasingly accessible. Through their application, a more thorough understanding of the role hibernation plays in the natural ecology of mammalian populations, and how these populations may be affected by climate change should be attainable.

J. E. Lane (✉)
Department of Biological Sciences, University of Alberta,
Edmonton, AB T6G 2E9, Canada
e-mail: u.columbianus@hotmail.com

T. Ruf et al. (eds.), *Living in a Seasonal World*,
DOI: 10.1007/978-3-642-28678-0_5, © Springer-Verlag Berlin Heidelberg 2012

5.1 Introduction

The phenologies of wild populations represent their timing of life history events and are assumed to have evolved so as to synchronize these occurrences with temporal abundances of resources (Both et al. 2009). These resource distributions are rapidly shifting worldwide, however, due to ongoing climate change (IPCC 2007) and phenological adjustments will thus be necessary in adapting populations to these new conditions. Not surprisingly, phenological adjustments are the most often cited ecological response to climate change (Parmesan 2006) and those species which have either not adjusted (Møller et al. 2008), or have exhibited an insufficient response to remain synchronized with their primary resources (Both et al. 2006) are more likely to be in decline.

Hibernation is widely hypothesized to be a seasonal adaptation to long-term energy and/or water shortages (Wang 1989; Humphries et al. 2003a, b; but see Lovegrove 2000; Liow et al. 2009). Yet, despite the important role that hibernation plays in the life cycle of many mammalian species, compared to phenological traits in other taxa (e.g., Przybylo et al. 2000; Charmantier et al. 2008) it has received substantially less attention in this context. We consequently have comparatively little understanding of how evolution has shaped natural hibernation expression and are limited in our ability to predict how populations of wild hibernators will be affected by climate change.

In this chapter, I will outline the sources and consequences of individual variation in hibernation phenology. Phenotypic plasticity represents one avenue by which individuals can adjust their phenology in response to variable climatic conditions (sensu Réale et al. 2003). In response to long-term directional change, however, it is thought that, because phenotypic plasticity has a theoretical limit, microevolution of phenological traits will be necessary (Visser 2008). Evolution by natural selection requires a combination of three fundamental ingredients: phenotypic variation, selection acting on this variation, and a heritable genetic basis underlying it (Endler 1986). While these features are often assumed in investigations of hibernation phenology, I will use concepts and theory from quantitative genetics and animal breeding to show how we can accurately quantify them. Although this data is currently limited, the framework presented in this chapter should allow for its collection from a diversity of species.

5.2 Intraspecific Variation and Phenotypic Plasticity

Whereas the focus in hibernation research, to date, has been on species-level patterns, if we are to understand how selection has shaped current hibernation phenologies and how these phenologies may respond to climate change, a tightened focus on individual variation is necessary. Individual variation in hibernation phenological traits is commonplace (e.g., Table 5.1) and its most striking

Table 5.1 Individual variation in five phenological traits from a wild mammal population (Coumbian ground squirrels, *Urocitellus columbianus*)

Trait	Mean (Julian days from 1-Jan; sample sizes noted in parentheses)	Annual range (± s.e.)
Adult female emergence date	116.9 (472)	19.1 ± 2.1
Adult female oestrous date	120.2 (444)	15.8 ± 2.1
Adult male emergence date	112.7 (195)	18.4 ± 2.4
Yearling female emergence date	123.1 (199)	17.2 ± 2.2
Yearling male emergence date	122.4 (212)	22.0 ± 2.7

Annual range indicates the difference in date between the first and last individual each year, averaged over a period of 18 years. Data from Lane et al. (2011)

examples are obtained by studying individuals across a range of environments. Black-tailed prairie dogs (*Cynomis ludovicianus*) are facultative hibernators resident throughout the great plains of North America (Lehmer et al. 2001). Although populations from a more benign environment (Colorado, USA) hibernate rarely (Lehmer et al. 2006), all focal individuals from a more northern population (in Saskatchewan, Canada) were shown to hibernate (Gummer 2005). Similarly, echidnas (*Tachyglossus aculeatus*) range throughout Australia and New Guinea and hibernation in the comparatively more benign climate of Kangaroo Island is more shallow, later and of shorter duration (Rismiller and McKelvey 1996) than in other locations (e.g., Tasmania; Nicol and Andersen 2002). There is also good reason to expect that the ecological consequences of this variation may be profound. For example, Landry-Cuerrier et al. (2008) showed that the energy expenditure across individual eastern chipmunks (*Tamias striatus*) expressing varying degrees of torpor varied by over threefold.

Davis (1976) hypothesized that physiological processes underlying hibernation are integrated through annual changes in the environment. Specifically, metabolism exhibits a free-running circannual cycle of approximately 11 months (Pengelley and Fisher 1963) and torpor occurs when the low phase of the metabolism cycle coincides with environmental energy shortages. This hypothesis was originally formulated to explain species-level variation among the Marmotini. However, comparison to other mammalian hibernators may shed light on the traits where we should expect to see individual variation as well as the levels of variation, and its environmental correlates. Whereas a pronounced circannual rhythm appears to limit hibernation to winter seasons in the Marmotini, other species are more flexible in their torpor patterns. Eastern chipmunks (Munro et al. 2008), edible dormice (*Glis glis*; Bieber and Ruf 2009) and pygmy possums (*Cercartetus nanus*; Geiser 2007) are all highly sensitive to environmental conditions and may hibernate for up to a year during resource shortages. I suggest, therefore, that Davis' hypothesis can be expanded. Relatively large hibernators (e.g., larger members of the Marmotini) are able to store more energy as fat prior to hibernation. In addition, the seasonality of more predictable environments was likely a useful 'guide' to the availability of resources. Over evolutionary time, circannual rhythms that matched this seasonality were thus likely to have been favored by selection. Larger hibernators and those occupying more predictable environments (Lovegrove 2000) should be more likely to have hibernation patterns heavily influenced by an innate circannual rhythm. Smaller species and populations from more stochastic environments, by comparison, should be more plastic.

In the wild, phenotypic plasticity has most often been assessed as a change in the mean value of a trait within a population (e.g., Przybylo et al. 2000) or as interpopulation differences in mean values of traits (e.g., Sheriff et al. 2011). However, phenotypic plasticity is inherently an individual-based phenomenon. Specifically, it is defined as the differential phenotypic expression of a genotype across an environmental gradient (Pigliucci 2001). For traits that are expressed multiple times throughout the lifetime, it is thus most appropriately evaluated using the reaction norm approach in which variation within an individual (or genotype) is compared against variation in the environmental trait of interest (Nussey et al. 2007). With repeated measurements of phenological traits from hibernators, levels of phenotypic plasticity can thus be estimated by using a specific type of mixed-effects model, the random-regression model:

$$y_{ij} = \mu + \beta E_j + p_i + p_{Ei}E_j + e_{ij}$$

In this model, the phenological trait, y (e.g., emergence date) is measured on individual i, at time j, and related to the environmental variable E. The fixed effects describe the population average response of y to variation in E: μ is the population mean in the average environment (i.e., the intercept if the environmental variable is mean-centered) and β is the population response to the environmental variation (i.e., the slope of y on E). The random effects then model the individual-specific deviations from the population response: p_i models consistent individual differences in the phenological trait (i.e., the variance in the intercepts of individuals; Fig. 5.1a) and p_{Ei} models individual differences in phenotypic plasticity (i.e., the variance in the slopes of individuals; Fig. 5.1b). For a more detailed description of the application of random-regression models to studies of life history variation in natural populations see Nussey et al. (2007).

A number of examples of phenotypic plasticity have been reported from wild populations (e.g., Kobbe et al. 2011) and perhaps the most convincing evidence comes from experimental studies that have manipulated resource abundance to directly examine its influence on hibernation expression (e.g., Humphries et al. 2003b). However, to my knowledge, no studies to date have quantified reaction norms for hibernation traits. Although such estimates undoubtedly require large sample sizes (Martin et al. 2011), appropriate dataset are increasingly available (e.g., Ozgul et al. 2010; Lane et al. 2011) and should provide for the ability to explicitly test a number of key hypotheses. For example, by statistically comparing reaction norm slopes across species/populations, one can test the hypothesis that those resident in more stochastic environments should display more plasticity in hibernation traits. Similarly, through appropriate phylogenetic comparisons, the relationship between phenotypic plasticity and body mass could be evaluated.

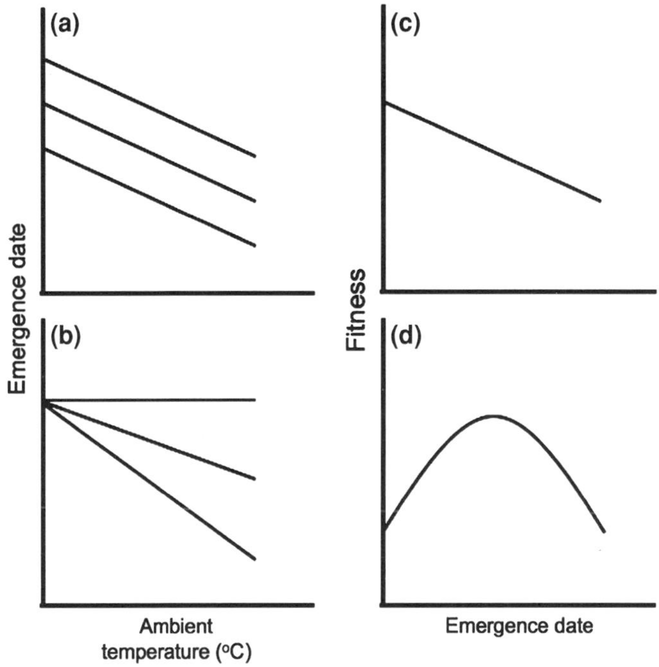

Fig. 5.1 Hypothesized relationships of the sources (**a**, **b**), and consequences (**c**, **d**) of individual variation in hibernation phenology. **a** represents the influence of an environmental variable (ambient temperature in this case) on variation in a hibernation phenological trait (emergence date). In this scenario, all individuals emerge earlier in response to warmer temperatures (they are phenotypically plastic) and individual intercepts differ, reflecting individual variation in mean emergence date which may, or may not, have a heritable genetic component. In **b** the slopes of the individuals differ, reflecting variation in their plastic response to the environmental predictor. **c** depicts negative selection on emergence date as individuals with increasingly delayed emergences experience diminished fitness. **d** depicts stabilizing selection as individuals with the mean emergence date have the highest fitness. Positive and disruptive selection would be represented by the reverse of **c** and **d**, i.e., respectively, a linear increase and a concave function

5.3 Selection

Gender-related variation in hibernation phenology has been described for a number of species (e.g., Young 1990; Kawimichi 1996). In many north temperate sciurids, for example, the emergence from hibernation of reproductive males precedes that of reproductive females (e.g., Murie and Harris 1982). This pattern (termed 'protandry'; Morbey and Ydenberg 2001) is hypothesized to have arisen due to differing selection pressures across the sexes (Michener 1983). Natural selection is presumed to favor female emergences that synchronize reproduction with the period of vegetative growth, while sexual selection favors earlier emergence of males so as to gain reproductive opportunity. However, an appropriate assessment of the adaptiveness of sex-specific emergence phenology, or indeed of much of hibernation phenology, has yet to be conducted because the strengths and directions of selection have never been calculated for any phenological traits.

Lande and Arnold (1983), using recognizable principles from multiple regression, developed the approach for measuring selection in wild populations that is most widely used today. Estimating selection in this way offers a number of advantages. First, standardized coefficients of selection are provided that can be compared across traits and taxa. Second, by analyzing multiple traits within a multiple-regression framework, direct estimates of selection are decomposed from the action of indirect selection on correlated traits. Third, by estimating the curvature in the trait-fitness relationship (by fitting quadratic functions), the action of directional, stabilizing, and disruptive selection can be evaluated (Fig. 5.1c, d). Finally, when coupled with estimates of the heritability of traits (see below), standardized selection estimates provide for an assessment of not only the direction, but also pace of microevolution (Falconer and MacKay 1996).

Numerous studies have since applied Lande and Arnold's (1983) analytical framework to investigate selection in wild populations (reviewed in Endler 1986; Kingsolver et al. 2001). However, the vast majority of these have been obtained for morphological traits (of the more than 1,500 estimates reviewed by Kingsolver et al. (2001), over 80% were for morphological traits). By comparison, only 13–17% were for life history/phenological traits and 61% of these were obtained from plants (no estimates for hibernation phenological traits were represented). To what extent the results from other traits and taxa may hold for hibernating mammals is thus currently unknown and there is a pressing need for selection estimates for hibernation traits. However, there are practical considerations to consider. For example, the median sample size represented in Kingsolver et al. (2001) was 134 and the authors' estimated that the probability with which one could reject a null hypothesis of no selection at the 95% level using this sample size was <50%. Nonetheless, I would suggest that the benefit gained from providing estimates of this sort should in many cases outweigh concerns over statistical significance.

5.4 Genetic Variation

In order for selection to lead to the evolution of a trait, shifts in the phenotypic distribution in one generation need to be passed reliably to the subsequent generations (Lynch and Walsh 1998). In other words, the trait must be heritable. Heritability refers to the proportion of phenotypic variance that has a genetic basis and is estimated by comparing measurements of phenotypic traits among individuals of known genetic relatedness (Lynch and Walsh 1998). Arguably the most intuitive way to do this is with a parent–offspring regression, in which the values of the phenotypic trait in the offspring are regressed against these values in one, or both, of their parents. The slope of the regression [or twice the slope if a mid-parent value (i.e., the mean of the maternal and paternal trait) is used] then represents the heritability of the trait. This technique has been used extensively but it suffers from a number of limitations. Specifically, by only taking advantage of a single genetic relationship (parent–offspring), it does not make efficient use of all available data (Kruuk 2004). In addition, for many species maternity is easier to identify than paternity and mother–offspring regressions are thus more commonly used than father–offspring regressions. However, maternal effects (i.e., non-genetic influences of a dam on her offspring) can lead to phenotypic similarity among kin and, unless explicitly accounted for, can confound estimates of heritability (Kruuk and Hadfield 2007). For mammals, this is especially problematic due to their prolonged periods of maternal care. For these reasons, a specific type of restricted maximum likelihood model, the 'animal model', is now used more often (Kruuk 2004), i.e.,

$$\mathbf{y} = X\mathbf{b} + \mathbf{Z}_1\mathbf{a} + \mathbf{Z}_2\mathbf{m} + \mathbf{Z}_3\mathbf{pe} + e$$

This model uses pedigree-derived measures of genetic relatedness and estimates of phenotypic similarity among all members of a population to estimate the variance components underpinning observed phenotypic variation. Specifically, **y** represents a vector of observed phenotypic values and **b** represents a vector of fixed effects. Importantly, because this is a type of mixed-effects model, variables such as age and sex can be controlled for by fitting them as fixed effects. **a**, **m**, and **pe** then represent, respectively, the vectors of additive genetic, maternal, and permanent environment effects and X and $\mathbf{Z}_{1\text{–}3}$ are the corresponding design matrices (Lynch and Walsh 1998; Kruuk 2004). The permanent environment effect arises due to individual differences other than those due to additive genetic or maternal effects (Kruuk and Hadfield 2007). Incorporating these additional random effects (**m** and **pe**), in the mixed-effects model framework, thus allows for a non-confounded estimate of the additive genetic variance. The heritability can then be estimated as the variance component associated with **a** (V_a) divided by the total phenotypic variance (V_p).

Despite its advantages, to my knowledge, our analysis of hibernation phenology in Columbian ground squirrels (*Urocitellus columbianus*) represents the only application of an animal model to a wild hibernator (Lane et al. 2011). In this population from Alberta, Canada the heritability of female emergence date was $h^2 = 0.22 \pm 0.05$ and the genetic correlation between male and female emergence dates was $r_G = 0.76 \pm 0.22$. In addition, the genetic correlation between female emergence date and oestrous date as $r_G = 0.98 \pm 0.01$. These results suggest that emergence date can evolve in response to selection and selection-induced changes will have corresponding effects across the sexes and traits.

5.5 Discussion and Future Directions

In this chapter, I have taken a perspective from evolutionary ecology to address the sources and consequences of individual variation in hibernation phenology. It is my view that, although these sorts of data are currently limited, appropriate datasets have been built (or are being built) and the analytical methodology with which to analyze them are readily accessible. By applying these frameworks, we should be able to gain a greater depth of insight into how evolution has shaped the hibernation phenotypes we currently observe, and how these phenotypes may respond to environmental change. Here, I outline three potential questions, incorporating both the theoretical and applied relevance of hibernation phenology, that could be addressed in future study.

How does selection act on variable torpor expression during hibernation? Torpid metabolic rates during hibernation average 5% of euthermic metabolic rate and can be <1% of the metabolic rate expressed during cold exposure (Geiser and Ruf 1995). These pronounced metabolic savings have led many authors to suggest that torpor expression should be maximized during hibernation (e.g., Harlow and Frank 2001). Others, however, have pointed to the ubiquitous arousals to euthermy during hibernation, as well as the ecological benefits of elevated energy expenditure, in general, to suggest that torpor entails non-trivial costs and an optimum level should thus balance energetic benefits against these costs (French 1988; Humphries et al. 2003a). Support for this hypothesis has been provided by the observation that animals fed supplemental food express less torpor (e.g., Humphries et al. 2003b) and the apparently polyphyletic loss of hibernation in southern populations of woodchucks (*Marmota monax*) and California ground squirrels (*Otospermophilus beecheyi*) (Davis 1976). The individual variation expressed in wild hibernator populations provides ideal raw material with which to test these hypotheses from an evolutionary standpoint. Specifically, by quantifying the strength and shape of selection on phenological traits (e.g., immergence and emergence dates, number and depth of torpor bouts) and coupling these measurements to estimates of the heritability of these traits, information can be gained as to the fitness consequences of individual variation and predictions can be made as to how they may evolve.

How will the phenologies of wild populations of hibernators respond to environmental change and will this change be sufficient to prevent population declines? In addition to providing for a more in depth appreciation of the evolutionary ecology of hibernation, an understanding of the phenotypic plasticity and microevolutionary potential of phenological traits is increasingly becoming of applied relevance. Their relevance to population dynamics has recently been shown in yellow-bellied marmots (*Marmota flaviventris*). Climate change has resulted in earlier emergence of the marmots, which has effectively lengthened the active season for the population and led to higher over-winter survival (Ozgul et al. 2010). However, this is the only investigation of its kind to date and predicting how populations of hibernators will respond to environmental change, more generally, clearly requires additional data. This is especially important as the environmental factors influencing phenological variation in hibernators (e.g., duration of snow cover) may be different for those species upon which most research has been conducted (i.e., ambient temperature for insectivorous birds), thus limiting our ability to generalize across taxa.

Are hypotheses to explain protandrous emergence from hibernation valid and, if so, how will shifting selection due to climate change alter these patterns? Enhanced degrees of protandry have recently been observed in some migratory bird populations and investigators have suggested that this is due to climate change (Møller 2004). Specifically, warming climates lead to more benign conditions during migration and, hence, lowered survival costs to early arrival. This relaxation of natural selection pressures then allows sexual selection to promote earlier male arrival. Thus, climate change could potentially influence not only the viability of populations, but also the social relationships within them. As many hibernators exhibit naturally protandrous emergence, climate change could conceivably further enhance this gender difference. A strong between-sex genetic correlation in Columbian ground squirrels suggests that such independent evolution of the sexes will be limited. However, estimates of natural and sexual selection on emergence dates of the two sexes have not been made in this, or any other species, thus rendering most predictions premature at this point.

In describing the focus of hibernation research over 30 years ago Davis (1976) stated: "Research on hibernating species often concerns primarily the physiological responses to low temperatures. While this topic is highly meritorious for its clinical and biochemical applications, it may perhaps neglect some aspects of adaptations to natural conditions that may provide clues to physiological processes". Although the intervening decades have yielded great advances in both hibernation physiology and evolutionary ecology, it is fair to argue that the current focus resembles that in 1976. I suggest, however, that a tightened focus on the genetic and environmental sources, as well as the fitness consequences, of variation in hibernation phenology will provide for a more thorough understanding of the ecological and evolutionary significance of natural hibernation expression.

Acknowledgments My research on hibernation phenology has been supported by the Royal Society of London and the Alberta Conservation Society, as well as, grants from the Agence Nationale de la Recherche of France (to Anne Charmantier) and the Natural Science and Engineering Research Council of Canada (to Stan Boutin).

References

Bieber C, Ruf T (2009) Summer dormancy in edible dormice (*Glis glis*) without energetic constraints. Naturwissenschaften 96:165–171

Both C, Bouwhuis S, Lessells CM, Visser ME (2006) Climate change and population declines in a long-distance migratory bird. Nature 441:81–83

Both C, van Asch M, Bijlsma RG, van den Burg AB, Visser ME (2009) Climate change and unequal phenological changes across four trophic levels: constraints or adaptations? J Anim Ecol 78:73–83

Charmantier A, McCleery RH, Cole LR, Perrins C, Kruuk LEB, Sheldon BC (2008) Adaptive phenotypic plasticity in response to climate change in a wild bird population. Science 320:800–803

Davis DE (1976) Hibernation and circannual rhythms of food consumption in marmots and ground squirrels. Q Rev Biol 51:477–514

Endler JA (1986) Natural selection in the wild. Princeton University Press, Princeton

Falconer DS, Mackay TFC (1996) Introduction to quantitative genetics, 4th edn. Longman, Harlow

French AR (1988) The patterns of mammalian hibernation. Am Sci 76:569–575

Geiser F (2007) Yearlong hibernation in a marsupial mammal. Naturwissenschaften 94:941–944

Geiser F, Ruf T (1995) Hibernation versus daily torpor in mammals and birds: physiological variables and classification of torpor patterns. Physiol Zool 68:935–966

Gummer DL (2005) Geographic variation in torpor patterns: the northernmost prairie dogs and kangaroo rats. PhD thesis. University of Saskatchewan, Saskatoon

Harlow HJ, Frank CL (2001) The role of dietary fatty acids in the evolution of spontaneous and facultative hibernation patterns in prairie dogs. J Comp Phys B 171:77–84

Humphries MM, Thomas DW, Kramer DL (2003a) The role of energy availability in mammalian hibernation: a cost-benefit approach. Physiol Biochem Zool 76:165–179

Humphries MM, Kramer DL, Thomas DW (2003b) The role of energy availability in mammalian hibernation: an experimental test in free-ranging eastern chipmunks. Physiol Biochem Zool 76:180–186

IPCC (2007) Summary for policymakers. In: Solomon S, Qin D, Manning M, Chen Z, Marquis M, Averyt KB, Tignor M, Miller HL (eds) Climate change 2007: the physical science basis. Contribution of working group I to the fourth assessment report of the intergovernmental panel on climate change. Cambridge University Press, Cambridge

Kawamichi M (1996) Ecological factors affecting annual variation in commencement of hibernation in wild chipmunks (*Tamias sibiricus*). J Mamm 77:731–744

Kingsolver JG, Hoekstra HE, Hoekstra JM, Berrigan D, Vignieri SN, Hill CE, Hoang A, Gilbert P, Beerli P (2001) The strength of phenotypic selection in natural populations. Am Nat 157:245–261

Kobbe S, Ganzhorn JU, Dausmann KH (2011) Extreme individual flexibility of heterothermy in free-ranging Malagasy mouse lemurs (*Microcebus griseorufus*). J Comp Physiol B 181: 165–173

Kruuk LEB (2004) Estimating genetic parameters in natural populations using the ‘animal model’. Philos Trans R Soc Lond B Biol Sci 359:873–890

Kruuk LEB, Hadfield JD (2007) How to separate genetic and environmental causes of similarity between relatives. J Evol Biol 20:1890–1903

Lande R, Arnold SJ (1983) The measurement of selection on correlated characters. Evolution 37:1210–1226

Landry-Cuerrier M, Munro D, Thomas DW, Humphries MM (2008) Climate and resource determinants of fundamental and realized metabolic niches of hibernating chipmunks. Ecology 89:3306–3316

Lane JE, Kruuk LEB, Charmantier A, Murie JO, Coltman DW, Buoro M, Raveh S, Dobson FS (2011) A quantitative genetic analysis of hibernation emergence date in a wild population of Columbian ground squirrels. J Evol Biol 24:1949–1959

Lehmer EM, Savage LT, Antolin MF, Biggins DE (2006) Extreme plasticity in thermoregulatory behaviors of free-ranging black-tailed prairie dogs. Physiol Biochem Zool 79:454–467

Lehmer EM, Van Horne B, Kulbartz B, Florant GL (2001) Facultative torpor in free-ranging black-tailed prairie dogs (*Cynomys ludovicianus*). J Mamm 82:551–557

Liow LH, Fortelius M, Lintulaakso K, Mannila H, Stenseth NC (2009) Lower extinction risk in sleep-or-hide mammals. Am Nat 173:264–272

Lovegrove BG (2000) Daily heterothermy in mammals: coping with unpredictable environments. In: Heldmaier G, Klingenspor M (eds) Life in the cold. Springer, Berlin, pp 29–40

Lynch M, Walsh B (1998) Genetics and the analysis of quantitative traits. Sinauer Associates, Sunderland

Martin JGA, Nussey DH, Wilson AJ, Réale D (2011) Measuring individual differences in reaction norms in field and experimental studies: a power analysis of random regression models. Methods Ecol Evol 2:362–374

Michener GR (1983) Spring emergence schedules and vernal behaviour of Richardson's ground squirrels: why do males emerge from hibernation before females? Behav Ecol Sociobiol 14: 29–38

Møller AP (2004) Protandry, sexual selection and climate change. Glob Change Biol 10:2028–2035

Møller AP, Rubolini D, Lehikoinen E (2008) Populations of migratory bird species that did not show a phenological response to climate change are declining. Proc Natl Acad Sci U S A 105:16195–16200

Morbey YE, Ydenberg RC (2001) Protandrous arrival timing to breeding areas: a review. Ecol Lett 4:663–673

Munro D, Thomas DW, Humphries MM (2008) Extreme suppression of aboveground activity by a food-storing hibernator, the eastern chipmunk (*Tamias striatus*). Can J Zool 86:364–370

Murie JO, Harris MA (1982) Annual variation of spring emergence and breeding in Columbian ground squirrels (*Spermophilus columbianus*). J Mamm 63:431–439

Nicol S, Andersen NA (2002) The timing of hibernation in Tasmanian echidnas: why do they do it when they do? Comp Biochem Physiol B 131:603–611

Nussey DH, Wilson AJ, Brommer JE (2007) The evolutionary ecology of individual phenotypic plasticity in wild populations. J Evol Biol 20:831–844

Ozgul A, Childs DZ, Oli MK, Armitage KB, Blumstein DT, Olson LE, Tuljapurkar S, Coulson T (2010) Coupled dynamics of body mass and population growth in response to environmental change. Nature 466:482–485

Parmesan C (2006) Ecological and evolutionary responses to recent climate change. Annu Rev Ecol Evol Syst 37:637–669

Pengelley ET, Fisher KC (1963) The effect of temperature and photoperiod on the yearly hibernating behavior of captive golden-mantled ground squirrels (*Citellus lateralis tescorum*). Can J Zool 41:1103–1120

Pigliucci M (2001) Phenotypic plasticity. John Hopkins University Press, Baltimore

Przybylo R, Sheldon BC, Merila J (2000) Climatic effects on breeding and morphology: evidence for phenotypic plasticity. J Anim Ecol 69:395–403

Réale D, McAdam AG, Boutin S, Berteaux D (2003) Genetic and plastic responses of a northern mammal to climate change. Proc R Soc Biol Sci Ser B 270:591–596

Rismiller PD, McKelvey MW (1996) Sex, torpor and activity in temperate climate echidnas. In: Geiser F, Hulbert AJ, Nicol SC (eds) Adaptations to the cold: tenth international hibernation symposium, University of New England Press, Armidale, pp 23–30

Sheriff MJ, Kenagy GJ, Richter M, Lee T, Tøien Ø, Kohl F, Buck CL, Barnes BM (2011) Phenological variation in annual timing of hibernation and breeding in nearby populations of Arctic ground squirrels. Proc R Soc Biol Sci Ser B 278:2369–2375

Visser ME (2008) Keeping up with a warming world: assessing the rate of adaptation to climate change. Proc R Soc Biol Sci Ser B 275:649–659

Wang LCH (1989) Ecological, physiological and biochemical aspects of torpor in mammals and birds. In: Wang LCH (ed) Comparative and environmental physiology. Animal adaptation to cold, vol 4. Springer, Berlin, pp 361–401

Young PJ (1990) Hibernating patterns of free-ranging Columbian ground squirrels. Oecologia 83:504–511

Chapter 6
Interrelationships Among Timing of Hibernation, Reproduction, and Warming Soil in Free-Living Female Arctic Ground Squirrels

Cory T. Williams, Michael J. Sheriff, Franziska Kohl, Brian M. Barnes and C. Loren Buck

Abstract The effects of climate change on hibernating species will depend, in part, on their responsiveness to environmental cues used to adjust the seasonal timing of annual events of hibernation and reproduction. Using long-term data collected from two arctic ground squirrel populations living 20 km apart in northern Alaska, we investigate the relationships between the timing of hibernation and reproduction and, in addition, the potential for change in soil temperatures to act as a proximate cue. Previously, we found that female ground squirrels living at the southern-most site, Atigun River, emerge from hibernation and give birth 13 days earlier than females at Toolik Lake. Here we show that timing of parturition was tightly linked to the termination of heterothermy and subsequent emergence from the hibernacula at both sites, whereas timing of entrance into hibernation was only weakly correlated with date of parturition in Toolik Lake females. Females ended heterothermy in spring coincident with rising soil temperatures from winter minima, but since average soil temperatures did not differ between the two sites, a single threshold in warming cannot explain the differences in timing of spring emergence and reproduction between the two populations. Earlier reproduction at Atigun is associated with earlier snowmelt, yet, how this is achieved and the relative importance of phenotypic plasticity versus genetic differences between the two populations will require further investigation.

C. T. Williams · C. Loren Buck (✉)
Department of Biological Sciences, University of Alaska Anchorage,
Anchorage, AK 99508, USA
e-mail: loren@uaa.alaska.edu

C. T. Williams
Department of Biological Sciences, University of Alberta, Edmonton,
AB T6G 2E9, Canada

M. J. Sheriff · F. Kohl · B. M. Barnes
Institute of Arctic Biology, University of Alaska Fairbanks,
902 N. Koyukuk Dr., Fairbanks, AK 99775, USA

T. Ruf et al. (eds.), *Living in a Seasonal World*,
DOI: 10.1007/978-3-642-28678-0_6,

6.1 Introduction

The arctic environment has long, cold winters and correspondingly short growing seasons that severely limit the time available each year for reproduction and growth of locally breeding animals. Climate warming is predicted to lengthen the growing season, particularly at high latitudes (Serreze and Francis 2006; Post et al. 2009), which provides increased foraging opportunities. However, biological interactions can also be disrupted due to intra- and inter-specific variability in the response to climate forcing (Høye et al. 2007). The response of species to changing seasonality depends on the rate at which populations evolve through natural selection and also on the ability of individuals to respond, via physiological mechanisms, to environmental cues that are indicative of changes in the growing season (Berteaux et al. 2004). Arctic migrants, for example, may not be capable of appropriately adjusting the timing of reproduction to match shifts in local resource status because cues used to fine-tune their migration may not reflect the conditions of the breeding grounds (e.g., Post and Forchhammer 2008).

As a year-round resident, arctic ground squirrels (*Urocitellus parryii*) overwinter on their breeding grounds. The capacity of individuals to respond adaptively to environmental change will depend on their phenotypic plasticity and the reliability of the cues used to align their annual timing of hibernation and reproduction with the local environment, since the environment within their underground hibernacula may not be reflective of changing conditions on the surface. As part of ongoing efforts to better understand how changes in seasonality impact the physiological and population ecology of arctic mammals, Sheriff et al. (2011) compared phenologies of hibernation and reproduction in arctic ground squirrels living at two sites in northern Alaska with different snow cover regimes. One study site located near the Atigun River, approximately 20 km south of the second site at Toolik Lake, becomes snow-free 27 days earlier in spring and becomes snow covered 17 days later in autumn. Timing of hibernation and reproduction of arctic ground squirrels at the Atigun site occurs 13 days earlier than at Toolik Lake, and we suggested that this variation in phenology likely reflects phenotypic plasticity generated via physiological mechanisms which provide individuals the ability to respond to changes in their environment (Sheriff et al. 2011).

Seasonal timing of hibernation and reproduction are under strong endogenous control in hibernating ground squirrels; however, the precise timing or fine-tuning of annual events reflects responsiveness to the local environment. The environmental cues used for hibernation and reproduction are not known, although snow cover and air temperature have previously been implicated as contributing factors in ground squirrels (Murie and Harris 1982; Michener 1992) and other hibernators (Ozgul et al. 2010). However, while squirrels are exposed to these factors following their spring emergence, information about current snow cover and ambient above-ground temperature is unlikely to be available to animals sequestered within hibernacula. Of cues available to hibernating ground squirrels, we propose that changes in soil temperature will be the most available and relevant to seasonal timing. During hibernation arctic ground squirrels are highly responsive to changes in soil temperature. Body temperature (T_b) during torpor closely tracks surrounding soil temperatures until the active layer freezes and cools below −3.0°C (e.g., Williams et al. 2011a). Mean (−8.9°C) and minimum (−25°C) soil temperatures across most of the heterothermic period are below minimum body temperature (T_b: −2.9°C) indicating squirrels are highly thermogenic and responsive to changes in soil temperature for most of the hibernation season (Barnes 1989; Buck and Barnes 1999a; Williams et al. 2012).

Here, we use the long-term data set described by Sheriff et al. (2011) to examine how the timing of the ending of heterothermy (torpor and arousal cycles) and emergence from hibernation relates to the timing of reproduction in arctic ground squirrels. Further, since there are overall differences in snow cover between the two sites, we examine whether they differ in either the average temperature of the soil or in the rate at which the soil warms as such differences may explain observed variance in the timing of spring emergence.

6.2 Methods

6.2.1 Study Species

With a body mass as great as 1.5 kg, *U. parryii* is the largest New World ground squirrel and is distributed across northern Alaska, Canada, and eastern Siberia. Their annual cycle includes a 3–5 month season of above-ground activity (Buck and Barnes 1999b; Williams et al. 2012), with breeding beginning in mid to late April shortly after females emerge from hibernation (Buck and Barnes 2003). Young are born in mid to late May, appear above ground shortly before weaning in mid to late June, and disperse 2–3 weeks after first appearing on the surface (Lacey 1991). Adult females enter hibernation as early as late July and young of the year and males enter in September. Animals spend the remaining 7–9 months of the year sequestered in frozen burrow systems (hibernacula), alternating between long (2–3 weeks) bouts of torpor and shorter (10–20 h) intervals of high body temperature during interbout arousal and euthermia (Buck et al. 2008). Hibernaculum depth is limited to ca. 1 m because the arctic tundra is underlain by a continuous permafrost layer that is an impenetrable barrier (Buck and Barnes 1999a).

6.3 Field Measurements

We determined the timing of annually recurring life-cycle events (termination of heterothermy, spring emergence, parturition, entrance into hibernation, and initiation of heterothermy) from patterns of core T_b recorded using data-loggers implanted in free-living arctic ground squirrels, as described in Williams et al. (2011b). Each spring, ground squirrels were implanted/explanted with abdominal temperature loggers (modified TidBit Stowaway model TBICU32-05 + 44, accuracy of ± 0.2°C, Onset Computer Corporation, Pocasset, MA) programmed to record T_b at 20-min intervals for up to 18 months (Long et al. 2007). Annual records of T_b were collected from 1999 to 2010 at a study site near Toolik Lake (68°38′ N, 149°38′ W) and from 2003 to 2010 at a site 20 km to the south near Atigun River (68°27′ N, 149°21′ W); physical and ecological site characteristics are described in Sheriff et al. (2011). Animal protocols were approved by the University of Alaska Fairbanks Institutional Animal Care and Use Committee and the Department of Defense Animal Care and Use Review Office.

Here, we restrict our analyses to the timing of hibernation, conception, and parturition in females known to have given birth; the day of parturition can be determined from an abrupt upward shift in T_b (Williams et al. 2011b), and the day of conception was estimated by subtracting an average gestation length of 25 days (Lacey 1991) from the date of parturition. Final sample sizes for the date females terminated heterothermy and emerged in spring were 22 (Toolik Lake) and 19 (Atigun River) and for dates of fall immergence and initiation of heterothermy were 24 (Toolik Lake) and 20 (Atigun River).

We recorded soil temperature once every 6 h at −1 m depth adjacent to 33 burrow entrances (17 Toolik; 16 Atigun) beginning in fall 2009 with data loggers (Hobo Temp II, Onset Computer Corporation; accuracy ± 0.16°C at the ice point).

6.4 Statistical Analyses

Relationships between the timing of annually recurring life-history events were examined using Pearson's product moment correlation analysis. As we were interested in the potential of conditions within the hibernacula to affect the timing of spring emergence, we restricted our analyses of soil temperature data from April 1 to May 20, a range which encompasses the earliest observed end date for heterothermy in females (April 13) and the latest observed date of emergence (May 18; Sheriff et al. 2011). We compared soil temperatures between sites using repeated-measures mixed models with average temperature (binned in 10-day increments) as the dependent variable. Use of mixed models permitted the inclusion of two burrows lacking T_h data for the final time interval. Means are reported ± SD.

6.5 Results

In females, the date of parturition was highly correlated with the date of termination of heterothermy (Fig. 6.1; both sites combined: $r = 0.87$, $P < 0.0001$; Toolik: $r = 0.77$, $P < 0.0001$; Atigun: $r = 0.88$, $P < 0.001$). The correlation

Fig. 6.1 Relationship between timing of parturition and (**a**) last day of heterothermy, (**b**) date of emergence, and (**c**) date of subsequent immergence into the hibernacula

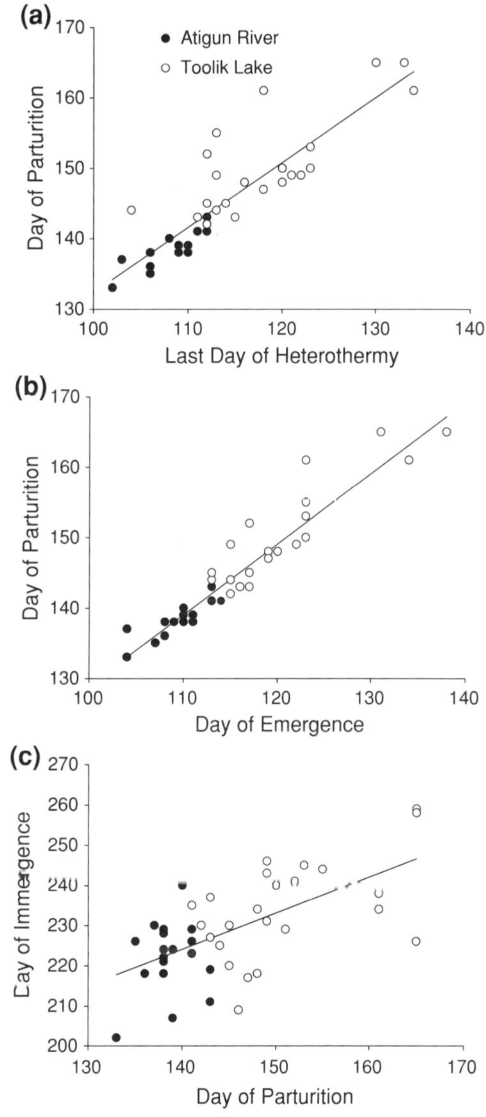

coefficient between dates of parturition and emergence was even higher (both: $r = 0.94$, $P < 0.0001$; Toolik: $r = 0.88$, $P < 0.0001$; Atigun: $r = 0.89$, $P < 0.001$), as the latency between emergence of females and subsequent courtship and impregnation is brief. Females emerged 2.2 ± 2.2 days (Atigun: 1.5 ± 0.9 days; Toolik: 2.9 ± 2.9 days) after they terminated heterothermy and were impregnated 4.0 ± 2.7 days following emergence (Atigun: 3.4 ± 1.5 days; Toolik 4.7 ± 3.3 days). The date of entrance into hibernation in fall was correlated with the date of parturition only in Toolik females (both: $r = 0.62$, $P < 0.0001$; Toolik $r = 0.53$, $P = 0.008$; Atigun $r = 0.25$, $P = 0.29$).

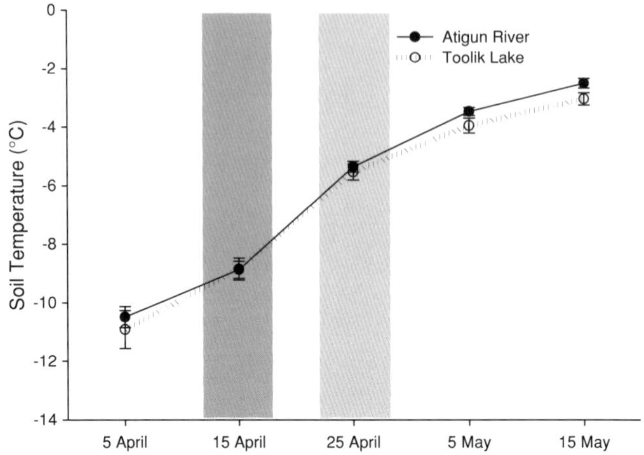

Fig. 6.2 Mean (± SE) soil temperatures, binned in 10-day increments, measured from burrows at a site near Toolik Lake (*dashed line*; $n = 17$) and Atigun River (*solid line*; $n = 16$) in 2010. *Dark and light gray-shaded* portions of the figure indicate the range of dates reproductive females terminated heterothermy at Atigun River and Toolik Lake in 2010, respectively

Across all years, females from Atigun were earlier than females from Toolik in the date they ended heterothermy (9 days earlier), emerged (11 days), gave birth (11 days), entered the hibernacula (11 days), and began heterothermy anew (12 days). In 2010, the year for which we have soil temperature data, females displayed a similar pattern of phenology as the average across all years: Atigun females ended heterothermy (10 days), emerged (11 days), gave birth (11 days), entered the hibernacula (11 days), and began heterothermy anew (12 days) earlier than their Toolik counterparts. Soil temperatures in 2010 became significantly warmer as spring progressed ($F_{4,30} = 174.8$; $P < 0.0001$), but did not differ between Atigun and Toolik ($F_{1,30} = 1.2$; $P = 0.3$) and the interaction between site and date was not significant ($F_{4,30} = 2.1$; $P = 0.1$; Fig. 6.2), indicating that it did not begin to warm sooner at Atigun. However, Atigun females ended heterothermy between 12 and 19 April 2010, an interval that coincided with when soil temperatures first began to warm from their winter minima. Toolik females ended heterothermy between 22 and 28 April 2010, as soil temperatures continued to warm.

6.6 Discussion

The ability of organisms to modulate seasonal timing of annually recurring life-history events to remain in synchrony with shifts in their environment depends, in part, on the environmental cues available and used to adjust timing. Previously, we reported that the phenology of hibernation and reproduction in free-living arctic ground squirrels varies greatly over a relatively small spatial scale and that this variation corresponds with differences in the duration of snow cover (Sheriff et al. 2011). Presently, we show that the timing of parturition in female arctic ground squirrels is tightly correlated with the date females end heterothermy and begin daily activity on the surface in spring. Females remain below ground after ending heterothermy for a short duration ($\sim$2 days), emerge, and then mate within $\sim$4 days. Considering that squirrels sequestered within their hibernacula presumably lack information about surface conditions, we examined whether soil warming might be the proximate cue used to adjust spring hibernation and thus reproductive phenology. Although warming of soil generally coincided with the timing of female emergence, we found no simple differences in patterns of soil temperatures between sites that explained the 12-day difference in dates of parturition. Whether females are utilizing other environmental cues to adjust their phenology remains unknown.

6.7 Timing of Reproduction

The timing of reproduction poses a fundamental challenge for animals that occupy regions characterized by short growing seasons. Although timing of hibernation and reproduction in ground squirrels are known to be under endogenous control (Kenagy 1980), there is also a likely interplay with cues from the local environment (Visser et al. 2010). If spring conditions occur earlier it will likely benefit animals to advance the timing of their breeding to either maximize the amount of time available for offspring growth and development or to match energy demand with an earlier availability of adequate food (Visser and Both 2005). However, animals can only do so if the cues they use to time these annual events are reliable indicators of the current environment and are anticipatory of future conditions.

The phenology of many migrant and hibernating species has advanced at some locales in response to warming, although the magnitude of the advance does not always match the shift in timing of resource availability (Inouye et al. 2000). For example, in caribou the earlier timing of migration in spring has not kept pace with the seasonal advancement in plant growth and this may be contributing to the decline in reproduction (Post and Forchhammer 2008). In arctic ground squirrels, the high correlation between the date of parturition and the date of emergence indicates females will only be able to appropriately adjust reproductive timing if they are able to terminate hibernation earlier when conditions warrant. Parturition was better correlated with the date of emergence than the end of heterothermy (Fig. 6.1), as some females remained below ground for an extended period (up to 12 days) after resumption of euthermia. These extended below ground periods were rare in females, and occurred only at Toolik, the site characterized by late and more variable snowmelt.

We also found that the timing of parturition was correlated with the subsequent date of entrance into hibernation, although this effect was relatively weak and driven solely by females at Toolik Lake that give birth late in the spring. Females that emerge, give birth, and wean later may be forced to slightly delay their impending hibernation until they gain sufficient mass for overwintering. Fitness consequences of delayed breeding remain unclear, as in late summer the ground remains snow-free and access to food does not appear to be limiting when adult females enter hibernation in late July and August. Late-borne offspring will encounter reduced foraging opportunities, however, if their active season extends through September when freezing temperatures and snow become widespread. Offspring born earlier in spring, in contrast, might be able to achieve a larger body mass which could improve over-winter survival rates (e.g., Ozgul et al. 2010). In other species, females that give birth later tend to have decreased weaning success (e.g. Réale et al. 2003; Broussard et al. 2008), which might explain the relatively weak relationship between timing of reproduction and fall immergence. If females that breed late do not wean their offspring they may enter hibernation sooner than expected having the ability to increase body mass earlier due to the loss of their litter. However, we were unable to determine whether females which bred late were successful or not in our study. We expect that weaning success along with timing of breeding may be major factors affecting the timing of immergence in females.

6.8 Phenotypic Plasticity Versus Genetic Variability

Whether differences in the timing of spring emergence, and consequently parturition, are driven by phenotypic plasticity in response to environmental cues or are reflective of genetic differences between the two populations remains unclear. Lane et al. (2011) demonstrated that timing of emergence from hibernation was a heritable trait in a population of Columbia ground squirrels (*Urocitellus columbianus*); however, the additive genetic variance component explained only about 20% of the total phenotypic variation in this trait. We suggest that some of the additional variance can be attributed to phenotypic plasticity in response to local conditions. One possibility is that females do respond to shifts in ambient temperature within the hibernacula (i.e., soil temperature), but the response to temperature differs between sites. This could occur if females from Toolik have a different soil temperature threshold that triggers the end of heterothermy relative to Atigun females. In this case interannual variation in timing of emergence and reproduction may be explained by variation in soil temperature within each site but not between sites. We are currently investigating this possibility.

Another possibility is that female arctic ground squirrels move out of the insulated "nest" chamber during brief arousal bouts of euthermia and assess snow conditions from burrow entrances. Buck et al. (2008) noted that the penultimate arousal interval is significantly longer than earlier intervals, which suggests that behavior and/or physiology is different during the brief interbout arousals later in hibernation. This shift in behavior/physiology may reflect an endogenously driven seasonal shift in sensitivity to environmental cues. Finally, it is possible that timing of female emergence is dictated primarily by date of entry into hibernation, as Sheriff et al. (2011) found no significant difference in the duration of hibernation between Toolik and Atigun females. However, there was substantial variation in the duration of hibernation among individuals (range: 200–271 days) making this an unlikely possibility. Patterns of hibernation are without a doubt under strong endogenous control but the degree to which females are capable of altering these patterns in response to environmental change, particularly with respect to the termination of hibernation, remains unclear. Therefore, the capacity of this species to respond to environmental changes predicted by climate models also remains uncertain.

Acknowledgments Funding was provided by the National Science Foundation (EF-0732763 and -0732755) to CLB and BMB and the US Army Medical Research and Materiel Command Grant #05178001 to BMB. We thank Trixie Lee and Melanie Richter for their invaluable help in gathering field data.

References

Barnes BM (1989) Freeze avoidance in a mammal: body temperatures below 0°C in an arctic hibernator. Science 244:1593–1595

Berteaux D, Réale D, McAdam AG, Boutin S (2004) Keeping pace with fast climate change: can arctic life count on evolution? Integr Comp Biol 44:140–151

Broussard DR, Dobson FS, Murie JO (2008) Previous experience and reproductive investment of female Columbian ground squirrels. J Mammal 89:145–152

Buck CL, Barnes BM (1999a) Temperatures of hibernacula and changes in body composition of arctic ground squirrels over winter. J Mammal 80:1264–1276

Buck CL, Barnes BM (1999b) Annual cycle of body composition and hibernation in free-living arctic ground squirrels. J Mammal 80:430–442

Buck CL, Barnes BM (2003) Androgen in free-living arctic ground squirrels: seasonal changes and influence of staged male–male aggressive encounters. Horm Behav 43:318–326

Buck CL, Breton A, Kohl F, Tøien Ø, Barnes BM (2008) Overwinter body temperature patterns in free-living arctic squirrels (*Spermophilus parryii*). In: Lovegrove BG, McKechnie AE (eds) Hypometabolism in animals: torpor hibernation and cryobiology. University of KwaZulu-Natal, Pietermaritzburg, pp 317–326

Høye TT, Post E, Meltofte H, Schmidt NM, Forchhammer MC (2007) Rapid advancement of spring in the High Arctic. Curr Biol 17:R449–R451

Inouye DW, Barr B, Armitage KB, Inouye BD (2000) Climate change is affecting altitudinal migrants and hibernating species. Proc Natl Acad Sci 97:1630–1633

Kenagy GJ (1980) Interrelation of endogenous annual rhythms of reproduction and hibernation in the golden-mantled ground squirrel. J Comp Physiol A 135:333–339

Lacey EA (1991) Reproductive and dispersal strategies of male Arctic ground squirrels (*Spermophilus parryii plesius*). PhD thesis, University of Michigan

Lane JE, Kruuk LEB, Charmantier A, Murie JO, Coltman DW, Buoro M, Raveh S, Dobson FS (2011) A quantitative genetic analysis of hibernation emergence date in a wild population of Columbian ground squirrels. J Evol Biol 24:1949–1959

Long RA, Hut RA, Barnes BM (2007) Simultaneous collection of body temperature and activity data in burrowing mammals: a new technique. J Wildl Manage 71:1375–1379

Murie JO, Harris MA (1982) Annual variation of spring emergence and breeding in Columbian ground squirrels (*Spermophilus columbianus*). J Mammal 63:431–439

Ozgul A, Childs DZ, Oli MK, Armitage KB, Blurnstein DT, Olson LE, Tuljapurkar S, Coulson T (2010) Coupled dynamics of body mass and population growth in response to environmental change. Nature 466:482–485

Post E, Forchhammer MC (2008) Climate change reduces reproductive success of an arctic herbivore through trophic mismatch. Phil Trans R Soc B 363:2367–2373

Post E, Forchhammer MC, Bret-Harte MS, Callaghan TV, Christensen TR, Elberling B, Fox AD, Gilg O, Hik DS, Høye TT, Ims RA, Jeppesen E, Klein DR, Madsen J, McGuire AD, Rysgaard S, Schindler DE, Stirling I, Tamstorf MP, Tyler NJC, van der Wal R, Welker J, Wookey PA, Schmidt NM, Aastrup P (2009) Ecological dynamics across the Arctic associated with recent climate change. Science 325:1355–1358

Réale D, McAdam AG, Boutin S, Berteaux D (2003) Genetic and plastic responses of a northern mammal to climate change. Proc Roy Soc B 270:591–596

Serreze MC, Francis JA (2006) The Arctic amplification debate. Clim Change 76:241–264

Sheriff MJ, Kenagy GJ, Richter M, Lee T, Tøien Ø, Kohl F, Buck CL, Barnes BM (2011) Phenological plasticity: variation in annual timing of hibernation and breeding in two nearby populations of arctic ground squirrels. Proc Roy Soc B 278:2369–2375

Visser ME, Both C (2005) Shifts in phenology due to global climate change: the need for a yardstick. Proc Roy Soc B 272:2561–2569

Visser ME, Caro SP, van Oers K, Shaper SV, Helm B (2010) Phenology, seasonal timing and circannual rhythms: towards a unified framework. Proc Roy Soc B 365:3113–3127

Williams CT, Goropashnaya AV, Buck CL, Fedorov VB, Kohl F, Lee TN, Barnes BM (2011a) Hibernating above the permafrost: effects of ambient temperature and season on expression of metabolic genes in liver and brown adipose tissue of arctic ground squirrels. J Exp Biol 214:1300–1306

Williams CT, Sheriff MJ, Schmutz JA, Kohl F, Tøien Ø, Buck CL, Barnes BM (2011b) Data logging of body temperatures provides precise information on phenology of reproductive events in a free-living arctic hibernator. J Comp Physiol B 181:1101–1109

Williams CT, Barnes BM, Buck CL (2012) Body temperature rhythms persist under the midnight sun but are absent during hibernation in free-living arctic ground squirrels. Biol Lett 8:31–34. doi:10.1098/rsbl.2011.0435

Chapter 7
Assessing the Effect of Climate Change on Hibernating Mammals Using Nonlinear Mixed Effects Method

István Németh

Abstract A hibernating lifestyle makes animals sensitive to the changing of the environmental temperature. Therefore, the effect of currently ongoing climate change might be considerable on these species. To assess this effect, we estimated the body mass change during hibernation under three different climate scenarios, using computational modeling. The chosen nonlinear mixed effects modeling technique was suitable to describe body mass change during hibernation. The proposed model predicted a decrease in spring emergence body mass in predicted (2070–2100) compared to control period (1960–1990). Probably this difference ($\sim$2 g) has an insignificant effect on the survival of hibernating animals. Such small disturbances can be compensated by the animals themselves or by the advantageous side effects of climate change (extended active period, enhanced primer production), but along with other disturbances, such as human activities (e.g., landscape using) might contribute to altering population dynamics.

Keywords Climate change · Ecophysiology · Hibernation · Ground dwelling sciurids · Population stochastic modeling

I. Németh (✉)
Developmental Drug Metabolism and Pharmacokinetics,
Gedeon Richter Plc. H-1103, Budapest, Gyömrői út 19-21, Hungary
e-mail: furgeurge@gmail.com

T. Ruf et al. (eds.), *Living in a Seasonal World*,
DOI: 10.1007/978-3-642-28678-0_7, © Springer-Verlag Berlin Heidelberg 2012

7.1 Introduction

In recent years, numerous studies support the ecological impact of current climate change (Parmesan and Yohe 2003; Walther et al. 2002). Changes induced by global warming are apparent at all levels of ecological organization, they appear in life history and phenology, population and community structure, and in geographical distribution of species (McCarty 2001). While some studies document and report the ongoing changes, others try to describe the potential outcome of the predicted climate change in the future (Inouye et al. 2000; Kausrud et al. 2008; Barbraud and Weimerskirch 2001). Several of the latter studies use the relationship between ecology and physiology for their forecast (Humphries et al. 2002; Barbraud and Weimerskirch 2001; Landry-Cuerrier et al. 2008). These ecophysiological modeling studies make use of the fact that certain life history or physiological properties sensitize organisms to effects of climate, and therefore these species have increased responsiveness to climate change (Hodkinson 1999).

As there is a close relationship between energy expenditure and environmental temperature, this makes animal species that hibernate sensible and vulnerable to climate change (Buck and Barnes 2000; Geiser 2004; Frank 2002). Decreased energy expenditure during hibernation affects the survivorship, maturation, and fecundity, via the exhaustible energy stores (Barnes 1984; Millesi et al. 1999a). Hibernating animals partially give up their control over their body temperature (Körtner and Geiser 2000). On one hand, it is favorable, e.g., to save energy, on the other hand, partial control offers only partial response to changing environment. Fat-storing hibernators are aphagic or severely hypophagic during the hibernation season (Davis 1976) and unresponsive to energy challenges that at other times of year promote increases in body mass (Dark et al. 1984, 1989).

Seasonal hibernation consists of two fundamentally different states. The heterothermic state, commonly called torpor, is the energy conservation state which generally lasts for 3–14 days (Heldmaier et al. 2004). In torpor state the body temperature of animals approaches the ambient temperature as long as this lowering of body temperature does not threaten the integrity of the organism (Barnes 1989; Carey et al. 2003). Via this mechanism a considerable amount of energy can be saved and the consumed energy does not exceed 10% of energy consumed in the normothermic state (Wang 1979; Geiser 1988). Heterothermic states are regularly interrupted by short ($\sim$1 day long) euthermic states (arousal), when animals recover their normal body temperature and normothermic metabolic rates (Heldmaier et al. 2004). In seasonal hibernators, torpor and arousal states are in close and clear relationship with ambient temperature, both in terms of energetics and phenology (Buck and Barnes 2000; Németh et al. 2009). This feature of hibernation renders hibernating animals a convenient subject for ecophysiological modeling studies (Landry-Cuerrier et al. 2008; Németh et al. 2009).

Here, we propose an ecophysiological model which estimates the population parameters as well as the individual deviations of the process that describes weight loss during hibernation. The aims of the study were to (1) fit and parameterize the model on previously measured laboratory data on European ground squirrels (*Spermophilus citellus*; Németh et al. 2009); (2) to make predictions about future consequences of global warming on hibernating animals based on published local climate change scenarios (Bartholy et al. 2008); and we also wanted to show that nonlinear mixed effects modeling is applicable for solving ecophysiological problems.

7.2 Materials and Methods

7.2.1 Laboratory Time Series Data

Fifteen European ground squirrels were live trapped from a free ranging population living on a grassy airport in Hungary (N 47°36′43″, E 19°08′40″), and were transferred to a climate-controlled room. The animals were housed individually in plastic containers (*h*: 8 cm, *d*: 16 cm). Nest materials were provided. Food and water were provided ad libitum. The animals were kept at 20°C and a light–dark (LD) cycle of 10:14 (short photoperiod conditions) until mid September when ambient temperature (T_a) was decreased to 15°C and the LD cycle turned to continuous dark. At this time food and water were removed to facilitate the beginning of hibernation. We kept three separate groups ($n = 5$ each) on different constant temperatures during the treatment (0, 5 and 9°C). The treatment lasted from 1 January to 10 April, when we gradually increased the ambient temperature and reintroduced the LD cycle. The body temperature of the animals was traced with a computerized data acquisition system (recorded on every 10 min) and body mass was recorded in 6–14 day intervals (Németh et al. 2009).

7.2.2 Estimation and Simulation

For the simulation of body mass change under artificial climate scenarios, the first step was to build and fit a model for the data measured in the laboratory. This model consists of two distinct parts. One describes the body mass change by using differential equations, the other describes the binary torpor–arousal cycles during hibernation. After model fitting, the obtained parameters of the model can be used for subsequent simulations.

7.2.3 Body Mass Change

Modeling of body mass change (BM) was carried out via the modeling of energy expenditure during hibernation. The unit of time axis was day of hibernation. The state of the animals on a given day could be torpor or arousal. Events on smaller time scales like warming-up period prior to arousals or cooling at the ends were disregarded. Such simplifications were necessary, to keep the number of estimable parameters to an optimum and to avoid gaining more, but meaningless information regarding accessible data (time, body mass, state). As modeling was carried out via energy expenditure, body mass was converted into energy by using a scaling parameter (37 kJ/g). Thus, the state variable of the model was the total amount of energy stored in the body of animals (E). Body mass of animal i measured on day t of hibernation can be described as,

$$\mathrm{BM}_{i,t} = \frac{E_{i,t}}{37\ \mathrm{kJ/g}} + \varepsilon_{i,t}$$

where $\varepsilon_{i,t}$ is the random intra-individual error (e.g., weighing error), and comes from a normal distribution, $\varepsilon_{i,t} \sim N(0, \sigma_{\mathrm{BM}})$.

In our model, energy expenditure depended on two variables. The first variable was the total amount of energy in the animal's body (that is, the body mass itself), because the smaller the body mass is, the less energy can be spent. The second was the ambient temperature. As the relationship between environmental temperature and energy expenditure during torpor and arousal states are essentially different from each other, the change in total energy in these states were included in the model separately. A nonlinear relationship between T_a and energy expenditure during torpor was included in the model as a second order polynomial of T_a, whereas this relationship during arousal was left to be linear. Hence, energy expenditure of an individual was described as,

$$\frac{\mathrm{d}E_i}{\mathrm{d}t} = \left\{ \begin{array}{ll} \text{if animal is in torpor,} & \left(E_i \times k_{i,2}\right) + \left(T_\mathrm{a} \times k_{i,3} + T_\mathrm{a}^2 \times k_{i,4}\right) \\ \text{if animal is in arousal,} & \left(E_i \times k_{i,5}\right) + \left(T_\mathrm{a} \times k_{i,6}\right) \end{array} \right\} \mathrm{kJ/day}.$$

Temperature was implemented on Kelvin-scale (K).

Energy amount at the beginning of the hibernation of an animal $\left(E_{i,t=0} = k_{i,1}\right)$ was also an estimated parameter. k_{ij}s in the equations above are the individual parameters which can be expressed as

$$k_{i,j} = k_j \times \exp(\eta_{i,j})$$

where k_js are the typical values in the population (population parameters) and $\eta_{i,j}$s refer to the difference between individuals' parameter values and the typical values in the population. Here, we used an exponential inter-individual error model by expression $\exp(\eta_{i,1})$.$\eta_{i,j}$s are the elements of a multivariate normal distribution with zero expected values.

7.2.4 Torpor–Arousal Cycles

Modeling of torpor–arousal cycles was carried out by a generalized linear model on binary time-series data of hibernating animals (arousal day–torpor day).

Relationship between the probability of an arousal day and time was characterized by a second order polynomial. Polynomial regression was made necessary by the fact that the time–arousal probability curve has a local minimum near the middle of hibernation (Körtner and Geiser 2000). Temperature was included in the model as a continuous fixed effect factor.

7.2.5 Simulating Body Mass Change

Body mass change under different climate scenarios was simulated over hypothetic hibernation periods. Daily temperature data (at +2 m) for projected (2070–2100) and control (1961–1990) periods were gained from PRUDENCE project public data (http://prudence.dmi.dk/). We have chosen three scenarios on the basis of published information about the region of interest, the Carpathian basin (Bartholy et al. 2008). Simulations were done by using these scenarios, ranging from conservative to extreme predictions. The chosen scenarios were developed by the Danish Meteorological Institute (DMI HS1 and DMI HC1), the Swiss Federal Institute of Technology (ETH HC A2 and ETH HC CTL), and The Royal Netherlands Meteorological Institute (KNMI HA2 and KNMI HC1), respectively. Air temperature data were used to compute average daily soil temperature data using a formerly published method (Strijkstra et al. 1999).

Hundred simulations were done for each scenario and period (overall 17,400 individual simulations). Simulated hibernation periods lasted from the beginning of October till the end of February, which means 150 days (every month consisting of 30 days). For simulating body mass changes under different climate scenarios and periods (control or projected) the immergence body mass was set to 400 g $\left(E_{i,t=0} = 400\ \text{g} \times 37\ \text{kJ/g}\right)$according to the literature (Millesi et al. 1999b).

7.2.6 Statistical Analysis

After completion of simulations, emergence body mass $(\text{BM}_{\text{em}} = \text{BM}_{\text{i,t=150}})$ was used to test the effect of different climate scenarios on body mass change during hibernation. ANOVA models were used to analyze the effect of climate change and overall effect of temperature.

Supportive soil temperature data analysis was performed for demonstrating differences among climate scenarios. In a nonlinear model, the year-round fluctuation of temperature was handled by sinusoid functions. (Pinheiro and Bates 2002; Pinheiro et al. 2008).

NONMEM software (7.2.0) with GFortran compiler (version 4.6) was used for population nonlinear mixed-effect modeling of body mass change during hibernation (Beal et al. 2011). ANOVA procedures and data management were performed with R, an open source software for statistical computing and modeling (R Development Core Team 2011).

7.3 Results

7.3.1 Modeling Body Mass Change During Hibernation

Evaluation of estimated parameters for body mass–time profiles was carried out by using data of simulated hibernations (Fig. 7.1). Results of 1,000 simulation were over-plotted on original data points to check the predictive power of the NONMEM model by visual inspection (Visual Predictive Check—VPC; Yano et al. 2001). As observations distribute equally both under and above the model predicted medians and 90% range of the simulations contain the majority of observations, hence we can conclude that the model reliably reproduced the variability of the observed data. The probability of the appearance of arousals was modeled as a function of ambient temperature and the time elapsed in hibernation. Both time ($P < 0.001$) and temperature ($P < 0.001$) affected significantly the probability of the appearance of arousals (Fig. 7.2).

7.3.2 Simulating Body Mass Change Under Different Climate Scenarios

Soil temperature data for control (1961–1990) and projected (2071–2100) periods for three different scenarios were derived from air temperature data gained from PRUDENCE project. Year-round oscillation of soil temperature was approximated with a sinusoid function. Soil temperature was approximately 4.0°C higher in the 2071–2100 period than in the 1961–1990 period ($P < 0.0001$, Fig. 7.3a). This increase in temperature was fairly similar among scenarios. Predicted rise in soil temperature were 4.0, 3.6, and 3.4°C for the DMI, KNMI, and ETH scenarios, respectively.

After completion of the 17,400 individual simulations, our model predicted that BM_{em} of populations was slightly affected by predicted climate change (Fig. 7.3b; $F1,17,396 = 90.7$, $P < 0.0001$). The prediction for population mean of simulated emergence body mass for period 1961–1990 is 167 g. This expected population BM_{em} decreased by 2 g at period 2071–2100 in comparison to period 1961–1990.

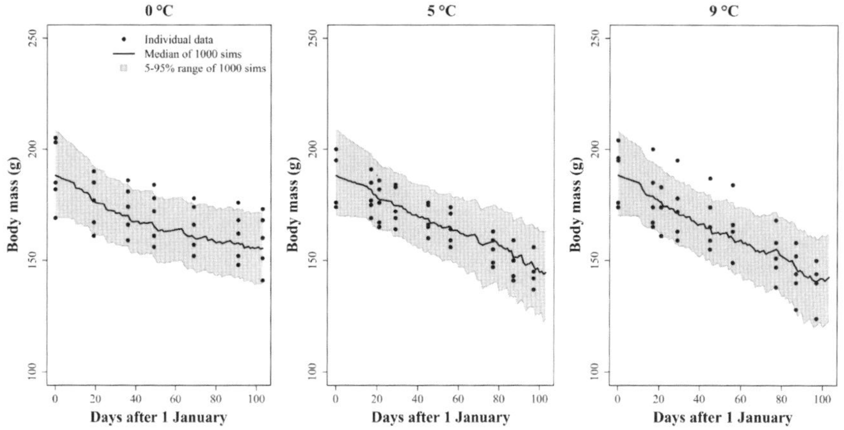

Fig. 7.1 Visual evaluation of the applied model. Body mass change data of 1,000 simulations were over-plotted on measured data. To facilitate evaluation, data on different ambient temperatures are plotted separately

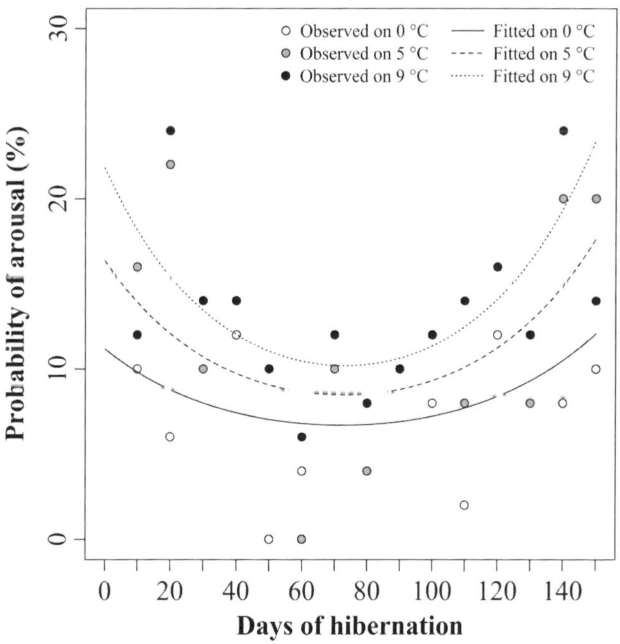

Fig. 7.2 Estimation of the effect of time and temperature on the appearance of arousals by using generalized linear modeling. Time was implemented as a second order polynomial, which reached its minimum near to the middle of the hibernation period. Measured time points (*circles*) represent the sums of individual data at every 10 days as a percentage (number of arousal days/ all days)

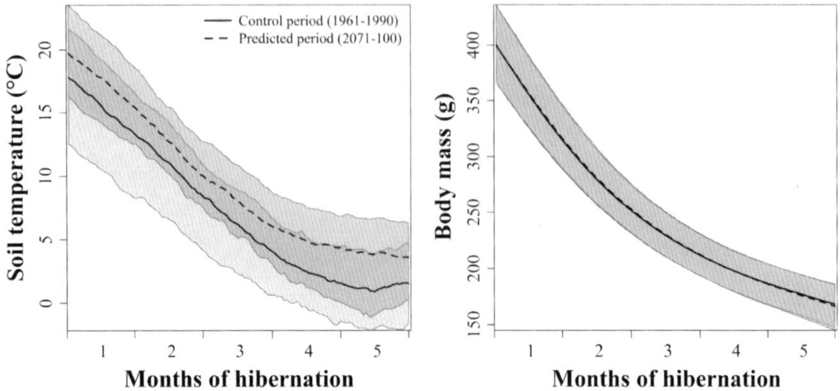

Fig. 7.3 Effect of climate change on the course of soil temperature (**a**) and body mass change (**b**) during the hibernation period. Soil temperature data of different scenarios (DMI, ETH, and KNMI, 29 years) were merged as one, as well as the simulated body mass data of different scenarios and years. *Lines* represent the median, *gray* areas represent the 90% range of the data

Difference among applied scenarios was also significant and comparable with the expected climate change effect (F2,17,396 = 179.9, $P < 0.0001$; ETH vs. DMI: 1.1 g, KNMI vs. DMI: 1.1 g, KNMI-ETH: −0.02 g).

7.4 Discussion

7.4.1 Evaluating the Performance of Applied Modeling Method

Ecophysiological models are based on structural models including ecologically relevant parameters (Humphries et al. 2002; Landry-Cuerrier et al. 2008). Our proposed method was to imply nonlinear mixed effects modeling in ecophysiology similar to pharmacology where it has been successfully applied for decades (Pillai et al. 2005). We used nonlinear mixed effects modeling technique to determine population parameters of partial differential equations relating to the energetics of hibernation and their inter- and intra-individual variations (Pinheiro and Bates 2002).

Model evaluation was carried out with visual predictive check (Yano et al. 2001), that confirms the reliability of the proposed partial differential equations and corresponding parameters. Accordingly, the chosen structural and random effect models are able to describe body mass change under various temperature conditions over hibernation. However, hibernation is a convenient modeling subject, because it takes place in a fairly isolated environment ($\sim$1 m underground). We would emphasize that some neglected aspects of hibernation (e.g., thermal conductance) in this model does not originate in the features of the used method, but in the simplicity of measured variables. Using additional data, such as oxygen consumption and body temperature together, would give the opportunity to build more sophisticated models.

7.4.2 Model Prediction on Future Climate and its Species–Specific Aspects

As some authors have pointed out, only a minority of organisms will/is not affected by climate change (Parmesan and Yohe 2003). Nearly all aspects of ecological organization, geographical distribution, community structures, phenology, and life history have been affected (McCarty 2001). The climatic effects on plants affect related animal species (Scholze et al. 2006; Kausrud et al. 2008). However, climate change also has its own effect on hibernating animals. For example, a change in the phenology of the active season (*Marmota flaviventris*), or a shift in the species distribution area (*Myotis lucifugus*) has been reported (Humphries et al. 2002; Inouye et al. 2000). Presumably, the majority of these effects on hibernators are expressed through the energetics of hibernation. Animals in torpor save considerable amounts of energy and no thermoregulation is required over a wide range of ambient temperatures and a fall in body temperature can lower the energy expenditure (Geiser 2004).

Relationship between environmental temperature and the energy requirement of hibernation is close and proven by many studies (Buck and Barnes 2000; Wang 1979). Our results indicated the effect of temperature as significant, but less pronounced on body mass change than would be expected. According to a study by Buck and Barnes (2000), the predicted ~4°C increase in year average soil temperature would cause a 12–20% decrease in emergence body mass. However, our results show an approximate 1–3% decrease in body mass at emergence from the same change in soil temperature. This departure probably results from the effect of temperature on the timing of arousals (Németh et al. 2009). Energy consumption is about 10 times higher in arousal than in torpor during a given time unit (Wang 1979). One day in arousal consumes the approximate energy saved in 10 days of torpor. In this sense, environmental temperature influences the overall energy consumption of hibernation via the timing of arousals rather than via its relationship to the rate of torpor metabolism (Buck and Barnes 2000).

Nevertheless, an effect which may seem negligible within a season, may have considerable effect on a population life span scale. In many species, timing of primiparity, maturity, and competitive skills of males, as well as reproductive output of females, are in close relationship with body weight (Millesi et al. 1999a; Neuhaus et al. 2004; Dobson and Michener 1995; Millesi et al. 1998; Barnes 1984).

7.4.3 Remarks on Conservation Strategies

Hibernation might have appeared along with the first representatives of euthermic vertebrates (Grigg et al. 2004). There have been numerous global changes, including climate changes, during these approximately 60 million years (Barrett 2003). Considering the past events in the evolutionary history of hibernating species helps us to discover the consequences of the present climate change on hibernating species. The squirrel family is a relatively young group (first fossils are approx. 34 million years old), but probably one of the groups which have the highest number of hibernators among its members (Sciuridae; Barclay et al. 2001). There have been many global changes during these approximately 30 million years as well. Climate changes and tectonic movement frequently occurred during the evolution of Sciuridae family. However, the major waves of these events coincide with the radiations of species rather than their extinctions (Mercer and Roth 2003). It seems probable that dispersion moving across ecological corridors, played an important role in those past radiations and should play an important role in the present as well (Thomas et al. 2004).

In conclusion, our model predicts the effect of climate change on energy expenditure during hibernation not to be considerable, compared to the effects (e.g., extended active period, increased primer production), which will act parallel to and compensate for the negative effects of climate change on the survival of seasonal hibernators. However, it is necessary to emphasize the importance of good landscape usage, in order to maintain the network of ecological corridors connecting suitable habitats.

Acknowledgments Climate change data have been provided through the PRUDENCE data archive, funded by the EU through contract EVK2-CT2001-00132. The author thanks the approval of publication for the ICON Plc, the licensor of NONMEM®. We especially thank Celeste Pongrácz and Orsolya Zeöld for their valuable comments and corrections.

References

Barbraud C, Weimerskirch H (2001) Emperor penguins and climate change. Nature 411(6834):183–186

Barclay RMR, Lausen CL, Hollis L (2001) What's hot and what's not: defining torpor in free-ranging birds and mammals. Can J Zool 79(10):1885–1890

Barnes BM (1984) Influence of energy stores on activation of reproductive function in male golden-mantled ground-squirrels. J Comp Physiol [B] 154(4):421–425

Barnes BM (1989) Freeze avoidance in a mammal: body temperatures below 0°C in an Arctic hibernator. Science 244(4912):1593–1595

Barrett P (2003) Palaeoclimatology: cooling a continent. Nature 421(6920):221–223

Bartholy J, Pongracz R, Gelybó G, Szabó P (2008) Analysis of expected climate change in the Carpathian basin using the PRUDENCE results. Időjárás 112:249–264

Beal S, Sheiner LB, Boeckmann A, Bauer RJ (2011) NONMEM User's Guides, version 7.2. Icon Development Solutions, Ellicott City

Buck CL, Barnes BM (2000) Effects of ambient temperature on metabolic rate, respiratory quotient, and torpor in an arctic hibernator. Am J Physiol-Reg I 279(1):R255–R262

Carey HV, Andrews MT, Martin SL (2003) Mammalian hibernation: cellular and molecular responses to depressed metabolism and low temperature. Physiol Rev 83(4):1153–1181

Dark J, Forger NG, Zucker I (1984) Rapid recovery of body mass after surgical removal of adipose tissue in ground squirrels. Proc Natl Acad Sci U S A 81(7):2270–2272

Dark J, Stern JS, Zucker I (1989) Adipose tissue dynamics during cyclic weight loss and weight gain of ground squirrels. Am J Physiol-Reg I 256(6):R1286–R1292

Davis DE (1976) Hibernation and circannual rhythms of food consumption in marmots and ground squirrels. Q Rev Biol 51(4):477

Dobson FS, Michener GR (1995) Maternal traits and reproduction in Richardsons ground-squirrels. Ecology 76(3):851–862

Frank CL (2002) Short-term variations in diet fatty acid composition and torpor by ground squirrels. J Mammal 83(4):1013–1019

Geiser F (1988) Reduction of metabolism during hibernation and daily torpor in mammals and birds—temperature effect or physiological inhibition. J Comp Physiol [B] 158(1):25–37

Geiser F (2004) Metabolic rate and body temperature reduction during hibernation and daily torpor. Annu Rev Physiol 66:239–274

Grigg Gordon C, Beard Lyn A, Augee Michael L (2004) The evolution of endothermy and its diversity in mammals and birds. Physiol Biochem Zool 77(6):982–997

Heldmaier G, Ortmann S, Elvert R (2004) Natural hypometabolism during hibernation and daily torpor in mammals. Resp Physiol Neurobi 141(3):317–329

Hodkinson ID (1999) Species response to global environmental change or why ecophysiological models are important: a reply to Davis et al. J Anim Ecol 68(6):1259–1262

Humphries MM, Thomas DW, Speakman JR (2002) Climate-mediated energetic constraints on the distribution of hibernating mammals. Nature 418(6895):313–316

Inouye DW, Barr B, Armitage KB, Inouye BD (2000) Climate change is affecting altitudinal migrants and hibernating species. Proc Natl Acad Sci USA 97(4):1630–1633

Kausrud KL, Mysterud A, Steen H, Vik JO, Ostbye E, Cazelles B, Framstad E, Eikeset AM, Mysterud I, Solhoy T, Stenseth NC (2008) Linking climate change to lemming cycles. Nature 456(7218):93–97

Körtner G, Geiser F (2000) The temporal organization of daily torpor and hibernation: circadian and circannual rhythms. Chronobiol Int 17(2):103–128

Landry-Cuerrier M, Munro D, Thomas DW, Humphries MM (2008) Climate and resource determinants of fundamental and realized metabolic niches of hibernating chipmunks. Ecology 89(12):3306–3316

McCarty JP (2001) Ecological consequences of recent climate change. Cons Biol 15(2):320–331

Mercer JM, Roth VL (2003) The effects of Cenozoic global change on squirrel phylogeny. Science 299(5612):1568–1572

Millesi E, Huber S, Dittami J, Hoffmann I, Daan S (1998) Parameters of mating effort and success in male European ground squirrels, *Spermophilus citellus*. Ethology 104(4):298–313

Millesi E, Huber S, Everts LG, Dittami JP (1999a) Reproductive decisions in female European ground squirrels: factors affecting reproductive output and maternal investment. Ethology 105(2):163–175

Millesi E, Strijkstra AM, Hoffmann IE, Dittami JP, Daan S (1999b) Sex and age differences in mass, morphology, and annual cycle in European ground squirrels, *Spermophilus citellus*. J Mammal 80(1):218–231

Németh I, Nyitrai V, Altbäcker V (2009) Ambient temperature and annual timing affect torpor bouts and euthermic phases of hibernating European ground squirrels (*Spermophilus citellus*). Can J Zool 87(3):204–210

Neuhaus P, Broussard DR, Murie JO, Dobson FS (2004) Age of primiparity and implications of early reproduction on life history in female Columbian ground squirrels. J Anim Ecol 73(1):36–43

Parmesan C, Yohe G (2003) A globally coherent fingerprint of climate change impacts across natural systems. Nature 421(6918):37–42

Pillai G, Mentré F, Steimer J-L (2005) Non-linear mixed effects modeling—from methodology and software development to driving implementation in drug development science. J Pharmacokinet Pharmacodyn 32(2):161–183

Pinheiro J, Bates D (2002) Mixed effects models in S and S-Plus. Springer, New York

Pinheiro J, Bates D, DebRoy S, Sarkar D, the R Core team (2008) nlme: linear and nonlinear mixed effects models. R package version 3.1-89

Scholze M, Knorr W, Arnell NW, Prentice IC (2006) A climate-change risk analysis for world ecosystems. Proc Natl Acad Sci USA 103(35):13116–13120

Strijkstra AM, Hut RA, Millesi E, Daan S (1999) Energy expenditure during hibernation in european ground squirrels (*Spermophilus citellus*). In: Strijkstra AM (ed) Periodic euthermy during hibernation in the european ground squirrel: causes and consequences. PhD thesis, University of Groningen, pp 31–56

R Development Core Team (2011) R: a language and environment for statistical computing, version 2.13.0. R Foundation for Statistical Computing, Vienna

Thomas CD, Cameron A, Green RE, Bakkenes M, Beaumont LJ, Collingham YC, Erasmus BFN, de Siqueira MF, Grainger A, Hannah L, Hughes L, Huntley B, van Jaarsveld AS, Midgley GF, Miles L, Ortega-Huerta MA, Townsend Peterson A, Phillips OL, Williams SE (2004) Extinction risk from climate change. Nature 427(6970):145–148

Walther GR, Post E, Convey P, Menzel A, Parmesan C, Beebee TJ, Fromentin JM, Hoegh-Guldberg O, Bairlein F (2002) Ecological responses to recent climate change. Nature 416(6879):389–395

Wang LC (1979) Time patterns and metabolic rates of natural torpor in the Richardson's ground squirrel. Can J Zool 57:149–155

Yano Y, Beal SL, Sheiner LB (2001) Evaluating pharmacokinetic/pharmacodynamic models using the posterior predictive check. J Pharmacokinet Pharmacodyn 28(2):171–192

Chapter 8
Impact of Climatic Variation on the Hibernation Physiology of *Muscardinus avellanarius*

Iris Pretzlaff and Kathrin H. Dausmann

Abstract Climate change will not only directly affect climatic parameters, such as temperature and precipitation, but ultimately also ecological and physiological parameters of animal species. Our study investigates the effects of variation in climatic conditions, as experienced by climate change, on hibernation physiology and energy balance of a small hibernating mammal, the hazel dormouse (*Muscardinus avellanarius*). To understand whether ambient conditions indeed impact thermoregulatory patterns (frequency of arousals) and energy budgets during hibernation, we measured skin temperature and metabolic rate of *M. avellanarius* under natural conditions during two winters of different strengths. Our results show that environmental conditions did significantly alter hibernation parameters in *M. avellanarius* at different levels. *M. avellanarius* did not compensate for changes in ambient temperature, but body temperature followed ambient temperature for most of the hibernation season (with skin temperature down to −2.9°C). Arousals occurred more frequently during warmer months, also after seasonal effects were eliminated, causing substantial additional energy expenditure. We conclude that hibernating small mammals such as *M. avellanarius* are more likely to deplete fat reserves before the end of hibernation when temperatures increase, due to generally increased body temperature (and thus energy expenditure) during hibernation, and an increase in arousal frequency.

Keywords Hibernation · Climate change · Arousal frequency · Energy budgets · *Muscardinus avellanarius*

I. Pretzlaff (✉) · K. H. Dausmann
Department of Animal Ecology and Conservation, Biocentre Grindel,
University of Hamburg, Martin-Luther-King-Platz 3,
20146 Hamburg, Germany
e-mail: iris.pretzlaff@gmx.de

T. Ruf et al. (eds.), *Living in a Seasonal World*,
DOI: 10.1007/978-3-642-28678-0_8,

8.1 Introduction

Balanced energy budgets are essential for any organism. Hibernating mammals without energy intake for about half a year have to be especially attentive to the allocation of energy reserves. Even small changes in daily energy expenditure will add up substantially over the months, severely affecting total energy balance. Small mammals are especially sensitive to changes in ambient temperature (T_a) (e.g., McNab 1971). Small hibernating mammals face the additional challenge of accumulating sufficient body fat stores before the onset of winter to last until the next spring season, and to economise with this storage sensibly, usually without the possibility to show distinctive behavioural adaptations when conditions change (e.g., change of hibernaculum). Due to the seasonality of their lifestyle, they are additionally especially sensitive to environmental parameters such as the duration and timing of vegetation periods (enabling seasonal fattening in autumn and the replenishment of depleted fat stores in spring). It is known that changes in T_a influence hibernation parameters (Geiser and Kenagy 1988; Speakman and Thomas 2003), but the mechanisms behind these changes and long-term consequences are still largely unclear.

The possibility to enter energy-saving hypometabolism (hibernation/daily torpor; Geiser and Ruf 1995) during periods of unfavourable conditions is considered to be a crucial adaptation to allow for the distribution of small mammals in the northern temperate zone with its seasonal, but predictable changes in environmental conditions and its more or less predictable fluctuations in food and water availability (Wojciechowski et al. 2007). Apart from some tropical species (Dausmann et al. 2004, 2005), hibernation is interrupted by periodic inter-bout arousals, short energetically expensive phases with general metabolism at high, euthermic levels. The function of these arousals is still under debate, but they seem to be necessary to keep up vital body functions at otherwise permanently low body temperature ($T_{b;}$ Willis 1982; Humphries et al. 2003; Geiser 2007). Torpor bout length and thus arousal frequency seems to be temperature-dependent, with increasing T_a increasing arousal frequency, at least in bats and other mammals (e.g., Geiser and Broome 1993; Humphries et al. 2002; Speakman and Thomas 2003). Climate change with generally increasing T_a and changes in a multitude of environmental parameters (e.g., duration and timing of vegetation periods) will affect all aspects of the life histories of animals. The ability to cope with these changes will be of vital importance determining range shifts or survival of animal species in the future. Humphries et al. (2002) showed that bioenergetics can provide the essential link between climate and biogeography needed to predict the consequences of climate change. Increasing temperatures during winter, even if differences are small, could change T_b during hibernation bouts, as well as arousal patterns, and therefore the pattern of thermoregulation and energy expenditure.

The aim of this study was to reveal to what extent and through which key parameters changes in environmental conditions (such as experienced in connection with climate change) directly influence patterns of thermoregulation and energy budgets during hibernation in a small hibernating mammal, by comparing winters of different strengths. Thermoregulation and energy expenditure are essential physiological control variables for survival and we wanted to know, if and how flexible *M. avellanarius* is able to regulate these control variables. We specifically aimed to answer the following questions: How does changing temperature affect T_b and thermoregulatory patterns during hibernation (arousal frequency)? What is the impact of changing temperatures during hibernation on energy expenditure [metabolic rate (MR)]? How does MR change during interbout arousals, and what are the energetic costs of additional arousals?

8.2 Materials and Methods

8.2.1 Study Animals and Animal Housing

We collected 13 *M. avellanarius* in Northern Germany by controlling nest tubes (Bright and Morris 1996) and nest boxes at the end of September 2009 and 2010 and released them at their sites of capture in early May 2010 and 2011, respectively. Thus, these animals had experienced their natural preparation period for the hibernation season in the wild. We transferred these 13 animals and their nests (both winters: 8 males, 5 females) into 13 wooden nest boxes that were readily accepted as sleeping and hibernation sites. Each nest box was isolated by an 8 cm organic insulation material to establish natural ambient hibernation conditions and each box was placed on the ground of an individual outdoor enclosure (0.75 m × 1.0 m × 0.5 m). Thus, *M. avellanarius* was kept under natural temperature and light regimes. Food and water was given ad libitum.

8.2.2 Temperature Measurements

We recorded skin temperature (T_{sk})—as a good indicator of T_b in small, curled-up hibernating mammals (Dausmann 2005)—every hour using small temperature data loggers (weetag lite®, Alpha Mach, Canada; weight: 1.1 g; resolution ± 0.5°C) that were fixed around the neck when the animals were torpid. We measured T_{sk} continuously from October until May ($N_{males} = 8$ and $N_{females} = 5$ for all months, winter 2009/2010 and 2010/2011). During the first winter three animals were presumably disturbed by shrews (*Sorex minutus*) at some point, and affected data were omitted from the analyses (thus $N_{males} = 5$ from December on). One female died in January 2010 ($N_{females} = 4$ from January on). T_a of the ground (under the leaf cover, the natural hibernation site of *M. avellanarius*) and nest temperature (T_{nest}) were also measured every hour, using temperature loggers (DS1922L Thermochron® iButtons, Dallas Semiconductor, USA; resolution ±0.5°C). We determined patterns of thermoregulation by identifying torpor bouts and arousal episodes. Torpor was defined as any time T_{skin} decreased below 30°C (T_{skin} of active, normothermic animals is around 36°C) and an arousal when T_{skin} increased above 30°C between torpor bouts, or when the difference $T_{skin}-T_{nest}$ was >30°C. The duration of an arousal was defined as the time period from when T_{skin} first began to increase from a torpor bout, ending in an arousal, to the time T_{skin} reached T_a again. In about 20% of the data, when monitored animals were not equipped with a functioning data logger, arousals were detected by comparing T_{nest} to T_a: whenever T_{nest} increased more than 4.5°C above $T_{a,}$ it can be safely assumed that the dormouse inside the nest was actively heating (Pretzlaff et al. 2010). Fitment of collars was checked monthly, without apparent disturbance to hibernating animals in deep torpor. During October and November, animals aroused occasionally after collar checks. But since this was not the case during deep torpor bouts, these arousals were considered as regularly occurring arousals.

8.2.3 Measurement of MR

We determined oxygen consumption of single dormice, as a measure of MR and thus of energy expenditure, continuously using open-flow respirometry. The wooden nest boxes served as metabolic chambers (volume approximately 1,725 cm^3), with the entrances serving as air inlets. A portable gas analyser (OxBox, designed and constructed by Thomas Ruf and Thomas Paumann, FIWI, University of Veterinary Medicine Vienna; Pretzlaff et al. 2010) was connected to the relevant nest box by an air-tight tube (Tygon®, Saint-Gobain Performance Plastics, France). The air was filtered from particles and dried with silica gel orange (Carl Roth®, Germany) after leaving nest boxes, before the air samples entered the gas analyser. The air was drawn at a flow rate of 50–80 l/h from the nest box and oxygen content was determined every minute. Consumption of oxygen by the animal was calculated as MR (ml O_2/h) per animal with the following equation (Heldmaier and Ruf 1992): MR (ml O_2/h) = Δvol%O_2* flow (l/h)* 10. To control for any drift of the oxygen sensor, we sampled reference air (baseline) of the ambient environment outside the nest boxes for 5 min every hour and adjusted values accordingly. Energy expenditure (kJ /day) was calculated by converting MR into units of energy. At the RQ of 0.85 (combustion of 50% fat and 50% carbohydrates; Dausmann et al. 2009), the caloric equivalent of oxygen consumption is 20.37 kJ/l O_2 (Heldmaier and Neuweiler 2004).

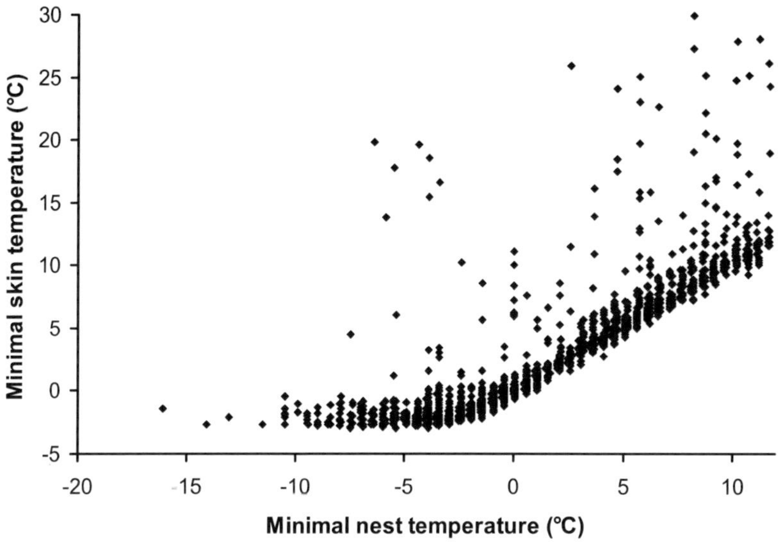

Fig. 8.1 Daily minimal skin temperature and daily minimal nest temperature, measured for 13 *M. avellanarius* during hibernation (October 2009 to April 2010)

8.2.4 Statistical Analyses

Statistical analyses were carried out using SPSS version 15.0 for Windows. Depending on whether data were normally distributed, we employed parametric or non-parametric tests (specified at each analysis). Data are presented as mean ± SD unless otherwise stated.

8.3 Results

Generally, *M. avellanarius* showed the typical pattern of T_b of small mammals during hibernation, including typical spontaneous periodical arousals, when held under natural conditions of photoperiod and T_a, with food and water given ad libitum. T_b started to drop in October and torpor bouts lengthened with increasing advancement into the winter and were most pronounced during December to February. After that, torpor bout duration started to decrease again until regular, normothermic T_b was adopted at the end of April. Average T_{skin} during torpor bouts was 2.6 ± 1.5°C ($N = 23$). Daily minimal T_{skin} followed minimal daily T_a and T_{nest}, respectively (Fig. 8.1). Remarkably, T_{skin} recorded by our loggers did drop regularly below 0°C, but seems to be defended just under −2°C (−1.85 to −2.9°C). T_{skin} remained constant at this threshold, even though T_{nest} and T_a did drop further substantially, confirming that the loggers were not merely measuring T_{nest}. To determine the value of this critical inflection point, we calculated the best fitting T_{skin}/T_{nest} regression equation for each individual by sequentially removing T_{skin} corresponding to the next lowest T_{nest}, following Corp et al. (1997). The best fit ($r^2 = 0.964 \pm 0.030$; $N = 13$) was obtained by exclusion of data below -2.3 ± 0.4°C, indicating that this is the threshold, below which T_b is actively defended.

Absolute minimal T_{nest} recorded in our study was −16.1°C, absolute minimal T_a was −19.5°C, absolute minimal T_{skin} was −2.9°C, and the longest torpor bout extended over 26 days.

Mean T_a during both winters was consistent with the long-term and accelerating warming trend. Between our two study years, winter 2009/2010 was colder than winter 2010/2011, which ranked among the top four winters in terms of extreme warm weather in the northern hemisphere over the last 63 years (Guirguis et al. 2011). To quantify climatic differences between the two study periods, we compared mean daily T_a of every month (October to April) between the first and the second winter (Table 8.1). Generally, temperatures decreased sooner during the first winter and stayed colder for a longer period of time. Mean daily T_a remained below 0°C throughout January (the coldest month) and February during winter 2009/2010, but only in December during winter 2010/2011. Mean daily T_a was significantly lower during the months of October, January, February and April of the first winter 2009/2010, significantly higher during the months of November and December (U-Test for every month, all $P < 0.001$), but did not differ significantly between the years during March (U-Test: $P = 0.471$).

To investigate adjustments of physiological parameters during hibernation to varying ambient conditions, we compared mean arousal frequencies per month between years. These did not differ significantly between the sexes, neither in winter 2009/2010 nor in winter 2010/2011 (t-test for all months, all $P > 0.05$). Thus, data of arousal frequency of females and males were pooled and analysed together. Since torpor bout length is generally dependent on the time within the season, we only compared the same months of different years with each other, and did not generally relate mean daily T_a to mean arousal frequency. This enabled us to untangle the effect of general changes of torpor bout durations over the course of the winter and the concomitant seasonal changes of temperature. We observed that mean arousal frequency per month differed significantly between winter 2009/2010 and winter 2010/2011 (Fig. 8.2). In most cases (except November 2009), *M. avellanarius* aroused significantly less frequent (and thus torpor bout duration was longer) during months that were significantly colder compared to the other year (t-test for every month, significance and N-numbers are indicated in Fig. 8.2). This difference was most pronounced in October and April, with a mean arousal frequency of 12.2 ± 4.5 in (the cooler) October 2009 compared to 29.1 ± 2.8 in (the warmer) October 2010 and 6.6 ± 1.9 in (the cooler) April 2010 compared to 22.9 ± 10.4 in (the warmer) April 2011. Only November did not yield a significant difference in mean arousal frequency, even though the mean average T_a did differ. During March (without a significant difference in mean T_a) mean arousal frequency was higher in the second year. Mean arousal frequency per month of the first winter 2009/2010 was lowest in January (3.4 ± 1.8 arousals/month) and

Table 8.1 Mean daily ambient temperature per month ± SD (°C) during winter 2009/2010 and winter 2010/2011

Year	October	November	December	January	February	March	April
2009/2010	7.1 ± 2.9	7.8 ± 2.0	0.6 ± 3.8	−3.5 ± 3.7	−1.1 ± 3.0	3.4 ± 5.4	7.0 ± 3.3
2010/2011	8.0 ± 3.0	4.4 ± 5.1	−3.8 ± 3.7	1.7 ± 3.9	1.0 ± 4.1	4.0 ± 4.7	11.4 ± 5.3
Significance	***	***	***	***	***	n.s.	***

Stars indicate significant differences: *** = $P < 0.001$, n.s = not significant

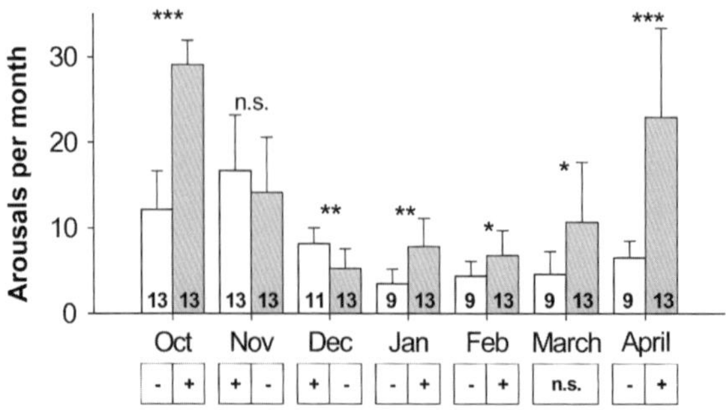

Fig. 8.2 Mean monthly arousal frequency per animal compared between winter 2009/2010 and winter 2010/2011; *stars* indicate significant differences: *** = $P < 0.001$, ** = $P < 0.01$, * = $P < 0.05$, n.s. = not significant; □ = winter 2009/2010, ▒ = winter 2010/2011, + = mean ambient temperature was significantly higher during the particular month, − = mean ambient temperature was significantly lower during the particular month, n.s. = mean ambient temperature did not differ significantly during the particular month, numbers in the *bars* represent *N* (animals)

during the second winter 2010/2011 in December (5.3 ± 2.3 arousals/month). Thus, during both winters, *M. avellanarius* aroused least frequently during the coldest month, regardless of which month this was. On the other hand, November was the warmest month in the winter 2009/2010, and this was also the month with the most frequent arousals (16.7 ± 6.5 arousals/month); April was the warmest month in the winter 2010/2011 and the month with the second frequent arousals (after October, the second warmest month) (22.9 ± 10.4 arousals/month). Not only torpor bout length, but also MR during torpor bouts, and thus energy expenditure during hibernation, was highly dependent on T_a, albeit in twofold directions. Generally, MR decreased with decreasing T_a, and concomitantly decreasing T_b. However, when T_b reached a certain threshold (at about −2.3°C T_{skin}), MR increased sharply, defending T_b (Fig. 8.3). Closely above this threshold increasing T_a had a positive effect on energy expenditure, i.e., MR decreased. Thus, there is an optimal range of T_a for optimised energy savings during hibernation; above and below this range energy expenditure increases.

To quantify additional energetic costs of additional arousals, we calculated the cost of one arousal, by means of energy expenditure, compared to remaining torpid throughout the same day. Mean hourly oxygen consumption and thus MR and energy expenditure of *M. avellanarius* were quite stable and low around 0.3 ml O_2/h × g (26 g) in the individual that remained torpid all 24 h. Mean hourly MR in the arousing animal (34 g) was comparable at the beginning of the day, but increased up to 2.9 ml O_2/h × g during the arousal that lasted for about 6 h. Integrated over the whole day, energy expenditure was 3.5 kJ/day (135 J/day × g) in the individual that remained torpid and it was 14.6 kJ/day (430 J/day × g) in the individual that aroused once. Thus, one arousal caused an additional energy

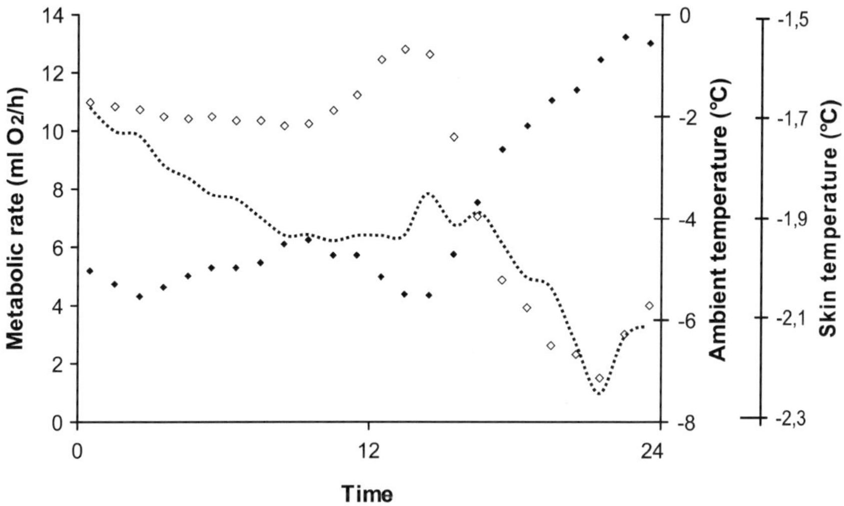

Fig. 8.3 Relationship between ambient temperature and energy expenditure. Shown is the torpid metabolic rate of a single *M. avellanarius* (body mass: 26 g) measured over 24 h during hibernation and ambient temperature. Skin temperature of this individual ranged between −2.2 and −1.7°C throughout the day. ◇ = mean hourly ambient temperature (°C); ■ = mean hourly torpid metabolic rate (ml O_2/h* animal), *dashed line* shows skin temperature

demand of about 200% compared to a day when the animal remained torpid at a mean T_a of −3.2°C.

8.4 Discussion

Hazel dormice, captured shortly before the onset of hibernation and exposed to natural environmental conditions during hibernation, showed the typical pattern of T_b and MR of small mammals during hibernation. Torpor bouts with minimal MR and T_b close to T_a were interspersed with periodic inter-bout arousals with euthermic MR and T_b. Torpor bout lengths changed over the season, and were longest in the middle of winter. However, environmental conditions did significantly alter hibernation parameters in *M. avellanarius*.

General T_b during hibernation as well as thermoregulatory patterns were significantly influenced by T_a, an effect that remained significant even after seasonal effects were eliminated. This effect of increased temperature on T_b, MR and thus energy expenditure was not uniform. T_b followed T_a and *M. avellanarius* aroused more often when temperatures (of the same month) were higher. Not surprisingly, MR generally increased with increasing T_a, and concomitantly increasing T_b (Humphries et al. 2002), leading to more energy expenditure the warmer it was. Studies on thermoregulation during torpor (e.g., Florant and Heller 1977) could make an internally fixed month-specific value of T_b during heterothermy feasible. This is obviously not the case in *M. avellanarius,* since the animals did not compensate differing T_a (e.g., by adjusting insulation capacities of their hibernacula) to match this possible month-specific value of T_b. Additionally, higher temperatures led to more frequent arousals (shorter torpor bouts), increasing energy demands. Generally, longest torpor bouts were found in the coldest month of each winter, regardless of which month this was. Thus, lengths of torpor bouts also do not seem to be an inherent, fixed variable for each month, but adjusted to prevailing T_a. When T_b (following T_a) dropped below a certain threshold (at about −2.3°C T_{skin}), T_b was defended and MR sharply increased. Given that this lower threshold was very stable despite a wide range of T_{nest} we conclude that the measurements of the loggers were not substantially influenced by T_{nest}, and effectively reflect T_b. Thus, *M. avellanarius* might be another mammal species found to utilise supercooling during hibernation (*Spermophilus parryii:* Barnes 1989). Closely above this threshold, increasing T_a led to a decrease of MR (in contrast to the usual increase of MR with increasing T_a). We attribute this to the decreased need to activate endogenous heat production for defence of T_b at this temperature level. Thus, *M. avellanarius* seems to be well adapted to low T_a during hibernation. Energy expenditure during torpor is stable and low within an (from an energetic point of view) optimal range of T_a in *M. avellanarius*. This result is consistent with studies on other hibernating mammals that demonstrated that energy expenditure increases and torpor bout length decreases when animals are exposed above or below the minimum T_b below which T_b is metabolically defended (Soivio et al. 1968; Heller and Hammel 1972; Geiser and Kenagy 1988).

During warmer winters, hibernating animals might be forced to heat up more often during more frequent arousals, utilising more energy additionally to the already generally higher energy expenditure at higher T_a (and thus T_b). These additional energetic costs might jeopardise winter survival of *M. avellanarius* and in the long-term survival of this species, if they exceed the amount of pre-hibernation energy accumulation (body fat reserves). This should be especially critical at the edge of distribution ranges (such as Northern Germany for *M. avellanarius*) because climatic conditions might already be suboptimal and thus more demanding.

The impact of T_a on arousal frequency and therefore torpor bout duration found in this study is consistent with that observed in laboratory studies in other hibernating mammals and bats at different constant T_a (Geiser et al. 1990; Geiser and Broome 1993). Furthermore, our findings emphasise the key role of arousals as the main factors of energy expenditure during hibernation. The energetic cost of a single arousal has been estimated to be as much as the amount of stored energy expended during 68 days of undisturbed torpor in bats (Thomas et al. 1990). The supplementary energetic cost of an arousal during the month of January (mid-hibernation) in *M. avellanarius* was about 200% of the energy expended during one day of continued torpor during the same time of that year.

Considering mean arousal frequencies, *M. avellanarius* aroused on average 48 times in the first, more severe winter 2009/2010 and 97 times in the second, milder winter 2010/2011. According to our estimation, the animals would have had to have eaten 51 additional fresh hazel nuts (about 51 g) to compensate for these additional arousals in the warmer winter. This amount corresponds to about 155% of body mass of well-fed individuals just before hibernation (33 g).

Moreover, our study showed that *M. avellanarius* can readily adjust the onset and end of hibernation, depending on the temperature of the relevant months. *M. avellanarius* aroused earlier from hibernation when the winter was shorter, indicating the ability to adapt to changing environmental conditions. Thus, they could potentially profit from food available through an advanced or prolonged vegetation period, caused by higher temperatures in spring and autumn. An earlier start of the vegetation period might be beneficial for *M. avellanarius* because reproduction could start earlier, prolonging the reproduction period and potentially making more reproduction efforts per year possible, which of course is directly relevant to improve individual fitness. Juškaitis (2008) observed females that successfully reared up to three litters during an unusually warm summer instead of one or two litters in a normal year. Longer feeding possibilities in autumn, on the other hand, might allow for increased fat stores, together with a shorter hibernation period, that needs to be fuelled. Similar flexibility to changing environmental conditions has been detected in birds that return earlier from or depart later to winter migration since the onset of global warming (Mills 2005). It is important to realise, that for successful utilisation of this additional time, other ecological factors must be matching as well, such as food availability (phenological mismatches) or the seasonal synchronisation of competitors (Juškaitis 2006), predator and prey, or parasite and host organism. Temporal mismatching between the common cuckoo (*Cuculus canorus*), a migratory brood parasitic bird, and its short-distance migrating hosts occurs already (Saino et al. 2009).

We have shown that *M. avellanarius* are not only flexibly responding in their thermoregulatory parameters to a generally mild or severe winter, but also more specifically to the climatic situation of each month. We conclude from our results that small mammals such as *M. avellanarius* hibernating at increased temperatures are more likely to deplete fat reserves before the end of hibernation. Ultimately, the extent of the effects of climate change on small hibernating mammals will depend on during which months and at what temperature level changes in T_a will be most prominent.

Acknowledgments The study has been carried out under licence from the "Landesamt für Naturschutz und Umwelt Schleswig–Holstein". We gratefully acknowledge their support and that of the local department of forestry. The "Deutsche Forschungsgemeinschaft (DFG)" provided financial support.

References

Barnes BM (1989) Freeze avoidance in a mammal: body temperatures below 0°C in an Arctic hibernator. Science 244:1593–1595

Bright PW, Morris PA (1996) Why are dormice rare? A case study in conservation biology. Mammal Rev 26:157–187

Corp N, Gorman ML, Speakman JR (1997) Seasonal variation in the resting metabolic rate of male wood mice *Apodemus sylvaticus* from two contrasting habitats 15 km apart. J Comp Physiol B 167:229–239

Dausmann KH (2005) Measuring body temperature in the field—evaluation of external vs. implanted transmitters in a small mammal. J Therm Biol 30:195–202

Dausmann KH, Glos J, Ganzhorn JU, Heldmaier G (2004) Hibernation in a tropical primate. Nature 429:825–826

Dausmann KH, Glos J, Ganzhorn JU, Heldmaier G (2005) Hibernation in the tropics: lessons from a primate. J Comp Physiol B 175:147–155

Dausmann KH, Glos J, Heldmaier G (2009) Energetics of tropical hibernation. J Comp Physiol B 179:345–357

Florant G, Heller H (1977) CNS regulation of body temperature in euthermic and hibernating marmots (*Marmota flaviventris*). Am J Physiol Regul Integr Comp Physiol 232(5): R203–R208

Geiser F (2007) Yearlong hibernation in a marsupial mammal. Naturwissenschaften 94:941–944

Geiser F, Broome LS (1993) The effect of temperature on the pattern of torpor in a marsupial hibernator. J Comp Physiol B 163:133–137

Geiser F, Kenagy GJ (1988) Torpor duration in relation to temperature and metabolism in hibernating ground squirrels. Physiol Zool 61(5):442–449

Geiser F, Ruf T (1995) Hibernation versus daily torpor in mammals and birds: physiological variables and classification of torpor patterns. Physiol Zool 68:935–966

Geiser F, Hiebert S, Kenagy GJ (1990) Torpor bout duration during the hibernation season of two sciurid rodents: interrelations with temperature and metabolism. Physiol Zool 63:489–503

Guirguis K, Gershunov A, Schwartz R, Bennett S (2011) Recent warm and cold daily winter temperature extremes in the Northern Hemisphere. Geophys Res Lett 38:L17701. doi:10.1029/2011GL048762

Heldmaier G, Neuweiler G (2004) Vergleichende Tierphysiologie. Springer, Heidelberg

Heldmaier G, Ruf T (1992) Body temperature and metabolic rate during natural hypothermia in endotherms. J Comp Physiol B 162:696–706

Heller HC, Hammel HT (1972) CNS control of body temperature during hibernation. Comp Biochem Physiol 41A:349–359

Humphries MM, Thomas DW, Speakman JR (2002) Climate-mediated energetic constraints on the distribution of hibernating mammals. Nature 418:313–316

Humphries MM, Kramer DL, Thomas DW (2003) The role of energy availability in mammalian hibernation: an experimental test in free-ranging eastern chipmunks. Physiol Biochem Zool 76(2):180–186

Juškaitis R (2006) Interactions between dormice (Gliridae) and hole-nesting birds in nestboxes. Folia Zool 55(3):225–236

Juškaitis R (2008) The common dormouse, *Muscardinus avellanarius*. Ecology, population structure and dynamics. Institute of Ecology of Vilnius University Publishers, Vilnius

McNab BK (1971) On the ecological significance of Bergman's rule. Ecology 52:845–854

Mills AM (2005) Changes in the timing of spring and autumn migration in North American migrant passerines during a period of global warming. Ibis 147:259–269. doi:10.1111/j.1474-919X.2005.00380.x

Pretzlaff I, Kerth G, Dausmann KH (2010) Communally breeding bats use physiological and behavioural adjustments to optimise daily energy expenditure. Naturwissenschaften 97:353–363. doi:10.1007/s00114-010-0647-1

Saino N, Rubolini D, Lehikoinen E, Sokolov LV, Bonisoli-Alquati A, Ambrosini R, Boncoraglio G, Møller AP (2009) Climate change effects on migration phenology may mismatch brood parasitic cuckoos and their hosts. Biol Lett 5:539–541. doi:10.1098/rsbl.2009.0312

Soivio A, Tähti H, Kristoffersson R (1968) Studies on the periodicity of hibernation in the hedgehog (*Erinaceus europaeus* L.). Ann Zool Fenn 5:224–226

Speakman JR, Thomas DW (2003) Physiological ecology and energetics of bats. In: Kunz TH, Fenton BM (eds) Bat ecology. University of Chicago Press, Chicago, pp 430–490

Thomas DW, Dorais M, Bergeron JM (1990) Winter energy budgets and cost of arousals for hibernating little brown bats, *Myotis lucifugus*. J Mammal 71:475–479

Willis JS (1982) The mystery of periodic arousal. In: Lyman CP, Willis JS, Malan A, Wang LCH (eds) Hibernation and torpor in mammals and birds. Academic Press, New York, pp 92–103

Wojciechowski MS, Jefimow M, Tegowska E (2007) Environmental conditions, rather than season, determine torpor use and temperature selection in large mouse-eared bats (*Myotis myotis*). Comp Biochem Physiol A 147:828–840

Chapter 9
Comparison of Variables of Torpor Between Populations of a Hibernating Subtropical/Tropical Bat at Different Latitudes

Clare Stawski

Abstract Torpor use differs among heterothermic species and data comparing populations of a single hibernating species at different latitudes could reveal how animals adapt to different climates. I investigated variables of torpor in free-ranging *Nyctophilus bifax*, a hibernating subtropical/tropical insectivorous microbat, during winter from a subtropical region and a tropical region. Mean torpor bout duration was significantly shorter and mean minimum skin temperature of torpid bats was significantly higher in the tropical population in comparison to the subtropical population. In both populations torpor bout duration was negatively correlated with ambient temperature and the slope of this relationship differed significantly ($P = 0.02$) between the populations when examined under the same thermal conditions. The differences found in these variables of torpor were most likely due to regional differences in weather and insect abundance and suggest that populations of heterothermic mammals can adapt or acclimatize to the local climate of the habitat that they occupy, even when thermal conditions are mild.

C. Stawski
Centre for Behavioural and Physiological Ecology, Zoology, University of New England, Armidale, NSW 2351, Australia

Present Address
C. Stawski (✉)
Institute of Environmental Sciences, Jagiellonian University, ul. Gronostajowa 7, 30-387 Kraków, Poland
e-mail: clare.stawski@gmail.com

T. Ruf et al. (eds.), *Living in a Seasonal World*,
DOI: 10.1007/978-3-642-28678-0_9, © Springer-Verlag Berlin Heidelberg 2012

9.1 Introduction

In most regions of the world, weather and temperature change daily and seasonally and generally influence food supply. During winter in temperate regions when ambient temperatures (T_a) are reduced or during the dry season in tropical regions, many plants cease to grow and insects reduce or halt activities. Decreases in food availability in turn will affect the energy balance of many animals. Endotherms are especially affected as they need large amounts of energy to maintain high, normothermic body temperatures (T_b), particularly at low T_a (Bartholomew 1982). However, many animals use a variety of energy saving mechanisms to overcome such periods and the most effective of these is torpor, which is characterized by controlled reductions in T_b, metabolic rate (MR) and water loss (Geiser 2004). Torpor is employed by a wide variety of animals and is especially prolific among small mammals (Geiser and Körtner 2010) due to their large surface area to volume ratio, which increases heat loss during cold exposure. Torpor use has been well documented in temperate species (e.g. Hall 1832; Hock 1951; Speakman and Thomas 2003; Geiser 2006; Turbill and Geiser 2008), but until recently relatively few quantitative data on torpor use in subtropical and tropical species were available (for review see Geiser and Stawski 2011).

Even fewer studies have compared variables of torpor between populations of a single species inhabiting different climates. Climate and weather patterns vary between regions and it is possible that the use of torpor may vary among populations of a species whose range encompasses various habitats. Generally, latitude and elevation do affect the variables and energetics of torpor (Geiser and Ferguson 2001; Fenn et al. 2009). For example, woodchucks (*Marmota monax*) increase the length of torpor bouts in the more northern latitudes of their range, i.e. where winters are colder (Zervanos et al. 2010). Different populations of big brown bats (*Eptesicus fuscus*) and eastern red bats (*Lasiurus borealis*) are also known to differ in the thermal physiology of torpor bouts (Dunbar and Brigham 2010). Therefore, populations of the same species appear to be able to adapt variables of torpor and thermal physiology to the local climate conditions of the habitat they occupy.

The thermal physiology of *Nyctophilus bifax*, a hibernating Australian microbat endemic to subtropical/tropical regions, differed between a subtropical and a tropical population primarily in regard to the T_a at which torpid bats defended their T_b (Stawski and Geiser 2011). This was suggested to be an adaptation to the climate of the region where each population resided. In the current analysis I compared the differences in variables of torpor specifically during winter between a tropical and a subtropical population of *N. bifax*. I predicted that torpor bout durations would be shorter and shallower during winter in the tropical region in comparison to the subtropical region. Further, I hypothesized that local weather conditions would affect variables of torpor of both subtropical and tropical *N. bifax*, but that these relationships would differ between the regions.

9.2 Materials and Methods

Field studies were undertaken during the Austral winter to obtain data on the differences in variables and patterns of torpor. This research was conducted at two sites (i) the subtropical Iluka Nature Reserve (29°24′S, 153°22′E) during July/August 2007 and (ii) ~2,000 km north at the tropical Djiru National Park (17°50′S, 146°03′E) during June 2008 and July/August 2009. Some of the data collected from these studies have been published previously (Stawski et al. 2009; Stawski and Geiser 2010), however, new data have been added and all data have been re-analyzed for this comparison. Throughout the study periods T_a was measured ~2 m off the ground in the shade with data loggers (±0.5°C, iButton thermochron DS1921G, Maxim Integrated Products, Inc., Sunnyvale, California, USA).

Mist nets were used to capture a total of eight *N. bifax* from the subtropical region and a total of 13 *N. bifax* from the tropical region. Once captured, a small patch of fur was removed from between the shoulder blades of each individual and a temperature-sensitive radio-transmitter (~0.5 g, LB-2NT, Holohil Systems Inc., Carp, Ontario, Canada) was attached to the exposed skin using a latex-based adhesive (SkinBond, Smith and Nephew United, Mount Waverley, NSW, Australia). Before attachment to the bat each transmitter was calibrated to the nearest 0.1°C in a water bath between 5 and 40°C against a precision thermometer. Each individual was released at the site of capture following this procedure. On the subsequent morning each individual was located via radio-tracking and their roost locations were marked with flagging tape. An antenna and a remote receiver/logger (Körtner and Geiser 2000) were placed within range of the bats' transmitter signal. This receiver/logger recorded the pulse interval of the transmitter signal once every 10 min and data from the receiver/loggers were downloaded every 2–5 days to prevent loss of data. Data were converted into skin temperatures (T_{skin}) of the bat and torpor bouts included any period of time that T_{skin} was <28°C for periods >30 min. This definition of torpor was used as it has previously been suggested that a T_b < 30°C is appropriate to define torpor (Barclay et al. 2001) and because it has often been shown that the difference between T_{skin} and T_b of a small torpid animal is <2°C.

Statistical analyses were conducted using Stastitixl (V 1.8, 2007) and null hypotheses were rejected if alpha was <0.05. The means of each individual were used to calculate means involving repeated measures and data were reported as means ±SD (n = the number of individuals or days, N = the number of observations). t tests were used to compare the two regions and ANCOVAs were used when comparing linear regressions. Data from the two winters at Djiru National Park (June 2008 and July/August 2009) were combined as no differences were found.

Table 9.1 Mean ± SD for the given environmental variables during winter for the tropical site ($n = 50$ days) and subtropical site ($n = 29$ days)

Region	Mean min T_a (°C)	Mean max T_a (°C)	Mean daily T_a fluctuations (°C)	Mean daily T_a (°C)	Mean nightly T_a (°C)
Tropical	16.2 ± 2.4	21.9 ± 1.7	5.8 ± 2.4	18.8 ± 1.6	17.7 ± 2.1
Subtropical	8.2 ± 2.2	17.4 ± 2.3	9.2 ± 1.9	12.4 ± 2.2	10.9 ± 2.4
P-value	<0.001	<0.001	<0.001	<0.001	<0.001

Significance of difference from two sample *t* tests

Table 9.2 Mean ± SD for the given variables for all free-ranging *N. bifax* during winter from the tropical and subtropical sites

Region	Body mass (g)	Longest individual bouts (h)	Mean torpor bout duration (h)	Daily minimum T_{skin} (°C)
Tropical	10.2 ± 0.7 (13)	11.4 ± 8.6 (13)	4.5 ± 3.1 (13, 210)	20.1 ± 3.1 (13, 102)
Subtropical	10.1 ± 1.3 (8)	83.7 ± 27.6 (8)	26.8 ± 11.3 (8, 114)	14.7 ± 2.4 (8, 147)
P-value	0.9	<0.001	<0.001	<0.001

Significance of difference from two sample *t* tests from the means of individuals. Body mass at capture. The brackets show: the number of individuals, the number of observations

9.3 Results

During winter at the tropical site T_a ranged from a minimum of 10.6°C to a maximum of 25.3°C; at the subtropical site T_a ranged from a minimum of 4.0°C to a maximum of 25.0°C. Mean minimum T_a, mean maximum T_a, mean daily (24 h) T_a, and mean nightly (sunset–sunrise) T_a were all significantly lower at the subtropical site in comparison to the tropical site (Table 9.1). However, the mean daily (24 h) fluctuations of T_a were significantly higher at the subtropical site in comparison to the tropical site (Table 9.1). The timing of sunset differed between the two field sites ($P < 0.001$) and at the subtropical site sunset occurred at 17:11 ± 00:04 h ($n = 29$) and at the tropical site sunset occurred at 17:52 ± 00:07 h ($n = 50$).

Body mass of individuals did not differ between the tropical and subtropical populations (Table 9.2). However, the length of the longest torpor bout recorded in a subtropical individual (128.5 h) was ~fourfold longer than the longest torpor bout recorded in a tropical individual (33.3 h) and the mean of all the individuals' longest bouts was ~sevenfold longer in the subtropical population compared to the tropical population (Table 9.2). Further, mean torpor bout duration was ~sixfold longer in the subtropical population in comparison to the tropical population (Table 9.2). Even when compared under similar thermal conditions (minimum T_a during each torpor bout in the range of 10.5–19.5°C), mean torpor bout duration was also significantly longer (~twofold) in the subtropical population (11.3 ± 8.0 h, $n = 8$, $N = 52$) in comparison to the tropical population (5.3 ± 4.8 h, $n = 13$, $N = 165$; $P < 0.001$). Mean daily minimum T_{skin} was 5.4°C lower in the subtropical population in comparison to the tropical population

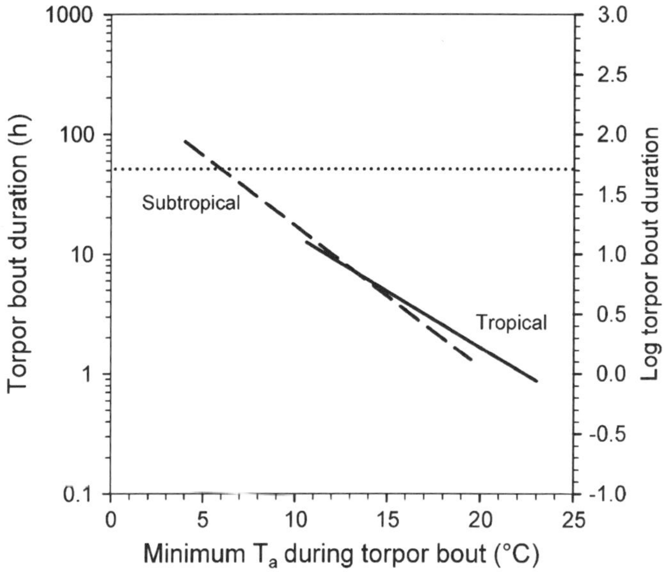

Fig. 9.1 The relationship between duration of torpor bouts ($\log_{10}$) and minimum T_a during winter at the tropical site (*solid line*; $\log_{10}$ TBD = 2.1–0.09[T_a°C]; $R^2 = 0.2$, $P < 0.001$, $F_{78,131} = 1.4$) and during winter at the subtropical site (*dashed line*; $\log_{10}$ TBD = 2.4–0.12[T_a°C]; $R^2 = 0.7$, $P < 0.001$, $F_{25,82} = 21.8$). The *dotted horizontal line* represents torpor bout durations of 48 h

(Table 9.2). However, when compared under similar thermal conditions (corresponding T_a in the range of 10.5–16.5°C), mean daily minimum T_{skin} did not differ between the subtropical population (16.9 ± 2.5°C, $n = 8$, $N = 45$) and the tropical population (16.7 ± 2.1°C, $n = 13$, $N = 46$; $P = 0.7$).

Torpor bout duration and the minimum T_a experienced during each torpor bout were negatively correlated in both the tropical population ($\log_{10}$ TBD = 2.1–0.09[T_a°C], $R^2 = 0.2$, $P < 0.001$, $F_{78,131} = 1.4$; Fig. 9.1) and the subtropical population ($\log_{10}$ TBD = 2.4–0.12[T_a°C], $R^2 = 0.7$, $P < 0.001$, $F_{25,82} = 21.8$; Fig. 9.1). For this relationship both slope ($P = 0.1$) and intercept ($P = 0.8$) did not differ between the populations. However, when this relationship was compared under similar thermal conditions (minimum T_a during each torpor bout in the range of 10.5–19.5°C) there was no common slope ($P = 0.02$) between the tropical and subtropical populations.

Daily minimum T_{skin} and the corresponding T_a during winter were positively correlated for both the tropical population (minimum T_{skin}[°C] = 2.4 + 1.0[T_a°C]; $R^2 = 0.5$, $P < 0.001$, $F_{61,39} = 3.0$; Fig. 9.2) and the subtropical population (minimum T_{skin}[°C] = 9.1 + 0.6[T_a°C]; $R^2 = 0.3$, $P < 0.001$, $F_{24,119} = 4.9$; Fig. 9.2). The slope differed significantly (P < 0.001) between the populations for this relationship. In contrast, there was a common slope ($P = 0.6$) and intercept ($P = 0.4$) between the populations when this relationship was compared under similar thermal conditions (corresponding T_a in the range of 10.5–16.5°C).

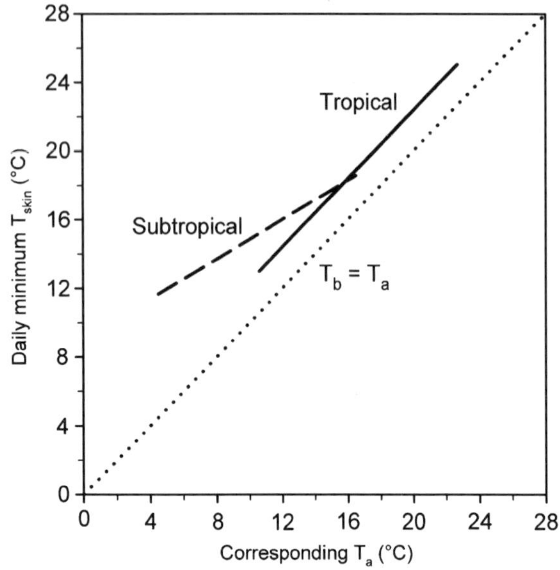

Fig. 9.2 The relationship between daily minimum T_{skin} and the corresponding T_a during winter at the tropical site (*solid line*; minimum T_{skin} [°C] = 2.4 + 1.0[T_a°C]; $R^2 = 0.5$, $P < 0.001$, $F_{61,39} = 3.0$) and during winter at the subtropical site (*dashed line*; minimum T_{skin} [°C] = 9.1 + 0.6[T_a°C]; $R^2 = 0.3$, $P < 0.001$, $F_{24,119} = 4.9$). The *dotted regression line* represents $T_b = T_a$

9.4 Discussion

The weather conditions recorded during winter differed significantly between the regions such that, as expected, in the northern tropical part of their range *N. bifax* were experiencing much milder winters in comparison to the southern subtropical part of their range. Further, during winter the number of insects captured during a night at the tropical site was significantly greater (~fourfold) than that captured at the subtropical site (Stawski 2010). These differences in weather patterns and food availability between the tropics and subtropics likely were responsible, as predicted, for differences in the variables of torpor between the subtropical and tropical populations of *N. bifax*. For example, mean torpor bout durations and the longest torpor bouts recorded during winter were significantly shorter in tropical *N. bifax* in comparison to subtropical *N. bifax*, likely a result of the more favorable foraging conditions in the tropical habitat. Nevertheless, as both tropical and subtropical *N. bifax* have to deal with a reduction in food availability during winter (Stawski 2010) and T_as that are frequently well below the TNZ in both regions, both populations show frequent use of torpor to conserve energy.

Similarly to temperate hibernating bats (Ransome 1971; Park et al. 2000; Rambaldini and Brigham 2008; Turbill and Geiser 2008), the amount of time an individual spent torpid was negatively affected by the minimum T_a experienced during each torpor bout in both the tropical and subtropical populations. The slope and intercept of this relationship did not differ between the populations, however, when this relationship was compared between the populations under similar thermal conditions the slopes did differ resulting in longer torpor bouts in subtropical individuals. While previous studies have shown that MR and temperature

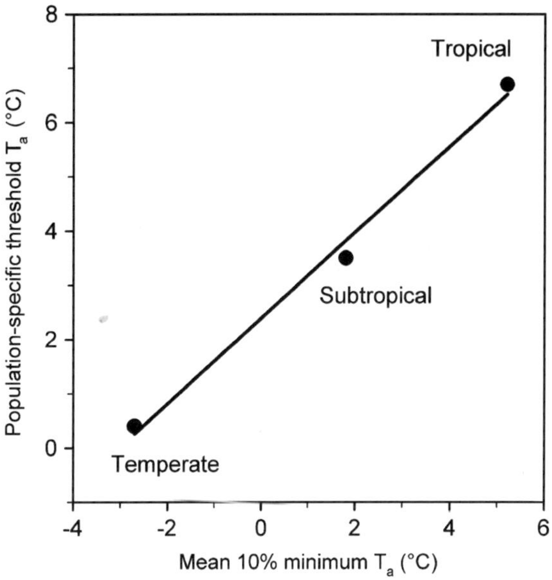

Fig. 9.3 Population-specific threshold T_a as a function of the mean of the lowest 10% of minimum T_a among temperate (*N. geoffroyi*, taken from Geiser and Brigham 2000), subtropical (*N. bifax*, taken from Stawski and Geiser 2011) and tropical (*N. bifax*, taken from Stawski and Geiser 2011) regions. This linear relationship is described by the following equation: population-specific threshold T_a (°C) = $-2.91 + 1.24T_a$ (°C), $r^2 = 0.9$ (Stawski and Geiser 2011)

influence torpor bout durations (French 1985; Geiser and Kenagy 1988), the data presented here suggest that there are also other factors that influence the length of torpor bouts. Some of these factors are ecological, for example, at a similar T_a there were twice as many insects captured during winter at the tropical site in comparison to the subtropical site (Stawski 2010), suggesting that tropical *N. bifax* would have more success foraging for food and therefore able to avoid prolonged torpor bouts at the same T_a when subtropical *N. bifax* cannot. Therefore, additionally to temperature and MR, the amount of food an individual obtains before becoming torpid or able to obtain after arousal may influence the length of a torpor bout (Geiser and Mzilikazi 2011).

Even though torpor bout duration varied between the populations, the lack of difference in MRs (Stawski and Geiser 2011) strongly suggests that like *N. bifax* from the subtropical population (Stawski et al. 2009), *N. bifax* from the tropical population are also able to enter prolonged bouts of torpor when conditions are unfavorable. Nevertheless, along with different torpor bout durations, the population-specific threshold T_a at which torpid individuals thermoregulate also varied between the populations, similarly to big brown bats (Dunbar and Brigham 2010). There was a positive relationship found between the population-specific threshold T_a of *Nyctophilus* species and the ambient conditions of the specific habitat they occupy (Fig. 9.3; Stawski and Geiser 2011), which also reveals that there are some long-term adaptations to the climate in which a population resides.

The higher daily minimum torpid T_{skin} in tropical *N. bifax* in comparison to subtropical *N. bifax* is likely due to the higher T_as experienced during the tropical winter in comparison to the subtropical winter, as under the same thermal conditions daily minimum torpid T_{skin} was similar between the populations. Further, under the same thermal conditions the slope of the positive relationship between daily minimum torpid T_{skin} and the corresponding T_a did not differ between the populations, but the slope did differ when all of the data was examined together. The slope of this relationship for the tropical population was similar to $T_b = T_a$, however, the regression line for the tropical population was ~2°C above $T_b = T_a$, which is the typical difference between T_{skin} and T_b. On the contrary, the slope of this relationship for the subtropical population was much shallower in comparison to both the tropical population and $T_b = T_a$. During the studies at both the subtropical and tropical field sites T_a was only measured in one location and not at each individual roost. Therefore, the shallower regression line for the subtropical population may be explained by the use of thermally buffered roosts, such as tree hollows, by the bats and that consequently they were experiencing much smaller daily T_a fluctuations within the roosts in comparison to the ambient conditions outside of the roost (Stawski et al. 2008). Conversely, it appears that individuals in the tropical population were most likely roosting in more open locations, such as under leaves, as their T_{skin} largely approximated the measured T_a (Stawski et al. 2008). Roost choice is important to hibernators because, while thermally buffered roosts will promote larger energy savings during prolonged torpor bouts as experienced by subtropical *N. bifax*, the large daily passive fluctuations of thermally unstable roosts can eradicate the large energetic costs of the arousal process for tropical *N. bifax* that arouse daily by passively re-warming (Chruszcz and Barclay 2002; Brack 2007; Turbill and Geiser 2008; Boyles and McKechnie 2010; Cory Toussaint et al. 2010).

My study has shown that even though the thermal physiology of *N. bifax* is largely similar throughout their range, there is some plasticity in torpor use that allows local populations to match the climatic and ecological conditions of the habitat they occupy. These variations most likely represent a response to the local energetic challenges faced by a population. The data presented here add to the limited knowledge that populations of the same species inhabiting different latitudes may vary in regard to torpor use and thermal physiology (Geiser and Ferguson 2001; Fenn et al. 2009; Dunbar and Brigham 2010; Zervanos et al. 2010). They demonstrate that research on a single population may lead to incorrect assumptions about the thermal biology and habitat usage of the species as a whole. Further, such data is particularly important for species with wide distribution ranges (Boyles et al. 2011) and may provide information on how animals respond to changes in the climate and ecology of their habitats.

Acknowledgments I thank Fritz Geiser for help and for constructive comments on the manuscript. For their help with this project I would also like to thank Alexander Foster, Gerhard Körtner, Anaïs LeBot, Alexander Riek, Margaret and Mike Stawski, and Christopher Turbill. This project was funded by an Australian Postgraduate Award, Bat Conservation International, Jagiellonian University, and the University of New England.

References

Barclay RMR, Lausen CL, Hollis L (2001) What's hot and what's not: defining torpor in free-ranging birds and mammals. Can J Zool 79:1885–1890

Bartholomew GA (1982) Body temperature and energy metabolism. In: Gordon MS (ed) Animal physiology: principles and adaptations. MacMillian Publishing Co. Inc., New York, pp 333–406

Boyles JG, McKechnie AE (2010) Energy conservation in hibernating endotherms: Why "suboptimal" temperatures are optimal. Ecol Model 221:1644–1647

Boyles JG, Seebacher F, Smit B, McKechnie AE (2011) Adaptive thermoregulation in endotherms may alter responses to climate change. Integr Comp Biol 51:676–690

Brack V Jr (2007) Temperatures and locations used by hibernating bats, including *Myotis sodalis* (Indiana bat), in a limestone mine: implications for conservation and management. Environ Manage 40:739–746

Chruszcz BJ, Barclay RMR (2002) Thermoregulatory ecology of a solitary bat, *Myotis evotis*, roosting in rock crevices. Funct Ecol 16:18–26

Cory Toussaint D, McKechnie AE, van der Merwe M (2010) Heterothermy in free-ranging male Egyptian free-tailed bats (*Tadarida aegyptiaca*) in a subtropical climate. Mamm Biol 75: 466–470

Dunbar MB, Brigham RM (2010) Thermoregulatory variation among populations of bats along a latitudinal gradient. J Comp Physiol B 180:885–893

Fenn AM, Zervanos SM, Florant GL (2009) Energetic relationships between field and laboratory woodchucks (*Marmota monax*) along a latitudinal gradient. Ethol Ecol Evol 21:299–315

French AR (1985) Allometries of the durations of torpid and euthermic intervals during mammalian hibernation: a test of the theory of metabolic control of the timing of changes in body temperature. J Comp Physiol B 156:13–19

Geiser F (2004) Metabolic rate and body temperature reduction during hibernation and daily torpor. Annu Rev Physiol 66:239–274

Geiser F (2006) Energetics, thermal biology, and torpor in Australian Bats. In: Zubaid A, McCracken GF, Kunz TH (eds) Functional and evolutionary ecology of bats. Oxford University Press, New York, pp 5–22

Geiser F, Brigham RM (2000) Torpor, thermal biology, and energetics in Australian long-eared bats (*Nyctophilus*). J Comp Physiol B 170:153–163

Geiser F, Ferguson C (2001) Intraspecific differences in behaviour and physiology: effects of captive breeding on patterns of torpor in feathertail gliders. J Comp Physiol B 171:569–576

Geiser F, Kenagy GJ (1988) Torpor duration in relation to temperature and metabolism in hibernating ground squirrels. Physiol Zool 61:442–449

Geiser F, Körtner G (2010) Hibernation and daily torpor in Australian mammals. Aust Zool 35:204–215

Geiser F, Mzilikazi N (2011) Does torpor of elephant shrews differ from that of other heterothermic mammals? J Mammal 92:452–459

Geiser F, Stawski C (2011) Hibernation and torpor in tropical and subtropical bats in relation to energetics, extinctions, and the evolution of endothermy. Integr Comp Biol 51:337–348

Hall M (1832) On hybernation. Philos T Roy Soc Lond 122:335–360

Hock RJ (1951) The metabolic rates and body temperatures of bats. Biol Bull 101:289–299

Körtner G, Geiser F (2000) Torpor and activity patterns in free-ranging sugar gliders *Petaurus breviceps* (Marsupialia). Oecologia 123:350–357

Park KJ, Jones G, Ransome RD (2000) Torpor, arousal and activity of hibernating greater horseshoe bats (*Rhinolophus ferrumequinum*). Funct Ecol 14:580–588

Rambaldini DA, Brigham RM (2008) Torpor use by free-ranging pallid bats (*Antrozous pallidus*) at the northern extent of their range. J Mammal 89:933–941

Ransome RD (1971) The effect of ambient temperature on the arousal frequency of the hibernating greater horseshoe bat, *Rhinolophus ferrumequinum*, in relation to site selection and the hibernation state. J Zool 164:353–371

Speakman JR, Thomas DW (2003) Physiological ecology and energetics of bats. In: Kunz TH, Fenton MB (eds) Bat Ecology. University of Chicago Press, Chicago, pp 430–492

Stawski C (2010) Ecology and thermal physiology of an insectivorous bat restricted to subtropical and tropical Australia. Dissertation, University of New England

Stawski C, Geiser F (2010) Seasonality of torpor patterns and physiological variables of a free-ranging subtropical bat. J Exp Biol 213:393–399

Stawski C, Geiser F (2011) Do season and distribution affect thermal energetics of a hibernating bat endemic to the tropics and subtropics? Am J Physiol 301:R542–R547

Stawski C, Turbill C, Geiser F (2008) Prolonged torpor use during winter by a free-ranging bat in subtropical Australia. In: Lovegrove BG, McKechnie AE (eds) Hypometabolism in animals: hibernation, torpor and cryobiology. University of KwaZulu-Natal, Pietermaritzburg, pp 353–360

Stawski C, Turbill C, Geiser F (2009) Hibernation by a free-ranging subtropical bat (*Nyctophilus bifax*). J Comp Physiol B 179:433–441

Turbill C, Geiser F (2008) Hibernation by tree-roosting bats. J Comp Physiol B 178:597–605

Zervanos SM, Maher CR, Waldvogel JA, Florant GL (2010) Latitudinal differences in the hibernation characteristics of woodchucks (*Marmota monax*). Physiol Biochem Zool 83:135–141

Chapter 10
The Other Functions of Torpor

Fritz Geiser and R. Mark Brigham

Abstract Although energy conservation by cold-climate adult endotherms in winter is often viewed as the main function of torpor, recent evidence suggests that this may not always be the case. We examined whether other functions of torpor may be equally or even more important in some instances. Torpor enhances fat storage during migration, apparently permits prolonged female sperm storage in bats, allows reproduction with limited or fluctuating food supply, and delays parturition until more favorable periods. Torpor appears to increase the efficiency of energy and nutrient use during development. Further, torpor reduces water requirements, appears to permit persistence during droughts, reduces the load of some parasites, permits co-existence of competing species, and also reduces the risk of predation and mammalian extinctions. Thus, the functions of torpor are complex and some of these appear to be not just proximate.

10.1 Introduction

It remains a widely held view that the ultimate function of torpor is energy conservation to increase the probability of survival despite limited food availability in winter by adult, non-reproductive mammals and birds living in cold climates (Lyman et al. 1982). To some extent this paradigm is related to the regions where

F. Geiser (✉)
Centre for Behavioural and Physiological Ecology, Zoology,
University of New England, Armidale NSW 2351, Australia
e-mail: fgeiser@une.edu.au

R. M. Brigham
Department of Biology, University of Regina, Regina SK S4S 0A2, Canada
e-mail: mark.brigham@uregina.ca

T. Ruf et al. (eds.), *Living in a Seasonal World*,
DOI: 10.1007/978-3-642-28678-0_10, © Springer-Verlag Berlin Heidelberg 2012

most of the classical work on torpor was conducted (northern Europe and North America) and where torpor often occurs predictably in winter when food availability decreases and ambient temperatures (T_a) are low (e.g. Hall 1832; Horvath 1878; Eisentraut 1929; Kayser 1939; Jaeger 1948; Lyman 1948).

The physiological variables measured in the context of torpor such as heart rate, rate of oxygen consumption, or carbon dioxide production that are often used as proxies for energy metabolism are variables relatively easy to measure. Their substantial decline during torpor is consistent with energy conservation being the ultimate function of torpor. In contrast, perceived proximate effects of torpor use such as predator avoidance, a prolonged life span, or other life history traits are often more obscure and therefore difficult to quantify, especially for animals under natural conditions. Nevertheless, evidence has emerged in recent years which taken together suggests that the functions of torpor may not be as straight forward as commonly thought. In some instances, the data suggest that energy conservation during torpor might in fact be a proximate, rather than the ultimate effect. Torpor is now known to occur in many mammals and birds including tropical and subtropical species (Dausmann et al. 2005; Stawski et al. 2009). Torpor is also used by reproductively active individuals as well as developing young. While many heterothermic species enter torpor mainly or only when energetically challenged (Wang 1978; Christian and Geiser 2007; Landry-Cuerrier et al. 2008) others display torpor in summer or in situations where they have access to food and exhibit no obvious energetic stress (Hill 1975; Geiser and Baudinette 1987; Nicol and Andersen 2002; Rimbaldini and Brigham 2008; Bieber and Ruf 2009; Stawski and Geiser 2010). This suggests that torpor likely has other functions besides energy conservation during periods of severe energy shortage or that animals have no reason to not use torpor. Our review is an attempt to synthesize what is known about some of these other functions and to examine whether these functions are more than simply proximate effects.

10.2 Torpor and Migration

An important difference in how small birds and terrestrial mammals deal with adverse conditions in winter often identified in text books is that because birds can fly, they commonly employ a behavioral option and migrate to avoid winter conditions. In contrast, because small terrestrial mammals cannot migrate long distances, they more often use a physiological option and employ torpor, characterized by pronounced reductions in metabolism and body temperature (T_b), to survive the winter. Clearly, the situation is more complex as bats can fly but many do not migrate and hibernate instead. Further, one of the first identified 'new functions of torpor' stems from evidence that migrating hummingbirds enter torpor not simply to reduce metabolic rate when energetically challenged, but as a means to enhance fat storage at night during the period of migration (Carpenter and Hixon 1988; Hiebert 1993). A free-ranging migratory hummingbird (*Selasphorus rufus*) used torpor for most of the night while roosting although it had enough stored fat to remain normothermic (Carpenter and Hixon 1988). Captive individuals of the same species enter the longest and most frequent bouts of spontaneous torpor during the autumnal migration period, the time of the year when they are at their greatest body mass and have amassed fat stores up to 50% of lean body mass. Such fat stores would easily allow for regulation of normothermic T_b in the cold (Hiebert 1993). These observations strongly suggest that torpor use is not simply an acute response to an energetic challenge, but rather a predictive strategy that anticipates a likely future energy-demanding event.

In non-migratory birds, torpor use, in addition to dealing with food shortages in winter, has been suggested to represent a key adaptation for maintaining a year round home range (Körtner et al. 2000). Tawny frogmouths (*Podargus strigoides*), an insectivorous species that has to deal with a substantial seasonal reduction of food and low T_a, regularly enter torpor in winter and remain in almost the identical part of the forest. Therefore, unlike migrating birds, they do not need to re-establish a new home range for breeding in early spring, which, as in hibernating terrestrial mammals, is likely to increase fitness.

10.3 Torpor and Reproduction

Although many endotherms undergoing reproduction appear to avoid torpor (Brigham 1992; Kissner and Brigham 1993; Barnes 1996; Mzilikazi and Lovegrove 2002), many others have been observed using torpor during various parts of the reproductive cycle. These include torpor during periods when females store sperm, during pregnancy, and during lactation. Sperm storage in female vertebrates is longest in ectothermic reptiles (up to several years) and up to 225 days in bats (Wimsatt 1960; Racey 1979; Birkhead and Møller 1993; Wang et al. 2008). In other mammals, sperm storage typically is 0.5–10 days, although 30 days has been reported in hares (*Lepus europaeus*) (Birkhead and Møller 1993). In bats, the longest periods of sperm storage (>100) days have been observed in cold-climate hibernating species (Racey 1979; Birkhead and Møller 1993, Wang et al. 2008). However, even in warm-climate species, including tropical bats, sperm storage lasting from several weeks to around 2 months have been observed (Wang et al. 2008). The enormous difference in duration of successful sperm storage in female bats in comparison to other mammals is likely related to some aspect of the extensive use of torpor. Bats frequently display torpor, including multiday torpor, even in tropical/subtropical areas (Stawski et al. 2009; Cory Toussaint et al. 2010; Liu and Karasov 2011), whereas most other mammals in which sperm storage is <1 month (Birkhead and Møller 1993) are homeothermic and a few (small marsupials and rodents) use daily torpor exclusively (Geiser and Ruf 1995).

Prolonged sperm storage in female bats has several implications (Racey 1979; Birkhead and Møller 1993). It often allows separation of the male and female reproductive cycle and likely facilitates optimal timing of reproduction. Female sperm storage represents insurance against not finding a partner in spring and enables females to synchronize births. So while the main function of torpor is likely to allow winter survival, it nevertheless also permits bats to cope with having to fertilize eggs soon after winter with enough time for the development of young, which can be critical especially in cold regions. Synchronization of births likely decreases predation and increases fitness.

Aside from enhancing sperm storage, torpor is also important after fertilization. Torpor during pregnancy has been recorded in echidnas (*Tachyglossus aculeatus*; Geiser 1996; Morrow and Nicol 2009), small marsupials (Frey and Fleming 1984; Geiser et al. 2005), tenrecs (*Microgale* spp.; Stephenson and Racey 1993), and vespertilionid and pteropodid bats (Racey 1973; Audet and Fenton 1988; Grinevich et al. 1995; Geiser 1996; Turbill and Geiser 2006; Stawski 2010). In many species, torpor during pregnancy is shorter and shallower than in non-reproductive females, but this is not always the case (Turbill and Geiser 2006). Although torpor during pregnancy appears to be mainly expressed during energetic stress, some mammals appear to use it to enhance fat storage (Geiser 1996). Pregnant mulgaras (*Dasycercus cristicauda/blythi*) enter spontaneous torpor (food available) under thermally mild conditions in captivity (Geiser and Masters 1994). During the period when torpor is expressed females increase body mass substantially. This is not due to the mass of the developing young because entire neonate marsupial litters weigh <0.5% of their mothers' body mass (Geiser and Masters 1994). Female mulgaras remain normothermic for a few days before giving birth and also do not enter torpor while lactating, which in marsupials is the energetically most demanding reproductive period. Although data are limited, female mulgaras in the wild in central Australia appear to use the same approach as captive animals with extensive use of torpor during pregnancy in winter (much deeper and longer than males at the same time of year) and homeothermy during lactation (Körtner et al. 2008). Thus, like for migration, it appears that torpor is not always a response to acute energy shortage or thermal stress, but rather an anticipatory strategy to reduce energy loss and enhance fat storage for future energy-demanding events.

In hoary bats (*Lasiurus cinereus*), entry into torpor by pregnant females during cold spells in spring appears to have an important function aside from energy conservation (Willis et al. 2006). Hoary bats use multiday torpor not only for survival, but also to slow embryonic growth which delays parturition until conditions are favorable for lactation and neonatal survival. Torpor has long been known to reduce growth rates and to prolong gestation in bats (Racey 1973), but this was typically viewed in a negative rather than a positive context.

Torpor during lactation is used by marsupial dunnarts (*Sminthopsis crassicaudata*; Morton 1978), sugar gliders (*Petaurus breviceps*; Geiser et al. 2008) and vespertilionid bats (Audet and Fenton 1988; Grinevich et al. 1995). Whereas in some bats, torpor during lactation appears to be shallower than during pregnancy, this is not the case in long-eared bats (*Nyctophilus* spp.; Turbill and Geiser 2006). In sugar gliders (*Petaurus breviceps*) torpor appears to be used exclusively during lactation and not pregnancy (Geiser et al. 2008). While it is known that torpor may be used during lactation, functions that go beyond simply saving energy to deal with limited food supply have not been identified, although conservation of water (see below) may be an important reason.

10.4 Torpor and Development

Many endotherms are altricial at hatching or birth. Altricial young lack insulation, are small, often uncoordinated, have a low capacity for heat production and a limited ability for physiological thermoregulation. Because they are well below adult size (in marsupials as little as $\sim 0.1\%$ or less) they have to continue to grow using imported nutrients, which are typically supplied by parents, and must use these efficiently. The largely poikilothermic thermoregulatory response (or perhaps better hypothermic response because heat loss exceeds heat production that is, however, well above that in ectotherms, Hill 1976) during early development permits the more or less exclusive investment of nutrients into growth. However, when the developing young becomes able to metabolically defend their T_b at low T_a at about 1/3 or 1/2 of adult size (or even at smaller sizes when huddling in nests or exposed to high T_a, Hill 1976) they face the problem of excessive heat loss that has to be compensated for by increased internal heat production via combustion of fuels to maintain a constant high T_b. It is likely that to minimize energy loss and to permit survival during adverse conditions or when parents are unable to supply enough food, some developing altricial young use torpor. Although this facet of torpor use has not received nearly enough scientific attention (it has not even been investigated in precocial young, which also face heat loss problems because of relatively large surface areas), it is currently known for five birds (chats *Ephthianura tricolor*, martins *Delichon urbica*, swifts *Apus apus*, mousebirds *Urocolius macrourus*, strom-petrels *Oceanodroma furcata*), four marsupials (dunnarts *Sminthopsis macroura*, antechinus *Antechinus stuartii*, *A. flavipes* and kowaris *Dasyuroides byrnei*), and three placental mammals (shrews *Crocidura russula*, hamsters *Phodopus sungorus*)(reviewed in Geiser 2008), and bats, *Eptesicus fuscus* (Hollis and Barclay 2008). In addition, reductions in core T_b by >5°C and even more in peripheral tissues have also been reported in developing king penguins (*Aptenodytes patagonicus*, Eichhorn et al. 2011). Torpor during development appears to have two major functions. The first is the traditionally accepted acute response permitting survival during adverse conditions and limited energy availability as for many adult heterotherms. However, torpor in developing young, although it likely slows growth, also has the potential to be used for diverting nutrients to enhance growth when the young are again normothermic (Giroud et al. 2012). Currently evidence supporting the hypothesis that this function of torpor is limited to a single study, but it certainly warrants further work.

10.5 Torpor and Water Conservation

Torpor is widely used by arid zone mammals and birds (Serventy and Raymond 1973; Warnecke et al. 2008; Doucette et al. 2011, 2012; Levy et al. 2011a). Lovegrove (2000) suggested that the unpredictability of food in environments such as deserts may be the reason for its extensive use, particularly on a daily basis.

Recent experimental data support this interpretation (Munn et al. 2010). However, use of torpor by desert dwellers and species from other regions may also function to reduce water loss. Free-ranging mouse-lemurs (*Microcebus murinus*) reduce rates of water turnover during torpor in comparison to individuals remaining normothermic (Schmid and Speakman 2009). Evaporative water loss of torpid cactus mice (*Peromyscus eremicus*) is only about 40% of that of normothermic individuals at the same T_a (MacMillen 1965). Similarly, for gerbils (*Gerbillus pusillus*) and marsupial dunnarts (*Sminthopsis macroura*) daily torpor reduces evaporative water loss to between 20 and 40% of that in normothermic individuals (Buffenstein 1985; Cooper et al. 2005). Extreme reductions in evaporative water loss of almost 90% have been reported for torpid bats (*Chalinolobus gouldii*) hibernating at T_a 10°C (Hosken and Withers 1997). Further, evaporative water loss of torpid honey possums (*Tarsipes rostratus*) during torpor was so low as to be undetectable (Withers et al. 1990).

Thus, torpor reduces water loss in mammals and is therefore ecologically important. It has been hypothesized that the large diversity of dasyurid marsupials in the arid zone of Australia, all of whom appear to express daily torpor, some on 100% of days in winter (Warnecke et al. 2008; Körtner and Geiser 2009), is one reason why these mammals persist in such a water poor environment. Although direct evidence for the initiation of torpor with the removal or restriction of water is more equivocal than for food restriction (but see Ibuca and Fukumura 1997), more extensive use of torpor in summer than winter in desert spiny mice (*Acomys russatus*; Levy et al. 2011a) suggests that water limitation rather than energy shortage may be the major cue for torpor use in this species in summer and thus the ultimate function of torpor may be for water rather than energy conservation.

10.6 Torpor and Drought

Cold northern winters are not the only periods of reduced food availability and have the advantage of lasting only about 6 months that can be bridged by fat or food storage and/or by appropriate use of torpor. In contrast, droughts, as for example during El Niño events, may last for a year or more, too long for survival on stored fat or food, at least in endotherms.

A recent study on free-ranging Australian owlet-nightjars (*Aegotheles cristatus*) showed for the first time that lack of rainfall in dry years can substantially affect the use and patterns of torpor in the wild (Doucette et al. 2012). In a dry year, owlet-nightjars in semi-arid central Australia used torpor more frequently than in a wet year (61% vs. 27%), torpor bout duration in the dry year was about twice as long as in the wet year and the minimum T_b in the dry year was 3.3°C lower than in the wet year. The variation in torpor variables between years was not strongly related to differences in T_a, but was best explained by availability of biomass (insects), which was less than half in the dry than the wet year because of reduced primary productivity. However, availability of water per se also may have affected torpor use (see above). Thus torpor seems to be an important adaptation to overcome droughts and persistence in areas that are subject to prolonged rather than seasonal shortages of food and water.

10.7 Torpor and Parasites

Although immune function during torpor appears to be suppressed (Prendergast et al. 2002; Boyles and Willis 2010), parasite loads, especially of some gastrointestinal parasites, decrease during hibernation in some animals (Chute 1964). This could be due to low T_b during hibernation being harmful to the parasite, slowed metabolic processes of the parasite, and seasonal changes in the chemical composition of the host because of hibernation that may be detrimental to the parasite (Chute 1964). Data for Alpine marmots (*Marmota marmota*) show that not all parasites respond in the same way (Callait and Gauthier 2000). A cestode (*Ctenotenia marmotae*) leaves the marmot intestine before/during hibernation and overwinters in an intermediate host, a mite. One nematode (*Ascaris laevis*) disappears from the marmot gut during hibernation. The only parasite that remains in the marmot gut during hibernation is a small nematode (*Citellina alpina*). This species, perhaps because of its small size and location in the marmot caecum is the only one known to persist in its host throughout hibernation (Callait and Gauthier 2000). Thus, it is correct that there is a 'self-curing' effect at the onset of hibernation that reduces the parasite load of this mammalian host. In contrast to marmots, all helminths (trematodes, cestodes, and nematodes) survived in the gut of bats (*Myotis lucifugus*) during hibernation (Coggins et al. 1982) even though T_b falls to near 0°C. This does not support the interpretation that parasites cannot cope with low temperatures nor that all parasites are removed during hibernation.

Ectoparasites of the bat *Miniopterus schreibersii*, which include bat flies, mites, and ticks, were present throughout the yearly cycle (Lourenco and Palmeirim 2008). However, these ectoparasites exhibited substantially reduced reproductive activities during the bats' hibernation period likely in response to reduced temperatures. For bats in tropical and subtropical regions where temperatures do not change dramatically with season, ectoparasites appear to reproduce year round (Marshall 1971). In contrast to marmot gut parasites, ectoparasites (mites) appear to increase in number immediately after hibernation when marmots visit other burrows and are more exposed to the mite than in winter because it lives in marmot nests (Arnold and Lichtenstein 1991).

Thus it appears that hibernation results in the reduction in some intestinal parasites and slows the growth of others. However, overall hibernation does not appear to be an effective approach to remove parasites. A reduction in some parasites is likely a by-product of low T_b during hibernation rather than its primary function.

10.8 Torpor and Inter-Specific Competition

Recent data suggest that torpor use is affected by inter-specific competition and may permit co-existence of competing species. Common spiny mice (*Acomys cahirinus*) in large outdoor enclosures competitively exclude their congener (*A. russatus*; golden spiny mice) from nocturnal activity forcing them to become diurnal (Levy et al. 2011b). This temporal partitioning allows the species to co-exist on a diet of arthropods in summer. In winter, when arthropod levels are low, both species rely on a largely vegetarian diet. Under these conditions, removal of common spiny mice reduced the duration of daily torpor in golden spiny mice whether food was supplemented or not. These results suggest that torpor may allow the co-existence of two competing species during periods of resource limitation and high energetic requirements (Levy et al. 2011b). Further, torpor use and the concomitant energy savings may allow subordinate species to occupy areas dominated by larger competitors. This may also be one reason for the large diversity of sympatric insectivorous/carnivorous marsupials (Dasyuridae) in the Australian arid zone (and perhaps other small desert mammals) despite the limited supply of food and water and the presence of introduced predators.

10.9 Torpor and Mammalian Extinctions

A traditional paradigm is that inactivity and immobility during torpor increase the risk of predation (Armitage 2004). This has been viewed as one of the major costs associated with torpor and selection pressure to minimize use of torpor. Contrary to this, recently published data show that torpor is used under mild conditions in summer without apparent energy stress. Non-reproductive captive dormice (*Glis glis*) in good condition enter a sequence of short torpor bouts in summer often after brief periods of activity following the final arousal from hibernation in spring (Bieber and Ruf 2009). The authors suggested that dormice may use torpor to actually avoid predation which may contribute to their high longevity. Field observations on subtropical bats support this contention. Long-eared bats (*Nyctophilus bifax*) in a subtropical, coastal region enter torpor frequently during summer (85% of observation days) and even do so on 38% of nights during their normal activity period (Stawski and Geiser 2010). Counter to more traditional predictions, bats in good condition (high body condition index) entered torpor more frequently, displayed longer torpor bouts and lower minimum T_bs than bats in poorer condition. Thus, it appears that these bats increased torpor use not because of food shortages or low energy stores, but likely to avoid exposure to predators during foraging when feeding is not required. This is exactly the opposite of what is traditionally predicted.

Predator avoidance may be one of the reasons why opportunistically heterothermic mammals are less threatened with extinctions (Liow et al. 2009) and have suffered fewer extinctions than their homeothermic relatives (Geiser and

Turbill 2009). Of the 61 worldwide confirmed extinctions of mammal species, over the past 500 years (American Museum for Natural History, Committee on Recently Extinct Organisms, http://creo.amnh.org), only four (6.5% of species) were likely heterothermic although the vast majority of these (approximately 85% of species) were within the size range of <10 kg in which deep torpor is expressed. Because only ~15% of extinct mammals weighed >10 kg, it is surprising that almost all extinct mammals (93.5%) were likely homeothermic (Geiser and Turbill 2009). Further, considering that most mammals (approximately 80%) are rodents, bats, insectivores, and marsupials and that many of these are known to be heterothermic, it is astonishing how few of the extinct mammals were likely heterothermic. It has been proposed that extinction risks in heterothermic mammals may be minimized by torpor use due to its enormous scope for adjusting energy requirements. This may allow long-term survival even under adverse environmental conditions and help individuals cope with habitat degradation and avoid introduced or native predators (Geiser and Turbill 2009; Liow et al. 2009). Thus, the use of torpor and the commonly prolonged life span of heterotherms (Turbill et al. 2011; Ruf et al. 2012) appear to have permitted opportunistically heterothermic mammals to deal with anthropogenic disturbances responsible for extinctions better than is the case for most other species.

In conclusion, our summary shows that torpor has implications for many facets of mammalian and avian biology that go well beyond energy conservation. Use of torpor is known to enhance fat storage during migration, apparently permits prolonged female sperm storage in bats, allows reproduction with limited or fluctuating food supply, and delays parturition until more favorable periods. Torpor also appears to increase the efficiency of energy and nutrient use during development, permits co-existence of competing species and survival of prolonged droughts, and reduce water requirements, parasite loads, and risk of mammalian extinctions. Future work, including data on daily torpor in free-ranging rodents that is currently almost entirely lacking, is likely to identify further 'other' functions of torpor.

Acknowledgments We thank Dick Hill, Chris Turbill and the students from RMB's laboratory for constructive comments. The work was supported by the ARC (FG) and NSERC (RMB).

References

Armitage KB (2004) Badger predation on yellow-bellied marmots. Am Midl Nat 151:378–387

Arnold W, Lichtenstein AV (1991) Ectoparasite loads decrease the fitness of Alpine marmots (*Marmota marmota*) but not the cost of sociality. Behav Ecol 4:36–39

Audet D, Fenton MB (1988) Heterothermy and the use of torpor by the bat *Eptesicus fuscus* (Chiroptera: Vespertilionidae): a field study. Physiol Zool 61:197–204

Barnes BM (1996) Relationship between hibernation and reproduction in male ground squirrels. In: Geiser F, Hulbert AJ and Nicol SC (eds) Adaptations to the Cold, 10th international hibernation symposium, University of New England Press, Armidale, pp 71–80

Bieber C, Ruf T (2009) Summer dormancy in edible dormice (*Glis glis*) without energetic constraints. Naturwissenschaften 96:165–171

Birkhead TR, Møller AP (1993) Sexual selection and the temporal separation of reproductive events: sperm storage data from reptiles, birds and mammals. Biol J Linn Soc 50:295–311

Boyles JG, Willis CKR (2010) Could localized warm areas inside cold caves reduce mortality of hibernating bats affected by white-nose syndrome? Front Ecol Environ 8:92–98

Brigham RM (1992) Daily torpor in a free-ranging goatsucker, the common poorwill (*Phalaenoptilus nuttallii*). Physiol Zool 65:457–472

Buffenstein R (1985) The effect of starvation, food restriction, and water deprivation on thermoreguation and average daily metabolic rates in *Gerbillus pusillus*. Physiol Zool 58: 320–328

Callait M-P, Gauthier D (2000) Parasite adaptations to hibernation in Alpine marmots (*Marmota marmota*). In: Heldmaier G and Klingenspor M (eds) Life in the cold. 11th international hibernation symposium, Heidelberg, Springer, pp 139–146

Carpenter FL, Hixon MA (1988) A new function of torpor: fat conservation in a wild migrant hummingbird. Condor 90:373–378

Christian N, Geiser F (2007) To use or not to use torpor? Activity and body temperature as predictors. Naturwissenschaften 94:483–487

Chute RM (1964) Hibernation and parasitism: recent developments and some theoretical consideration. Mammalian hibernation 2. Ann Acad Sci Fenn A, 4 Biol 71:113–122

Coggins JR, Tedesco JL, Rupprecht CE (1982) Seasonal changes and overwintering of parasites in the bat, *Myotis lucifugus* (Le Conte), in a Wisconsin hibernaculum. Am Midl Nat 107: 305–315

Cooper CE, McAllan BM, Geiser F (2005) Effect of torpor on the water economy of an arid-zone marsupial, the striped-faced dunnart (*Sminthopsis macroura*). J Comp Physiol B 175:323–328

Cory Toussaint D, McKechnie AE, van der Merwe M (2010) Heterothermy in free-ranging male Egyptian free-tailed bats (*Tadarida aegyptiaca*) in a subtropical climate. Mamm Biol 75: 466–470

Dausmann KH, Glos J, Ganzhorn JU, Heldmaier G (2005) Hibernation in the tropics: lessons from a primate. J Comp Physiol B 175:147–155

Doucette LI, Brigham RM, Pavey CR, Geiser F (2011) Roost type influences torpor use by Australian owlet-nightjars. Naturwissenschaften 98:845–854

Doucette LI, Brigham RM, Pavey CR, Geiser F (2012) Prey availability affects daily torpor by free-ranging Australian owlet-nightjars (*Aegotheles cristatus*). Oecologia. doi:10.1007/s00442-011-2214-7

Eichhorn G, Groscolas R, Le Glaunec G, Parisel C, Arnold L, Medina P, Handrich Y (2011) Heterothermy in growing king penguins. Nat Commun 2:435. doi:10.1038/ncomms1436

Eisentraut M (1929) Beobachtungen über den Winterschlaf der Haselmaus (*Muscardinus avellanarius* L.). Z Säugetierkd 4:213–239

Frey H, Fleming MR (1984) Torpor and thermoregulatory behaviour in free-ranging feathertail gliders (*Acrobates pygmaeus*) (Marsupialia: Burramyidae) in Victoria. In: Smith AP, Hume ID (eds) Possums and gliders. Surrey Beatty and Australian Mammal Society, Sydney, pp 393–401

Geiser F (1996) Torpor in reproductive endotherms. In: Geiser F, Hulbert AJ and Nicol SC (eds) Adaptations to the cold. 10th international hibernation symposium, University of New England Press, Armidale, pp 81–86

Geiser F (2008) Ontogeny and phylogeny of endothermy and torpor in mammals and birds. Comp Biochem Physiol A 150:176–180

Geiser F, Baudinette RV (1987) Seasonality of torpor and thermoregulation in three dasyurid marsupials. J Comp Physiol B 157:335–344

Geiser F, Masters P (1994) Torpor in relation to reproduction in the Mulgara, *Dasycercus cristicauda* (Dasyuridae: Marsupialia). J Thermal Biol 19:33–40

Geiser F, Ruf T (1995) Hibernation versus daily torpor in mammals and birds: physiological variables and classification of torpor patterns. Physiol Zool 68:935–966

Geiser F, Turbill C (2009) Hibernation and daily torpor minimize mammalian extinctions. Naturwissenschaften 96:1235–1240

Geiser F, McAllan BM, Brigham RM (2005) Daily torpor in a pregnant dunnart (*Sminthopsis macroura* Dasyuridae: Marsupialia). Mamm Biol 70:117–121

Geiser F, Christian N, Cooper CE, Körtner G, McAllan BM, Pavey CR, Turner JM, Warnecke L, Willis CKR, Brigham RM (2008) Torpor in marsupials: recent advances. In: Lovegrove BG, McKechnie AE (eds) Hypometabolism in animals: torpor, hibernation and cryobiology, 13th international hibernation symposium, University of KwaZulu-Natal, Pietermaritzburg, pp 297–306

Giroud S, Turbill C, Ruf T (2012) Torpor use and body mass gain during pre-hibernation in late-born juvenile garden dormice exposed to food shortage. In: Ruf T, Bieber C, Arnold W, Millesi E (eds) Living in a seasonal world: thermoregulatory and metabolic adaptations, Springer, Heidelberg

Grinevitch L, Holroyd SL, Barclay RMR (1995) Sex differences in the use of daily torpor and foraging time by big brown bats (*Eptesicus fuscus*) during the reproductive season. J Zool Lond 235:301–309

Hall M (1832) On hybernation. Trans Roy Soc Lond B 122:335–360

Hiebert SM (1993) Seasonality of daily torpor in a migratory hummingbird. In: Carey C, Florant GL, Wunder BA, Horwitz B (eds) Life in the cold: ecological, physiological and molecular mechanisms. Westview, Boulder, pp 25–32

Hill RW (1975) Daily torpor in *Peromyscus leucopus* on an adequate diet. Comp Biochem Physiol A 51:413–423

Hill RW (1976) The ontogeny of homeothermy in neonatal *Peromyscus leucopus*. Physiol Zool 49:292–306

Hollis L, Barclay RMR (2008) Developmental changes in body temperature and use of torpor by the big brown bat (*Eptesicus fuscus*). In: Lovegrove BG, McKechnie AE (eds) Hypometabolism in animals: torpor, hibernation and cryobiology, 13th international hibernation symposium, University of KwaZulu-Natal, Pietermaritzburg, pp 361–372

Horvath A (1878) Beitrag zur Lehre über den Winterschlaf. Verh phys med Ges Würzburg 12:139–198

Hosken DJ, Withers PC (1997) Temperature regulation and metabolism of an Australian bat, *Chalinolobus gouldii* (Chiroptera: Vespertilionidae) when euthermic and torpid. J Comp Physiol B 167:71–80

Ibuka N, Fukumura K (1997) Unpredictable deprivation of water increases the probability of torpor in the Syrian hamster. Physiol Behav 62:551–556

Jaeger EC (1948) Does the poorwill "hibernate"? Condor 50:45–46

Kayser C (1939) Exchanges respiratoires des hibernants réveillés. Ann Physiol Physicochim Biol 15:1087–1219

Kissner KJ, Brigham RM (1993) Evidence for the use of torpor by incubating and brooding common poorwills, *Phalaenoptilus nuttallii*. Ornis Scand 24:333–334

Körtner G, Geiser F (2009) The key to winter survival: daily torpor in a small arid zone marsupial. Naturwissenschaften 96:525–530

Körtner G, Brigham RM, Geiser F (2000) Winter torpor in a large bird. Nature 407:318

Körtner G, Pavey CR, Geiser F (2008) Thermal biology, torpor and activity in free-living mulgaras in arid zone Australia during the winter reproductive season. Physiol Biochem Zool 81:442–451

Landry-Cuerrier M, Munro D, Thomas DW, Humphries MM (2008) Climate and resource determinants of fundamental and realized metabolic niches of hibernating chipmunks. Ecology 89:3306–3316

Levy O, Dayan T, Kronfeld-Schor N (2011a) Adaptive thermoregulation in golden spiny mice: the influence of season and food availability on body temperature. Physiol Biochem Zool 84:175–184

Levy O, Dayan T, Kronfeld-Schor N (2011b) Interspecific competition and torpor in golden spiny mice: two sides of the energy-acquisition coin. Int Comp Biol 51:441–448

Liow LH, Fortelius M, Lintulaakso K, Mannila H, Stenseth NC (2009) Lower extinction in sleep-or-hide mammals. Am Nat 173:264–272

Liu J-N, Karasov WH (2011) Hibernation in warm hibernacula by free-ranging Formosan leaf-nosed bats, *Hipposideros terasensis*, in subtropical Taiwan. J Comp Physiol B 181:125–135

Lourenco S, Palmeirim JM (2008) Which factors regulate the reproduction of ectoparasites of temperate-zone cave-dwelling bats? Parasitol Res 104:127–134

Lovegrove BG (2000) Daily heterothermy in mammals: coping with unpredictable environments. In: Heldmaier G and Klingenspor M (eds) Life in the cold: 11th international hibernation symposium, Springer, Heidelberg, pp 29–40

Lyman CP (1948) The oxygen consumption and temperature regulation in hibernating hamsters. J Exp Zool 109:55–78

Lyman CP, Willis JS, Malan A, Wang LCH (1982) Hibernation and torpor in mammals and birds. Academic Press, New York

MacMillen RE (1965) Aestivation in the cactus mouse *Peromyscus eremicus*. Comp Biochem Physiol 16:227–247

Marshall AG (1971) The ecology of *Basilia hispida* (Diptera: Nycteribiidae) in Malaysia. J Anim Ecol 40:141–154

Morrow G, Nicol SC (2009) Cool sex? Hibernation and reproduction overlap in the echidna. PLoS ONE 4(6):e6070. doi:10.1371/journal.pone.0006070

Morton SR (1978) Torpor and nest-sharing in free-living *Sminthopsis crassicaudata* (Marsupialia) and *Mus musculus* (Rodentia). J Mammal 59:569–575

Munn AJ, Kern P, McAllan BM (2010) Coping with chaos: unpredictable food supplies intensify torpor use in an arid-zone marsupial, the fat-tailed dunnart (*Sminthopsis crassicaudata*). Naturwissenschaften 97:601–605

Mzilikazi N, Lovegrove BG (2002) Reproductive activity influences thermoregulation and torpor in pouched mice, *Saccostomus campestris*. J Comp Physiol B 172:7–16

Nicol SC, Andersen NA (2002) The timing of hibernation in Tasmanian echidnas: why do they do it when they do? Comp Biochem Physiol B 131:603–611

Prendergast BJ, Freeman DA, Zucker I, Nelson JR (2002) Periodic arousal from hibernation is necessary for initiation of immune response in ground squirrels. Am J Physiol 282: R1054–R1062

Racey PA (1973) Environmental factors affecting the length of gestation in heterothermic bats. J Reprod Fert Suppl 19:175–189

Racey PA (1979) The prolonged storage and survival of spermatozoa in Chiroptera. J Reprod Fert 56:391–402

Rimbaldini DA, Brigham RM (2008) Torpor use by free-ranging pallid bats (*Antrozous pallidus*) at the northern extent of their range. J Mammal 89:933–941

Ruf T, Bieber C, Turbill C (2012) Survival, aging, and life-history tactics in mammalian hibernators. In: Ruf T, Bieber C, Arnold W, Millesi E (eds) Living in a seasonal world: thermoregulatory and metabolic adaptations, Springer, Heidelberg

Schmid J, Speakman JR (2009) Torpor and energetic consequences in free-ranging grey mouse lemurs (*Microcebus murinus*): a comparison of dry and wet forests. Naturwissenschaften 96: 609–620

Serventy V, Raymond R (1973) Torpidity in desert mammals. Aust Wildl Heritage 14:2233–2240

Stawski C (2010) Torpor during the reproductive season in a free-ranging subtropical bat, *Nyctophilus bifax*. J Thermal Biol 35:245–249

Stawski C, Geiser F (2010) Fat and fed: frequent use of summer torpor in a subtropical bat. Naturwissenschaften 97:29–35

Stawski C, Turbill C, Geiser F (2009) Hibernation by a free-ranging subtropical bat (*Nyctophilus bifax*). J Comp Physiol B 179:433–441

Stephenson PJ, Racey PA (1993) Reproductive energetics of the Tenrecidae (Mammalia: Insectivora). II. The shrew-tenrecs, *Microgale* spp. Physiol Zool 66:664–685

Turbill C, Geiser F (2006) Thermal biology of pregnant and lactating female and male long-eared bats, *Nyctophilus geoffroyi* and *N. gouldi*. J Comp Physiol B 176:165–172

Turbill C, Bieber C, Ruf T (2011) Hibernation is associated with increased survival and the evolution of slow life histories among mammals. Proc Roy Soc B 278:3355–3363

Wang LCH (1978) Energetics and field aspects of mammalian torpor: the Richardsons's ground squirrel. In: Wang LCH, Hudson JW (eds) Strategies in cold. Academic Press, New York, pp 109–145

Wang Z, Liang B, Racey PA, Wang Y-L, Zhang S-Y (2008) Sperm storage, delayed ovulation, and menstruation of the female Rickett's big-footed bat (*Myotis ricketti*). Zool Stud 47: 215–221

Warnecke L, Turner JM, Geiser F (2008) Torpor and basking in a small arid zone marsupial. Naturwissenschaften 95:73–78

Willis CKR, Brigham RM, Geiser F (2006) Deep, prolonged torpor by pregnant, free-ranging bats. Naturwissenschaften 93:80–83

Wimsatt WA (1960) Some problems of reproduction in relation to hibernation in bats. In: Lyman CP, Dawe AR (eds) Mammalian hibernation. Bull Mus Comp Zool, Cambridge, pp 248–270

Withers PC, Richardson KC, Wooller RD (1990) Metabolic physiology of euthermic and torpid honey possums, *Tarsipes rostratus*. Aust J Zool 37:685–693

Chapter 11
Survival, Aging, and Life-History Tactics in Mammalian Hibernators

Thomas Ruf, Claudia Bieber and Christopher Turbill

Abstract Hibernation is commonly viewed as an adaptation that simply allows animals to survive periods of food shortage and climatically harsh conditions. Here, we review accumulating evidence suggesting that hibernation is part of a specific "slow-paced" mammalian life-history tactic that is associated with increased survival, retarded physiological aging, increased maximum longevity, low rates of fecundity, and long generation times. We argue that these traits can be explained if the primary function of hibernation—at least in many species—is the reduction of extrinsic mortality risks, namely predation, under environmental conditions that are not life-threatening, but do not favor reproduction. According to this view, hibernation is but one element of a life-history strategy that maximizes fitness by bet-hedging, i.e., reducing the risk of losing offspring by spreading lifetime reproductive effort over a number of temporally separated bouts. Further, increased survival and spreading of reproductive bouts should allow hibernators to produce young at times when climate and food resources are optimal for the rearing of offspring.

11.1 Introduction

Hibernation and torpor are commonly considered adaptations that allow animals to endure periods of food or water shortage and/or climatically harsh conditions (e.g., Aloia and Raison 1989). This is probably the primary function of hibernation

T. Ruf (✉) · C. Bieber · C. Turbill
Department of Integrative Biology and Evolution,
University of Veterinary Medicine, Vienna, Austria
e-mail: thomas.ruf@vetmeduni.ac.at

C. Turbill
School of Science and Health,
University of Western Sydney, Sydney, Australia

T. Ruf et al. (eds.), *Living in a Seasonal World*,
DOI: 10.1007/978-3-642-28678-0_11,

in certain mammals. For instance, herbivorous montane marmots or arctic ground squirrels in snow-covered or extremely cold habitats would almost certainly perish during winter if prevented from hibernation in underground burrows. Other hibernators, however, share habitats with non-hibernators of similar size and lifestyle. For example, the arboreal, granivorous edible dormouse (*Glis glis*) hibernates for 8 months (Bieber and Ruf 2009), whereas—also granivorous—tree squirrels, living in the same woods, do not. There are more species, such as mouse lemurs in Madagascar, that hibernate through seasons that may be unfavorable for reproduction, but hardly seem life-threatening in terms of climatic conditions or lack of food and water resources (Dausmann et al. 2004). These observations raise the question whether there are other selective forces, apart from energy constraints, that have contributed to the evolution of hibernation.

Here, we review recent evidence suggesting that one of the major functions of hibernation is predator avoidance. In this case, the principal benefit of hibernation would be temporal retreat into protective hibernacula or nests, while energy saving via metabolic depression would be only a secondary requirement, enabling the animals to remain inactive and hidden. If predator avoidance indeed was a major function of hibernation, this would lead to several predictions: (1) hibernation should be characterized by high survival probability, compared to the active season; (2) hibernation should be more common among small mammals, because extrinsic mortality (particularly predation) becomes increasingly important as body weight declines; (3) hibernators should have longer maximum lifespans (indicating they age slower) than non-hibernators of similar size, because evolutionary theories of aging predict that investment of resources into delaying senescence will be worthwhile only if the chances of dying from extrinsic cause are sufficiently low; and (4) survival rates and maximum longevity should be highest among mammals that combine hibernation with other means of predator avoidance, such as body armor or the ability to fly. Below, we discuss the evidence for the proposed link between hibernation and predation, and its consequences for life history tactics of hibernators.

11.2 Survival During Hibernation

There are surprisingly few studies in which survival rates of free-living mammals have been quantitatively determined during both the active and hibernation season. In 32 paired seasonal comparisons of 19 species, however, Turbill et al. (2011a) recently found (using phylogenetically informed statistics) that mean monthly adult survival rates over winter hibernation (97.0 $\pm$ 3.3% SD) were significantly higher than during the summer active season (84.5 $\pm$ 13.6%). These data probably even underestimated survival rates during hibernation, because "vanished" animals are typically classified as hibernating whereas they actually may have died prior to or shortly after hibernation (but outside of the last or first trapping period, respectively). Thus, despite contrary beliefs (e.g., Schwartz et al. 1998), hibernation is an extremely safe period in the life of mammals.

High survival rates during hibernation could be predicted to affect life history tactics and lead to a "slow" pace of life, compared to non-hibernators. This is because the increase in annual survival rate owing to seasonal dormancy allows animals to (i) optimally time reproduction to the most favorable environmental conditions (even if these seasonal periods are relatively short, or occur only in intermittent years) and (ii) improve survival of progeny by spreading offspring over more reproductive bouts. In other words, hibernators apparently increase their fitness by allocating their reproductive effort over more summer seasons, which requires increased longevity (i.e., a low risk of extrinsic mortality and slow rate of aging). Such a bet-hedging strategy is predicted to evolve under conditions of low adult mortality (Charnov and Schaffer 1973; Stearns 1992), and also appears to have contributed to the evolution of diapause in insects (e.g., Bradford and Roff 1993). Among mammalian hibernators, a prime example for such a tactic is the edible dormouse, which also shows lower rates of mortality during hibernation than during all other seasons, and this pattern can be found in different populations across Europe (Lebl et al. 2011). This species inhabits deciduous woods and is highly adapted to year-to-year fluctuations in high-caloric food, namely beechnuts and acorn (Bieber 1998; Pilastro et al. 2003; Ruf et al. 2006). Remarkably, entire dormouse populations skip reproduction in years when trees synchronously fail to reproduce seeds. In those years, dormice appear to retreat into underground burrows to aestivate throughout summer in addition to the obligatory winter hibernation. Hence, the animals can spend more than 10 months per year in dormancy (Bieber and Ruf 2009). We have argued that this prolonged inactivity in underground hibernacula serves to effectively protect the animals from their main predators, i.e., nocturnal birds of prey. This behavior could explain why, in areas in which high seed availability is rare, dormouse populations can reach an average age of >9 years (Pilastro et al. 2003), which is much higher than in a typical rodent of its size (~ 100 g). Protection from predation may well be the primary cause for high over-winter survival rates in many hibernators, both because they typically retreat into burrows, nests, or at least sheltered locations, and also owing to dormancy. As noted earlier, and even experimentally confirmed (e.g., Brown 1970), torpid animals further minimize the risk of being detected by predators because they are motionless, emit minimal noise and body odor, and do not respond to external stimuli ("freeze", a form of passive defense commonly used to avoid a lethal attack).

11.3 Hibernation and Aging

A basic tenet of evolutionary theories of aging, in particular of the disposable soma theory (Kirkwood 1977), is that animals can allocate energy either to reproduction early in life, or towards repairing wear and damage to tissues, thus prolonging life. One important prediction of this theory is that allocating energy to maintenance of the body should be beneficial only if rates of (early adult) survival are reasonably high. Therefore, whether or not it pays to invest in physiological mechanisms that attenuate aging will largely depend on rates of external mortality, such as death caused by predation (Kirkwood 1977; Ricklefs 2008). This would explain why animals that have evolved effective strategies to avoid predation also reach higher maximum longevities (a useful proxy for rates of aging, Ricklefs 2010; Turbill and Ruf 2010). An excellent example for this phenomenon is the evolution of flight, which strongly decreases the risk of predation (Ricklefs 1998). Flight is believed to explain why both birds and bats have evolved the physiological capacity to age much slower than non-flying mammals of similar size (e.g., Munshi-South and Wilkinson 2010). Bats were also the first mammalian order for which a positive effect of hibernation on maximum longevity was demonstrated (Wilkinson and South 2002). Surprisingly, it was shown only recently that small hibernators in general reach significantly higher maximum lifespans than non-hibernating mammals (Turbill et al. 2011a).

As illustrated in Fig. 11.1, this effect is only present in small mammals ($< \sim 1$ kg) and becomes increasingly more prominent as body weight decreases. The phylogenetically informed statistical analysis by Turbill et al. (2011a) showed that longevity was highest among hibernating bats, but significantly elevated even in non-bat small hibernators. For a typical hibernator weighing 50 g for instance, maximum longevity was 50% higher than in a similar sized non hibernator (when belonging to the order Chiroptera or not was taken into account), which suggests that rates of aging are approximately halved in hibernators. These analyses also indicated an effect of hibernation on other key life history traits. Small hibernators reproduce at slower rates, mature at older ages and have longer generation times compared with similar-sized non-hibernators (Turbill et al. 2011a). Hence, an increase in survival among hibernating mammals is associated with the co-evolution of a suite of traits indicative of relatively slow life histories. A possible effect of latitude was tested in these analyses but was not retained as an explanatory variable in any of the final models. Hence, the brevity of the breeding season, in itself, does not appear to be an overriding influence causing the evolution of slow life histories among small mammals at high latitudes (where the majority of hibernating species occur). As mentioned before, a major selective force leading to these traits might be the advantages of "bet-hedging", that is, the spreading of reproductive bouts over more active seasons with varying environmental conditions. Also, increased survival and spreading of reproductive bouts should allow hibernators to produce young at times when climate and food resources are optimal for the rearing of offspring, beyond what is possible for shorter-lived, non-hibernating seasonal breeders. However, even among hibernators, small species with short life expectancy may be forced to reproduce at suboptimal times of the active season, largely depending on when they were born themselves (Bieber et al. 2011).

The analysis by Turbill et al. (2011a) also indicated that effects of hibernation and other means of predator avoidance appear to be additive. In particular, the combination of hibernation with the ability to flight is associated with extremely high longevity in bats, among which even very small species such as the 7 g Brandt's bat (*Myotis brandti*) can—in the wild—reach a maximum lifespan of

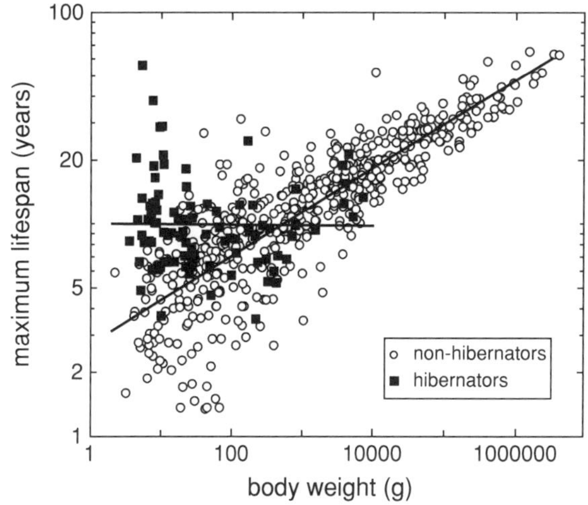

Fig. 11.1 Maximum longevity as a function of body weight in hibernating and non-hibernating mammals ($n = 652$ species; excluding winter-denning mammals with body weights above 10 kg). To easy visualization of the effects, the graph shows partial residual plots from a linear mixed effects model using phylogenetic classification (order and family) as random effects, which are removed from the plot. The effects of hibernation are principally identical whether this or more advanced methods (i.e., tree-based analyses) are applied (for details on data and methods see Turbill et al. 2011a)

41 years (Munshi-South and Wilkinson 2010). Another stand-out example is the short-beaked echidna, which is well-protected by a coat of spines, in addition to its extensive use of torpor, and accordingly has high rates of survival and an exceptionally long maximum lifespan of up to 45 years in the wild, and slow reproductive rate for its size (Augee et al. 2006; Nicol and Andersen 2007). We predict that similar additive effects exist for the combination of hibernation with other traits that reduce predation and increase longevity, such as an arboreal lifestyle (Shattuck and Williams 2010).

11.4 Body Size

Why is it that positive effects of hibernation on maximum longevity—which incidentally represents a good proxy for average longevity (Turbill and Ruf 2010)—should be limited to small mammals? If, as we argue, these effects are due to lowered rates of extrinsic mortality, namely predation, one would indeed expect a relation to body weight: First, large mammals have fewer predators and extrinsic mortality declines as body weight increases (Owen-Smith and Mills 2008; Ricklefs 2008). Second, if it is true that reduced mortality is largely due to the retreat of hibernators into underground burrows, this avenue of predator avoidance would be unavailable for many large mammals for morphological reasons alone, at least as far as access to self-dug hibernacula is concerned. In fact, declining benefits of hibernation as a means to evade predators may well help to explain why hibernation appears to become exceedingly rare as body size increases (Geiser and Ruf 1995).

This is not meant to imply that predation risks are entirely negligible for larger species. Large mammals, precisely because they typically cannot retreat to safe hide-outs, may in fact face another trade-off between the costs and benefits of torpor-like states. There is increasing evidence showing that large ungulates, for instance, undergo states of profound hypometabolism during winter that are accompanied by only slight decreases in core body temperature, but strong decreases in peripheral tissue temperatures (e.g., Arnold et al. 2004; Turbill et al. 2011b). It has long been known that mammals appear to counteract a loss of peripheral tissue function at low temperatures, namely in the extremities, by remodeling of cell membranes (Aloia and Raison 1989; Arnold et al. 2011). However, it is easy to envision that in spite of these adjustments, low tissue temperatures during hypometabolic states, especially in leg muscles, will still impair locomotor abilities and hence the capability to outrun predators. This may be another reason for the apparent lack of "deep hibernation" among large mammals. Not surprisingly then, profound hypometabolism—but still only moderate hypothermia—seems to be restricted to large mammals that have no (non-human) predators, such as bears (e.g., Tøien et al. 2011).

For small mammals there are, on the other hand, numerous reasons why hibernation and a "slow pace of life" may not maximize fitness. These include very high predation pressure that may prevent animals from even living to the hibernation season, non-seasonal environments that allow continuous reproduction (even if only occasionally, c.f., Blank and Ruf 1992), or specialized morphological adaptations that prevent digging of hibernacula. An example for the latter case maybe the genus *Lepus*: Hares are specialized fast runners that escape predators by flight using an extremely efficient hopping locomotion driven by powerful hindleg muscles (Garland 1983). This running mode is associated, however, with light-weight, relatively fragile forelegs that appear unsuited for the digging of burrows, which indeed are lacking in this genus. These factors may have contributed to the apparent absence of torpor and hibernation in hares, even though many hare species live in strongly seasonal habitats. Finally the absence of hibernation may be due to phylogenetic constraints. Analyzing the data set of Turbill et al. (2011a) we found that hibernation, even though taxonomically widespread, has a substantial phylogenetic signal (Blomberg's $K = 0.59$, $P < 0.0001$), meaning that its presence or absence is non-randomly distributed among clades. Consequently, the absence of hibernation in certain taxon may be simply due to a loss of this trait (if it is ancestral) or its lack of evolution within the respective clade.

11.5 Possible Mechanisms of Retarded Aging

As outlined above, high survival rates during the dormant state arguably are a prerequisite for hibernators to invest in mechanisms that delay senescence (although it remains to be seen whether these mechanisms may be mere by-products of the torpid state). Well-known evidence for the existence of such mechanisms comes from a study by Lyman et al. (1981) who found a positive correlation between longevity and the time spent in hibernation in Turkish hamsters, "suggesting that the process of aging is slowed during hibernation". This observation is in accordance with one of the oldest theories of aging, the "rate of living theory" (ROL), which suggests that a gram of tissue spends approximately the same amount of energy before an animal dies (Prinzinger 2005; Rubner 1908). As hibernation lowers energy expenditure, the ROL theory would indeed predict retarded aging and hence a prolonged maximum lifespan in hibernators. However, it has been recognized that there are severe problems with the ROL theory in its simple form, most obviously because comparisons made across classes (for example, between birds and mammals) contradict its expectations, and even within taxa there is high variability in the mass-specific energy expenditure per lifespan (reviewed in Speakman 2005).

A currently more widely accepted refinement of the ROL theory focuses not on metabolism or energy expenditure, in itself, but on a role of oxidative stress in aging. Specifically, aging has been linked to oxidative damage to DNA and other macromolecules caused by mitochondrial production of reactive oxygen species (ROS) (review in Kregel and Zhang 2007). However, it appears that the effects of ROS are complicated and that a minor oxidative stress causes an increase in maintenance and repair mechanisms that actually retards aging (e.g., Ristow and Schmeisser 2011). Several studies have suggested that ROS production may be increased during certain phases of the hibernation cycle, namely the sudden shift in energy consumption during arousal and interbout euthermia, which appears to be compensated by increased levels of antioxidants in hibernators (reviewed in Brown et al. 2011). A comprehensive recent study on ground squirrels, however, has demonstrated that overall, hibernation likely reduces oxidative stress. This is because hypometabolism at low body temperature reduces mitochondrial ROS production, and hibernators apparently avoid bursts in ROS production during arousal (Brown et al. 2011). As noted by these authors (Brown et al. 2011) a reduction in oxidative stress could help to explain previous observations that hibernation seems to contribute to longevity.

One of the detrimental consequences of oxidative stress is its effect on facilitating the loss of telomeres, i.e., DNA sequences that cap and protect the ends of eukaryotic chromosomes, by disrupting the balance between their shortening during cell division and their repair by the enzyme telomerase. Critically short telomeres and relatively high rates of telomere shortening have been linked to reduced future survival rates (Monaghan 2010). If hibernation is associated with decelerated physiological aging one would expect that this is reflected by lower rates of telomere shortening during torpor. As recently demonstrated by Turbill et al. (2011c) this is indeed the case. Although these results were obtained in Djungarian hamsters undergoing daily torpor, we hypothesize that they probably apply to deep hibernators as well. There are likely two factors that attenuate telomere shortening during torpor: First, if further studies confirm the data by Brown et al. (2011) on reduced oxidative stress during hibernation, it is clear that this would decrease ROS-induced telomere shortening; second, as telomere shortening occurs during cell divisions only, Arrhenius effects on biochemical reactions including those involved in mitosis will inevitably slow down telomere loss at low body temperature (Koizumi et al. 1992). Irrespective of the mechanism, however, the results by Turbill et al. (2011c) support the hypothesis that torpor is associated with a physiological state of increased somatic maintenance and retarded aging.

To avoid a possible misunderstanding: we do not intend to imply that torpor and hibernation are always entirely beneficial physiological states. As pointed out previously (e.g., Humphries et al. 2003), torpor may well involve risks and trade-offs. Whereas it doubtlessly helps to conserve energy and appears to attenuate senescence, it also bears its own risks such as dysfunction of the heart at low body temperature (Ruf and Arnold 2008), a possible impairment of immune function (Prendergast et al. 2002), or even an increased vulnerability to predators in case torpid animals are in fact routed out by predators (Michener 2004). Still, in spite of these risks winter dormancy generally appears to be a much safer state than above ground activity at any other season (Turbill et al. 2011a).

11.6 Conclusions

Taken together, the current evidence suggests that the role of hibernation as a strategy of predator avoidance, and its association with a slow-paced life history strategy may have been underestimated in the past. There may well be other functions of torpor (see also Geiser and Brigham 2012) that deserve more attention. Thus, a shift of the focus away from viewing torpor and hibernation as energy saving mechanisms alone seems a promising avenue for further research.

Acknowledgments This study was supported by the city of Vienna, the province of Lower Austria, and the Austrian Science Fund (Austrian Science Fund [FWF], Project P20534-B17).

References

Aloia RC, Raison JK (1989) Membrane function in mammalian hibernation. Biochmica et Biophysica Acta 988:123–146

Arnold W, Ruf T, Reimoser S, Tataruch F, Onderscheka K, Schober F (2004) Nocturnal hypometabolism as an overwintering strategy of red deer (*Cervus elaphus*). Am J Physiol Regul Integr Comp Physiol 286(1):R174–R181

Arnold W, Ruf T, Frey-Roos F, Bruns U (2011) Diet-independent remodeling of cellular membranes precedes seasonally changing body temperature in a hibernator. PLoS One 6(4):e18641

Augee ML, Gooden B, Musser A (2006) Echidna: extraordinary egg-laying mammal. Cisro Publishing, Melbourne

Bieber C (1998) Population dynamics, sexual activity, and reproduction failure in the fat dormouse (*Myoxus glis*). J Zool (Lond) 244:223–229

Bieber C, Ruf T (2009) Summer dormancy in edible dormice (*Glis glis*) without energetic constraints. Naturwissenschaften 96:165–171

Bieber C, Juskaitis R, Turbill C, Ruf T (2011) High survival during hibernation affects onset and time of reproduction. Oecologia (published online). doi:10.1007/s00442-011-2194-2197

Blank JL, Ruf T (1992) Effect of reproductive function on cold tolerance in deer mice. Am J Physiol 263:R820–R826

Bradford MJ, Roff DA (1993) Bet hedging and the diapause strategies of the cricket *Allonemobius fasciatus*. Ecology 74:1129–1135

Brown LN (1970) Population dynamics of the western jumping mouse (*Zapus princeps*) during a four-year study. J Mammalogy 51:651–658

Brown JCL, Chung DJ, Belgrave KR, Staples JF (2011) Mitochondrial metabolic suppression and reactive oxygen species production in liver and skeletal muscle of hibernating thirteen-lined ground squirrels. Am J Physiol Regul Integr Comp Physiol 302:R15–R28

Charnov EL, Schaffer WM (1973) Life history consequences of natural selection: Cole's result revisited. Am Nat 107:791–793

Dausmann KH, Glos J, Ganzhorn JU, Heldmaier G (2004) Hibernation in a tropical primate. Nature 429:825–826

Garland T Jr (1983) The relation between maximal running speed and body mass in terrestrial mammals. J Zool (Lond) 199:157–170

Geiser F, Ruf T (1995) Hibernation versus daily torpor in mammals and birds: physiological variables and classification of torpor patterns. Physiol Zool 68:935–966

Geiser F, Brigham RM (2012) The other functions of torpor. In: Ruf T, Bieber C, Arnold W, Millesi E (eds) Living in a seasonal world: thermoregulatory and metabolic adaptations. Springer, Heidelberg

Humphries MM, Thomas DW, Kramer DL (2003) The role of energy availability in mammalian hibernation: a cost-benefit approach. Physiol Biochem Zool 76:165–179

Kirkwood TB (1977) Evolution of ageing. Nature 170:201–204

Koizumi A, Tsukada M, Wada Y, Masuda H, Weindruch R (1992) Mitotic activity in mice is supressed by energy restriction-induce torpor. J Nutr 122(7):1446–1453

Kregel KC, Zhang H (2007) An integrated view of oxidative stress in aging: basic mechanisms, functional effects, and pathological considerations. Am J Physiol Regul Integr Comp Physiol 292:R18–R36

Lebl K, Bieber C, Adamík P, Fietz J, Morris P, Pilastro A, Ruf T (2011) Survival rates in a small hibernators, the edible dormouse: a comparison across Europe. Ecography 34(4):683–692

Lyman CP, O'Brien RC, Greene GC, Papafrangos ED (1981) Hibernation and longevity in the Turkish hamster *Mesocricetus brandti*. Science 212:668–670

Michener GA (2004) Hunting techniques and tool use by North American badgers preying on Richardson's ground squirrels. J Mammalogy 85(5):1019–1027

Monaghan P (2010) Telomeres and life histories: the long and the short of it. Ann NY Acad Sci 1206:130–142

Munshi-South J, Wilkinson GS (2010) Bats and birds: exceptional longevity despite high metabolic rates. Ageing Res Rev 9(2010):12–19

Owen-Smith N, Mills MGL (2008) Predator prey size relationships in an African large-mammal food web. J Anim Ecol 77:173–183

Pilastro AG, Marin G, Tavecchia G (2003) Long living and reproduction skipping in the fat dormouse. Ecology 84:1784–1792

Prinzinger R (2005) Programmed ageing: the theory of maximal metabolic scope. EMBO 6: 14–19

Prendergast BJ, Freeman DA, Zucker I, Nelson RJ (2002) Periodic arousal from hibernation is necessary for initiation of immune responses in ground squirrels. Am J Physiol Regul Integr Comp Physiol 282:R1054–R1082

Ricklefs RE (1998) Evolutionary theories of aging: confirmation of a fundamental prediction, with implications for the genetic basis and evolution of life span. Am Nat 152:24–44

Ricklefs RE (2008) The evolution of senescence from a comparative perspective. Funct Ecol 22:379–392
Ricklefs RE (2010) Insights from comparative analyses of aging in birds and mammals. Aging Cell 9:273–284
Ristow M, Schmeisser S (2011) Extending lifespan by increasing oxidative stress. Free Radical Biol Med 51:327–336
Rubner M (1908). Das Problem der Lebensdauer und seine Beziehung zu Wachstum und Ernährung. Oldenberg Verlag, München
Ruf T, Arnold W (2008) Effects of polyunsaturated fatty acids (PUFAs) on hibernation and torpor: a review and hypothesis. Am J Physiol Regul Integr Comp Physiol 294:R1044–R1052
Ruf T, Fietz J, Schlund W, Bieber C (2006) High survival in poor years: life history tactics adapted to mast seeding in the edible dormouse. Ecology 87:372–381
Schwartz OA, Armitage KB, Van Vuren D (1998) A 32-year demography of yellow-bellied marmots (*Marmota flaviventris*). J Zool (Lond) 246:337–346
Shattuck MR, Williams SA (2010) Arboreality has allowed for the evolution of increased longevity in mammals. Proc Natl Acad Sci 107(10):4635–4639
Speakman JR (2005) Body size, energy metabolism and lifespan. J Exp Biol 208(1717–1730): 1717
Stearns SC (1992) The evolution of life histories. Oxford University Press, Oxford
Tøien Ø, Blake J, Edgar DM, Grahn DA, Heller HC, Barnes BM (2011) Hibernation in black bears: independence of metabolic suppression from body temperature. Science 331:906–909
Turbill C, Ruf T (2010) Senescence is more important in the natural lives of long- than short-lived mammals. PLoS One 5(8):e12019
Turbill C, Bieber C, Ruf T (2011a) Hibernation is associated with increased survival and the evolution of slow life histories among mammals. Proc Biol Sci 278(1723):3355–3363
Turbill C, Ruf T, Mang T, Arnold W (2011b) Regulation of heart rate and rumen temperature in red deer: effects of season and food intake. J Exp Biol 214:963–970
Turbill C, Smith S, Deimel C, Ruf T (2011c) Daily torpor is associated with telomere length change over winter in Djungarian hamsters (Published online before print). doi:10.1098/rsbl.2011.0758
Wilkinson GS, South JM (2002) Life history, ecology and longevity in bats. Aging Cell 1:124–131

Chapter 12
Does Age Matter? Effects of Age on Hibernation Patterns in Edible Dormice (*Glis glis*)

Claudia Bieber and Thomas Ruf

Abstract We investigated the impact of age on patterns of hibernation in the edible dormouse. Multi-model inference showed that changes in hibernation duration, measured in 34 dormice implanted with temperature loggers, were best explained by effects of age. On average, each year of life shortened the hibernation duration by 15.17 days. While we did not observe significant changes in the timing of hibernation onset, every additional year of life led to an earlier mean termination of hibernation by 11.33 days. Since age did not influence torpor bout duration, arousal frequency or the duration of inter-bout euthermia, we conclude that the observed effects are not caused by ageing processes (e.g. physiological incapability). We discuss the observed age effects in relation to the unique life-history pattern in this species. Dormice invest into reproduction only in years with mast seeding. Thus, it may take some years until conditions are suitable for reproduction. While long hibernation periods predictably increase survival, an early termination of hibernation may provide an advantage in the establishment of high quality territories, which in turn may increase fitness. With increasing age, dormice are likely to favour investment into reproduction instead of maximising survival probability.

C. Bieber (✉) · T. Ruf
Department of Integrative Biology and Evolution,
University of Veterinary Medicine Vienna, Vienna, Austria
e-mail: claudia.bieber@vetmeduni.ac.at

T. Ruf et al. (eds.), *Living in a Seasonal World*,

DOI: 10.1007/978-3-642-28678-0_12,

12.1 Introduction

Dormice (*Glis glis*) are highly specialised seed predators and restrict reproduction to years of seed masting of trees (Bieber 1998; Bieber and Ruf 2009a; Fietz et al. 2004, 2005; Schlund et al. 2002). Dormice seem to compensate for the low number of reproductive bouts in their life with a generally high life expectancy, reaching population means of up to 9 years (Pilastro et al. 2003). While the probability to survive a reproductive (masting) year for an adult dormouse is low, survival probability in years with reproduction skipping is very high (Lebl et al. 2011a; Pilastro et al. 2003; Ruf et al. 2006). This is the more astonishing since food availability is lower in these years (Fietz et al. 2005). One pathway leading to this effect is most likely the extensive use of hibernation in this species. Even in the temperate climate dormice hibernate on average for 8 months (Vietinghoff-Riesch 1960) and further have the ability to enter summer dormancy (Bieber and Ruf 2009b). Recently it was shown that small hibernators in general show a slower life-history pace (lower yearly reproductive output, higher life expectancy) compared to non-hibernating mammals of the same body size (Turbill et al. 2011). Also, capture–recapture data on edible dormice have shown that survival rates in this species are highest during hibernation (Lebl et al. 2011a). Therefore, we have hypothesised that dormice may increase hibernation duration in years with reproductive skipping to maximise survival in those years (Bieber and Ruf 2009b).

The hypothesis that certain mammals use the ability to hibernate not just (or even mainly) to cope with unsuitable and harsh climatic conditions but as a tool to increase survival via predator avoidance has not received much attention in the past and requires further investigation (Bieber and Ruf 2009b). Up to now we lack detailed knowledge of hibernation patterns in this context. In the present study we aimed to investigate whether the age of an animal influenced hibernation pattern and duration. We hypothesised that young dormice should favour to maximise hibernation duration and hence survival, while older animals (with a lower chance to face another opportunity to reproduce in a masting year) should shorten hibernation duration in order to improve reproductive competitiveness in the subsequent spring.

12.2 Materials and Methods

12.2.1 Animals and Climate

Dormice (*Glis glis*, colony established in 1996) were held in mixed groups year-round in three outdoor enclosures (6 $\times$ 4 $\times$ 3.5 m each) between 2005 and 2007. The dormouse enclosures were located at 362 m a.s.l. in Vienna, Austria (48°13′N; 16°16′E). Animals were captured in their nest-boxes (sleeping sites) once a week during their active season. During weekly checks body mass was recorded to the nearest 1.0 g. While the animals occupied the offered nest boxes during the active season, all dormice exclusively used underground burrows dug by the animals for hibernation. Dormice were fed ad libitum with rodent chow (Altromin 1314 FORTI).

Mean air temperature (recorded at the close-by weather station; "Hohe Warte", Vienna, Austria, 48°15′N and 16°22′E, 203 m a.s.l., http://www.zamg.ac.at/) during the study period was 11.1°C. The mean monthly air temperature during the core hibernation period (October–April) in 2005/2006 was 4.2 ± SEM 2.1°C and in 2006/07 8.4 ± 1.4°C. IButtons (DS1922L, Maxim/Dallas) were used to record burrow temperature in the enclosures (buried 15 cm below ground).

12.2.2 Implantation of Temperature Loggers

Wax-coated iButtons (DS1922L, Maxim/Dallas) were implanted for measurements of subcutaneous temperature (t_{subc}) in the lateral area of the thorax, caudal of the scapula in 24 dormice (13 males, 11 females; for further details see Bieber and Ruf 2009a). One week after implantation dormice were released to their groups in the outdoor enclosures. Subcutaneous temperature was recorded at approximately hourly (3,650 s) intervals to cover 1 year. Out of 24 dormice 10 were implanted in two subsequent years (resulting in 34 data sets, 17 in 2005/2006 and 17 in 2006/07). Implantations were carried out between June and August 2005, explantation and implantation/replacement between May and August 2006, final explantation between May and July 2007. Since all implanted animals were born in our enclosures (birth season between June and early August) we were able to determine the exact age for each animal. Per definition animals reached age one after termination of their first hibernation season and stayed in this age class until termination of their second hibernation season and so on. We recorded t_{subc} in five animals at age 1 (one female, four males), 12 at age 2 (two females ten males), 12 at age 3 (seven females, five males) and five at age 4 (five females, 0 males). We did not implant juvenile dormice and therefore did not record body temperature data from the first hibernation season in a dormouse's life.

12.2.3 Statistics

We analysed t_{subc} data using program R (R Development Core Team 2011). We defined t_{subc} of 25°C as the threshold between entry and leaving the state of torpor/euthermia. Onset of hibernation was defined by a $t_{subc} < 25$°C for more than 24 h. Hibernation was considered as terminated when an animal remained at $t_{subc} > 25$°C for more than 72 h. We then computed minima, maxima and mean profiles for torpor bout duration, duration of inter-bout euthermia (IBE) and number of arousals for each recorded hibernation period.

To test simple hypotheses, we used ANOVA tables of linear regression models for repeated measurements.

We used lme models [*l*inear *m*ixed *e*ffect models, R-package lme4 (Bates et al. 2011)] to analyse the effects of age, sex and body mass on hibernation duration, onset and end of hibernation, mean torpor bout duration, mean IBE duration and number of arousals. Beside the fixed effects (age, sex and body mass) we also used the factors animal and year as random effects in all modelling procedures. Thus, we were able to adjust for intra-individual variation (repeated measurements of the same animal) and for variation between the years (e.g. climatic variations).

We used the multi-model inference approach [R-package MuMIn (Barton 2011)] to determine relative importance of predictor variables. We started with a global set of linear models including all combinations of additive and two-way interaction effects (if data were sufficient) of the above-mentioned parameters and created a table ranked by minimising the value of *A*kaikes *I*nformation *C*riterion (AICc) (Akaike 1973), corrected for small sample size. We also calculated Akaike weights, which represent the relative likelihood of each model given the data and the entire set of models (Burnham and Anderson 2002). To identify influential parameters, we made robust inferences based on a set of supported models (ΔAICc < 10). We could then estimate the relative explanatory strength of each parameter while simultaneously accounting for uncertainty in model selection. We calculated model-averaged coefficients for all parameters by averaging the estimates, weighted by multiplication with that model's Akaike weight, over the set of supported models (ΔAICc < 10) including each parameter. We also derived the *P*-value associated with each coefficient (i.e. not dependent on a single model). The relative importance of all parameters was calculated by summing the Akaike weights of each model containing that parameter.

If not stated otherwise all mean values were given $\pm$ standard error of the mean (SEM).

12.3 Results

We recorded an average hibernation duration of 5590.9 h $\pm$ 105.88 h (equivalent to 232.9 days $\pm$ 4.41 days or 7.64 months). Mean hibernation duration was longer in 2005/2006 (5831.1 h $\pm$ 136.98 h, 242.9 days) than in 2006/2007 (5350.7 h $\pm$ 142.36 h, 222.9 days). Maximum hibernation duration was recorded in 2005/2006 as 6575.04 h (equivalent to 273.9 days or 8.98 months; female w02/084F), minimum hibernation duration was 4293.5 h (equivalent to 178.9 days or 5.87 months) in female w02/3294 in 2006/2007. Animals entered hibernation on average on 14th September (day 257 $\pm$ 2.75, range: day 234–291) and emerged on 5th May (day 125 $\pm$ 3.51 d, range: day 87–173). In 2005/2006 dormice entered hibernation earlier (2005/2006: day 248.6 $\pm$ 3.66, 2006/2007: day 265.5 $\pm$ 2.99). Since we used year as a random effect in our modelling procedure, we corrected for these annual differences (see below).

Table 12.1 Relative importance and *P*-values for the model-averaged coefficient of the parameters age, sex and body mass prior to hibernation, for the response variables hibernation duration, onset of hibernation and end of hibernation

Response variables	Relative parameter importance and *P*-value for model-averaged coefficient		
	Age	Sex	Body mass[a]
Hibernation duration (h)	**0.87**	0.42	0.20
	0.002**	n.s.	n.s.
Onset of hibernation (day)	0.21	0.32	0.41
	n.s.	n.s.	n.s.
End of hibernation (day)	**0.81**	**0.74**	0.21
	0.011*	**0.019***	n.s.

Shown are robust values considering all models with ΔAICc $\leq$ 10. Values with high relative importance and correspondingly significant coefficients are printed in bold. Asterisks indicate significance levels; n.s. = non significant

[a] Body mass prior to hibernation (last capture event per year)

Dormice showed on average 28.8 $\pm$ 1.26 arousals per hibernation period with a mean IBE duration of 6.1 h $\pm$ 0.25 h (mean minimum 2.0 h, mean maximum 26.7 h). Mean torpor bout duration was 192.6 h $\pm$ 9.10 h (mean minimum 20.6 h, mean maximum 487.4 h). Taken together animals spent on average 175.9 h $\pm$ 10.64 h ($\sim$3%) during a hibernation period in euthermia and 5415.1 h $\pm$ 106.19 h ($\sim$97%) in hypothermia.

Mean body mass prior to the onset of hibernation was 208.4 g $\pm$ 6.97 g (range: 100–291 g). Prior to hibernation males had a significantly higher mean body mass (220.9 g $\pm$ 7.53 g) than females (192.5 g $\pm$ 11.62 g; $F_{1,22} = 6.29$, $P = 0.0200$). We did not observe significant body mass differences between age classes ($F_{1,8} = 0.004$, $P = 0.953$).

12.3.1 Factors Influencing Hibernation Patterns

Variation in hibernation duration was best explained by the parameter age alone (best model weight 0.37). All models with a ΔAICc $\leq$ 2.56 (first three models) included this parameter. Considering all models with a ΔAICc $\leq$ 10 we computed a high relative importance and coefficient significance for this parameter (Table 12.1). Every year of life shortened the hibernation duration by 364.1 h (equivalent to 15.17 days, Fig. 12.1).

None of the parameters age, sex and body mass prior to hibernation contributed to explain the variation in the onsets of hibernation in the best model (model weight 0.27). Consequently, their relative importance, especially that of the parameter age, was low (Table 12.1).

Variation in the end of hibernation was best explained (best model weight 0.53) by the parameter combination age + sex. Both predictors showed a high relative

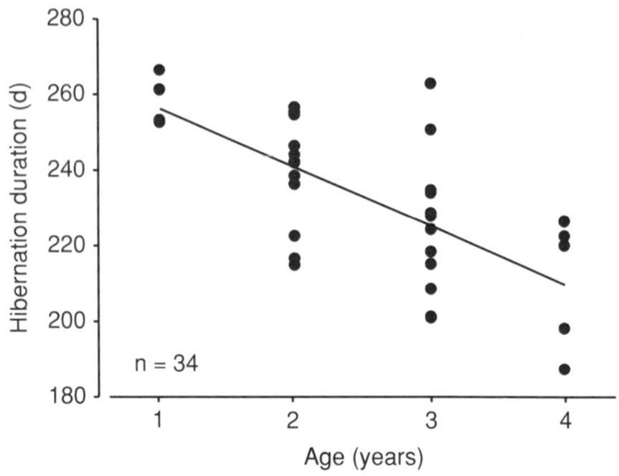

Fig. 12.1 Influence of age on the hibernation duration in edible dormice. Partial residual plot from the best linear mixed model using age as a fixed effect, and individual as well as study year as random effects

importance and coefficients differed significantly from 0 (all models ΔAICc $\leq$ 10, Table 12.1). Every additional year of life led to an earlier termination of hibernation by 271.9 h (equivalent to 11.33 days, Fig. 12.2). Males terminated hibernation on average 464.0 h (equivalent to 19.33 days) earlier than females.

The parameter age did not, however, contribute to the observed variation in the number of arousals, mean IBE duration or mean torpor bout duration. The overall relative parameter importance for age for these response variables was 0.17, 0.43 and 0.17, respectively. In neither case the model-averaged coefficient for the parameter age was significant.

12.4 Discussion

We observed a strong effect of age on hibernation duration in dormice. Body mass, however, did not contribute to explain the observed differences. The longest mean hibernation duration was observed in yearling dormice (256.6 days) and decreased with each additional year of life (Fig. 12.1). However, we did not consider the first hibernation season (as a juvenile) in our evaluation. This is because it has been shown earlier that mean hibernation duration in juveniles is significantly shorter than in older animals (Bieber and Ruf 2004). As dormice are born extremely late in the active season (August, Vietinghoff-Riesch 1960) juveniles delay the onset of hibernation as much as possible presumably to maximise body mass gain prior to their first hibernation. Accordingly, several field studies showed that juvenile dormice are the last age group to enter hibernation (Bieber and Ruf 2009a; Schlund et al. 2002; Vietinghoff-Riesch 1960). Because of this known bias in the onset of hibernation we omitted juvenile dormice from our experiment.

Hibernation onset was not affected by the age of an animal. However, our analysis revealed that the change in hibernation duration was entirely caused by significant differences in the termination of hibernation. A four-year-old animal

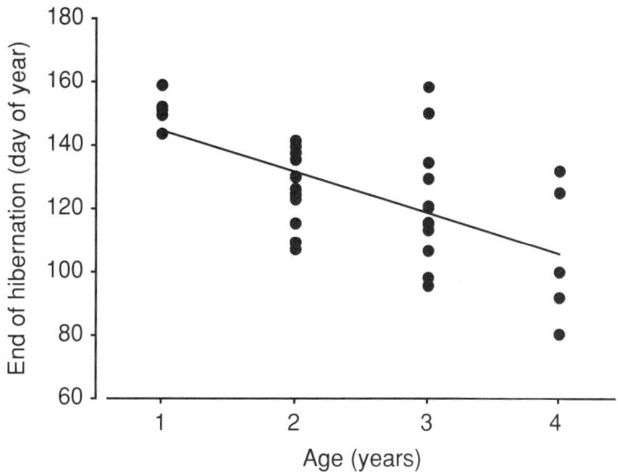

Fig. 12.2 Influence of age on the termination of hibernation in edible dormice. Partial residual plot from the best linear mixed model using age and sex as fixed effects, and individual as well as study year as random effects

emerged about 30 days earlier than a two-year old dormouse. Clearly, these time differences are large enough to cause notable effects on both reproductive success and survival probability. While early emergence is likely to have positive effects on establishing a good breeding territory, a possible disadvantage is an increased mortality risk early in the active season. In the Vienna forest the survival probability of edible dormice is highest during hibernation (98%, CI 97–100%) and lowest early in the active season (77%, CI 74–80%; Lebl et al. 2011a). The same pattern was observed in a Europe-wide study, and was attributed to increased predation risk in spring (Lebl et al. 2011a). To prolong hibernation, therefore, seems an effective strategy to increase survival probability. Recently it could be shown that hibernation affects life-history strategies in mammals in general. Compared to non-hibernating species, small hibernators showed a higher life expectancy leading to a 'slow' pace of life (Turbill et al. 2011). Edible dormice, for example, may reach a mean lifespan of 9 years (Pilastro et al. 2003). Thus, it is likely that age effects on life-history decisions should be particularly pronounced in this small mammal.

Due to the extremely close adaptation of dormice to the availability of pulsed resources, such as beech seeds, (Bieber 1998; Fietz et al. 2005; Ruf et al. 2006) one should expect that a trade off between investment into reproduction and optimising survival is influenced by an animals' age. Since dormice invest into reproduction only when seeds are available they may face several non-masting years prior to their first reproductive bout. Therefore, young animals should maximise survival—via prolonged hibernation—because there is a high likelihood for them to experience another masting year in their lifetime. Older animals, however, have little chance to experience another masting year with optimal food resources. Thus, as age progresses dormice should become increasingly more risk prone and terminate hibernation earlier to invest into reproduction.

Seeding in beech trees is, however, not always an all-or-nothing decision. In the Vienna woods we also observed intermediate masting years. In these years, which are actually more frequent than full masting years, only some of the trees

(25–50%) produce seeds (Lebl et al. 2011a). Under these conditions the occupation of suitable breeding ranges seems to be a major advantage. Occupying areas with older beech trees (with higher reserves to produce seeds even under suboptimal conditions) significantly increases reproductive success in dormice (Lebl et al. 2011b). Hence, older animals with slim chances to live to another full masting year should increase their lifetime reproductive success by investment into reproduction, especially in these intermediate mast years.

Further, a feeding experiment in the Vienna woods demonstrated that only yearling females increase their investment into reproduction in the area, where surplus food (sunflower seeds) was provided in nest boxes. Adult females (beyond their second hibernation season), however, did not show this effect (Lebl et al. 2010). It seems at least possible that older females occupied territories of higher quality (higher seed availability) and therefore did not respond to the availability of additional food. As in the present study, the observed age effects were not caused by differences in body mass in the feeding experiment (Lebl et al. 2010). However, based on our present findings we would conclude in the study of Lebl et al. (2010) that older animals, due to their smaller chance to survive another year, invested more into reproduction than younger animals. Whether social hierarchy or increased experience in older females plays a role in this context might be an interesting question for future studies.

We did not observe age effects (considering animals between ages 1 and 4) on the number of arousals, the mean IBE duration or mean torpor bout duration. Thus, there is no evidence that the observed age effects on hibernation duration could have been caused by senescence (e.g., decreasing ability to cope with the metabolic state of hibernation). Moreover, the lifespan investigated in the present study (up to age 4) does not reach the age of a truly "old" animal [maximum lifespan in our enclosure housed population is 9 years (two 9-year-old animals are still alive), CB unpublished data].

Our data indicated that in addition to age, sex had a significant influence on the termination of hibernation. The fact that males emerged significantly before females underlines observations from several field studies (Bieber and Ruf 2004). The reason underlying this difference could be related to spermatogenesis, which requires euthermia (Barnes et al. 1986) and can take more than 2 months to be completed (Heller and Clermont 1963). Alternatively, the early occupation of good territories may be even more important for males than for females. Whatever the causes, the similarity between field and enclosure data in this respect suggest that the age effects observed here were not biased by investigating animals in captivity. Since our dormice were kept year-round in outdoor enclosures and hibernated below ground in self-dug hibernacula, we are confident that our results reflect the natural pattern. This is further supported by the natural seasonal pattern of immergence and emergence observed here. However, since we provided the animals with food throughout the investigation period, we cannot rule out that varying environmental conditions may mask age effects in field studies. This underlines the need for the combination of field studies with experimental settings to uncover adjustments of physiological responses to life-history strategies.

Acknowledgments We thank P. Steiger, K. Außerlechner, C. Skerget for their help with data collection and W. Zenker, F. Balfanz, C. Beiglböck, C. Walzer for implantation of iButtons. We declare that all experiments in this study comply with the current laws of Austria in which they were performed. We thank F. Knauer for his support with the package MuMIn. This project was supported by the city of Vienna, the province of Lower Austria and the Austrian Science Fund (FWF, Project P20534-B17).

References

Akaike H (1973) Information theory as an extension of the maximum likelihood principle. In: Petrov BN, Csádki F (eds) Second international symposium on information theory. Akadémiai Kiadó (Budapest), Tsahkadsor, pp 267–281

Barnes BM, Kretzmann M, Licht P, Zucker I (1986) The influence of hibernation on testis growth and spermatogenesis in the golden-mantled ground squirrel, *Spermophilus lateralis*. Biol Reprod 35:1289–1297

Barton K (2011) MuMIn: Multi-model inference. R package version 1.0.0. http://CRAN.R-project.org/package–MuMIn

Bates D, Maechler M, Bolker B (2011) lme4: Linear mixed-effects models using S4 classes. R package version 0.999375-39. http://CRAN.R-project.org/package=lme4

Bieber C (1998) Population dynamics, sexual activity, and reproduction failure in the fat dormouse (*Myoxus glis*). J Zool Lond 244:223–229

Bieber C, Ruf T (2004) Seasonal timing of reproduction and hibernation in the edible dormouse (*Glis glis*). In: Barnes BM, Carey HV (eds) Life in the cold: evolution, mechanism, adaptation, and application. Twelfth international hibernation symposium. Institute of Arctic Biology, University of Alaska, Fairbanks, pp 113–125

Bieber C, Ruf T (2009a) Habitat differences affect life history tactics of a pulsed resource consumer, the edible dormouse (*Glis glis*). Popul Ecol 51:481–492

Bieber C, Ruf T (2009b) Summer dormancy in edible dormice (*Glis glis*) without energetic constraints. Naturwissenschaften 96:165–171

Burnham KP, Anderson DR (2002) Model selection and multimodel inference, 2nd edn. Springer, New York

Fietz J, Pflug M, Schlund W, Tataruch F (2005) Influences of the feeding ecology on body mass and possible implications for reproduction in the edible dormouse (*Glis glis*). J Comp Physiol 175:45–55

Fietz J, Schlund W, Dausmann KH, Regelmann M, Heldmaier G (2004) Energetic constraints on sexual activity in the male edible dormouse (*Glis glis*). Oecologia 138:202–209

Heller CG, Clermont Y (1963) Spermatogenesis in man: an estimate of its duration. Science 140:184–186

Lebl K, Kürbisch K, Bieber C, Ruf T (2010) Energy or information? The role of seed availability for reproductive decisions in edible dormice. J Comp Physiol 180:447–456

Lebl K et al (2011a) Survival rates in a small hibernator, the edible dormouse: a comparison across Europe. Ecography 34:683–692

Lebl K, Rotter B, Kürbisch K, Bieber C, Ruf T (2011b) Local environmental factors affect reproductive investment in female edible dormice. J Mammal 92:926–933

Pilastro A, Tavecchia G, Marin G (2003) Long living and reproduction skipping in the fat dormouse. Ecology 84:1784–1792

R Development Core Team (2011) R: a language and environment for statistical computing. R Foundation for Statistical Computing, Vienna. Version 2.13.1. http://www.R-project.org

Ruf T, Fietz J, Schlund W, Bieber C (2006) High survival in poor years: life history tactics adapted to mast seeding in the edible dormouse. Ecology 87:372–381

Schlund W, Scharfe F, Ganzhorn JU (2002) Long-term comparison of food availability and reproduction in the edible dormouse (*Glis glis*). Mammal Biol 67:219–223

Turbill C, Bieber C, Ruf T (2011) Hibernation is associated with increased survival and the evolution of slow life histories among mammals. Proc R Soc B Biol Sci 278:3355–3363

Vietinghoff-Riesch AFv (1960) Der Siebenschläfer. In: Müller-Using D (ed) Monographien der Wildsäugetiere. Gustav Fischer Verlag, Jena, pp 1–196

Chapter 13
Sex and Seasonality: Reproduction in the Echidna (*Tachyglossus aculeatus*)

Stewart C. Nicol and Gemma E. Morrow

Abstract We studied seasonality in free-ranging echidnas (*Tachyglossus aculeatus*) in Tasmania near the most southern part of their range. Both sexes showed a large seasonal variation in body mass associated with hibernation and reproduction. Male echidnas entered hibernation in mid-February (late summer) and females 1 month later. Not all reproductively mature adults mated every year: in non-reproductive years both sexes hibernated for approximately 6 months, becoming active in spring when ecosystem productivity was increasing and reliable. In reproductive years, males aroused from hibernation in early winter, and sought out females. Matings began before females had completed hibernation, and females re-entered hibernation between matings and sometimes when pregnant. This timing of mating ensures that maximum growth rate of the young coincides with the period of greatest ecosystem productivity, while female torpor through the mating period minimizes energy expenditure during the time of lowest food availability.

13.1 Introduction

The short-beaked echidna (*Tachyglossus aculeatus*) is the most widespread native mammal in the Australasian ecozone. At the most northerly part of their range in New Guinea, echidnas occur within a few hundred kilometers of the equator, where seasonal changes in day length are minimal. For the most southerly populations in Tasmania (43° S) day length varies from 15 h in summer to 9 h in winter. Compared with equivalent northern hemisphere locations, seasonal

S. C. Nicol (✉) · G. E. Morrow
School of Zoology, University of Tasmania, Hobart, Australia
e-mail: S.C.Nicol@utas.edu.au

T. Ruf et al. (eds.), *Living in a Seasonal World*,
DOI: 10.1007/978-3-642-28678-0_13, © Springer-Verlag Berlin Heidelberg 2012

temperature variations in Tasmania are relatively small. At our Tasmanian field site (Lovely Banks, 42°27′S), mean summer maximum and minimum are 24 and 10°C, and winter maximum and minimum are 11 and 2°C. Despite these mild winters, echidnas in Tasmania and in parts of mainland Australia hibernate. In a previous study (Nicol and Andersen 2002) we investigated the timing of hibernation in Tasmanian echidnas, and compared them with data from other Australian sites. We suggested that echidna hibernation was part of low energy strategy that was of particular value in the Australian climate, which is characterized by high year-to-year rainfall variability, and very variable productivity. We also showed that reproductively active animals aroused from hibernation in mid-winter to mate, and suggested that this allowed young to be raised at the time of maximum food availability. At the time of this study, we had relatively little information on mating and lactation and in this chapter we review and update our previously published data, adding new information on reproduction and ecosystem productivity to provide a comprehensive overview of echidna seasonality at the southernmost part of their range, and its relationship to reproduction.

13.2 Materials and Methods

The study was carried out on a 12 km^2 site on a grazing property in the Tasmanian southern midlands 50 km north of Hobart, Tasmania. The study site consists of improved and native pasture with areas of *Eucalyptus amygdalina* woodland on sandstone (Harris and Kitchener 2005). The site has variable topography with altitudes ranging from 200 to 400 m above sea level with numerous sandstone outcrops, caves, creeks, and gullies. Echidnas found while driving slowly around the property were caught by hand, weighed, and scanned with a hand-held RFID reader (Destron Fearing, MN, USA). Echidnas that had not been previously tagged had a passive transponder tag (LifeChip, Destron Fearing, MN, USA) implanted subcutaneously. Sex was determined by the presence or absence of a spur on the ankle. Juvenile echidnas of both sexes have a sheath covered spur, which in males loses the sheath to become an adult spur, while females lose the spur entirely. Echidnas were deemed to be reproductively mature if they were found in a mating group, or by the presence of a palpable penis bulge in males, and in females by a developing pouch, and entry into a nursery burrow.

In any year up to 22 echidnas had a RF tracking transmitter (Bio-Telemetry Tracking, St Agnes, South Australia) glued to the spines of the lower back allowing them to be located using a hand-held receiver. Body mass, reproductive condition and any reproductive activity, evidence of hibernation, and other details of the location and animal activity were recorded in a database. Blood samples were taken from the rostral sinus for measurement of reproductive hormones. In the period 1996–2007, five male and 16 female echidnas were surgically implanted with temperature data loggers (Stowaway Tidbit, Onset Computer Corporation, MA, USA) and in the period 2007–2011, 21 echidnas (5 M, 5F) were fitted with external temperature loggers (Thermochron iButtons, DS1922L; Maxim/Dallas Semiconductor, TX, USA) which were glued to the tracking transmitter. Times of initial entry into hibernation and final arousal were determined from the temperature records. Echidnas with implanted temperature loggers were considered to have entered hibernation when T_b fell below 20°C and subsequent periods of euthermia did not exceed 2 days. For echidnas with external loggers entry was taken to be the time at which the daily variation in recorded temperature decreased to <5°C (see Fig. 13.2). For echidnas with internal loggers final arousal was considered to have occurred when T_b remained above 30°C without further hypothermic bouts in that year. For echidnas with external loggers final arousal was considered to have occurred when normal daily temperature variation returned and persisted. Hibernation duration was the time between initial entry and final arousal.

From 2007 onwards reproductive status of echidnas was checked by taking cloacal swabs (Morrow and Nicol 2009). Microscopic examination of retrieved cells indicated reproductive status of females, and the presence of sperm showed recent mating. If the mating resulted in fertilization there were characteristic changes in reproductive tract cytology. The date of egg-laying was determined from the characteristic changes of the internal or external temperature loggers (Nicol and Andersen 2006; Morrow and Nicol 2009). For males, a swab of the penis was taken to check for the presence of viable sperm.

To allow comparison between echidnas of different body size long-term mean masses were calculated for echidnas for which we had more than 3 years of data, and the measured mass converted to a percentage of the long-term mean (mass%). Statistical analyses were performed using the software package Statistica 6.1 (Statsoft, Tulsa, Oklahoma). Values shown are means ± standard error.

13.3 Results

Between February 1996 and October 2011, 254 adult echidnas (115 female, 97 male) were tagged in the study area. Of the 50 echidnas tagged during the first year of the study, five were known to be alive in 2011, while 13 were known to be dead. Of the remainder, 11 were not seen after their first capture, while eight were seen in the last 5 years of the study. For all echidnas for which we had multiyear temperature records allowing unambiguous assessment of reproductive status, males were reproductively active in nine of 12 yearly records (75%), while females were reproductively active in 30 of 54 (55%) yearly records. Body mass of both sexes followed a strong seasonal cycle, with significant differences between masses in reproductive and non-reproductive years (Fig. 13.1). For both males and females there was highly significant effects of month (males: $F_{11,299} = 14.0$, $P < 0.0001$; females: $F_{11,374} = 23.5$, $P < 0.0001$) and reproductive status (males: $F_{1,299} = 11.1$, $P < 0.001$; females: $F_{1,374} = 15.5$, $P < 0.001$). The apparent increases seen in Fig. 13.1 in mass in August for non-reproductive females, and in

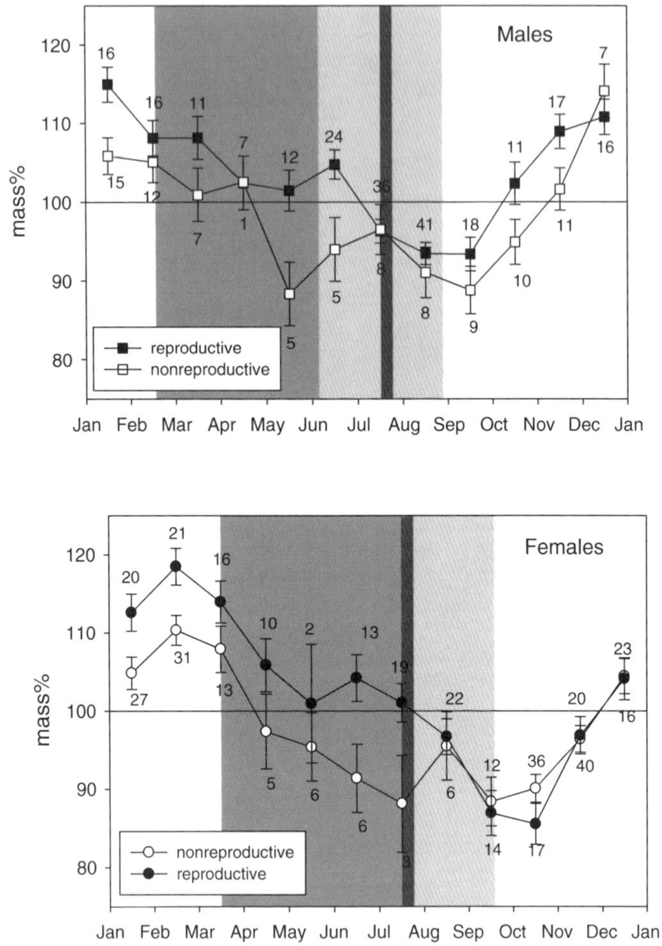

Fig. 13.1 Annual variation in body mass for reproductive and non-reproductive adult male and female echidnas. *Dark bar* shows mean mating time ±SE. *Dark shaded* area—mean hibernation period for reproducing individuals; *light shaded* area—additional hibernation period for non-reproducing individuals

July, for non-reproductive males, were not statistically significant (Posthoc unequal N HSD test).

Main effects ANOVAs for ID and reproductive status of males and females showed no effect of ID on date of entry, or final arousal, or duration of hibernation, so data from different years from individual animals were treated as being statistically independent. Factorial ANOVA for the effects of sex and reproductive status on entry date showed significant differences between males and females ($F_{1,60} = 20.9$, $P < 0.0001$), but no effect of reproductive status

($F_{1,60} = 0.79$, $P = 0.38$). Mean entry day for males was 17 February ± 3.61, ($n = 14$), and for females 18 March ± 2.29, ($n = 50$). There were significant effects of both sex and reproductive status on final arousal date: sex $F_{1,70} = 26.3$, $P < 0.0001$; reproductive status $F_{1,70} = 122.6$, $P < 0.0001$. The earliest final arousal was for reproductive males (6 June ± 5.9, $n = 13$), then reproductive females (21 July ± 3.6, $n = 32$), non-reproductive males (29 August ± 10.3, $n = 4$), and then non-reproductive females (19 September ± 4.1, $n = 25$). Final arousal dates of non-reproductive males and females were not significantly different (Unequal N HSD $P = 0.45$). Duration of hibernation did not differ between sexes ($F_{1,53} = 1.26$, $P = 0.267$), but differed significantly with reproductive status ($F_{1,53} = 83.0$, $P < 0.0001$). Mean hibernation duration for reproductively active animals was 118.4 ± 4.0 days ($n = 35$) and for non-reproducing echidnas 186.1 ± 4.4 ($n = 22$). Example temperature records from reproductive and non-reproductive male and female echidnas are shown in Fig. 13.2. Before the start of the hibernation period a number of echidnas showed periods during which T_b fell below 25°C. The average time of the first of these events was 25.6 ± 2.4 days before hibernation entry ($n = 24$, range 51–9.5). There was no difference in the relative dates for these prehibernatory events between males and females ($t_{27} = 0.58$, $P = 0.56$). For males, the minimum time between the end of hibernation and mating was 25 days, and the maximum was 85 days, giving a mating period of 40 days.

Mean fertilization date for females was July 22 ± 4.2 days ($n = 13$, range 1 July–4 October), and mean egg-laying date was August 16 ± 3.7 days ($n = 29$, range 24 July–28 August). There was a negative relationship between the number of days between final arousal date and the time from arousal to egg-laying ($y = 49.1–0.53x$, $r^2 = 0.28$, $F_{1,19} = 7.57$, $P < 0.02$, where x is day of the year), i.e., later arousing females had a shorter time before egg-laying. (Fig. 13.3a). In the period 2008–2011 five females re-entered hibernation after becoming pregnant, (mean duration of torpor 15.8 ± 6.3 days, range 3–50 days, median = 10 days). Data from these females, and from three others from previous years where the temperature records suggests post-fertilization hibernation are shown as shaded squares in Fig. 13.3a. Data from the intensively monitored animals (2008–2010) show a close relationship between gestation length and the period of post-fertilization hibernation ($y = 1.16x + 21.0$, $r^2 = 0.99$, $F_{1,9} = 972$, $P < 0.0001$; Fig. 13.3b). If the highest point on the graph was omitted the regression was very similar and still significant ($y = 0.90 + 21.7$, $r^2 = 0.99$, $F_{1,8} = 107$, $P < 0.0001$). There was no clear relationship between the date of fertilization and the length of post-fertilization hibernation, but the latest date at which a pregnant female re-entered hibernation was 25 July. Three females whose temperature loggers showed that they lost their egg or young during the period when the mother is confined to the nursery burrow re-entered hibernation. For two animals these hibernation periods were relatively brief (5 and 7 days in late September and early October), but the other animal re-entered hibernation on 31 July, and did not resume euthermia until 10 November, 102 days later.

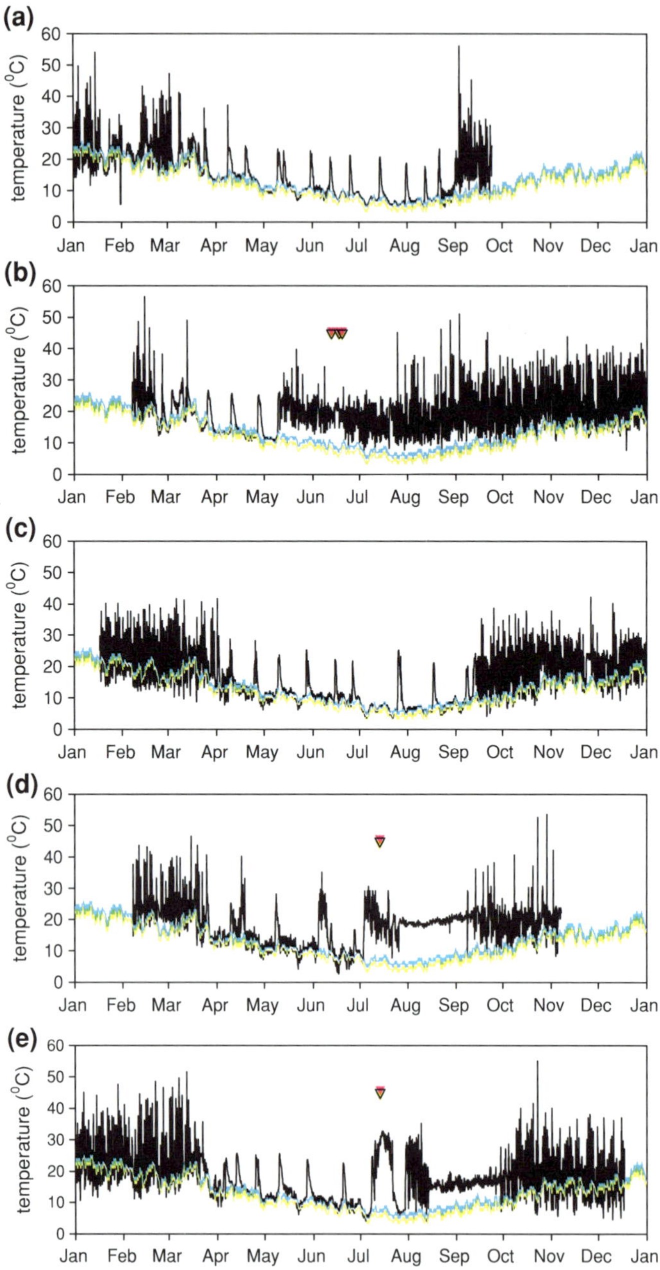

Fig. 13.2 External iButton records from five adult echidnas during 2008. **a** Non-reproducing male, **b** reproducing male, **c** non-reproducing female, **d** reproducing female, **e** reproducing female which re-entered hibernation after fertilization. *Gray line* shows soil temperature at 20 cm recorded at Bureau of Meteorology station approximately 2 km from the field site. *Triangles* indicating observed matings for the male (**b**) and estimated fertilization dates for the females (**d**, **e**)

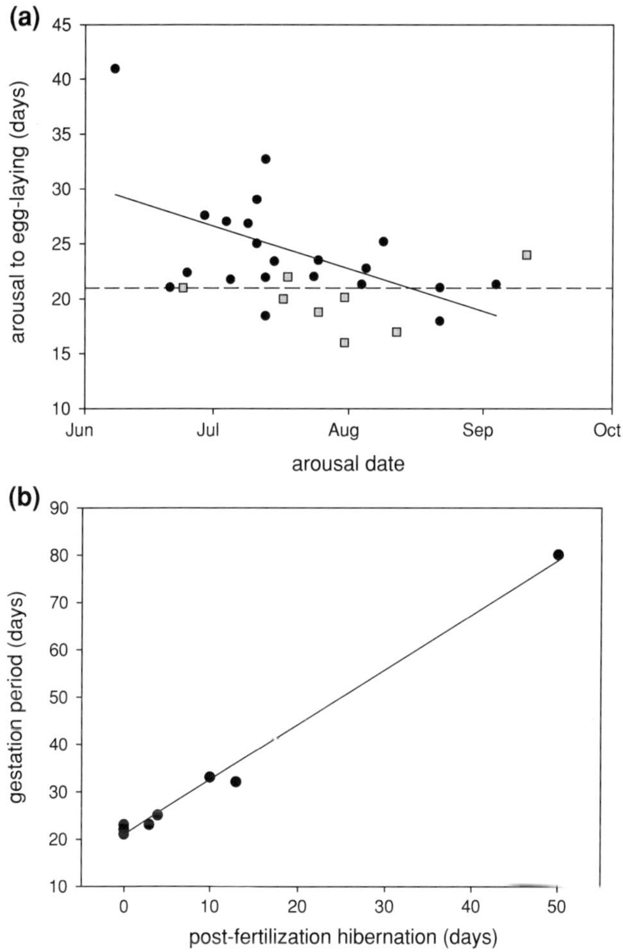

Fig. 13.3 Factors affecting date of egg-laying in the echidna. **a** Time from final arousal from hibernation to egg-laying. *Squares* indicate data from animals that re-entered hibernation after fertilization. *Line* is least squares regression fitted to remaining data. ($y = 49.1–0.53x$, $r^2 = 0.28$, where x is day of the year). The *dashed line* indicates the accepted gestation length, 21 days. **b** Gestation period as a function of the length of post-fertilization hibernation. *Line* is least squares regression ($y = 1.16x + 21.0$, $r^2 = 0.99$)

13.4 Discussion

Tasmanian echidnas are strongly seasonal. Both sexes show a significant annual cycle of body mass, related to hibernation and reproduction (Fig. 13.1). Figure 13.4 shows a graph of mean monthly C_3 pasture productivity for the Lovely Banks field site, calculated using the pasture simulation model EcoMod (Johnson

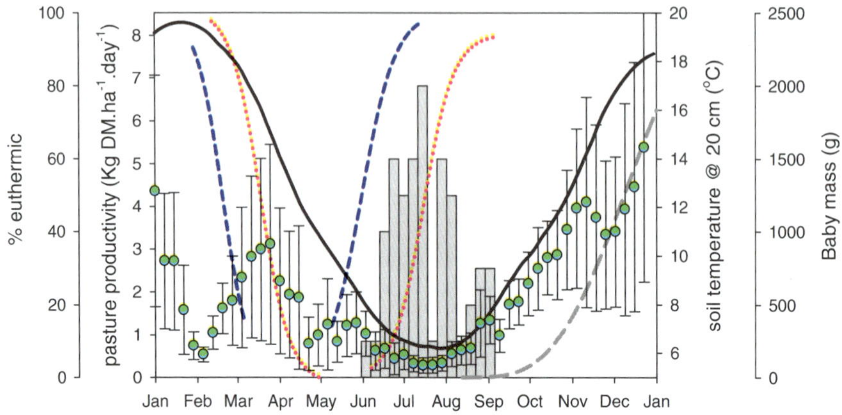

Fig. 13.4 Seasonality at the Tasmanian field site. *Circles* show estimated mean monthly C_3 productivity for pasture (Kg DM ha^{-1} day^{-1}) at Lovely Banks based on climate data from the site during the course of this study. *Bars* show standard error. *Solid line*: soil temperature at a depth of 20 cm. This measure closely approximates the temperature that hibernating echidnas are exposed to (Nicol and Andersen 2007b). *Short dashes*: percentage of male echidnas that are euthermic, *dotted line*: percentage of euthermic female echidnas. *Curves* are logistic regressions fitted to data from this study. *Gray bars*: distribution of mating groups. *Long dashed line*: average growth curve for young. Pasture productivity modeling by Mike Perrring and Mark Hovenden (School of Plant Science, University of Tasmania). Growth of young from Nicol and Andersen (2002), and mating data from Morrow and Nicol (2009)

et al. 2008). This model uses climatic data from the site, including temperature, day length, rainfall, and evapotranspiration records for the last 15 years. Superimposed on this are curves showing the proportion of male and female echidnas that are active (data from this study), the occurrence of mating groups (data from Morrow et al. 2009), and the growth of young (data from Nicol and Andersen 2007a), as well as soil temperature at 20 cm. This temperature closely approximates the temperature that hibernating echidnas are exposed to (Nicol and Andersen 2007b). The diet of echidnas at this site consists almost entirely of ants (approximately two-thirds of dietary volume) and the underground larvae of cockchafer beetles or moths pasture grubs (Sprent, unpublished). Ant abundance is closely related to plant productivity (Siemann 1998), and C_3 productivity calculated in this model can be used as a proxy for abundance of potential prey items of echidnas. During the period January–April productivity is quite unreliable, and it is very low when echidnas begin entering hibernation, although temperatures are quite high. Prehiberatory hypothermic bouts occurred as early as January 7, but only occurred when there was a drop in ambient temperature, while stable torpor only occurred if T_b remained below 16–17°C, or temperature on the external logger remained below 14–15°C. These data imply that males are physiologically prepared for hibernation by mid-summer.

In non-reproducing years, echidnas hibernate for 6 months, arousing when ecosystem productivity begins to increase in spring. For these echidnas, activity is confined to the most productive time of the year rather than the warmest.

Reproductively active male echidnas finish hibernation two weeks before the winter solstice while ambient temperatures are still falling (Figs. 13.2b, 13.4). At this stage the testes are near their maximum size, having increased in size before entry into hibernation (Morrow unpublished). This prehibernatory testicular recrudescence is likely to be initiated by decreasing day length with the response mediated by melatonin (Scherbarth and Steinlechner 2010). Field observations show that some males leave the hibernaculum almost immediately after they become euthermic, and may feed extensively and increase in weight during the approximately 25 days before they begin mating. Males then seek out females, most of which are still hibernating. Mating occurs at the coldest, least productive time of the year, and for males the 40 day mating period appears to be the most energetically demanding part of the year, as this is when they show their maximum mass loss. As well as coping with low ambient temperatures, males compete with other males for matings. We have recorded up to four males in mating groups (Morrow et al. 2009), while on Kangaroo Island up to 11 males have been observed with a single female (Rismiller 1992). After mating, males may guard the female for several days, before looking for another mating opportunity, which means they will have few opportunities for feeding. By October, mating is over, their testes have regressed, and feeding and weight gain are maximal.

In this echidna population, the majority of females are still hibernating when found by males. The gestation period for echidnas is 20–24 days (Morrow et al. 2009) (but see below), but only eight of 28 egg-laying events (29%) occurred more than 24 days after the final arousal (Fig. 13.3a). In the period 2007–2010, when females were closely monitored during the mating period, none aroused from hibernation and moved from their hibernaculum before being mated. It is not clear what the T_b of females is during these matings. All seem to coincide with periodic arousals, but the presence of a male in the hibernaculum could initiate rewarming. We never found fresh sperm in the tract of a female whose T_b had not rewarmed above 15°C, although a male may have been with her for much or all of the rewarming period. We have found mating groups where the female had a deep cloacal temperature below 25°C and fresh sperm in her reproductive tract, and a female with a cloacal temperature of 26°C was observed being pursued by a group of males, so females may not be at the normal euthermic temperature of 32°C when mating occurs. Of eleven pregnancies only three females did not re-enter hibernation after mating, and of the eight that re-entered hibernation six were pregnant. The negative relationship between time from arousal and egg-laying and final arousal date (Fig. 13.3a) shows that late arousing females become pregnant very quickly, while very early matings are less likely to result in fertilization.

Although Rismiller and McKelvey (2000) claim that females mate only once, cloacal swabbing for sperm shows that females mate many times with several males (Morrow et al. 2009; Morrow and Nicol 2009). Perhaps more surprisingly, females will continue to mate when pregnant, right up to the time that they enter the nursery burrow (Morrow unpublished). Up until late July pregnant females may re-enter hibernation. Figure 13.3b shows that the gestation period increases in direct relationship to the length of post-fertilization hibernation, indicating that embryonic development is completely arrested during maternal torpor. All of this suggests that female echidnas would "prefer" to not come out of hibernation until spring, and only do so when they have adequate energy reserves and become pregnant.

Females minimize energy expenditure during winter by hibernating until disturbed by a male (or males), and re-entering hibernation after mating. If fertilization has occurred relatively early in the mating season (before 25 July) females will re-enter hibernation even if pregnant. After approximately 18 days of euthermic gestation, which may be broken by periods of hibernation, the female digs a nursery burrow and moves into it, blocking the entrance behind her, and laying her single egg into her pouch 3 days later (Morrow unpublished). Although they may re-enter torpor when pregnant, once the egg has been laid females maintain a much more constant T_b than usual. Although mean T_b does not vary significantly during the burrow period and immediately after (mean $T_b = 31.2 \pm 0.35$, $n = 36$), while the mother is incubating the egg the SD for T_b is only 0.53, compared with 0.89 in the nursery burrow before egg-laying, 0.62 while the young is in the pouch, and 1.54 when the mother begins to leave the nursery burrow ($F_{3,32} = 33.7$, $P < 0.0001$, (data from Nicol and Andersen 2006). Maintaining such a constant T_b during egg incubation is likely to be very energetically expensive, and for an animal which is notable for the way it minimizes energy expenditure this indicates that at this stage of development of the young minimizing temperature variability is essential. The egg hatches after about 10–11 days of incubation, but the mother remains in the burrow for a total of approximately 37 days (Morrow et al. 2009), and she may have had virtually no opportunity to feed before entering the burrow. The lactation period for mothers at our field site is approximately 150 days (Morrow unpublished), with young being weaned in late December to early January. During these 150 days the young is completely dependent on the mother for food, and grows from approximately 0.6 g at hatching to about 1.5 kg at weaning (Fig. 13.4). This gives the mother $<$3 months after weaning the young to gain weight before entering hibernation. By comparison, Kangaroo Island echidnas, which mate at the same time as Tasmanian echidnas, but do not show deep prolonged hibernation have a lactation period of about 207 days (Morrow et al. 2009).

Although approximately only one in four young are raised to weaning female echidnas mate on average roughly every second year, with only females which are in good body condition before entering hibernation being likely to mate in the next winter. By contrast, only 25% of adult males hibernate through the mating season, implying that reproduction is less energetically demanding for males than for females. For males, maximum energy expenditure occurs over about 40 days during the middle of winter, when food availability is at its lowest. Food availability increases after the mating period, and males are able to attain maximum weight and be prepared for hibernation by January. Females attempt to minimize energy expenditure during the mid-winter mating period by re-entering torpor when possible. Males must be euthermic to seek out females, and compete with other males, but females can remain torpid until found by a male. Re-entering torpor when pregnant will delay egg-laying, but once the egg is laid the mother must expend a considerable amount of energy to maintain a constant T_b during this critical period. Figure 13.4 shows that by mating at the coldest time of the year echidnas ensure that maximum growth of young coincides with maximum ecosystem productivity. Despite the fact that Tasmanian winters are relatively mild, Tasmanian echidnas show a strongly seasonal pattern of energy expenditure consistent with their overall low energy life history.

Acknowledgments We thank all those people who have assisted with collecting data in the field, particularly Niels Andersen, Rachel Harris and Jenny Sprent, and the McShane family for allowing us continued access to their property. We are indebted to Mike Perring and Mark Hovenden for the productivity modeling. This work was carried out under permit from the Tasmanian Department of Primary Industries, Water and Environment, and the University of Tasmania Animal Ethics Committee, and complies with the Tasmanian and the Australian Code of Practice for the Care and Use of Animals for Scientific Purposes (2004). We are grateful for financial support from the Australian Research Council, the University of Tasmania Institutional Research Grants Scheme, the National Geographic Committee for Research and Exploration, and the Holsworth Wildlife Research Endowment.

References

Harris S, Kitchener A (2005) From forest to Fjaeldmark: descriptions of Tasmania's vegetation. Department of Primary Industries, Water and Environment, Hobart

Johnson IR, Chapman DF, Snow VO, Eckard RJ, Parsons AJ, Lambert MG, Cullen BR (2008) DairyMod and EcoMod: biophysical pasture-simulation models for Australia and New Zealand. Aust J Exp Agric 48:621–631. doi:10.1071/EA07133

Morrow G, Andersen NA, Nicol SC (2009) Reproductive strategies of the short-beaked echidna—a review with new data from a long-term study on the Tasmanian subspecies (*Tachyglossus aculeatus setosus*). Aust J Zool 57:275–282. doi:10.1071/ZO09037

Morrow G, Nicol SC (2009) Cool sex? Hibernation and reproduction overlap in the echidna. PLoS One 4:e6070

Nicol SC, Andersen NA (2002) The timing of hibernation in Tasmanian echidnas: why do they do it when they do? Comp Biochem Physiol B Biochem Mol Biol 131:603–611

Nicol SC, Andersen NA (2006) Body temperature as an indicator of egg-laying in the echidna, *Tachyglossus aculeatus*. J Therm Biol 31:483–490. doi:10.1016/j.jtherbio.2006.05.001

Nicol S, Andersen NA (2007a) The life history of an egg-laying mammal, the echidna (*Tachyglossus aculeatus*). Ecoscience 14:275–285. doi:10.2980/1195-6860(2007)14[275:TLHOAE]2.0.CO;2

Nicol SC, Andersen NA (2007b) Cooling rates and body temperature regulation of hibernating echidnas (*Tachyglossus aculeatus*). J Exp Biol 210:586–592. doi:10.1242/jeb.02701

Rismiller PD (1992) Field observations on Kangaroo Island echidnas (*Tachyglossus aculeatus multiaculeatus*) during the breeding season. In: Augee ML (ed) Platypus and echidnas. R Zool Soc NSW, Mosman, pp 101–105

Rismiller PD, McKelvey MW (2000) Frequency of breeding and recruitment in the short-beaked echidna, *Tachyglossus aculeatus*. J Mammal 81:1–17

Scherbarth F, Steinlechner S (2010) Endocrine mechanisms of seasonal adaptation in small mammals: from early results to present understanding. J Comp Physiol B 180:935–952. doi:10.1007/s00360-010-0498-2

Siemann E (1998) Experimental tests of effects of plant productivity and diversity on grassland arthropod diversity. Ecology 79:2057–2070. doi:10.1890/0012-9658(1998)079[2057:etoeop]2.0.co;2

Chapter 14
Sex Differences in Foraging Behaviour, Body Fat and Hibernation Patterns of Free-Ranging Common Hamsters

Carina Siutz, Marc Pluch, Thomas Ruf and Eva Millesi

Abstract In this study, we investigated foraging behaviour, body fat and hibernation patterns in adult common hamsters (*Cricetus cricetus*). In addition, we compared the proportion of body fat prior to hibernation between adult and juvenile individuals. We investigated a population of free-ranging common hamsters in an urban area in southern Vienna, Austria. We applied capture-mark-recapture techniques and focal-animal-sampling for behavioural recordings. Body fat was calculated by integrating morphometric parameters. Body temperature changes were recorded during winter using subcutaneously implanted data loggers. We calculated the time individuals spent in torpor, the number and length of torpor bouts. The results revealed sex differences in foraging behaviour, in that males were feeding mainly above ground, whereas females almost exclusively cached food. These potential differences in energy allocation prior to winter were reflected in body fat at the end of the active season. Adult males had a higher proportion of body fat than females. Although juveniles had lower body fat than adults, similar sex differences were found. During winter adult males spent more time in torpor than females, indicating differing overwintering strategies depending on the individual potential to allocate internal or external energy reserves.

C. Siutz (✉) · M. Pluch · E. Millesi
Department of Behavioural Biology, University of Vienna,
Althanstrasse 14, 1090 Vienna, Austria
e-mail: carina.siutz@univie.ac.at

T. Ruf
Department of Integrative Biology and Evolution,
University of Veterinary Medicine Vienna, Savoyenstrasse 1a,
1160 Vienna, Austria

T. Ruf et al. (eds.), *Living in a Seasonal World*,
DOI: 10.1007/978-3-642-28678-0_14,

14.1 Introduction

Endothermic animals living in seasonal environments are severely challenged during winter as food shortage is paralleled by increasing thermoregulatory costs due to low ambient temperatures. Therefore, animals have to prepare energy reserves to survive the winter period. Energy can be stored either internally as body fat (Barnes 1984; Buck and Barnes 1999; Humphries et al. 2003; Kunz et al. 1998) or externally as food hoards (Day and Bartness 2003; Grulich 1986; Humphries et al. 2003; Niethammer 1982). The advantages of the latter strategy are the potential time-savings in foraging and the broader potential of external energy accumulation (French 2000). However, the two energy accumulation strategies can be combined, depending on fat-storing capacity (Humphries et al. 2003; Lindstedt and Boyce 1985; Lovegrove 2000), reproductive effort, the availability of storable food items, and the size and fatty acid composition of food hoards (Florant 1998; Geiser and Kenagy 1993; Munro and Thomas 2004; Munro et al. 2005). The common hamster (*Cricetus cricetus*) is known for its high plasticity in overwintering strategies and reproductive performance (Franceschini-Zink and Millesi 2008; Franceschini et al. 2007; Grulich 1986; Wassmer and Wollnik 1997). Individual hibernation patterns of common hamsters under semi-natural conditions ranged from regular deep torpor bouts alternating with short euthermic periods (arousals) during the winter, to extended euthermic phases in the hibernaculum lasting for several weeks (Wassmer 2004; Wassmer and Wollnik 1997; Wollnik and Schmidt 1995). Some individuals even remained active throughout the winter.

In contrast to many other hibernators, common hamsters have an expanded reproductive period lasting from April to August. Females can raise up to 3 litters per season and the more litters a female produces, the more pups are weaned (Franceschini-Zink and Millesi 2008; Grulich 1986). During the breeding period males frequently change their burrows when searching for oestrous females, resulting in large home ranges and high fluctuation rates of individuals present in the population (Kayser and Stubbe 2003). Testes regression occurs during July/ August and thereafter, males start to prepare for hibernation before immerging into their hibernacula during September (Franceschini and Millesi 2005; Lebl and Millesi 2008). Female common hamsters are more philopatric and have smaller home ranges than males (Kayser and Stubbe 2003). Litter sizes range between one and nine pups (mean $\pm$ SD 4.3 $\pm$ 2.1; Franceschini-Zink and Millesi 2008). In females, the length of the postbreeding period used to prepare for winter varied with maternal effort (Franceschini-Zink and Millesi 2008). Individuals that successfully raised three litters weaned their last one in September and had only a short time period until immergence.

Juveniles emerge from the natal burrow at an age of about 3 weeks and are weaned shortly thereafter (Kayser and Stubbe 2003; Seluga et al. 1996). Juvenile hamsters terminate surface activity during October (Franceschini and Millesi 2005; Schmelzer and Millesi 2008).

In common hamsters, two foraging behaviours, feeding above ground and food caching can be distinguished. When performing the latter, the animals fill their cheek pouches and carry the food into their burrows (Eibl-Eibesfeldt 1958; Niethammer 1982).

Differences in reproductive effort and foraging tactics could cause individual variation in body condition, particularly in body fat prior to the hibernation period. Since adult males and females differ in their spatial behaviour and the duration of the breeding period, differences in energy allocation and overwintering strategies could be expected. Juveniles have to invest both in structural growth and preparation for hibernation in a limited time span. Thus it can be assumed that body fat in juveniles at the end of the active season may be below adult levels.

In this study we investigated foraging behaviour in adult common hamsters and calculated the proportion of body fat shortly before winter for both adults and juveniles. Body temperature recordings enabled us to compare the time spent torpid in the hibernaculum between adult males and females.

14.2 Methods

14.2.1 Field Techniques

We investigated a population of free-ranging common hamsters in an urban area in southern Vienna, Austria, where hamsters inhabited the green areas (2.7 ha) surrounding buildings. We applied capture-mark-recapture techniques using Tomahawk live traps baited with peanut butter. From the traps, the hamsters were released into black, cone-shaped cotton sacks laterally equipped with Velcro fasteners, which enabled an investigation of an individual without anaesthesia. Traps were checked at 20-min intervals and after the investigation (lasting for 5–10 min) the animals were released in front of their burrows (Franceschini et al. 2007). For permanent identification we subcutaneously implanted a transponder (PIT tag, Data Mars) at first capture. For distant recognition hamsters were fur marked in individual patterns using commercial hair dye. We documented sex and reproductive status (Franceschini-Zink and Millesi 2008) and distinguished between adults (hibernated at least once) and juveniles (born in the current season; before their first hibernation period). At each capture, we measured body mass (±1 g), head, tibia, and food length (±0.1 mm each). In females, parturition was indicated by rapid body mass loss and increasing teat size, partly with milk remains (Franceschini-Zink and Millesi 2008). We recorded the number of litters in each female and determined litter size by counting all juveniles emerging from a female's breeding burrow. We individually marked (transponder and fur mark) each pup to exclude double counts. Capture and handling procedures and parameters recorded in juveniles were the same as in adults.

14.2.2 Foraging Behaviour

Behavioural observations were carried out only on adult individuals using focal-sampling techniques. Feeding and food caching were recorded in 14 males and 24 females for 15 min/individual several times throughout the season. Food caching was defined as collecting food above ground (clearly recognisable by increasing sizes of cheek pouches), entering the burrow with full cheek pouches and leaving the burrow again with empty cheek pouches within a few minutes. The proportion of feeding and food caching was calculated for each individual.

14.2.3 Body Fat

Body fat was calculated using the morphometric parameters body mass, head, tibia, and food length. We have validated this method for the common hamster based on total body fat analysed in carcasses of dead found animals (died within 12 h before collection). Body fat was determined by lipid extraction using Soxhlet apparatus and petroleum ether as solvent. Total body fat was calculated as percentage of total wet mass. We performed a multiple regression model using percentage fat as response variable and body mass, head, tibia, and foot length as predictor variables ($p = 2.86^{-08}$; $R^2 = 83\%$; $n = 11$). By applying this model we were able to predict the proportion of body fat with a mean absolute error of 1.12% from measured values (Pearson's correlation, $r = 0.91$). This method enables the calculation of body fat in free-raging hamsters non-invasively.

Prehibernation body fat was determined in both adults and juveniles at last capture within 1 week before the individuals' immergence into their hibernacula (between 26 August and 5 October).

14.2.4 Hibernation Patterns

Body temperature changes in adult male and female hamsters in their hibernacula were monitored during winter. Temperature data loggers (I-buttons, coated in Elvax ethylene vinyl acetate resins, DuPont and paraffin, gas-sterilised) were implanted subcutaneously in the neck region. This method is less invasive compared to intraperitoneal implantations and was already applied successfully in this species. Body temperature was recorded at 150-min intervals until recapture in spring. The implantation was done in a veterinary clinic using isoflurane anaesthesia. Shortly thereafter, when the animals had recovered from anaesthesia, they were released at their burrows. The burrows were monitored during winter (open/closed) and checked at daily intervals starting in mid March. In spring, the hamsters were recaptured and the data loggers were removed. None of the recaptured animals had lost the data logger in spring. Overwinter survival rates of implanted individuals did not differ from untreated ones of same age and sex. We calculated the time spent torpid (*d*) by counting consecutive days in which body temperature continuously decreased to a minimum of ≤15°C until it had reached ≥30°C again. These periods of hypothermia between two arousals were defined as torpor bouts. Hibernation duration was defined as the time span (*d*) between the onset of the first and the termination of the last torpor bout. We calculated the time the animals hibernated (*d*) as percentage of the total time spent in the hibernaculum between immergence in autumn and vernal immergence in the subsequent year.

14.2.5 Statistics

Statistical analysis was performed in R 2.9.2 (R Development Core Team 2009). Normality of data distribution was tested using Shapiro–Wilk tests. In heterogenic data, we applied Wilcoxon-tests for related and Mann–Whitney U-tests for independent samples. For group comparisons we applied Student's t tests for unpaired samples in normally distributed data. If not stated otherwise, results are shown as means ± SD.

14.3 Results

14.3.1 Foraging Activity

We monitored foraging behaviour in 14 adult males and 24 adult females throughout the active season (Fig. 14.1). In males, the proportion of feeding above-ground was significantly higher than that of food caching (Wilcoxon-test, $p = 0.007$, $n = 14$). Females, however, almost exclusively cached food above-ground (Wilcoxon test, feeding vs. caching, $p < 0.000$, $n = 24$). Foraging behaviours differed significantly between the sexes (Fig. 14.1). Males showed more feeding behaviour than females (Mann–Whitney U-test, $p < 0.000$, $n = 14/24$), whereas in females the proportion of food caching was significantly higher compared to males (Mann–Whitney U-test, $p < 0.000$, $n = 14/24$; Fig. 14.1).

14.3.2 Body Fat

We calculated the proportion of body fat in 9 adults (4 males, 5 females; note that sample size in adults decreased during the season due to dispersal and/or death of individuals) and 24 juveniles (13 males, 11 females) shortly before the individuals

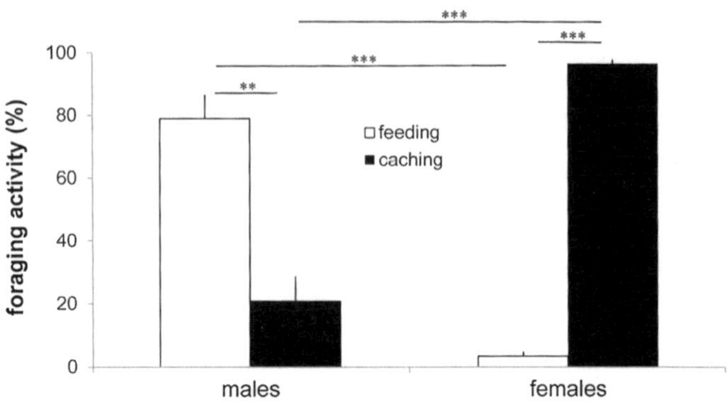

Fig. 14.1 Proportion of foraging activities feeding and food caching above-ground in adult males ($n = 14$) and females ($n = 24$) (means ± SE)

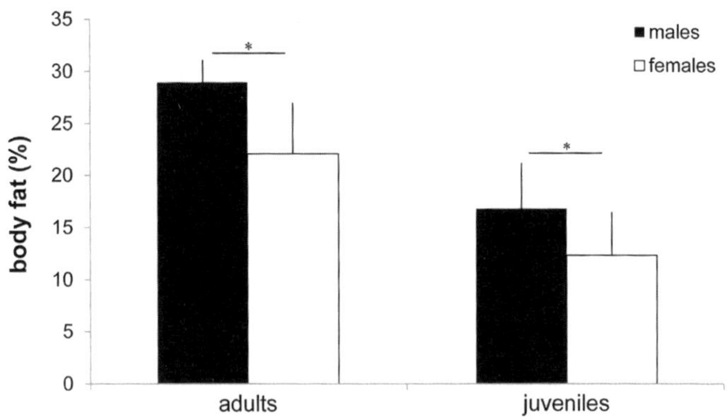

Fig. 14.2 Body fat contents in adult males ($n = 4$) and females ($n = 5$) as well as in juvenile males ($n = 13$) and females ($n = 11$) shortly before immergence into the hibernaculum

immerged into their hibernacula (Fig. 14.2). The results revealed that during the pre-hibernation period adult males had a higher proportion of body fat than females (t test, $p = 0.033$, n(m/f) = 4/5). Juveniles had less body fat than adults (t test, $p < 0.000$, n(adults/juveniles) = 9/24) and, similar to adults, males had significantly more body fat before immergence into the hibernaculum than females (t test, $p = 0.018$, n(m/f) = 13/11; Fig. 14.2).

14.3.3 Hibernation Patterns

None of the focal individuals left its hibernaculum between immergence in autumn and emergence in spring. This time period was significantly longer in males than in females (males: $187 \pm 5.6d$, $n = 5$; females: $175 \pm 6.3d$, $n = 4$, t test, $p < 0.05$).

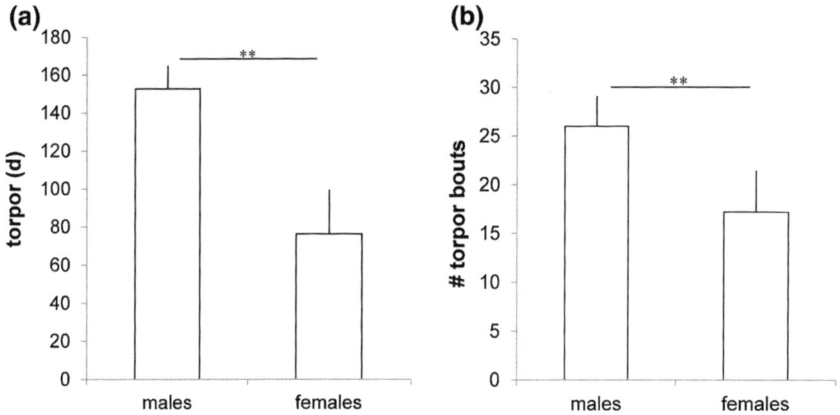

Fig. 14.3 Days spent in torpor (**a**) and number of torpor bouts (**b**) in adult male ($n = 4$) and female ($n = 5$) common hamsters in their natural hibernacula during the entire hibernation period (Sept/Oct—Mar/Apr)

The sex difference in the allocation of body fat was reflected by the hibernation patterns of the focal animals. Males hibernated for 82% (± 3.6, $n = 5$) of the time in the hibernaculum compared to only 51.7% (± 11.7, $n = 4$) in females (t test, $p = 0.001$). Correspondingly, adult males spent more time in torpor (t test, $p = 0.003$, $n = 5/4$) and showed more torpor bouts (t test, $p = 0.009$, $n = 5/4$) during winter than adult females (Fig. 14.3). Hence, females were euthermic for about half of the days in the hibernaculum. Average torpor bout lengths did not differ between the sexes (males: 87.6 ± 2.7 h, females: 85.8 ± 4.5 h; t test, $p > 0.05$).

14.4 Discussion

In this study, we aimed at investigating foraging behaviour, body fat and hibernation patterns in adult common hamsters. Adult males mainly fed on plants above-ground and were only rarely observed to cache food. This is not surprising because males moved far distances in the study area particularly during the breeding season to search for females in oestrus and frequently changed their burrows. Building up food stores could be expected in males after the breeding period when they had to adapt a burrow for hibernation. However, even during this period, the proportion of caching behaviour remained rather low. This indicates that at least in the studied males, building up body fat reserves was the preferred strategy to accumulate energy reserves for the winter. Males terminated reproductive activity earlier in the season than females, which in some cases had to care for their offspring until late September (Franceschini-Zink and Millesi 2008; Lebl and Millesi 2008). Hence, adult males had more time to store body fat and seemed to rely less on food caches. This is supported by the higher proportion of body fat in adult males compared to females shortly before immergence. Although the sample size is rather low, all studied males spent about 80% of the winter season hibernating, showing regular patterns of alternating torpor bouts and arousal episodes. Males immergence into their hibernacula earlier than females and thus stayed underground between autumn and spring for a longer period than females. This difference has been observed in all field seasons (Franceschini and Millesi 2005; Schmelzer and Millesi 2008). Hibernation in males started shortly after immergence into the hibernaculum. During the last 3 weeks before vernal emergence the males remained euthermic, probably to activate the reproductive system. Like many other hibernators, males emerge with scrotal testes and elevated androgen levels (Barnes et al. 1988; Michener 1992; Millesi et al. 1998; Nelson et al. 1990). During this last period in the burrow food caches could be important.

Adult females changed their burrows only occasionally and were predominantly observed caching food above-ground throughout the active season, indicating that food stores also served as energy reserve during gestation and lactation. In addition, food caching could shorten the time spent above ground because hamsters can transport larger amounts of food in their cheek pouches (e.g., 49 g cherries) than could be consumed in the same time. This foraging behaviour could reduce predation risk, and in case of lactating females, minimise the time the juveniles are left alone in the burrow.

The differences in fattening and hibernation could either indicate a sexual dimorphism in overwintering strategies in common hamsters or simply reflect an animal's condition at the end of the active season. Previous reproductive output and hence, maternal effort varied among individual females (Franceschini-Zink and Millesi 2008), which could affect the mothers' condition in early autumn. Females with low reproductive output (i.e., zero or one litter) might be in a better condition prior to winter than females with high output (i.e., two or three litters), which have lactated until September. Although the sample size in this preliminary data set is too small to draw general conclusions, we assume that the time and resources available to prepare for hibernation are reflected in the proportion of body fat prior to winter and could lead to the sex differences in overwintering behaviour. This is supported by body temperature patterns in females with low reproductive output resembling that of males. Females with high previous maternal effort seem to rely on their food stores in the burrow to accumulate internal energy reserves that might be necessary to trigger hibernation onset. Similarly, in European ground squirrels (*Spermophilus citellus*) breeding females started to hibernate about 4 weeks later than females that were prevented from breeding and hence, could invest more time in pre-hibernation fattening (Millesi et al. 2008).

Sex differences in the use of torpor have been documented in some rodent species (e.g., Richardson's ground squirrels, *Spermophilus richardsonii*, Michener 1992; eastern chipmunks, *Tamias striatus*, Munro et al. 2005; eastern woodchucks, *Marmota monax*, Zervanos and Salsbury 2003), but in these cases males had shorter hibernation periods than females, which was mainly attributed to longer euthermic periods in males prior to vernal emergence. In common hamsters, however, we suggest that body condition prior to winter affects overwintering strategies. Fat-dependent sex differences in hibernation patterns have been documented, but in these species females accumulated more fat stores and hibernated for longer periods than males. In grey mouse lemurs (*Microcebus murinus*) females had higher body mass and tail fat stores than males at the onset of the dry season and correspondingly, were inactive for a longer period than males (Schmid 1999). Likewise, brown mouse lemurs (*Microcebus rufus*) that increased body fat shortly before the dry season entered torpor, whereas other individuals showed no changes in body fat and remained active (Atsalis 1999). Female jerboas (*Jaculus orientalis*) showed an autumnal body mass increase in contrast to males and spent more time in torpor than males (El Ouezzani et al. 2011).

Due to small sample size at the end of the season, we were not able to statistically analyse potential relationships between previous reproductive output and body fat shortly before immergence.

Finally, we also calculated the proportion of body fat in juvenile common hamsters prior to winter to determine whether the sex difference in body fat is detectable before the first hibernation period. Juveniles have to prepare not only for hibernation, but must also invest energy in structural growth. Accordingly, they had less body fat than adults at the end of the active season. Analogous to adults, juvenile males had more internal fat stores than females prior to winter, although these differences were absent while the litters were still lactated. None of the investigated female juveniles reproduced in the year of birth. Males developed scrotal testes during the first weeks of life, and although reproductive activity cannot be excluded, successful matings seem rather unlikely when considering the presence of larger adult males. Male juvenile common hamsters seem to grow and accumulate body fat at higher rates than females demonstrating that the sexual dimorphism found in adult common hamsters (Niethammer 1982) becomes apparent during the first active season.

In conclusion, our data indicate that, in contrast to females, adult males use the post-breeding period mainly to accumulate internal energy reserves. Future studies should focus on relationships among body fat, quantity and quality of food stores, and hibernation patterns.

Acknowledgments This study was supported by the Faculty of Life Sciences, University of Vienna. We thank Minh Hien Le for her assistance in the lab.

References

Atsalis S (1999) Seasonal fluctuations in body fat and activity levels in a rain-forest species of mouse lemur, *Microcebus rufus*. Int J Primatol 20:883–910

Barnes BM (1984) Influence of energy stores on activation of reproductive function in male golden-mantled ground squirrels. J Comp Physiol B 154:421–425

Barnes BM, Kretzmann M, Zucker I, Licht P (1988) Plasma androgen and gonadotropin levels during hibernation and testicular maturation in golden-mantled ground squirrels. Biol Reprod 38:616–622

Buck CL, Barnes BM (1999) Annual cycle of body composition and hibernation in free-living arctic ground squirrels. J Mammal 80:430–442

Day DE, Bartness TJ (2003) Fasting-induced increases in food hoarding are dependent on the foraging-effort level. Physiol Behav 78:655–668

Eibl-Eibesfeldt I (1958) *Cricetus cricetus* (L.), Hamstern I. In: Wolf G (ed) Encyclopaedia cinematographica. Inst. f. d. Wiss. Film, Göttingen

El Ouezzani S, Janati IA, Magoul R, Pévet P, Saboureau M (2011) Overwinter body temperature patterns in captive jerboas (*Jaculus orientalis*): influence of sex and group. J Comp Physiol B 181:299–309

Florant GL (1998) Lipid metabolism in hibernators: the importance of essential fatty acids. Am Zool 38:331–340

Franceschini C, Millesi E (2005) Reproductive timing and success in common hamsters. In: Proceedings of the 12th Meeting of the International Hamsterworkgroup, Strasbourg, France

Franceschini-Zink C, Millesi E (2008) Reproductive performance in female common hamsters. Zoology 111:76–83

Franceschini C, Siutz C, Palme R, Millesi E (2007) Seasonal changes in cortisol and progesterone secretion in common hamsters. Gen Comp Endocrinol 152:14–21

French AR (2000) Interdependency of stored food and changes in body temperature during hibernation of the eastern chipmunk, *Tamias striatus*. J Mammal 81:979–985

Geiser F, Kenagy GJ (1993) Dietary fats and torpor patterns in hibernating ground squirrels. Can J Zool 7:1182–1186

Grulich I (1986) The reproduction of *Cricetus cricetus* (Rodentia) in Czechoslovakia. Acta Sci Nat Brno 20:1–56

Humphries MM, Thomas DW, Kramer DL (2003) The role of energy availability in mammalian hibernation: a cost-benefit approach. Physiol Biochem Zool 76:165–179

Kayser A, Stubbe M (2003) Untersuchungen zum Einfluss unterschiedlicher Bewirtschaftung auf den Feldhamster *Cricetus cricetus* (L.), einer Leit-und Charakterart der Magdeburger Börde. In: Stubbe M, Stubbe A (eds) Tiere im Konflikt 7. Martin-Luther-Universität Halle-Wittenberg, Halle

Kunz TH, Wrazen JA, Burnett CD (1998) Changes in body mass and fat reserves in pre-hibernating little brown bats (*Myotis lucifugus*). Ecoscience 5:8–17

Lebl K, Millesi E (2008) Yearling male common hamsters and the trade-off between growth and reproduction. In: Millesi E, Winkler H, Hengsberger R (eds) The common hamster (*Cricetus cricetus*): perspectives on an endangered species. Biosystematics and ecology series 25. Austrian Academy of Sciences Press, Vienna

Lindstedt SL, Boyce MS (1985) Seasonality, fasting endurance, and the body size in mammals. Am Nat 125:873–878

Lovegrove BG (2000) Daily heterothermy in mammals: coping with unpredictable environments. In: Heldmaier G, Klingenspor M (eds) Life in the cold. Springer, Berlin

Michener GR (1992) Sexual differences in over-winter torpor patterns of Richardson's ground squirrels in natural hibernacula. Oecologia 89:397–406

Millesi E, Divjak A, Strauss A (2008) Seasonal timing and pre-hibernation fattening in breeding and non-breeding European ground squirrels. In: Lovegrove BG, McKechnie AE (eds) Hypometabolism in animals: hibernation, torpor and cryobiology. Interpak Books, Pietermaritzburg

Millesi E, Huber S, Dittami JP, Hoffmann IE, Daan S (1998) Parameters of mating effort and success in male European ground squirrels, *Spermophilus citellus*. Ethology 104:298–313

Munro D, Thomas DW (2004) The role of polyunsaturated fatty acids in the expression of torpor by mammals: a review. Zoology 107:29–48

Munro D, Thomas DW, Humphries MM (2005) Torpor patterns of hibernating eastern chipmunks *Tamias striatus* vary in response to the size and fatty acid composition of food hoards. J Anim Ecol 74:692–700

Nelson RJ, Badura LL, Goldman BD (1990) Mechanisms of seasonal cycles of behavior. Annu Rev Psychol 41:81–108

Niethammer J (1982) *Cricetus cricetus* (Linnaeus, 1758)—Hamster (Feldhamster). In: Niethammer J, Krapp F (eds) Handbuch der Säugetiere Europas, Bd 2/I. Akademische Verlagsgesellschaft Wiesbaden, Germany

R Development Core Team (2009) R: a language and environment for statistical computing. R Foundation for Statistical Computing, Vienna, Austria, ISBN 3-900051-07-0, http://www.R-project.org

Schmelzer E, Millesi E (2008) Activity patterns in a population of European hamsters (*Cricetus cricetus*) in an urban environment. In: Proceedings of the 11th Meeting of the International Hamsterworkgroup, Budapest, Hungary

Schmid J (1999) Sex-specific differences in activity patterns and fattening in the grey mouse lemur (*Microcebus murinus*) in Madagascar. J Mammal 80:749–757

Seluga K, Stubbe M, Mammen U (1996) Zur Reproduktion des Feldhamsters (*Cricetus cricetus* L.) und zum Ansiedlungsverhalten der Jungtiere. Abh Ber Mus Heineanum 3:129–142

Wassmer T (2004) Body temperature and above-ground patterns during hibernation in European hamsters (*Cricetus cricetus* L.). J Zool (Lond) 262:281–288

Wassmer T, Wollnik F (1997) Timing of torpor bouts during hibernation in European hamsters (*Cricetus cricetus* L.). J Comp Physiol B 167:270–279

Wollnik F, Schmidt B (1995) Seasonal and daily rhythms of body temperature in the European hamster (*Cricetus cricetus*) under semi-natural conditions. J Comp Physiol B 165:171–182

Zervanos SM, Salsbury CM (2003) Seasonal body temperature fluctuations and energetic strategies in free-ranging eastern woodchucks (*Marmota monax*). J Mammal 84:299–310

Chapter 15
Summer Torpor and Sexual Segregation in the Subtropical Bat *Rhinopoma microphyllum*

Eran Levin, Amos Ar, Yoram Yom-Tov and Noga Kronfeld-Schor

Abstract During the past few years it has become clear that torpor and hibernation are not limited to high latitudes or cold climates. We studied summer roost selection and torpor patterns of a subtropical greater mouse-tailed bat, *Rhinopoma microphyllum*, on the edge of its distribution range. During summer, these bats exhibited complete sexual segregation, with males inhabiting a higher and cooler region. Both sexes inhabited warm and dry caves (28–32°C on average). Using radio telemetry, we measured body (skin) temperature and foraging patterns of both sexes during summer. Lactating females remained normothermic during the day and performed longer foraging bouts during the night, while both males and non-lactating females performed shallow torpor, and performed short foraging bouts during the night. We suggest that these differences result from the different energetic challenges of the two sexes during summer, which may contribute to sexual segregation in bats.

15.1 Introduction

During the past decade, it became evident that torpor is not limited to high latitudes, but also occurs in subtropical and tropical mammals, and in response to diverse environmental conditions including low ambient temperature, food and water availability, and ecological interactions (Carey et al. 2003; Cossins and Barnes 1996; Dausmann et al. 2004; Geiser and Stawski 2011; Levy et al. 2011a, b; Lovegrove and Génin 2008; Lovegrove and Raman 1998; Schmid et al. 2000; Schmid and Speakman 2000; Turbill et al. 2003). Here, we studied the summer roost

E. Levin (✉) · A. Ar · Y. Yom-Tov · N. Kronfeld-Schor
Department of Zoology, Tel Aviv University, 69978 Tel Aviv, Israel
e-mail: levinere@post.tau.ac.il

T. Ruf et al. (eds.), *Living in a Seasonal World*,
DOI: 10.1007/978-3-642-28678-0_15, © Springer-Verlag Berlin Heidelberg 2012

selection and thermal physiology of males and females of a subtropical insectivore bat, the greater mouse-tailed bat (*Rhinopoma microphyllum*) in northern Israel, which is the northern edge of its range. During summer, these bats exhibited complete sexual segregation, with males inhabiting a higher and cooler region.

Animal populations inhabiting the margins of the species distribution are interesting physiologically, since they may preserve important character combinations for adaptation to extreme environmental conditions (Hampe and Petit 2005; Lesica and Allendorf 1995). Thermal physiology and distribution constraints of these edge populations are expected to be of high importance for the response of species to climate change.

Greater mouse-tailed bats are medium-sized insectivore bats (25 g) inhabiting dry and warm regions of the "old world"—from North Africa, east to Thailand, and Sumatra (Altringham 1996). Greater mouse-tailed bats are well adapted to arid (Schliter and Qumsiyeh 1996) or subtropical environments (Kulzer 1965). The tolerance of these bats to relatively low temperatures is limited, and it was assumed that they are unable to perform deep torpor or hibernation (Kulzer 1965). During summer, greater mouse-tailed bats have a specialist diet preference, based mainly on beetles and queens of carpenter ants (*Camponotus felah*). From July, greater mouse-tailed bats accumulate significant amounts of fat, sometimes reaching 50% of their body mass (Levin et al. 2009), which is similar (and even higher) to that reported during the pre-hibernatory period in temperate bats (Kronfeld-Schor et al. 2000; Kunz et al. 1998).

We studied the thermal physiology of both sexes of the greater mouse-tailed bats during summer, trying to understand the significance of the geographic segregation between sexes. Based on our results, we suggest that the sexual segregation in these bats is related to different thermal preferences between males and reproductively active females.

15.2 Methods

15.2.1 Study Area

The study was conducted in the northern Jordan valley, Israel (32° 46′N 35° 32′E). The climate in this area is subtropical, with a dry and hot summer (mean daily minimal and maximal temperatures in July: 22.8 and 37.5°C, respectively) and a moderate winter (mean daily minimal and maximal temperatures in February: 9.2 and 19.2°C, respectively; Israel Meteorological Service).

15.2.2 Radio Telemetry

During summer 2007 (23.7–10.8) and 2008 (26.6–1.8), a total of 38 greater mouse-tailed bats (19 females and 19 males) were tagged with radio-transmitters for monitoring skin temperature and activity. During summer 2007 we tagged 10 males and 10 lactating females. During summer 2008 no reproductive females were observed in any of the greater mouse-tailed bat populations in Israel, so 9 non-lactating females and 9 males were tagged. We used BD2-CT radio transmitters (Holohil Systems, Canada) weighing 0.9 g for females and 1.1 g for males, within the accepted range of less than 5% of body weight (Aldridge and Brigham 1988; Bontadina et al. 2002). The hair between the shoulder blades was clipped and transmitters were attached using medical glue (Skin-Bond cement, Smith and Nephew United Inc., Largo, Florida, USA).

Two RX900 stationary receiver-data loggers (Televilt International, Sweden) were positioned near the entrance of the male and female roosts and a di-pole antenna was fixed inside each roost. Receivers were powered by external 12 V batteries that were charged during the day using solar panels (Shell Solar Industries, USA). The receivers were programmed to scan the relevant transmitter frequencies, allocating 45 s for each frequency (so each individual was scanned every ~15 min). An individual was considered to be foraging if no data were logged for more than 30 min. Skin temperature (T_{sk}) was obtained using the transmitter pulse rate, as measured and logged by the stationary receivers in the roosts. We used T_{sk} as an approximation of T_b based on studies showing that in small mammals T_{sk} is closely correlated to T_b (Audet and Thomas 1996; Daniel et al. 2010). The activity temperature of each individual (T_{ac}) was calculated from the last T_{sk} reading before bats emerged for their evening foraging bout (Kelly et al. 2007). Maximum skin temperature (T_{max}) was calculated from the average of 30 maximum T_{sk} measurements during the radio tracking period. In both colonies, ambient temperature (T_a) was recorded every 90 min using a temperature-data logger (HOBO, Onset, USA) positioned on the roost ceiling.

All procedures were carried out under permit no 25057/2006 from the Israel Nature Reserves and Park Authority (NPA).

15.3 Results

15.3.1 Summer Activity Rhythms and Body Temperature

As expected, the bats exhibited complete sexual segregation:

Males: Average daily T_a in the male colony during summer ranged between a minimum of 29.2 ± 1°C around 5:00 to 32.9 ± 0.8°C around 15:00 in summers 2007–2008 (Fig. 15.1a). A total of 138 full days of T_{sk} recordings were obtained from ten males (five in each season, Table 15.1). The other nine tagged individuals apparently moved to an unknown nearby colony after being caught and tagged, and were recorded foraging in the area. A single male remained in the males' roost every night after the evening foraging bout, but left the roost every morning before sunrise. The T_{sk} patterns of males in both summers were similar (Fig. 15.1c); after returning from a foraging bout T_{sk} fell almost immediately, and moderately rose at

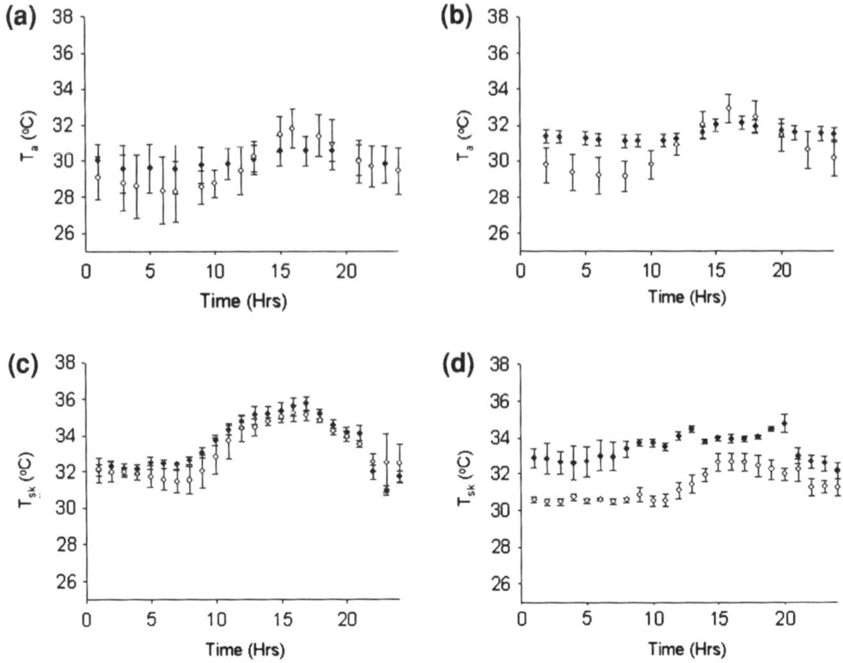

Fig. 15.1 Daily roost temperature cycles (±SD) of males (**a**) and females (**b**) and T_{sk} cycles (±SD) of males (**c**) and females (**d**) during summers 2007 (*filled dots*) and 2008 (*empty dots*)

Table 15.1 Foraging bout length (SD) and skin temperature (SD) of males and females greater mouse-tailed bats in summers 2007–2008 (t test, $*P < 0.05$)

Season	Sex	N	Min. foraging time (min)	Max. foraging time (min)	Av. foraging time (min)	T_{max}	T_{ac}
2007	M	5	41 (7)	272 (151)	117 (38)	38.1 (0.8)	34.4 (0.6)
	F	5	144 (138)*	457 (158)	306 (154)*	35.4 (0.5)*	35.3 (0.7)
2008	M	5	37 (8)	466 (148)	164 (70)	36.9 (0.8)	34.6 (0.7)
	F	4	240 (113)*	278 (103)	240 (113)	35.5 (0.9)*	34.5 (1)

noon (from 7:30 to 15:00), with a peak of T_{sk} above T_a around 15:30. After this peak, T_{sk} moderately decreased with T_a until bats actively re-warmed before leaving the cave to forage (Fig. 15.2).

Females: T_a in the female roost was similar to that of male roost, ranging between minimum of 29.6 ± 1.3°C around 5:00 to 30.5 ± 0.8°C around 15:00 (Fig. 15.1b). In response to trapping and transmitter attachment most females left the roost (carrying their pups with them during summer 2007) and were located using a directional antenna in another colony, 3.5 km northeast from the original maternity roost. We positioned the receiver-data logger in this roost for the two recording seasons. A total of 34 full days of T_{sk} recording were obtained from nine females (five individuals in 2007 and four in 2008, see Table 15.1). T_{sk} of lactating females (recorded during 2007) were significantly higher than recorded in the non-

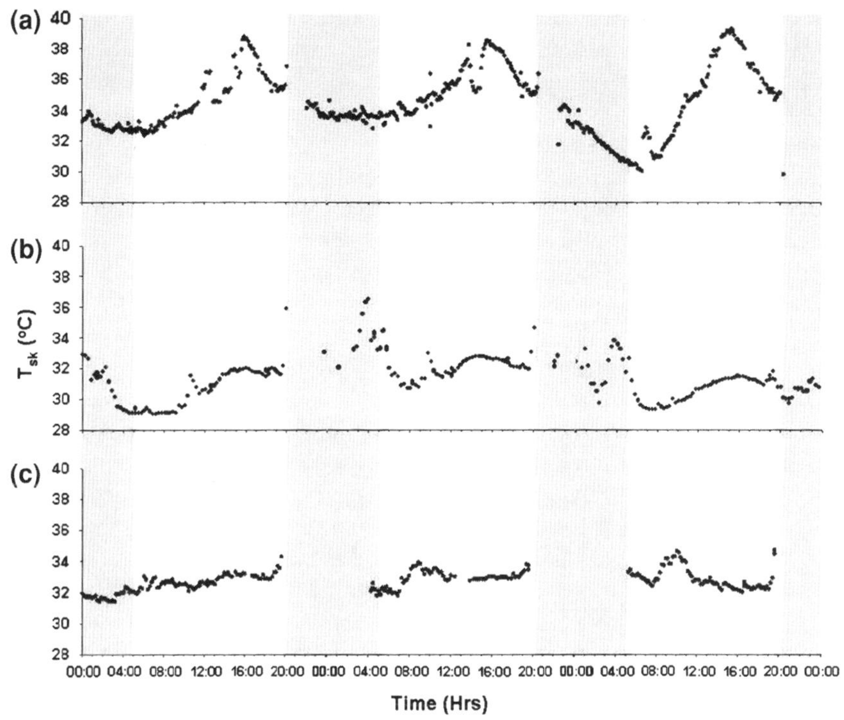

Fig. 15.2 T_{sk} cycles during 3 days in a male (**a**), a non-lactating female (**b**) and a lactating female (**c**) greater mouse-tailed bat. The empty parts are when bats left the cave to forage. Night time is *gray shaded*

lactating females (average $33.4 + 0.7°C$, $n = 5$ vs. $31.2 \pm 0.8°C$, $n = 4$), t test, $t = 9.4$, $P < 0.001$, Figs. 15.1d, 15.2b, c). In contrast to males, females tended to shift roosts, which resulted in a smaller sample size of T_{sk} recordings compared to males (34 vs. 138). When females left the roost to forage (lactating and non-lactating), they spent a significantly longer time outside the roost than males (Table 15.1). T_{ac} of females and males was not significantly different, but T_{max} was significantly higher in females during both summers (38.1 vs. 35.4°C on 2007 and 36.9 vs. 35.5°C on 2008, t test, $P < 0.05$, Table 15.1).

15.4 Discussion

We found that male and female greater mouse-tailed bats are "thermophile" bats that segregate throughout summer. During this time females are pregnant, give birth, and lactate, and both sexes accumulate pre-hibernatory body fat that is used as an energy source during winter hibernation, which they perform in relatively high T_a (Levin et al. 2009).

In northern Israel, both sexes of greater mouse-tailed bats inhabit subtropical habitats with high T_a and rich food sources. However, males roost at higher and cooler altitudes (average of 4°C difference in minimal daily temperature, [Israel meteorological service]). Nevertheless, we found no significant differences in summer roost T_a between males and females. Similar sexual segregation along an altitudinal gradient was observed in several temperate bat species (Senior et al. 2005). A common explanation for such segregation is that the bigger females exclude the smaller males to less productive regions. However, greater mouse-tailed bats males are at least 25% heavier than females (Levin et al. 2009) and therefore do not fit this pattern.

One possible variable differing between the male and female summer roost sites is the T_a outside the roost, which is about 4°C lower at the male roost area (18°C vs. 22°C). It is possible that some physiological or morphological differences, such as body mass or thermoregulatory abilities, differ between males and females, allowing males to be active during colder nights. In a recent paper, Humphries and Careau (2011) tested the hypothesis of activity-thermoregulatory heat substitution, which suggests that heat produced during activity in the cold reduces the energy expenditure for thermoregulation. They found that the scope for heat substitution increases with animals' body size and intensity of the activity (Humphries and Careau 2011). As T_a drops some of the heat produced by activity is used for thermoregulation, up to a point where heat produced by activity equals the amount needed for thermoregulation. Below this equilibrium point the animal has to start inversing energy in thermoregulation. Being larger than the females, it is expected that T_a at this equilibrium point, which is optimal in terms of energy usage efficiency, will be lower for greater mouse tail bat males. Moreover, it was suggested that in many mammalian species, including humans, there are thermoregulatory differences between males and females: females have lower T_{sk}, lower cooling rate, and reduced metabolic reaction to cold compared to males (McArdle et al. 1992; Vierck et al. 2008). These differences between sexes were shown to affect behavioral and physiological aspects such as aggregation in females (Terrien et al. 2010), or differences in geographic dispersal patterns between sexes (Cryan and Wolf 2003). Physiological, morphological, and behavioral differences between sexes (hormones, lactation, water balance, exercise capacity, adipose tissue to muscle ratio, body size, sociality, etc.) have the potential to contribute to these different ecological preferences (Kaciuba-Uscilko and Grucza 2001). These differences may exist also in greater mouse-tailed bats and contribute to the summer sexual segregation.

Acknowledgments This research was supported by THE ISRAEL SCIENCE FOUNDATION (grant No. 232/08). We are grateful to Dr. Amit Dolev of the Society for the Protection of Nature in Israel for good advice and help in the field.

References

Aldridge HDJN, Brigham RM (1988) Load carrying and maneuverability in an insectivorous bat: a test of the 5% "rule" of radio-telemetry. J Mammal 69:379–382

Altringham JD (1996) Bats biology and behavior. Oxford University Press, London

Audet D, Thomas DW (1996) Evaluation of the accuracy of body temperature measurements using external radio transmitters. Can J Zool 74:1778–1781

Bontadina F, Schofield H, Naef-Daenzer B (2002) Radio-tracking reveals that lesser horseshoe bats (*Rhinolophus hipposideros*) forage in woodland. J Zool 258:281–290

Carey HV, Andrews MT, Martin SL (2003) Mammalian hibernation: cellular and molecular responses to depressed metabolism and low temperature. Physiol Rev 83:1153–1181

Cossins AR, Barnes B (1996) Hibernation—Southern discomfort. Nature 382:582–583

Cryan PM, Wolf BO (2003) Sex differences in the thermoregulation and evaporative water loss of a heterothermic bat, *Lasiurus cinereus*, during its spring migration. J Exp Biol 206:3381–3390

Daniel S, Korine C, Pinshow B (2010) The use of torpor in reproductive female Hemprich's long-eared bats (*Otonycteris hemprichii*). Physiol Biochem Zool 83:142–148

Dausmann KH, Glos J, Ganzhorn JU, Heldmaier G (2004) Physiology: hibernation in a tropical primate. Nature 429:825–826

Geiser F, Stawski C (2011) Hibernation and torpor in tropical and subtropical bats in relation to energetics, extinctions, and the evolution of endothermy. Integr Comp Biol 51(3):337–348

Hampe A, Petit RJ (2005) Conserving biodiversity under climate change: the rear edge matters. Ecol Lett 8:461–467

Humphries MM, Careau V (2011) Heat for nothing or activity for free? Evidence and implications of activity-thermoregulatory heat substitution. Integr Comp Biol 51:419–431

Kaciuba-Uscilko H, Grucza R (2001) Gender differences in thermoregulation. Curr Opin Clin Nutr Metab Care 4:533–536

Kelly EJ, Mason M, Stoffberg S, Jacobs DS (2007) Thermoregulation in two free-ranging subtropical insectivorous bat species: *Scotophilus* species (Vespertilionidae). Can J Zool 85:883–890

Kronfeld-Schor N, Richardson C, Silvia BA, Kunz TH, Widmaier EP (2000) Dissociation of leptin secretion and adiposity during prehibernatory fattening in little brown bats. Am J Physiol 279:R1277–R1281

Kulzer E (1965) Temperaturregulation bei Fledermausen (*Chiroptera*) aus Verschiedenen Klimazonen. Zeitschrift Fur Vergleichende Physiologie 50:1–34

Kunz TH, Wrazen JA, Burnett CD (1998) Changes in body mass and fat reserves in pre-hibernating little brown bats (*Myotis lucifugus*). Ecoscience 5:8–17

Lesica P, Allendorf FW (1995) When are peripheral populations valuable for conservation? Conserv Biol 9:753–760

Levin E, Yom-Tov Y, Barnea A (2009) Frequent summer nuptial flights of ants provide a primary food source for bats. Naturwissenschaften 96:477–483

Levy O, Dayan T, Kronfeld-Schor N (2011a) Adaptive thermoregulation in golden spiny mice: the influence of season and food availability on body temperature. Phys Biochem Zool 84:175–184

Levy O, Dayan T, Kronfeld-Schor N (2011b) Interspecific competition and torpor in golden spiny mice: two sides of the energy acquisition coin. Integr Comp Biol 51(3):441–448

Lovegrove B, Génin F (2008) Torpor and hibernation in a basal placental mammal, the Lesser Hedgehog Tenrec (*Echinops telfairi*). J Com Physiol B 178:691–698

Lovegrove BG, Raman J (1998) Torpor patterns in the pouched mouse (*Saccostomus campestris; Rodentia*): a model animal for unpredictable environments. J Com Physiol B 168:303–312

McArdle W, Toner M, Magel J, Spinal R, Pandolf K (1992) Thermal responses of men and women during cold-water immersion: influence of exercise intensity. Eur J Appl Physiol 65:265–270

Schliter A, Qumsiyeh M (1996) *Rhinopoma microphyllum*. Mammal Sp 542:1–5

Schmid J, Ruf T, Heldmaier G (2000) Metabolism and temperature regulation during daily torpor in the smallest primate, the pygmy mouse lemur (*Microcebus myoxinus*) in Madagascar. J Com Physiol B 170:59–68

Schmid J, Speakman JR (2000) Daily energy expenditure of the grey mouse lemur (*Microcebus murinus*): a small primate that uses torpor. J Com Physiol B 170:633–641

Senior P, Butlin RK, Altringham JD (2005) Sex and segregation in temperate bats. Proc R Soc B 272:2467–2473

Terrien J, Perret M, Aujard F (2010) Gender markedly modulates behavioral thermoregulation in a non-human primate species, the mouse lemur (*Microcebus murinus*). Physiol Behav 101:469–473

Turbill C, Kortner G, Geiser F (2003) Natural use of heterothermy by a small, tree roosting bat during summer. Physiol Biochem Zool 76:868–876

Vierck CJ, Acosta-Rua AJ, Rossi HL, Neubert JK (2008) Sex differences in thermal pain sensitivity and sympathetic reactivity for two strains of rat. J Pain 9:739–749

Chapter 16
Heterothermy in Caprimulgid Birds: A Review of Inter- and Intraspecific Variation in Free-Ranging Populations

R. M. Brigham, A. E. McKechnie, L. I. Doucette and F. Geiser

Abstract Caprimulgid birds represent one of the best studied endotherm taxa in terms of the occurrence of heterothermy in free-ranging populations. In this chapter, we review currently available data on heterothermy in this group, and examine potential ecological correlates of these responses. To date, heterothermic responses have been examined in at least one population of each of six species, ranging in body mass (M_b) from 40 to 450 g and occurring in habitats from deserts to mesic woodlands. Patterns of heterothermy vary from infrequent, shallow bouts to periods of uninterrupted torpor lasting several days, during which body temperature may be reduced below 5°C. Overall levels of heterothermy, as quantified using a recently proposed metric, do not show statistically significant relationships with M_b nor with ecological variables such as minimum air temperature or habitat aridity. Nevertheless, it is striking that the two most heterothermic species recorded to date, the Common Poorwill and the Freckled Nightjar, both inhabit arid habitats. Moreover, the former species remains the only bird known to hibernate. Within species, patterns of heterothermy may vary considerably among populations, with lunar cycles, temporal fluctuations in insect abundance and roost site characteristics being among the ecological determinants of heterothermy.

R. M. Brigham
Department of Biology, University of Regina, Regina, SK S4S 0A2, Canada
e-mail: mark.brigham@uregina.ca

A. E. McKechnie (✉)
DST/NRF Centre of Excellence at the Percy FitzPatrick Institute,
Department of Zoology and Entomology, University of Pretoria,
Private Bag X20, Hatfield, 0028, South Africa
e-mail: aemckechnie@zoology.up.ac.za

L. I. Doucette · F. Geiser
Centre for Behavioural and Physiological Ecology, Zoology,
University of New England, Armidale, NSW 2351, Australia

T. Ruf et al. (eds.), *Living in a Seasonal World*,
DOI: 10.1007/978-3-642-28678-0_16, © Springer-Verlag Berlin Heidelberg 2012

16.1 Introduction

Approximately 65% of extant endotherms are birds, yet heterothermic responses in this taxon have received considerably less attention than in mammals (Geiser and Ruf 1995; McKechnie and Mzilikazi 2011). Nevertheless, the available evidence suggests that heterothermy is widespread among birds, with approximately 100 species representing 11 orders (including the Caprimulgiformes, Coraciiformes, Coliiformes, Apodiformes, Trochiliformes, Columbiformes and Passeriformes) known to facultatively reduce body temperature (T_b) below normothermic levels (i.e. rest-phase T_b observed at thermoneutral environmental temperatures) (McKechnie and Lovegrove 2002). Nearly all these species employ either daily torpor (substantial reductions in T_b accompanied by lethargy and reduced responsiveness), or shallower reductions of rest-phase T_b during which behavioural responses to stimuli are retained (Prinzinger et al. 1991); hibernation has been reported for only one species, the Common Poorwill.

In terms of heterothermy, the order Caprimulgiformes is one of the best studied avian taxa. Traditionally, the Caprimulgiformes consist of the frogmouths, potoos, Oilbird, nightjars, eared-nightjars and owlet-nightjars (Holyoak 2001), although Sibley and Ahlquist (1990) placed these families within the Strigiformes alongside the owls, and recent work has raised the possibility that the owlet-nightjars may in fact be more closely related to swifts than to the other caprimulgid families (Hackett et al. 2008; Mayr 2002). Although there are avian taxa (such as the hummingbirds) for which data on heterothermy are available for more species than is the case for the caprimulgids, what sets the latter group apart in terms of our current knowledge of avian heterothermy is the number of detailed studies involving free-ranging populations (Table 16.1). Thus, data on the caprimulgids present a valuable opportunity to investigate the ecological correlates of avian heterothermy in inter- and intraspecific analyses, and understand the significance of these responses for wild birds in their natural habitats.

Several ecological factors are thought to be broadly responsible for the widespread use of heterothermy in caprimulgids. First, most species are nocturnal aerial insectivores, and the availability of prey is thus strongly dependent on variables such as air temperature and rainfall (Ashdown and McKechnie 2008; Jetz et al. 2003; Rydell et al. 1996). Second, caprimulgids are visually orientating predators, and foraging opportunities are severely reduced during periods with low ambient light levels (Ashdown and McKechnie 2008; Jetz et al. 2003). Third, the foraging activity of caprimulgids may be constrained by the risk of predation (Brigham et al. 1999; Woods and Brigham 2008). Since caprimulgids are a cosmopolitan group that occur on all continents except Antarctica, inhabit habitats ranging from tropical rainforest to deserts and vary in body mass (M_b) by more than an order of magnitude (<40–600 g, Holyoak 2001), this taxon offers rich opportunities to test predictions concerning ecological determinants of heterothermy.

The data available in 2002 regarding ecological correlates of caprimulgid heterothermy were qualitatively reviewed by Brigham et al. (2006). However, the

Table 16.1 Summary of attributes of heterothermic responses by seven species of Caprimulgiform birds

Species	Body mass (g)	Use	Period	Min. T_b (°C)	Duration	Timing	Location[a]	References
Common Poorwill (*Phalaenoptilus nuttallii*)	40–45	Common (all year)	Night/day	3–5	h-days	All but summer	AZ, BC, SK N. America	French (1993); Brigham (1992); Woods (2002); Woods and Brigham (2004)
Australian Owlet-nightjar (*Aegotheles cristatus*)	40–50	Common	Day	20	4–8 h	Winter (all year)	NSW, NT Australia	Brigham et al. (2000); Doucette et al. (2011, 2012)
Whippoorwill (*Caprimulgus vociferus*)	50–55	Rare	Day	20	3–8 h	Spring, fall (spring-fall)	ON, SD N. America	Hickey (1993); Lane et al. (2004)
European Nightjar (*Caprimulgus europaeus*)[b]	54–82	Rare	Day	7	9–12 h	Spring, fall (spring-fall)	Finland	Peiponen (1965, 1966)
Freckled Nightjar (*Caprimulgus tristigma*)	73–81	Common	Night/day	10	2–8 h	Winter (winter)	NC, NW South Africa	McKechnie et al. (2007); Smit et al. (2011)
Common Nighthawk (*Chordeiles minor*)	75–85	V. rare	Night/day	25	<8 h	Fall (spring-fall)	BC, SK N. America	Firman et al. (1993); Fletcher et al. (2004)
Tawny Frogmouth (*Podargus strigoides*)	350–550	Common	Night	29	3–11 h	Winter (winter)	NSW Australia	Körtner et al. (2000, 2001)

Use is a qualitative assessment of the frequency of heterothermy based on the number of days it was recorded. *Period* represents the most common time of the 24h cycle for heterothermy. *Min* T_b is the lowest body or skin temperature recorded by any individual during the study(-ies). *Duration* is the length of typical heterothermic bouts. *Timing* is the period of the year when torpor was recorded followed by the period when the study occurred in parentheses. *Location* refers to the province or state and/or country where the study(-ies) took place

[a] *AZ* Arizona, *BC* British Columbia, *SK* Saskatchewan, *NSW* New South Wales, *NT* Northern Territory, *ON* Ontario, *SD* South Dakota, *NC* Northern Cape, *NW* Northwest Province

[b] Data collected under semi-natural conditions

subsequent decade has seen an increase in the number of studies on free-ranging populations, and has also seen new data on intraspecific variation in heterothermy among populations that differ in factors such as habitat aridity. Moreover, a recently proposed comparative metric [Heterothermy index (HI), Boyles et al. 2011] provides a quantitative index of overall levels of heterothermy, allowing for statistical analyses that were previously not available. HI is calculated using an equation similar to that for standard deviation, and increases with increasing depth and/or duration of heterothermic bouts (Boyles et al. 2011). Values range from 0°C when T_b is perfectly stable, to approximately 40°C in a continuously hibernating animal in which T_b is 40°C lower than normothermic T_b. Our aims are to review currently available data on heterothermy in free-ranging caprimulgids, and examine potential ecological correlates of these responses. As the relatively small number of species investigated (and the large gap in M_b between the Tawny Frogmouth and the other species) imposes constraints on the statistical power of these analyses, we intend this to be an exploratory analysis aimed at developing hypotheses that need to be addressed in future.

16.2 Materials and Methods

16.2.1 Heterothermy Data

We surveyed the literature for studies on thermoregulation in free-ranging caprimulgids, and obtained data on variables such as M_b, minimum T_b and study location (Table 16.1). For context we also include information about European nightjars which have been the focus of some field work, albeit in semi-natural conditions with birds deprived of food (Peiponen 1965, 1966). Data from this species were not included in the HI analysis. For each study population, we calculated an average HI value (Boyles et al. 2011) from T_b or skin temperature (T_{skin}) traces, with a mean daily HI value calculated for each individual for which data were available, and an overall mean then calculated for each study population (Table 16.2). For each species, HI was calculated for the daily period during which heterothermic bouts typically occurred, rather than the full 24 h (Table 16.2). We opted to use this approach for two reasons. First, in many studies, data collection was biased towards periods of heterothermy and far fewer data were available from normothermic birds (one reason being that normothermic birds were active and away from roost sites). Second, the HI integrates T_b or T_{skin} both below and above the modal value, and we wanted to minimise the likelihood of HI values being inflated by artificially high normothermic T_{skin} values resulting from solar irradiance on backpack-mounted transmitters during the day. In the case of the Common Poorwill, an average HI value for hibernating birds was calculated by extracting and digitising data from Fig. 8.6 in Woods (2002), and modal normothermic T_b was taken as the average maximum T_b achieved during spontaneous arousal by shaded individuals (33.4°C; Woods 2002).

Table 16.2 Heterothermy index (HI), body mass and environmental variables [aridity index (Q, Emberger 1955) and mean daily minimum air temperature during the coldest month of the year (data from www.worldclimate.com)] for eight caprimulgid taxa

Species/population	Body mass (g)	Aridity index (Q)	Minimum T_a (°C)	HI (n)	Source
Common Poorwill	42.5	212	3.6	20.29[a]	Woods (2002)
Australian Owlet-Nightjar (Alice Springs)	42.3	212	4	4.87 (5)	Doucette et al. (2011, 2012)
Australian Owlet-Nightjar (Armidale)	50.1	1071	0.4	5.40 (6)	Doucette et al. (2011), Brigham et al. (2000)
Whippoorwill[b]	52.5	636	3.2	4.04 (8)	Lane et al. (2004)
Freckled Nightjar (western)	73	224	6.8	7.73 (6)	Smit et al. (2011)
Freckled Nightjar (eastern)	81	677	2.9	4.86 (5)	McKechnie et al. (2007)
Common Nighthawk[b]	83	417	0.5	2.50 (3)	Fletcher et al. (2004)
Tawny Frogmouth	453	1071	0.4	3.42 (8)	Körtner et al. (2001)

[a] Calculated from average trace

[b] Not included in analyses of relationships between HI and environmental variables

16.2.2 Climate Variables

For each population studied, the following historical climate data were obtained from the nearest weather station (www.worldclimate.com): mean annual precipitation (MAP, mm), mean maximum air temperature of the hottest month (T_{max}, °C) and mean minimum air temperature of the coldest month (T_{min}, °C). These data were used to calculate the aridity index $Q = MAP/((T_{max} + T_{min})(T_{max} - T_{min})) \times 1000$ (Emberger 1955), following Tieleman et al. (2003).

16.2.3 Statistical Analyses

The small number of taxa involved here makes it impossible to reliably test for a phylogenetic signal within our data set (Blomberg et al. 2003). Because all the species in this analysis are within the same order, and because of the exploratory nature of this analysis, we analysed the data using conventional (i.e. non-phylogenetically independent) statistics only.

16.3 Results and Discussion

16.3.1 Interspecific Variation

The caprimulgid species for which thermoregulation has been investigated in free-ranging populations have all been found to exhibit reductions in T_b below normothermic levels. However, the extent of heterothermy varies widely among

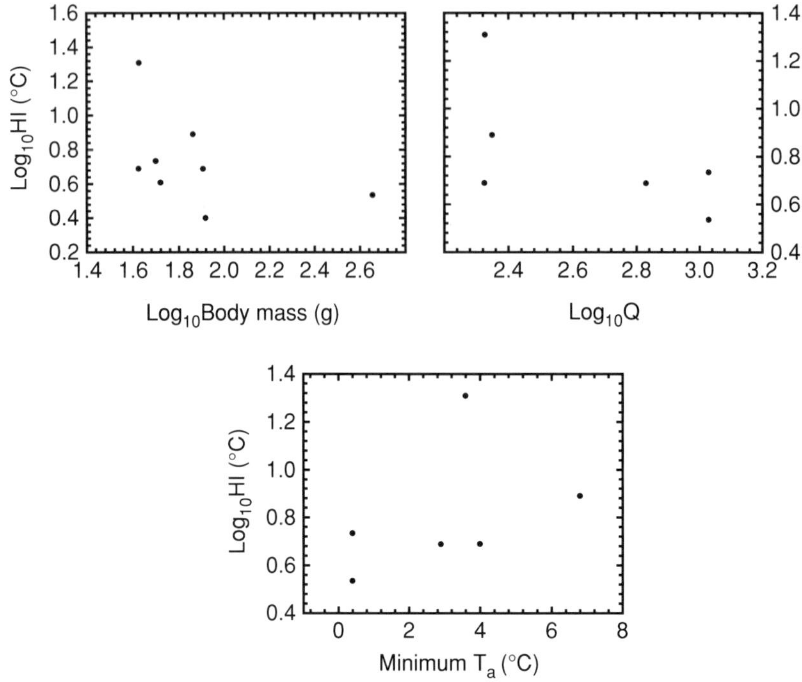

Fig. 16.1 Heterothermy index values for caprimulgids were not significantly related to M_b, aridity index (Q) nor the mean daily minimum air temperature (T_a) for the coldest month of the year. The latter two analyses included only winter studies on non-migratory populations

species, with HI values ranging from 2.5°C in Common Nighthawks to 20.3°C in Common Poorwills (Table 16.2). This variation reflects large differences in the frequency, depth and duration of heterothermic bouts; the former species employed heterothermy infrequently and reduced T_b by only 5–15°C (Fletcher et al. 2004), whereas the latter species exhibited deep, continuous hibernation during winter and reduced T_b to as low as 3°C (Woods 2002, Woods and Brigham 2004). European Nightjars also employ deep torpor, but we have not included this species in our quantitative analysis as the only data available originate from captive birds held under semi-natural conditions (Peiponen 1965, 1966).

The caprimulgid species investigated to date vary approximately tenfold in M_b from 40–450 g (Tables 16.1, 16.2). There does not appear to be any consistent relationship between M_b and HI; neither a linear nor logarithmic regression model fitted to $\log_{10}M_b$ and $\log_{10}$HI data yielded a significant fit (linear: $F_{1,6} = 1.450$, $P = 0.274$, logarithmic: $F_{1,6} = 1.723$, $P = 0.235$; Fig. 16.1). Thus, we compared HI values directly (as opposed to using residuals) with climatic variables for further analyses.

In addition to M_b, the species/populations for which data are available vary widely in the latitude and the climatic characteristics of their habitat (Table 16.2). For the six non-migratory taxa for which heterothermy has been examined during winter (i.e. excluding Common Nighthawks and Whippoorwills), $\log_{10}$HI did not show a significant relationship with either average daily minimum temperature during the coldest month of the year ($F_{1,4} = 0.905$, $P = 0.395$; Fig. 16.1) nor with aridity ($\log_{10}Q$, $F_{1,4} = 2.661$, $P = 0.178$; Fig. 16.1). Despite the latter non-significant result, we would argue that habitat aridity may in fact be the variable to pursue further as a climatic correlate of interspecific variation in heterothermy in caprimulgid birds. It is noteworthy, for instance, that the two taxa with the highest HI values in our data set (Common Poorwills, HI = 20.3°C and western Freckled Nightjars, HI = 7.7°C) both occur in desert habitats. Possible correlations between heterothermy and factors such as gender and foraging mode were reviewed by Brigham et al. (2006), with no clear patterns evident.

The Common Poorwill remains the only species known to employ both daily and seasonal bouts of heterothermy (i.e. short bouts of torpor as well as hibernation (Brigham 1992; Woods 2002; Woods and Brigham 2004), and $T_b < 5$°C has been reported in this species, the lowest for any bird. Jaeger's (1948, 1949) early contention that this species is the only bird capable of hibernation has thus been supported by subsequent work (French 1993; Woods and Brigham 2004).

In addition to variation in the depth, duration and frequency of heterothermy, the timing of bouts also varies considerably among species. For example, Australian owlet-nightjars in New South Wales, Australia use torpor mostly during the morning (rest phase), whereas Tawny Frogmouths at the same site most commonly enter torpor at night (active phase) after an evening foraging bout (Brigham et al. 2000; Körtner et al. 2000, 2001). There is evidence that some species employ passive re-warming near midday, and may re-enter torpor in the afternoon prior to activity (Doucette et al. 2012; Körtner et al. 2001; McKechnie et al. 2007; Woods 2002).

Heterothermy appears to be used by both sexes in most species, although the data available to directly evaluate sex differences are extremely limited. Data for Common Poorwills, Whippoorwills and Australian owlet-nightjars during the reproductive season suggest that heterothermic responses, while uncommon, do nevertheless occasionally occur (Kissner and Brigham 1993; Csada and Brigham 1994; Brigham et al. 2000; Woods 2002; Lane et al. 2004). However, the reproductive period typically encompasses the warmest parts of the year when insect prey tends to be most common, and furthermore most studies have been conducted during periods outside the reproductive season. It is commonly argued that the decrease in T_b that accompanies reduced metabolic rates is incompatible with incubation and brooding, but this has not been evaluated directly.

A priori, T_a would seem a likely candidate variable to predict the occurrence of heterothermy, since the availability of aerial insects is strongly temperature-dependent (Ashdown and McKechnie 2008; Rydell et al. 1996). Whereas some published studies have suggested there is an association between T_a and the occurrence of heterothermy in caprimulgids (Brigham 1992; Csada and Brigham 1994;

Körtner et al. 2001), other studies suggest the opposite. For instance, heterothermy in Freckled Nightjars at an arid winter-rainfall site in western South Africa was much more strongly correlated with lunar cycles (i.e. light availability) than T_a, although the latter was nevertheless a significant predictor of heterothermy in a colder, drier year (Smit et al. 2011). On the other hand, several studies have revealed considerable variation among individuals at any given time at a particular site, with some individuals entering deep torpor while others remain normothermic. Thus, all that can reliably be concluded from the data currently available is that the ecological determinants of heterothermy vary across taxa.

16.3.2 Intraspecific Variation

All the caprimulgid species discussed above, with the exceptions of the European Nightjar and Tawny Frogmouth, have been investigated at more than one location (Table 16.3), and variation in the environmental correlates and patterns of heterothermy within species has the potential to shed considerable light on the proximate and ultimate determinants of these responses.

In Common Nighthawks and Whippoorwills, some populations exhibited heterothermy whereas others did not (Table 16.3). In the remaining species, however, populations differed in the type and/or environmental correlates of heterothermy. In South Africa, Freckled Nightjars at a winter-rainfall site exhibited patterns of winter heterothermy that were strongly correlated with lunar phase. As the lunar cycle progressed towards full moon, heterothermy became less pronounced, and then increased in frequency and depth as nights became progressively darker around new moon (Smit et al. 2011). Moreover, entry into and re-warming from torpor was tightly linked to moonset and moonrise respectively (Smit et al. 2011). In contrast, conspecifics at a cooler, summer-rainfall site (Rustenburg) did not show any obvious relationship between winter heterothermy and lunar phase, and moreover heterothermy was exhibited by some individuals but not others (McKechnie et al. 2007). In spite of substantial evidence that owlet-nightjars, poorwills and Whippoorwills are able to increase activity (and presumably foraging) during periods of the lunar cycle with more light (Brigham and Barclay 1992; Brigham et al. 1999, Mills 1986; Woods and Brigham 2008), there is no clear evidence that any of these three species adjust heterothermic responses in response to lunar conditions.

Common Poorwills investigated at four sites exhibited broadly similar patterns of heterothermy (Table 16.3). Data collected from spring through autumn for a considerable number of individuals suggest that heterothermic responses are commonly employed by both sexes after foraging bouts at dusk (except during the breeding season), and with no obvious effect of T_a, insect availability or the lunar cycle (Brigham 1992; Csada and Brigham 1994; Woods 2002; Woods and Brigham 2004). Prolonged, multi-day torpor has been recorded during winter only at a site north of Tucson. The latter population is the only non-migratory

Table 16.3 Summary of intraspecific variation in heterothermy in caprimulgid birds

Species	Sites	Migrant?	Variation in heterothermy
Common Nighthawk	Okanagan Valley	Migrant	No evidence of heterothermy, despite poor conditions and some individuals starving to death[a]
	Cypress Hills	Migrant	Short, shallow torpor bouts[b]
Whippoorwill	Ontario	Migrant	No evidence of heterothermy[c]
	South Dakota	Migrant	Occasional torpor bouts[d]
Freckled Nightjar	Rustenburg	Non-migrant	Frequent torpor by some individuals but not others during winter; no obvious relationship to lunar cycles[e]
	Namaqualand	Non-migrant	Frequent torpor by all individuals during winter; strongly related to lunar cycles[f]
Common Poorwill	Okanagan Valley	Migrant	Frequent deep torpor (typically <8 h duration) during summer[g]
	Cypress Hills	Migrant	Frequent deep torpor (typically <8 h duration) during summer[h]
	30 km N of Tucson	Non-migrant	Frequent deep torpor (typically <8 h duration) during summer, hibernation during winter[i,j]
	100 km SE of Tucson	Migrant	Frequent deep torpor (typically <8 h duration) during summer[i,j]
Australian Owlet-Nightjar	Armidale	Non-migrant	Frequent torpor during winter[k,l]
	Ormiston Gorge	Non-migrant	Frequent torpor during winter; torpor is significantly more frequent in individuals roosting in tree hollows compared to rock crevices[l,m]
	Alice Springs (irrigated site)	Non-migrant	Frequent torpor during winter; torpor is significantly more frequent in individuals roosting in tree hollows compared to rock crevices[l,m]

[a] Firman et al. (1993), [b] Fletcher et al. (2004), [c] Hickey (1993), [d] Lane et al. (2004), [e] McKechnie et al. (2007), [f] Smit et al. (2011), [g] Brigham (1992), [h] Csada and Brigham (1994), [i] Woods (2002), [j] Woods and Brigham (2004), [k] Brigham et al. (2000), [l] Doucette et al. (2012), [m] Doucette et al. (2011)

population so far investigated, raising the question of whether migratory tendency is a factor in the expression of heterothermy. The fact that heterothermic patterns during spring, summer and autumn are similar in migratory populations in Canada and Arizona suggests little influence, but it remains unknown whether, like the non-migratory Arizona population, these birds use multi-day torpor on their wintering grounds.

Perhaps the most instructive data set for a species studied at multiple locations is for the Australian owlet-nightjar, which has been the focus of projects conducted near Armidale in northern New South Wales (Brigham et al. 2000) and at two sites 150 km apart near Alice Springs in the Northern Territory (NT, Doucette et al. 2011; Doucette et al. 2012). One NT site is natural (Ormiston Gorge, MacDonnell Ranges), whereas the other (Alice Springs Desert Park) has been irrigated to promote re-establishment of native vegetation. At all three sites, data are available from more than 1 year for a number of individuals. Individuals from all three populations are sedentary. Ecologically, there are differences between the populations. In NSW, owlet-nightjars roost and nest in the cavities of intact trees and stumps in a small remnant of relatively moist forest. In contrast, at both NT sites, birds use both tree cavities and rock crevices (Doucette et al. 2011). Different individuals seem to have preferences for the type of roost they use with some using only one type and some both (Doucette et al. 2011). Torpor occurs during winter (May–September) at all study sites and the depth, duration and frequency are similar in an overall sense. However, roost site choice by birds in the NT is a highly important predictor of torpor use. Birds roosting in rock crevices are buffered from fluctuations in T_a and use torpor much less (about half as often) than birds roosting in tree cavities which were less buffered from daily variation in T_a. Perhaps most interesting, birds roosted in tree cavities on about 2/3rds of tracking days, which may imply a benefit of tree roosts where torpor is used more. Doucette et al. (2011) suggested that a reduced risk of predation and lower costs of arousal because of the ability to passively re-warm by birds roosting in trees may be among the reasons. Doucette et al. (2012) noted that the three habitats of their study differed in annual T_a, rainfall and arthropod abundance. They found that torpor frequency, depth and duration were greatest during years and at sites with lower arthropod abundance, although there was a strong correlation between arthropod abundance and environmental variables. These data provide evidence that direct measures of food availability can be used to predict heterothermic responses.

16.4 Emerging Questions and Future Directions

The steadily increasing body of data on caprimulgid heterothermy has provided key insights into the ecological significance of this phenomenon, both within this taxon and in endotherms in general. These data, however, also raise numerous questions regarding the ecological determinants of these responses. Why, for instance, do many caprimulgids exhibit only shallow heterothermy and short bout lengths? Some species, such as Australian Owlet-Nightjars and Tawny Frogmouths, exhibit "W"-shaped patterns of T_b consisting of two short bouts interrupted by a re-warming phase, a pattern that is presumably less energetically efficient than a single, longer bout. Similarly, variation between and within species in patterns of heterothermy remains poorly understood—why do some individuals at a particular site enter deep torpor while others remain normothermic?

We would argue that one of the most productive future lines of research on caprimulgid heterothermy would involve experimental manipulations of energy balance. Caprimulgids are not generally suitable subjects for laboratory studies, and the focus would need to be on free-ranging individuals. Energy supply can be experimentally increased using artificial light; Woods and Brigham (2004) showed that patches of high insect density created by erecting lights in the territories of poorwills led to changes in patterns of heterothermy. This approach could readily be used for many species to explore the role of prey availability on patterns of heterothermy. Artificially increasing energy demand is more challenging, but could be achieved by increasing heat loss rates via removal of patches of feathers, or by increasing costs of locomotion via the clipping of flight feathers (e.g. Tieleman et al. 2008) or the attachment of additional mass (e.g. Hughes and Rayner 1991). Such studies could greatly increase our understanding of the factors responsible for variation in heterothermy among individuals within a species.

A key gap in our current understanding of heterothermy in caprimulgids concerns the energetic consequences of heterothermy; while there is little question that heterothermy reduces energy and water demands, the savings associated with these responses under natural conditions have yet to be quantified. Caprimulgids may prove relatively tractable subjects for measurements of energy expenditure using doubly labeled water—they are generally territorial, respond readily to play-back and can be caught using a variety of methods, factors that make it easier to re-capture birds after 24 or 48 h (Speakman 1997). Alternatively, heart rate telemetry may also prove a feasible approach for quantifying energy savings associated with heterothermy, although this method requires careful laboratory calibrations of heart rate—oxygen consumption relationships (Butler et al. 2004). Measurements of metabolic rate and water turnover will provide vital information for understanding when and why individuals use heterothermy.

While descriptive studies may not appear to be particularly imaginative, investigations of the occurrence of heterothermy in more caprimulgid species, particularly in tropical regions (such as South America and southeast Asia) have the potential to significantly improve our understanding of ecological determinants of heterothermy in this group, particularly if they involve species that are carefully selected on the basis of their habitat characteristics. Furthermore, such studies will be vital for answering the interesting question of whether the Common Poorwill is indeed the only avian hibernator. More work is also required during reproductive periods, as well as establishing when heterothermic responses first appear in young birds, particularly since foraging efficiency is often lower in juvenile birds compared with adults (Weathers and Sullivan 1989) and juvenile caprimulgids might a priori be expected to experience more severe energy stress.

Finally, a key set of questions that need to be answered before we can gain a better understanding of caprimulgid heterothermy concerns the costs of these responses. Although increased vulnerability to predation is widely accepted as being one of the key ecological costs of heterothermy, at least for animals occupying exposed roost sites, virtually no data other than anecdotes are available. Studies comparing heterothermy among caprimulgid populations that vary in predation risk and/or experimental manipulations of perceived predation risk could provide key insights into ecological determinants of these responses.

Acknowledgments We thank Gerhard Körtner, Jeff Lane and Quinn Fletcher for providing us with raw data for HI analyses.

References

Ashdown RAM, McKechnie AE (2008) Environmental correlates of Freckled Nightjar activity in a seasonal, subtropical habitat. J Orn 149:615–619

Blomberg SP, Garland T, Ives AR (2003) Testing for phylogenetic signal in comparative data: behavioral traits are more labile. Evolution 57:717–745

Boyles JG, Smit B, McKechnie AE (2011) A new comparative metric for estimating heterothermy in endotherms. Physiol Biochem Zool 84:115–123

Brigham RM (1992) Daily torpor in a free-ranging goatsucker, the common poorwill (*Phalaenoptilus nuttallii*). Physiol Zool 65:457–472

Brigham RM, Barclay RMR (1992) Lunar influence on foraging and nesting activity of common poorwills (*Phalaenoptilus nuttallii*). Auk 109:315–320

Brigham RM, Gutsell RCA, Wiacek RS, Geiser F (1999) Foraging behavior in relation to the lunar cycle by Australian Owlet-nightjars (*Aegotheles cristatus*). Emu 99:253–261

Brigham RM, Körtner G, Maddocks TA, Geiser F (2000) Seasonal use of torpor by free-ranging Australian owlet-nightjars (*Aegotheles cristatus*). Physiol Biochem Zool 73:613–620

Brigham RM, Woods CP, Lane JE, Fletcher QE, Geiser F (2006) Ecological correlates of torpor use among five caprimulgiform birds. Acta Zool Sinica 52 (suppl):401–404

Butler PJ, Green JA, Boyd IL, Speakman JR (2004) Measuring metabolic rate in the field: the pros and cons of the doubly labelled water and heart rate methods. Funct Ecol 18:168–183

Csada RD, Brigham RM (1994) Reproduction constrains the use of daily torpor by free-ranging common poorwills (*Phalaenoptilus nuttallii*) (Aves: Caprimulgidae). J Zool Lond 234:209–216

Doucette LI, Brigham RM, Pavey CR, Geiser F (2011) Roost type influences torpor use by Australian owlet-nightjars. Naturwissenschaften 98:845–854

Doucette LI, Brigham RM, Pavey CR, Geiser F (2012) Prey availability affects daily torpor by free-ranging Australian owlet-nightjars (*Aegotheles cristatus*). Oecologia. doi:10.1007/s00442-011-2214-7

Emberger L (1955) Afrique du Nord-Ouest. In: UNESCO (ed) Plant ecology: reviews of research. UNESCO, Paris, pp 219–249

Firman MC, Brigham RM, Barclay RMR (1993) Do free-ranging common nighthawks enter torpor? Condor 95:157–162

Fletcher QE, Fisher RJ, Willis CKR, Brigham RM (2004) Free-ranging common nighthawks use torpor. J Therm Biol 29:9–14

French AR (1993) Hibernation in birds: comparisons with mammals. In: Carey C, Florant GL, Wunder BA, Horwitz B (eds) Life in the cold. Westview Press, Boulder, pp 43–53

Geiser F, Ruf T (1995) Hibernation versus daily torpor in mammals and birds: physiological variables and classification of torpor patterns. Physiol Zool 68:935–966

Hackett SJ, Kimball RT, Reddy S, Bowie RCK, Braun EL, Braun MJ, Chojnowski JL, Cox WA, Han K-L, Harshman J, Huddleston CJ, Marks BD, Miglia KJ, Moore WS, Sheldon FH, Steadman DW, Witt CC, Yuri T (2008) A phylogenomic study of birds reveals their evolutionary history. Science 320:1763–1768

Hickey MBC (1993) Thermoregulation in free-ranging whip-poor-wills. Condor 95:744–747

Holyoak DT (2001) Nightjars and their allies: the Caprimulgiformes. Oxford University Press, Oxford

Hughes PM, Rayner JMV (1991) Addition of artificial loads to long-eared bats *Plecotus auritus*: handicapping flight performance. J Exp Biol 161:285–298

Jaeger EC (1948) Does the poor-will hibernate? Condor 50:45–46

Jaeger EC (1949) Further observations on the hibernation of the poor-will. Condor 51:105–109

Jetz W, Steffen J, Linsenmair KE (2003) Effects of light and prey availability on nocturnal, lunar and seasonal activity of tropical nightjars. Oikos 103:627–639

Körtner G, Brigham RM, Geiser F (2000) Winter torpor in a large bird. Nature 407:318

Körtner G, Brigham RM, Geiser F (2001) Torpor in free-ranging tawny frogmouths (*Podargus strigoides*). Physiol Biochem Zool 74:789–797

Lane JE, Brigham RM, Swanson DL (2004) Daily torpor in free-ranging whip-poor-wills (*Caprimulgus vociferus*). Physiol Biochem Zool 77:297–304

Mayr G (2002) Osteological evidence for paraphyly of the avian order Caprimulgiformes (nightjars and allies). J Orn 143:82–97

McKechnie AE, Ashdown RAM, Christian MB, Brigham RM (2007) Torpor in an Afrotropical caprimulgid, the Freckled Nightjar (*Caprimulgus tristigma*). J Avian Biol 38:261–266

McKechnie AE, Lovegrove BG (2002) Avian facultative hypothermic responses: a review. Condor 104:705–724

McKechnie AE, Mzilikazi N (2011) Heterothermy in Afrotropical birds and mammals: a review. Int Comp Biol 51:349–363

Mills AM (1986) The influence of moonlight on the behavior of goatsuckers (Caprimulgidae). Auk 103:370–378

Peiponen VA (1965) On hypothermia and torpidity in the nightjar (*Caprimulgus europaeus* L.). Ann Acad Sci Fenn (Ser A IV. Biol) 87:1–15

Peiponen VA (1966) The diurnal heterothermy of the nightjar (*Caprimulgus europaeus* L.). Ann Acad Sci Fenn (Ser A IV. Biol) 101:1–35

Prinzinger R, Preßmar A, Schleucher E (1991) Body temperature in birds. Comp Biochem Physiol 99A:499–506

Rydell J, Entwistle A, Racey PA (1996) Timing of foraging flights of three species of bats in relation to insect activity and predation risk. Oikos 76:243–252

Sibley CG, Ahlquist JE (1990) Phylogeny and classification of birds. Yale University Press, New Haven

Smit B, Boyles JG, Brigham RM, McKechnie AE (2011) Torpor in dark times: patterns of heterothermy are associated with the lunar cycle in a nocturnal bird. J Biol Rhythms 26:241–248

Speakman JR (1997) Doubly labelled water: theory and practise. Chapman & Hall, New York

Tieleman BI, Dijkstra TH, Klasing KC, Visser GH, Williams JB (2008) Effects of experimentally increased costs of activity during reproduction on parental investment and self-maintenance in tropical house wrens. Behav Ecol 19:949–959

Tieleman BI, Williams JB, Bloomer P (2003) Adaptation of metabolic rate and evaporative water loss along an aridity gradient. Proc R Soc B 270:207–214

Weathers WW, Sullivan KA (1989) Juvenile forgaging proficiency, parental effort, and avian reproductive success. Ecol Monogr 59:223–246

Woods CP (2002) Ecological aspects of torpor use and inactivity during winter by common poorwills. Ph.D. thesis, University of Regina, Regina

Woods CP, Brigham RM (2004) The avian enigma: "hibernation" by common poorwills (*Phalaenoptilus nuttalli*). In: Barnes BM, Carey HV (eds) Life in the cold: evolution, mechanisms, adaptation and application 12th International Hibernation Symposium, pp 129–138

Woods CP, Brigham RM (2008) Common poorwill activity and calling behavior in relation to moonlight and predation. Wilson J Ornithol 120:505–512

Part II
Physiology of the Torpid State

Chapter 17
The Brain at Low Temperature: Tau Hyperphosphorylation Dynamics in Hibernation Torpor

A. S. Boerema, J. N. Keijser, H. R. Bouma, E. A. van der Zee and A. M. Strijkstra

Abstract Mammalian torpor is associated with neuronal tau protein hyperphosphorylation. This process is fully reversed upon rewarming to euthermy. Not much is known about the hyperphosphorylation dynamics during cooling and rewarming to and from torpor. In this study we show that there is a negative relation between brain temperature and the amount of tau hyperphosphorylation in the cortex of Syrian hamsters. This relation was found to be nonlinear: the fastest changes in the hyperphosphorylation state of the tau protein occurred around brain temperatures of ~27°C. The amount of hyperphosphorylation did not substantially increase further with the time spent in torpor. In mice, reversible hyperphosphorylation could also be detected during torpor-like hypothermia at 21°C, but was not present during torpor-like hypothermia at an ambient temperature of 30°C. These results suggest that tau hyperphosphorylation is not only passively connected with brain temperature, but is actively regulated. The results argue against a need for periodic euthermy to reverse hyperphosphorylation of tau in the

A. S. Boerema · H. R. Bouma · A. M. Strijkstra
Unit of Chronobiology, Centre for Behaviour and Neuroscience (CBN),
University of Groningen, Groningen, The Netherlands

A. S. Boerema · J. N. Keijser · E. A. van der Zee
Unit of Molecular Neurobiology, Centre for Behaviour and Neuroscience (CBN),
University of Groningen, Groningen, The Netherlands

H. R. Bouma · A. M. Strijkstra
Department of Clinical Pharmacology, University Medical Centre Groningen,
Groningen, The Netherlands

Present Address:
A. S. Boerema (✉)
Department of Nuclear Medicine and Molecular Imaging,
University Medical Centre Groningen, Groningen, The Netherlands
e mail: a.s.boerema@umcg.nl

T. Ruf et al. (eds.), *Living in a Seasonal World*,
DOI: 10.1007/978-3-642-28678-0_17,

brain. Alternatively we hypothesise that the phosphorylation of the tau protein may play an active role in the regulation of the neuronal metabolism, facilitating the entrance and or maintenance of the torpid state.

17.1 Hibernation and Torpor in Mammals

Hibernation is an overwintering behaviour observed in many mammalian species (Heldmaier et al. 2004). During hibernation and torpor mammals are subject to the most extreme fluctuations in metabolism and body temperature (T_b) observed in homeotherms. During torpor metabolic rate can drop to values as low as 1–2% of the basal metabolic rate (BMR) (Lyman 1948; Heldmaier and Ruf 1992), whereas during rewarming from torpor the metabolic rate is much higher than the BMR (see also Fig. 17.1). In deep hibernation T_b ranges between euthermic (36–38°C) values and values of 1–2°C above the ambient temperature (Lyman and O'Brien 1974; Heldmaier and Ruf 1992). In mammals, even T_bs below 0°C have been measured in Arctic ground squirrels (Barnes 1989) and in European ground squirrels (Strijkstra 1999, p 102), although thorax and head temperatures did not show sub-zero temperatures. During rewarming, hibernators seem to prioritize rewarming the brain. There are big differences between body and brain temperature in arousal (Fig. 17.1), which have to be taken into account when analysing temperature-dependent brain parameters.

17.2 Tau Protein Hyperphosphorylation Dynamics in Mammalian Torpor

Neuronal protein hyperphosphorylation is mainly known for its association with pathological and neurodegenerative phenomena, such as Alzheimer's disease (AD) in humans. In torpor tau hyperphosphorylation also occurs, but it is fully reversed upon rewarming to euthermic conditions. So far, tau hyperphosphorylation has been observed in Syrian hamsters and ground squirrels during deep torpor (Arendt et al. 2003; Härtig et al. 2007; Su et al. 2008; Stieler et al. 2011), in black bears during continuous torpor (Stieler et al. 2011) and in Djungarian hamsters during daily torpor (Boerema et al. 2008). Torpor can also be induced in mice. When mice have to work hard for their food, they show daily torpor (Schubert et al. 2010; Hut et al. 2011). This workload induced torpor in mice is also caused by the food shortage (Boerema 2012). A torpor-like hypometabolic state can also be induced in mice pharmacologically by injection of 5′-adenosine-monophosphate (5′AMP) (Zhang et al. 2006; Swoap et al. 2007). Both workload induced torpor and a 5′AMP induced torpor-like hypo metabolic state in mice are also associated

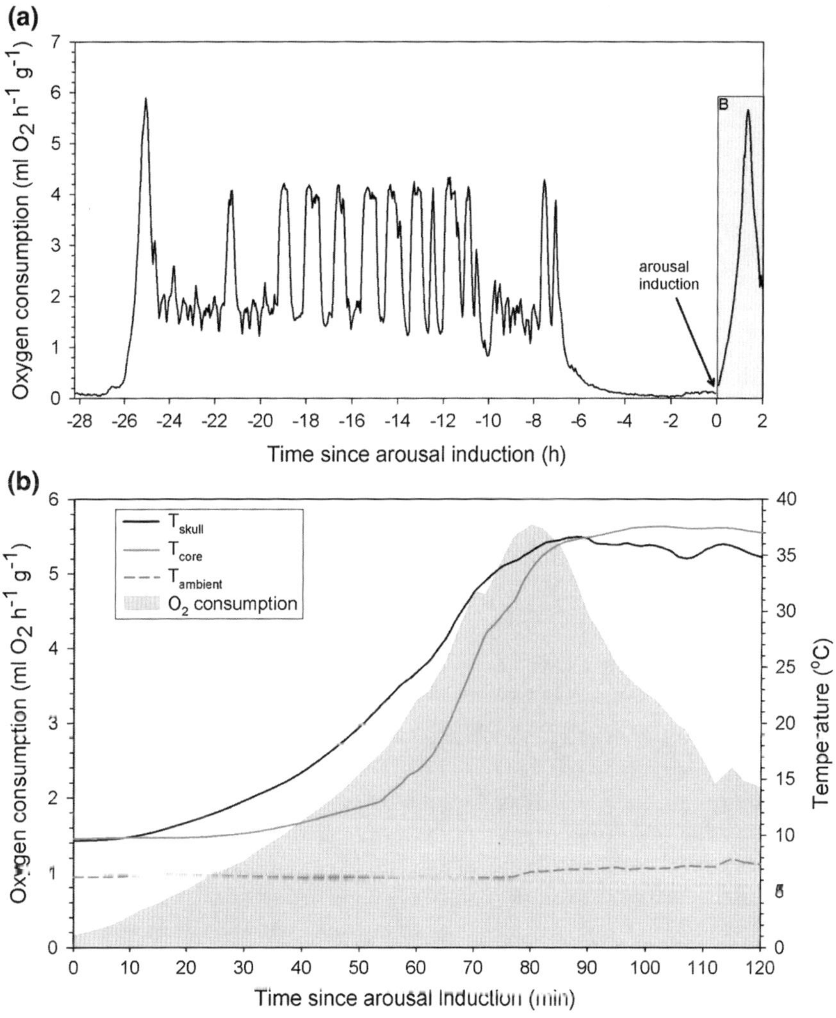

Fig. 17.1 Example of the oxygen consumption of a Syrian hamster during torpor and arousal from torpor (**a**). The second arousal is shown in detail, together with thermocouple recordings of skull temperature (T_{skull}; *black line*), rectal body temperature (T_{core}; *grey line*) and ambient temperature (T_{ambient}; *dashed line*) in (**b**). The filled *grey area* is the oxygen consumption as measured by indirect calorimetry in an open flow system. Body temperature curves were smoothed by a running mean with a window of 10 min prior to plotting

with tau protein hyperphosphorylation (Boerema 2012). In mice the tau hyperphosphorylation is also temperature dependent. We detected reversible tau hyperphosphorylation in mice during 5′AMP induced torpor-like hypothermia at an ambient temperature of 21°C, but it was not present in 5′AMP induced torpor-like hypothermia at an ambient temperature of 30°C (Fig. 17.2).

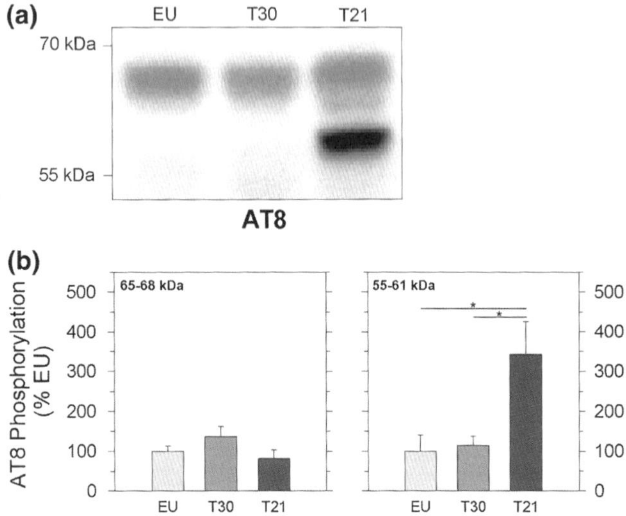

Fig. 17.2 **a** Shows representative examples of mice in all 3 groups, stained for tau hyperphosphorylation with the AT8 antibody. **b** Shows the optical density (OD) analysis of the hyperphosphorylation of the tau protein in two molecular weight ranges: 65–68 kDa (**b**, *left*) and 55–61 kDa (**b**, *right*). Average tau hyperphosphorylation with s.e.m. is plotted. Prior to averaging, OD's were corrected for the amount of protein loaded by using the β-tubulin-3 signal and corrected for the amount of total tau protein present within each molecular weight range. The T_{21} mice show a significantly higher expression of AT8 hyperphosphorylation in the 55–61 kDa weight range (EU vs. T_{21}: $q = 6.136$, $p < 0.001$; T_{21} vs. T_{30}: $q = 5.380$, $p < 0.001$)

In Syrian hamsters we further explored hyperphosphorylation dynamics associated with the transitions from cold to warm, and from warm to cold, during hibernation. We measured body and (estimates of) brain temperature during cooling and rewarming from torpor. We sampled hamsters at a range of brain temperatures during cooling and rewarming and analysed the brains for tau hyperphosphorylation. The fastest changes in tau hyperphosphorylation load in the cortex occurred around a brain temperature of ~27°C, during both cooling and rewarming (Fig. 17.3). After the initial increase around ~27°C the total amount of hyperphosphorylation only marginally increased with time spent in torpor, which was also observed by (Stieler et al. 2011). This makes it unlikely that the degree of hyperphosphorylation represents an accumulating cost of torpidity provoking a need for de-phosphorylation in periodic euthermy.

17.3 A Role for Tau Hyperphosphorylation in Torpor Physiology?

One of the characteristics of AD is a reduced glucose uptake in the brain (Jagust et al. 1991). Altered glucose metabolism leading to hypothermia is also associated with tau hyperphosphorylation in mice (Planel et al. 2004). Gong et al. (2006) have

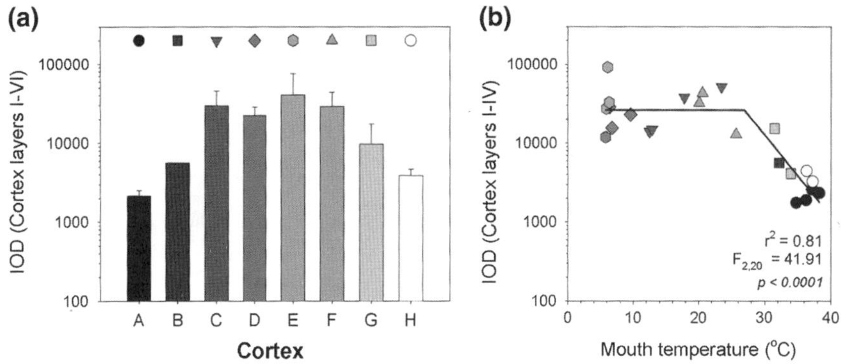

Fig. 17.3 Integrated optical density (IOD) measurements of the staining for hyperphosphorylated tau with the AT8 antibody and the relation with mouth temperature in layers I–VI of the cortex. The *left* shows an overview of the average IOD for each experimental group with the standard deviation of the mean. The letters below the bars correspond to the experimental groups. The symbols correspond to the symbols for the individual data points in the right panel. The *right* shows the measured IOD for each individual, plotted against mouth temperature during sampling. The *fitted curve* represents the outcome of the two phase regression model

recently offered a novel hypothesis on the relation between glucose metabolism, tau hyperphosphorylation and neurofibrillary degeneration as observed in AD. These authors hypothesised that decreased glucose metabolism leads to decreased tau *O*-GlcNAcylation, which increases the hyperphosphorylation of tau. Protein *O*-GlcNAcylation is a form of protein Glycosylation, similar to phosphorylation and abundant in eukaryotic organisms (Hart 1997). It is an important regulatory process. Modifications in the gene coding for the *O*-GlcNAc transferase enzyme are incompatible with life (Shafi et al. 2000).

The tau protein is known to be subject to both *O*-GlcNAcylation and phosphorylation. Increased *O*-GlcNAcylation negatively affects the amount of tau hyperphosphorylation and vice versa (Liu et al. 2004). The production of the enzyme necessary for protein *O*-GlcNAcylation is directly dependant on the glucose available (Gong et al. 2006), providing a direct link between hyperphosphorylation and metabolism.

Because of reduced transcription and translation during torpor (van Breukelen and Martin 2001, 2002), post-translational protein modifications are important for physiological regulation in torpor. Apart from phosphorylation, several other forms of post-translational regulation, such as protein SUMOylation (Lee et al. 2007), Ubiquitylation and Acetylation (Storey 2010) have been measured in torpor. Torpor entrance is a well-orchestrated cell biological process (Carey et al. 2003). In the brain, however, torpor entry may cause a discrepancy between energy production and metabolic demands of neurons. Metabolic rate decreases rapidly at torpor entry, but T_b goes down much more slowly, by passive cooling. The electrical activity of cortical neurons is associated with the T_b. Frequencies in the cortical EEG slow down with a constant Q_{10} of ~2.5 (Deboer et al. 1994;

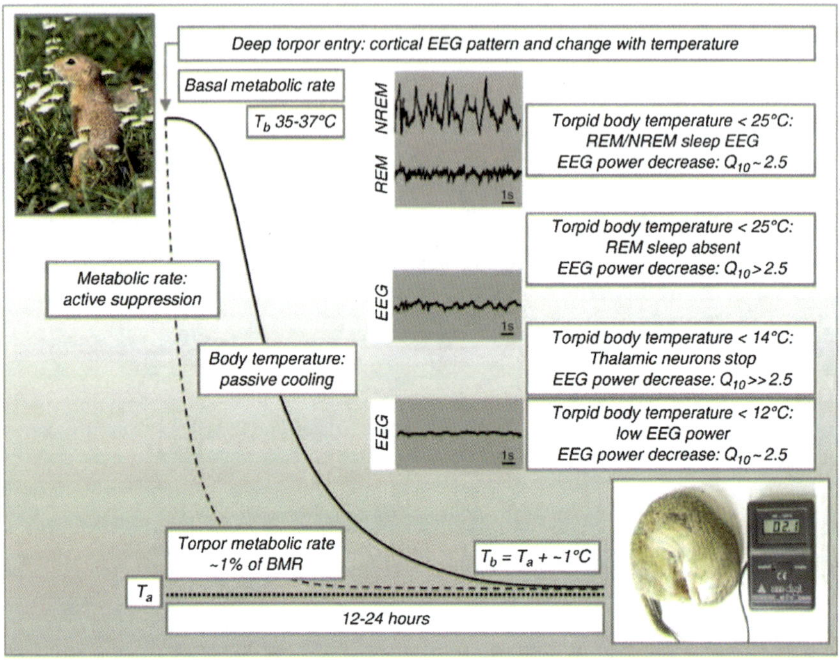

Fig. 17.4 Schematic representation of deep torpor entry from an euthermic T_b to a deep torpor T_b just over ambient temperature (T_a) in the European ground squirrel. Active metabolic suppression (*dotted line*) from basal metabolic rates at euthermia is followed by a passive T_b decrease (*solid line*). The 10 s EEG traces initially resemble Non-REM and REM-sleep; REM-sleep is lost below 25°C. At lower temperatures the EEG shows progressive changes in power with different rates, indicating regulated activity changes ($Q_{10} > 2$–3) and passive changes ($Q_{10} \sim 2$–3), ultimately resembling a flat line below ~10°C. Figure adapted from (Strijkstra 2009, p. 1834)

Deboer and Tobler 1995; Strijkstra et al. 2000). The power of the EEG signal shows stepwise changes with significant deviations from a Q_{10} of ~2.5 around 25 and 14°C (Strijkstra 2006). These steps can be associated with the disappearance of REM-sleep around 25°C and the silencing of thalamic neurons around 14°C (Krilowicz et al. 1988). The effects are summarised in Fig. 17.4.

The discrepancy between glucose availability and the glucose demands of electrically active neurons may result in low brain glucose levels early during torpor entry. This could affect the balance between phosphorylation and *O*-GlcNAcylation of the tau protein, tilting the balance in favour of phosphorylation. The temperature of ~27°C at which the largest changes in tau hyperphosphorylation occur during torpor entry in hamsters fit this view. The apparent cholinergic specificity of tau hyperphosphorylation in the basal forebrain in hamsters (Härtig et al. 2007) also fits in. The cholinergic neuron associated REM-sleep is one of the first electrical characteristics to disappear from the EEG during cooling, between 30 and 25°C (Deboer et al. 1994; Strijkstra 2006).

Hyperphosphorylation of tau negatively affects the binding capacity of tau to microtubules and thereby also negatively influences axonal vesicle transport (reviewed in: Johnson and Stoothoff 2004). This reduces the capability of neurons to perform neurotransmission. In this scenario, tau hyperphosphorylation could play an active role in early torpor entry by reducing neurotransmission capability of cholinergic neurons, effectively decreasing the metabolic demands of the brain.

Instead of leading to arousals, tau hyperphosphorylation together with tau *O*-GlcNAcylation may therefore have a function in the entrance and/or maintenance of the state of torpor in mammalian hibernation.

17.4 Brief Methods

Torpor-like hypothermia was induced by ip injection of 7.5 μmol 5′AMP/g body mass in male C57BL/6 J mice. Brains were collected during topor-like hypothermia at an ambient temperature of 21°C (T_{21}; $n = 5$) and 30°C (T_{30}; $n = 5$) or in euthermia following 5′AMP induced torpor-like hypothermia at 21°C (EU; $n = 5$). Cortex lysates were prepared as described previously (Boerema et al. 2008). Proteins were separated on 4–20% gradient Nu Page Novex Bis–Tris mini gels (Invitrogen, Breda, the Netherlands) using 57 μg of total protein per well and were subsequently transferred to a 0.2 μm PVDF-membrane, using the I-blot dry blotting system (Invitrogen, Breda, the Netherlands). Immunodetection was performed by overnight incubation at 5°C, with the AT8 antibody (MN-1020; Pierce; 1:1000). Detection of bound primary antibodies was performed with (SC-2005; Santa Cruz; 1:10000). Detection of bound secondary antibody was done by ECL, using Super Signal West Dura (Pierce, Rockford, IL, USA) diluted 1:1 in TBS (pH 8.0) in a Biorad Chemidoc XRS + imaging system.

Brains of 23 Syrian hamsters were sampled during summer euthermy (A; $n = 4$), during torpor entry with a brain temperature above 30°C (B; $n = 1$) and below 30°C (C; $n = 4$), after 24–30 h of torpor (D; $n = 3$), late in torpor after >84 h of torpor (E; $n = 4$), early during arousal from torpor with a brain temperature below 30°C (F; $n = 3$), late during arousal from torpor with a brain temperature above 30°C (G; $n = 2$) and late (>8.5 h) during periodic euthermy (H; $n = 2$). All hamsters were sampled based on measurements of body, brain and/or skull temperature obtained with a combination of radio telemetry (T_b) and thermocouples (brain and skull temperature). Mouth temperature taken at the moment of sacrifice confirmed the correct state. The AT8 staining was performed according to the methods described by (Härtig et al. 2007) with a primary antibody dilution of 1:100. OD was measured by placing a region of interest (ROI) on the slice. Each OD was corrected for non-specific background staining, by measuring an area in the corpus callosum in the same slice and subtracting the OD from this area from the ROI. The Leica Qwin software (Leica, Cambrigde, UK) was used to analyse the OD in the regions of interest. Staining intensity was determined, using an analysis macro specifically detecting lineated structures. In this procedure both the coverage area and the OD contain relevant information; therefore the AT8 staining levels were expressed using IOD. The IOD was multiplied with the percentage coverage of the ROI detected by the macro to correct for the size of the complete ROI.

The relation between mouth temperature and the intensity of the AT8 staining was analysed by Continuous two phase regression, a method for detecting physiological thresholds, or stepwise changes in physiological parameters (Nickerson et al. 1989). Detailed methods are described in (Boerema 2012).

References

Arendt T, Stieler J, Strijkstra AM, Hut RA, Rudiger J, Van der Zee EA, Harkany T, Holzer M, Hartig W (2003) Reversible paired helical filament-like phosphorylation of tau is an adaptive process associated with neuronal plasticity in hibernating animals. J Neurosci 23:6972–6981

Barnes BM (1989) Freeze avoidance in a mammal: body temperatures below 0 degree C in an Arctic hibernator. Science 244:1593–1595

Boerema AS (2012) The brain at low temperature: neuronal and behavioural dynamics in mammalian hibernation and torpor. PhD thesis, University of Groningen

Boerema AS, Steinlechner S, Van der Zee EA, Keijser JN, Stieler J, Strijkstra AM (2008) Reversible hyperphosphorylation of the microtubule-associated protein tau during daily torpordaily torpor in Djungarian hamsters. In: Lovegrove BG, McKechnie AE (eds) Hypometabolism in animals: hibernation, torpor and cryobiology. Pietermartizburg, South Africa, pp 151–156

van Breukelen F, Martin SL (2001) Translational initiation is uncoupled from elongation at 18 degrees C during mammalian hibernation. Am J Physiol Regul Integr Comp Physiol 281:R1374–R1379

van Breukelen F, Martin SL (2002) Reversible depression of transcription during hibernation. J. Comp. Physiol. B. Biochem Syst Environ Physiol 172:355–361

Carey HV, Andrews MT, Martin SL (2003) Mammalian hibernation: cellular and molecular responses to depressed metabolism and low temperature. Physiol Rev 83:1153–1181

Deboer T, Franken P, Tobler I (1994) Sleep and cortical temperature in the Djungarian hamster under baseline conditions and after sleep deprivation. J Comp Physiol A 174:145–155

Deboer T, Tobler I (1995) Temperature dependence of EEG frequencies during natural hypothermia. Brain Res 670:153–156

Gong C-X, Liu F, Grundke-Iqbal I, Iqbal K (2006) Impaired brain glucose metabolism leads to Alzheimer neurofibrillary degeneration through a decrease in tau O-GlcNAcylationtau. J Alzheimers Dis 9:1–12

Hart GW (1997) Dynamic O-linked glycosylation of nuclear and cytoskeletal proteins. Annu Rev Biochem 66:315–335

Härtig W, Stieler JT, Boerema AS, Wolf J, Schmidt U, Weissfuss J, Bullmann T, Strijkstra AM, Arendt T (2007) Hibernation model of tau phosphorylation in hamsters: selective vulnerability of cholinergic basal forebrain neurons—implications for Alzheimer's disease. Eur J Neurosci 25:69–80

Heldmaier G, Ortmann S, Elvert R (2004) Natural hypometabolism during hibernation and daily torpordaily torpor in mammals. Respir Physiol Neurobiol 141:317–329

Heldmaier G, Ruf T (1992) Body temperature and metabolic rate during natural hypothermia in endotherms. J Comp Physiol B Biochem Syst Environ Physiol 162:696–706

Hut RA, Pilorz V, Boerema AS, Strijkstra AM, Daan S (2011) Working for food shifts nocturnal mouse activity into the day. PLoS ONE 6:e17527

Jagust WJ, Seab JP, Huesman RH, Valk PE, Mathis CA, Reed BR, Coxson PG, Budinger TF (1991) Diminished glucose transport in Alzheimer's disease: dynamic PET studies. J Cereb Blood Flow Metab 11:323–330

Johnson GVW, Stoothoff WH (2004) Tau phosphorylation in neuronal cell function and dysfunction. J Cell Sci 117:5721–5729

Krilowicz BL, Glotzbach SF, Heller HC (1988) Neuronal activity during sleep and complete bouts of hibernation. Am J Physiol Regul Integr Comp Physiol 255:R1008–R1019

Lee Y-ja, S-ichi Miyake, Wakita H, McMullen DC, Azuma Y, Auh S, Hallenbeck JM (2007) Protein SUMOylation is massively increased in hibernation torpor and is critical for the cytoprotection provided by ischemic preconditioning and hypothermia in SHSY5Y cells. J Cereb Blood Flow Metab 27:950–962

Liu F, Iqbal K, Grundke-Iqbal I, Hart GW, Gong C-X (2004) O-GlcNAcylation regulates phosphorylation of tau: a mechanism involved in Alzheimer's disease. Proc Natl Acad Sci USA 101:10804–10809

Lyman CP (1948) The oxygen consumption and temperature regulation of hibernating hamsters. J Exp Zool 109:55–78

Lyman CP, O'Brien RC (1974) A comparison of temperature regulation in hibernating rodents. Am J Physiol 227:218–223

Nickerson DM, Facey DE, Grossman GD (1989) Estimating physiological thresholds with continuous two-phase regression. Physiol Zool 62:866–887

Planel E, Miyasaka T, Launey T, Chui D-H, Tanemura K, Sato S, Murayama O, Ishiguro K, Tatebayashi Y, Takashima A (2004) Alterations in glucose metabolism induce hypothermia leading to tau hyperphosphorylation through differential inhibition of kinase and phosphatase activities: implications for Alzheimer's disease. J Neurosci 24:2401–2411

Schubert KA, Boerema AS, Vaanholt LM, de Boer SF, Strijkstra AM, Daan S (2010) Daily torpor in mice: high foraging costs trigger energy-saving hypothermia. Biol Lett 6:132–135

Shafi R, Iyer SPN, Ellies LG, O'Donnell N, Marek KW, Chui D, Hart GW, Marth JD (2000) The O-GlcNAc transferase gene resides on the X chromosome and is essential for embryonic stem cell viability and mouse ontogeny. Proc Nat Acad Sci 97:5735–5739

Stieler JT, Bullmann T, Kohl F, Tøien O, Brückner MK, Härtig W, Barnes BM, Arendt T (2011) The physiological link between metabolic rate depression and tau phosphorylation in mammalian hibernation. PLoS ONE 6:e14530

Storey KB (2010) Out cold: biochemical regulation of mammalian hibernation—a mini-review. Gerontology 56:220–230

Strijkstra AM (1999) Periodic euthermy during hibernation in the European ground squirrel: causes and consequences. PhD Thesis, University of Groningen

Strijkstra AM (2006) Good and bad in the hibernating brain. J British Interplanet Soc 59:119–123

Strijkstra AM (2009) Hibernation in Encyclopedia of neuroscience Berlin. Springer, Heidelberg, pp 1831–1836

Strijkstra AM, de Boer T, Daan S (2000) Slowing of sigma and theta EEG frequencies during entry into natural hypothermia in European ground squirrels. Sleep Wake Res Neth 11:114–119

Su B, Wang X, Drew KL, Perry G, Smith MA, Zhu X (2008) Physiological regulation of tau phosphorylation during hibernation. J Neurochem 105:2098–2108

Swoap SJ, Rathvon M, Gutilla M (2007) AMP does not induce torpor. Am J Physiol Regul Integr Comp Physiol 293:R468–R473

Zhang J, Kaasik K, Blackburn MR, Lee CC (2006) Constant darkness is a circadian metabolic signal in mammals. Nature 439:340–343

Chapter 18
The Hibernation-Related Peptide TSKY Acts as a Neuroprotector in Cultured Pond Snail Neurons

Ludmila I. Kramarova, Natalya A. Ivlicheva, Rustam H. Ziganshin, Alexey A. Andreev and Edith N. Gakhova

Abstract The effect of addition to the incubation/preservation medium of the peptide Thr-Ser-Lys-Tyr (TSKY) isolated earlier from the brain of the ground squirrel *Spermophillus undulatus* on the survival of cultured neurons from the pond snail *Lymnaea stagnalis* was investigated. An increase in cell viability in culture (determined by the trypan blue) was observed after the treatment of the brain with TSKY at different temperatures (22–24°C and 4–6°C). The number of neurons with neurites after the peptide treatment was markedly lower at 22–24°C compared to control. This might indicate that the peptide could decrease the metabolic and functional activity of neurons and decline cell death. Our studies revealed that TSKY did not change the survival of neurons in culture isolated from the pond snail frozen–thawed brain. TSKY is not a cryoprotective factor at extra low temperatures (−196°C) but it exerts a neuroprotective effect at normal (22–24°C) and low (4–6°C) temperatures.

18.1 Introduction

It is generally assumed that extra low temperature and subzero preservation of biological materials is associated with a number of cryoprotectants: dimethyl sulfoxide (DMSO), glycerol, some carbohydrates and salts, alcohols, glycoproteins, etc. They preserve the morphological integrity of cells and tissues during

L. I. Kramarova (✉)
Institute of Theoretical and Experimental Biophysics RAS, Pushchino, Russia
e-mail: luda_kramarova@rambler.ru

N. A. Ivlicheva · A. A. Andreev · E. N. Gakhova
Institute of Cell Biophysics RAS, Pushchino, Russia

R. H. Ziganshin
Shemyakin—Ovchinnikov Institute of Bioorganic Chemistry RAS, Moscow, Russia

T. Ruf et al. (eds.), *Living in a Seasonal World*,
DOI: 10.1007/978-3-642-28678-0_18,

freezing; in some cases, they prevent tissue degradation during thawing (Meryman 2007). Cryoprotectants might trigger the processes that support the viability of cells and tissues. Besides the involvement of nontoxic and effective cryoprotectants, the physiological state of cells to be cryopreserved or transplanted is of paramount importance, and a preconditioning of cells before long-term cryopreservation provides a basis for a successful survival of cells and tissues at low and extra low temperatures (Kramarova et al. 1994; Drew et al. 2001; Terry et al. 2006; Ross and Drew 2006).

Scientists first conceived the idea to use endogenous factors produced in organisms in the state of hibernation for preservation of cells, tissues, and organs in the early 1990s (Chien et al. 1991; Oeltgen et al. 1996; Kramarova et al. 1994). It is well known that hibernators in the state of hibernation are protected from cold, ischemia, muscle diseases and atrophy, various infections, cancer, and even ionizing radiation (Smith and Grenan 1951; Kemper and Ruben 1982; Wang 1988; Drew et al. 2001; Cotton and Harlow 2010). In addition, a number of neuroprotective adaptations in the state of hibernation, such as a decreased metabolism, immunosuppression, hypocoagulation, and increased concentration of antioxidants were described (Drew et al. 2001; Ross and Drew 2006). The nerve tissue of hibernators is more cold-resistant than that of non-hibernators. A method was developed to preserve an isolated brain of adult hibernating animals for about 3 days, after which some of brain areas were proven to maintain structure-specific electrical activity (Pakhotin and Pakhotina 1994). An elevated tolerance to hypoxia and hypoglycemia was demonstrated in hippocampus slices from hibernating ground squirrels (Frerichs and Hallenbeck 1998). It becomes clear that the processes and principles of organization that allow hibernators to survive at subzero body temperatures for days and weeks should be comprehensively studied to find and develop a model system appropriate for short- and long-term cell preservation.

Various well-known peptides have been considered as the major factors in circannual adaptation, seasonal preparation, and the onset of hibernation (Beckman et al. 1981; Wang 1988; Kramarova et al. 2009; Nurnberger et al. 1991). The involvement of endogenous peptides in the control of respiration, thermoregulation, regulation of DNA synthesis, cardiovascular responses, cardioprotection, blood pressure, electrical-related activity, sexual activity, and the interaction of number of peptides with many hormonal systems were studied extensively on hibernating and non-hibernating species, and provided a strong basis for the use of these peptides for the long-term organ and tissue preservation (Chien et al. 1991; Oeltgen et al.1996; Govindaswami et al. 2000; Ross and Drew 2006). It was shown that some opioid peptides (DADLE, morphine, DAMGO, DPDPE, U-69593, and Deltorphine-D_{variant}) would most likely have neuroprotective effects, including low-temperature-induced cell death (Borlongan et al. 2000; Tamura et al. 2006; Govindaswami et al. 2008).

In the past three decades, we have isolated and identified new peptides that could induce hibernation-specific effects in homeothermic and heterothermic animals and participate in preparation, induction, or regulation of the state of natural hypobiosis. A search for such peptides in the brain of ground squirrels and Yakutian horses has been performed, and about 30 new peptides were isolated. Some of them were capable of regulating inward potential-dependent calcium currents in cardiac myocytes of rats and ground squirrels and possessed cardiotropic, hypometabolic, and hyperthermic activities (Ziganshin et al. 1994; Kramarova et al. 2009). Recently, state-dependent effects of these peptides on the neuronal activity of some structures in brain slices of the ground squirrel were shown (Zenchenko et al. 2000; Popova et al. 2003).

It has been shown earlier that TSKY is one of the promising peptides that may participate in regulation of many processes. It can enhance the inward potential-dependent Ca^{2+} current in rat cardiac myocytes (Ziganshin et al. 1994), modulate the electrical activity of neurons in rat brain slices (Zenchenko et al. 2000; Popova et al. 2003), decrease the heart rate of chicken embryos (Kramarova et al. 1996), and even is able to delay the arousal of ground squirrels from hibernation (Ignat'ev et al. 2001).

In this study, we investigated the effect of peptide TSKY on cell viability of isolated and cultured neurons from the brain of the pond snail. We studied dose-dependent and temperature-dependent effects of this peptide and its influence on formation of ice microparticles (IMP). We also examined the survival of the whole mollusk brain and the ability of isolated neurons to regenerate neurites in culture after freeze–thawing of the brain in liquid nitrogen.

18.2 Materials and Methods

18.2.1 Neuron Preparation

Fresh-water pond snails (*Lymnaea stagnalis*) were collected near Pushchino, Moscow region, in September–October and maintained in an aquarium at 22–24°C (Fig. 18.1a) where they were able to reproduce. Experiments were carried out during November–March. For all experiments, 3- to 6-month-old animals with a shell length of 30–40 mm were used. The central ring ganglia of the nervous system including buccal ganglia were isolated. To obtain this preparation, animals were first deshelled and placed in anesthetic solution (Bell et al. 2007).

18.2.2 Neuroprotective Experimentation

A pool of isolated ganglia (Fig. 18.1b) was incubated for 60 min in Liebowitz medium (20% L-15, Sigma) with TSKY at different concentrations (1×10^{-5} M, 1×10^{-6} M, and 1×10^{-7} M) and temperatures (22–24 and 4–6°C). The brains incubated in medium without the peptide were used as control. The peptide TSKY

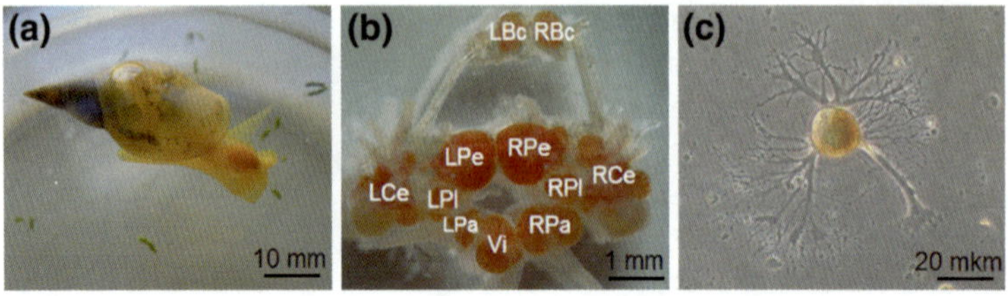

Fig. 18.1 **a** A photo of the pond snail *Lymnaea stagnalis*. **b** The central nerve system of the pond snail with different paired ganglia [right (R) and left (L)]. L(R)BcG—buccal ganglia; L(R)CeG—cerebral ganglia; L(R)PeG—pedal ganglia; L(R)PlG—leural ganglia; L(R)PaG—parietal ganglia; ViG—the visceral ganglion. **c** A single neuron with neurites isolated from paired ganglia after 24 h in culture

was isolated and identified from the brain of ground squirrels (*Spermophillus undulatus*) in the state of hibernation, and then it was synthesized by classical methods in solution (Ziganshin et al. 1994). Dissociated neurons from the mollusk brain were counted, inoculated in culture medium consisting of 20% L-15, gentamicin (20 μg/ml), NaCl (90 mM), KCl (5 mM), $CaCl_2$ (2 mM), $MgCl_2$ (1.5 mM), Tris–HCl (0.25 mM), pH 7.8 at 22–24°C for 24 h, and then counted again (Kostenko et al. 1981). The integrity of cellular membranes was estimated with trypan blue. Neuronal degenerative processes were also evaluated by the early cytoplasmic vacuolization.

18.2.3 Cryopreservation Experimentation

The regeneration ability of neurons was estimated by the formation of neurites in culture after 24 h or after freezing/thawing (500°C/min, in presence of 2 M DMSO) of mollusk ganglia and further culturing of neurons also for 24 h (Fig. 18.1c) (Gakhova et al. 1989; Ivlicheva et al. 2004). We studied the formation of IMP in: (1) culture medium for neurons (NCM) [20% Leibowitz medium, (L-15), Sigma]; (2) NCM + 2 M DMSO; (3) NCM + TSKY (10^{-5} M); (4) NCM + DMSO + TSKY (10^{-5} M). IMP were assessed under a microscope, model Orthoplan (Zeiss, Germany). The Fuchs-Rosenthal chamber was used for sample freezing. The samples under investigation were 10 μl droplets of cryoprotective agents having a standard thickness of 200 μm. The temperature was measured directly in the freezing sample by an ATT-2006 thermometer (Aktakom, Taiwan), with the copper/constantan microthermocouple of diameter 0.1 mm. When the temperature of a sample reached −130°C in liquid nitrogen vapor, it was placed in liquid nitrogen. The formation of IMP was performed by an NV-35 video recorder (NV group, Russia) at the temperatures from +24 to −196°C. The image acquisition was done at −196°C.

18.2.4 Statistical Analysis

The results are presented as means ± SE from the indicated number of experiments. The statistical significance of difference was analyzed by the Student's unpaired *t*-test.

18.3 Results and Discussion

We examined the dose-dependent effects of TSKY on the viability of neurons isolated from the central nervous system of the fresh-water pond snail *Lymnaea stagnalis* after 24 h in culture (Fig. 18.1).

We found that, after the incubation of mollusk ganglia for 60 min at different concentrations of TSKY (1×10^{-5} M, 1×10^{-6} M, and 1×10^{-7} M) and different temperatures (22–24 and 4–6°C), dissociated neurons remain viable after 24 h in culture and are capable of forming neurites, typical for mollusk neurons (Fig. 18.1c).

18.3.1 Incubation of the Isolated Mollusk Brain with Peptide TSKY at 22–24°C

After the incubation of mollusk ganglia for 60 min at different concentrations of TSKY (1×10^{-5} M, 1×10^{-6} M, and 1×10^{-7} M), the total number of viable neurons in culture after 24 h insignificantly increased compared with control and averaged 172.2% at 1×10^{-5} M, 154.4% at 1×10^{-6} M, 157.2% at 1×10^{-7} M (Fig. 18.2a).

The number of neurons with neurites under the same conditions and at the same concentrations of TSKY significantly decreased as compared with control, and averaged 30.8% at 1×10^{-5} M, 29.17% at 1×10^{-6} M, and 13.3% at 1×10^{-7} M (Fig. 18.3a).

18.3.2 Incubation of the Isolated Mollusk Brain with TSKY at 4–6°C

A short-term hypothermic incubation (60 min, 4–6°C) of the mollusk brain at different concentrations of TSKY (1×10^{-5} M, 1×10^{-6} M, and 1×10^{-7} M) increases the total number of neurons in culture as compared with control and averages 357.5% at 1×10^{-5} M, 322.5% at 1×10^{-6} M, and 205% at 1×10^{-7} M (Fig. 18.2b). Thus, the low temperatures and the treatment of molluscan ganglia with TSKY have a profound positive effect on the viability of neurons in culture.

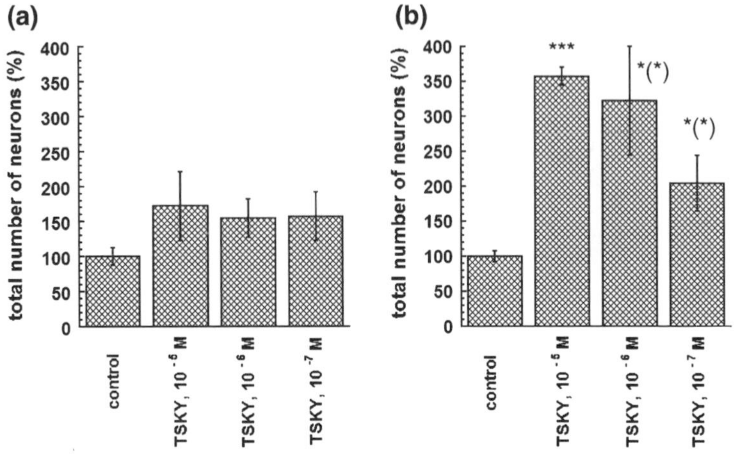

Fig. 18.2 Dose-dependent effect of TSKY on the viability of neurons after 24 h in culture. Incubation (60 min) of the mollusk brain at different concentrations of TSKY: 1×10^{-5} M, 1×10^{-6} M, and 1×10^{-7} M at 22–24°C (**a**) and at 4–6°C (**b**). Incubation of the brain without the peptide was taken as a control. The data shown are the means ± SE from three independent cell culture experiments. The statistical significance of difference was analyzed by the Student's unpaired t-test. *(*), *** indicate a significant effect of TSKY; $p < 0.025$; $p < 0.0001$, respectively

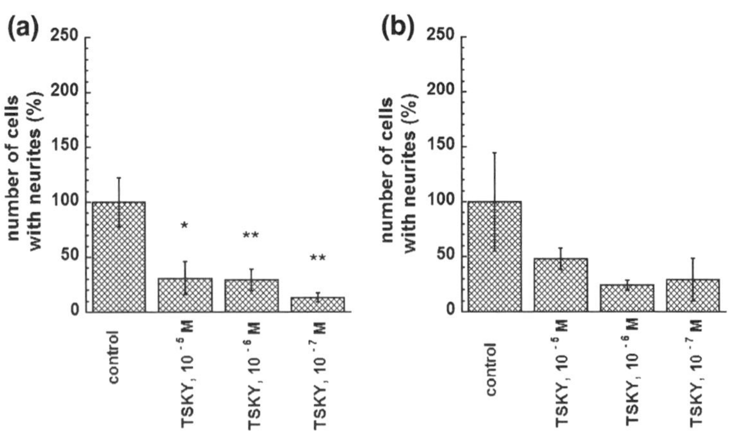

Fig. 18.3 Dose-dependent effect of TSKY on the viability of neurons with neurites at different temperatures. Incubation (60 min) of the mollusk brain at different concentrations of TSKY: 1×10^{-5} M, 1×10^{-6} M, and 1×10^{-7} M at 22–24°C (**a**) or at 4– 6°C (**b**). The incubation of the brain without the peptide was taken as a control. The data shown are the means ± SE from three independent cell culture experiments. The statistical significance of difference was analyzed by the Student's unpaired t-test. *, ** indicate a significant effect of TSKY; ($p < 0.05$), $p < 0.001$, respectively

The formation of neurons with neurites in the same conditions and at the same concentrations of TSKY differed statistically insignificantly from control (presumably, because of the scattering in the control data) and averaged 48% at 1×10^{-5} M, 24% at 1×10^{-6} M, and 29% at 1×10^{-7} M (Fig. 18.3b).

18.3.3 Freezing-Thawing of Isolated Mollusk Brain

It can be suggested that TSKY would affect the viability of cells after freeze–thawing. Therefore, the mollusk brain after incubation for 60 min at 22–24°C in a medium containing 2 M DMSO and TSKY at a concentration of 1×10^{-5} M was frozen to a temperature of liquid nitrogen. However, after the thawing of the brain, the total number of isolated neurons and the number of neurons with neurites in culture after 24 h did not change relative to the control. The results of this experiment indicate that TSKY does not act as a cryoprotectant.

18.3.4 Formation of Ice Microparticles

The observation of IMP formation at −196°C confirms this (Fig. 18.5a–c). It is well known that the formation of IMP, along with other negative factors (osmotic shock, pH shift, recrystallization during thawing, etc.), markedly reduces the cellular integrity and survival. The size and configuration of IMP affect the degree of mechanical cell damage. Figure 18.5a demonstrates that the freezing of culture medium for pond snail neurons (NCM) to −196°C induced the formation of well-defined four-sided IMP of rectangular and sharp-corned configuration. Similar shape and size have IMP after freezing of distilled water and physiological solution for sturgeon sperm (Andreev et al. 2009). The freezing of NCM with 2 M DMSO induced the formation of well-defined round and six-sided IMP (Fig. 18.5c). The freezing of NCM with TSKY at a concentration of 1×10^{-5} M (Fig. 18.5b) led to the formation of microparticles of similar shape as in Fig. 18.5a with rare inclusions of six-sided ice particles. The addition of TSKY (1×10^{-5} M) to 2 M DMSO and NCM (Fig. 18.5d) slightly changed the form of IMP shown in Fig 18.5c. As demonstrated above (Figs. 18.4 and 18.5), TSKY is not a cryoprotective factor at extra low temperatures (−196°C), but it definitely exerts a neuroprotective effect at normal (22–24°C) and low (4–6°C) temperatures.

Thus, we showed that the treatment of the mollusk brain with the TSKY peptide at 4–6°C increases considerably (approximately 3.5 times) the number of living neurons in culture compared with the control (Fig. 18.2b). The effect of TSKY is most clearly pronounced at the concentration of 10^{-5} M. At the same time, the number of neurons forming neurites at 24 h of culture after treatment with the peptide TSKY at different concentrations was significantly less than in the control at a temperature of 22–24°C (Fig. 18.3a) and did not significantly differ from control upon incubation at a low temperature (4–6°C) (Fig. 18.3b).

Earlier we found that the concentration of peptide TSKY isolated from the brain of ground squirrels in the hibernation state was approximately twofold higher than that in the brain of summer-active animals. This tetrapeptide was capable of substantially inhibiting the heart rate in isolated 36 h chicken embryos (Kramarova et al. 1996). It was also shown that the cooling of mice preliminarily injected with the peptide TSKY induced prolonged hypothermia in animals (Ignat'ev et al. 2001).

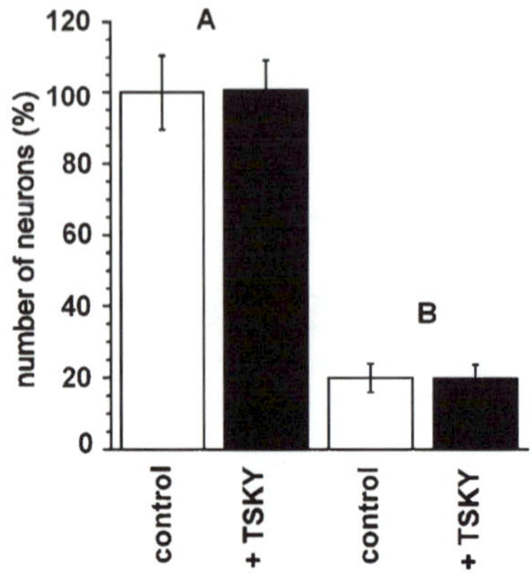

Fig. 18.4 Effect of TSKY on the viability of dissociated neurons in culture after freezing–thawing of mollusk ganglia. The mollusk brains were incubated with TSKY (1×10^{-5} M). The incubation of the brain without the peptide was taken as a control. **a** The total number of neurons. **b** The number of neurons with neurites. Data shown are the means ± SE from six independent cell culture experiments

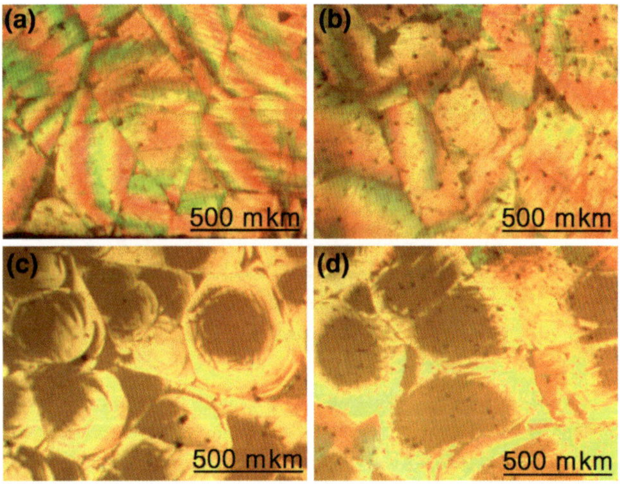

Fig. 18.5 Ice microparticles in different cryoprotective medium at −196°C: **a** NCM (20% Leibovitz (L-15) medium), **b** NCM + TSKY (1×10^{-5} M), **c** NCM + DMSO (2 M), **d** TSKY (1×10^{-5} M) + NCM + DMSO (2 M)

These data, together with the results of this study, lead us to suggest that the treatment of the *Lymnaea* brain with this peptide can trigger a transition of neurons to a new physiological state (probably, hypobiosis/hypometabolism); the neurons can stay in this state for a prolonged period of time (unpublished data) without loss of viability but with a weakening of the regeneration properties. Therefore, there arises the next problem, how to induce the exit or "arousal" of cells from such a hypometabolic state. The identification of endogenous neuroprotective factors in tissues of animals that survive cold temperatures will make it possible to turn to a search for novel effective and safe methods of low temperature conservation of neurons and tissues and the elucidation of mechanisms of their resistance to the cold.

This work was supported by the Russian Foundation for Basic Research (project no. 10-04-01319-a).

References

Andreev AA, Gakhova EN, Pashovkin TN (2009) Congelation of cryoprotective solutions and cryopreservation of fish sperm. Biofiz (Russ) 54(5):869–875

Beckman AL, Llados-Eckman C, Stanton TL, Adler MW (1981) Physical dependence on morphine fails to develop during the hibernating state. Science 212(4502):1527–1529

Bell HJ, Inoue T, Shum K, Luk C, Syed NI (2007) Peripheral oxygen-sensing cells directly modulate the output of an identified respiratory central pattern generating neuron. Eur J Neurosci 25(12):3537–3550

Borlongan CV, Su TP, Wang Y (2000) Treatment with delta opioid peptide enhances in vitro and in vivo survival of rat dopaminergic neurons. NeuroReport 11(5):923–936

Chien S, Oeltgen PR, Diana JN, Shi X, Nilekani SP, Salley R (1991) Two-day preservation of major organs with autoperfusion multiorgan preparation and hibernation induction trigger. J Thorac Cardiovasc Surg 102(2):224–234

Cotton CJ, Harlow HJ (2010) Avoidance of skeletal muscle atrophy in spontaneous and facultative hibernators. Physiol Biochem Zool 83(3):551–560

Drew KL, Rice ME, Kuhn TB, Smith MA (2001) Neuroprotective adaptations in hibernation: therapeutic implications for ischemia-reperfusion, traumatic brain injury and neurodegenerative diseases. Free Radic Biol Med 31(5):563–573

Frerichs KU, Hallenbeck JM (1998) Hibernation in ground squirrels induces state and species-specific tolerance to hypoxia and aglycemia: an in vitro study in hippocampal slices. J Cereb Blood Flow Metab 18(2):168–175

Gakhova EN, Chekurova NR, Kislov AN, Veprintsev BN (1989) Giant neurons save the viability after deep freezing of pond snail brain *Lymnaea stagnalis*. Kriobiologiya (Russ) 1:19–23

Govindaswami M, Sanchez JA, Wedge J, Langstone MD, Bishop PD, Bruce DS, Oeltgen PR (2000) Opioid-like hibernation factors provide protection to the ischemic myocardium. In: Heldmaier G, Klingenspor M (eds) Life in the cold. Springer, Berlin

Govindaswami M, Brown SA, Yu J, Zhu H, Bishop PD, Kindy MS, Oeltgen PR (2008) Delta 2-specific opioid receptor agonist and hibernating woodchuck plasma fraction provide ischemic neuroprotection. Acad Emerg Med 15(3):250–257

Ivlicheva NA, Dmitrieva EV, Kostenko MA, Gakhova EN (2004) Cryopreserved snail neurons are capable of morphological differentiation in vitro. Biofiz (Russ) 49(4):710–714

Ignat'ev DA, Mikhaleva II, Sosulina LIu, Zakarian AA, Sukhova GS, Ziganshin RKh (2001) TSKY peptide inhibits the ground squirrel recovery after hibernation. Zh Evol Biokhim Fiziol (Russ) 37(3):201–205

Kemper GB, Ruben RL (1982) Effect of 7,12-dimethylbenz(a)anthracene on the integument of the hibernating and nonhibernating 13-lined ground squirrel. Comp Biochem Physiol C 73(2):445–450

Kostenko MA, Musienko VS, Smolikhina TI (1981) Effects of pH value and calcium content on morphological differentiation of mollusk neurons in culture. Tsitolog (Russ) 23(7):779–787

Kramarova LI, Gakhova EN, Karnaukhov VN (1994) Hypomethabolic factors of hibernating animals as a potential regulators of cell resistance to low and extra low temperatures. Biofiz Living Cell (Russ) 6:47–55

Kramarova LI, Ziganschin RH, Timoshenko MA, Mikhaleva II, Karnaukhov VN (1996) The peptide Thr-Ser-Lys-Tyr isolated from the brain of hibernating ground squirrels (*Spermophilus undulatus*) decreases the heart rate of chicken embryos. In: Geiser F, Hulbert AJ, Nicol SC (eds) Adaptations to the cold. University of New England Press, Armidale

Kramarova LI, Ziganshin RKh, Gakhova EN (2009) Endogenous hypometabolic-hypothermic factors and their possible application to life in the cold. Bioorg Khim (Russ) 35(5):597–609

Meryman HT (2007) Cryopreservation of living cells: principles and practice. Transfusion 47(5):935–945

Nurnberger F, Lee TF, Jourdan ML, Wang LC (1991) Seasonal changes in methionine-enkephalin immunoreactivity in the brain of a hibernator, *Spermophilus columbianus*. Brain Res 547(1):115–121

Oeltgen PR, Horton ND, Bolling SF, Su TP (1996) Extended lung preservation with the use of hibernation trigger factors. Ann Thorac Surg 61(5):1488–1493

Pakhotin PI, Pakhotina ID (1994) Preparation of isolated perfused ground squirrel brain. Brain Res Bull 33(6):719–721

Popova IY, Kokoz YM, Zenchenko KI, Vinogradova OS (2003) The paradoxically high reactivity of septal neurons in hibernating ground squirrels to endogenous neuropeptides is lost after chronic deafferentation of the septum from the preopticohypothalamic areas. Comp Biochem Physiol A Mol Integr Physiol 135(3):383–402

Ross AP, Drew KL (2006) Potential for discovery of neuroprotective factors in serum and tissue from hibernating species. Mini Rev Med Chem 6(8):875–884

Smith F, Grenan MM (1951) Effect of hibernation upon survival time following whole-body irradiation in the marmot (*Marmota monax*). Science 113(2946):686–688

Tamura Y, Monden M, Shintani M, Kawai A, Shiomi H (2006) Neuroprotective effects of hibernation-regulating substances against low-temperature-induced cell death in cultured hamster hippocampal neurons. Brain Res 1108(1):107–116

Terry C, Dhawan A, Mitry RR, Lehec SC, Hughes RD (2006) Preincubation of rat and human hepatocytes with cytoprotectants prior to cryopreservation can improve viability and function upon thawing. Liver Transpl 12(1):165–177

Wang LCH (1988) Mammalian hibernation: an escape from the cold. In: Gilles R (ed) Advances in comparative and environmental physiology. Springer, Berlin

Zenchenko KI, Kokoz YM, Ivanov VT, Ziganshin RH, Vinogradova OS (2000) State-dependent effects of some neuropeptides and neurotransmitters on neuronal activity of the medial septal area in brain slices of the ground squirrel, *Citellus undulatus*. Neuroscience 96(4):791–805

Ziganshin RH, Svieryaev VI, Vaskovsky VN, Mikhaleva II, Ivanov VT, Kokoz YuM, Alekseev AE, Korystova AF, Sukhova GS, Emelyanova TG, Usenko AB (1994) Bioactive peptides isolated from the brain of hibernating ground squirrel. Bioorg Chem 20(8):899–918

Chapter 19
The Torpor-Arousal Cycle is Controlled by an Endogenous Clock

André Malan

Abstract This chapter aims at elucidating the control of the torpor-arousal cycle of mammalian hibernators, based on recent progress in circadian biology. The circadian system is now viewed as a complex network of central and peripheral clocks. In euthermy, the whole system synchronizes to a master, light-driven clock, located in the suprachiasmatic nuclei of the hypothalamus (SCN). It is proposed that in a torpor bout another, non temperature-compensated clock, the torpor-arousal clock, takes over as the master clock. According to the Arrhenius law, the endogenous period of the torpor-arousal clock will expand as temperature decreases. The corresponding subjective time will then diverge from astronomical time. Body temperature recordings of 17 complete hibernation seasons of five species have been analyzed. In astronomical time, torpor bout length (TBL) varied up to fivefold, but the corresponding durations in subjective time were remarkably constant. In all cases, the animal would simply arouse at a constant subjective time given by the torpor-arousal clock. In the frame of modern circadian biology, this suggests a much broader role of the circadian system in the control of the torpor-arousal cycle.

Keywords Circadian clock · Torpor-arousal cycle · Torpor bout length

Abbreviation

SCN	Suprachiasmatic nuclei of the hypothalamus
T_a	Ambient temperature
T_b	Body temperature
TBL	Torpor bout length

A. Malan (✉)
Institute for Cellular and Integrative Neurosciences, CNRS and University of Strasbourg, 5 rue Blaise Pascal, 67084 Strasbourg, France
e-mail: malan@inci-cnrs.unistra.fr

T. Ruf et al. (eds.), *Living in a Seasonal World*,
DOI: 10.1007/978-3-642-28678-0_19, © Springer-Verlag Berlin Heidelberg 2012

19.1 The Circadian System

During the hibernation season, mammalian hibernators alternate between prolonged bouts of torpor and short arousals in a pseudo-periodic fashion. The control of this torpor-arousal cycle (Malan 2008) is still unexplained. Could an answer be provided by circadian biology, which has kept expanding very rapidly over the past few years?

The molecular mechanism of the master endogenous clock of mammals, located in the suprachiasmatic nuclei (SCN) of the hypothalamus, has now been largely elucidated. Circadian clocks based on a similar gene machinery have been found in a variety of brain sites and of peripheral organs, e.g., liver, lung, pancreas, kidney, retina (Mendoza and Challet 2009; Pezuk et al. 2010). The SCN clock is synchronized by the photoperiod, and synchronizes in turn the other clocks, generating the overall circadian rhythm of the body.

This light-driven circadian system controls a wide variety of processes in the body.

When the SCN clock has been eliminated by lesions or by genetic manipulation, other clocks may take over. Arrhythmic, SCN-lesioned euthermic animals can anticipate the 24-h periodicity of food access by the so-called food-entrainable oscillator (Challet and Mendoza 2010; Feillet et al. 2008; Mistlberger 2011). This oscillator controls a multiplicity of sites involved in metabolism (Mistlberger 2011). In the organs, substrate-driven local clocks control cellular processes. In the liver the local clock controls the synthesis of at least 20% of the soluble proteins. These include rate-limiting factors in vital pathways, such as urea formation and carbohydrate metabolism (Reddy et al. 2006). A complex interplay exists between central and local clocks, with a tight crosstalk between the clock and metabolic circuitries (Asher and Schibler 2011; Huang et al. 2011). With its organization and ubiquitous distribution within the body the circadian system seems to be well suited for a polycentric control of the torpor-arousal cycle.

19.2 Requirements for a Torpor-Arousal Clock

Could then torpor bout length (TBL) be controlled by a circadian clock? TBL varies with body temperature (T_b). Therefore the torpor-arousal clock controlling it cannot be the SCN clock, which is temperature-compensated to keep its endogenous period close to the day–night period of 24 h. When it is expressed at torpor T_b, SCN period remains close to 24 h (Ruby 2003). The SCN-controlled circadian period of *tau* mutant golden hamsters significantly differs from that of wild types, yet their TBL is nearly identical (Oklejewicz et al. 2001). While torpid in a dark burrow, a hibernator may not need to synchronize on the photoperiod.

The torpor-arousal clock must be able to substitute the SCN as the main clock. The food-entrainable oscillator fulfills this condition. Its eventual temperature compensation has never been evidenced and may be lacking. It may not need it since it uses no astronomical Zeitgeber. Whether or not torpor-arousal clock is the same as the food-entrainable oscillator, the lack of temperature compensation does not need to be constitutive. It might also result from epigenetic control, possibly seasonal. Epigenetic gene control occurs in hibernation (Morin and Storey 2009). Clock genes present multiple control sites by sirtuins or micro-RNAs (Asher et al. 2008; Zhang et al. 2009). Lipid utilization resulting from fasting and acting via PPARα and FGF21-NPY is a possible epigenetic control factor of the seasonal induction of torpor (Ishida 2009; Melvin and Andrews 2009).

19.3 How Would a Temperature-Dependent Circadian Clock Affect Torpor Bout Length?

The equation describing the temperature effects on the period of the torpor-arousal clock has been given elsewhere (Malan 2010). It has been derived from the Arrhenius law governing enzyme reaction rates, based on published data on enzymes from hibernators and on a detailed set of observed TBL's in three ground squirrel species (Twente et al. 1977; Malan 2010). The endogenous period of the torpor-arousal clock can be predicted for any value of T_b. As T_b decreases, the torpor-arousal clock will run more slowly. At 6°C, for instance, a full cycle of the clock will take about 330 h instead of 24. The subjective time given by the clock will thus be slowed by a factor of 330/24 = 13.75. In the brain or tissues of the animal, 13.75 min will be shown as only one minute by the internal clock. As T_b decreases, subjective time will widely diverge from the astronomical time.

What TBL will result? Let us calculate the subjective time shown on the torpor-arousal clock by the end of a torpor bout, over which T_b may have been quite variable. To this end, the total duration of the bout is divided in equal short astronomical time intervals. The subjective time elapsed during each interval is derived from the equation with the corresponding T_b. By summing up the subjective times for all intervals, one gets the time that reads on the torpor-arousal clock when the arousal occurs (Malan 2010). The time courses of T_b in 17 full hibernation seasons from five species of hibernators (three species of ground squirrels, a marsupial, and a monotreme) have been analyzed. Torpor bouts shorter than 4 days were excluded to avoid the transients at the extremities of the hibernation season, as well as intercurrent short bouts associated with cold defense (Buck and Barnes 2000; Ortmann and Heldmaier 2000) or spurious disturbances.

On the astronomical time scale, TBL was highly variable (Fig. 19.1). In all cases, however, TBL was accurately predicted as a constant subjective time, while actual durations in astronomical time exhibited up to fivefold variations. For instance, in an Anatolian ground squirrel (Gür et al. 2009) TBL ranged 4–20 days in astronomical time against 38.4 ± 1.2 h in subjective time (Fig. 19.1b); in a mountain pigmy possum (Körtner et al. 1998) TBL ranged 4–13 days astronomical time against 17.4 ± 1.4 h in subjective time (Fig. 19.1a); in a female European ground

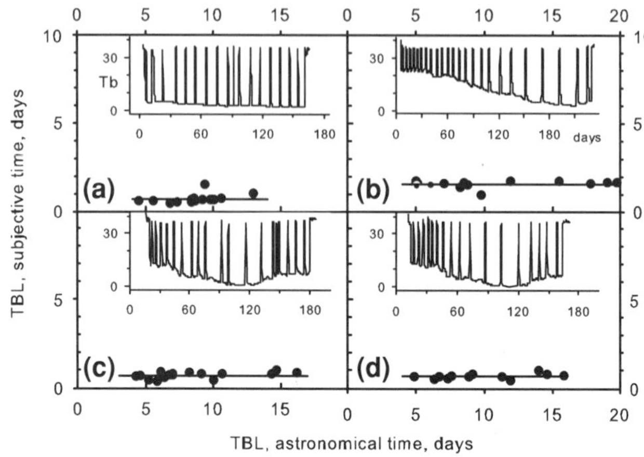

Fig. 19.1 Torpor bout length (TBL), in subjective time versus TBL in astronomical time. Data plotted for a full hibernation season in representative individuals of (**a**) mountain pigmy possum (Körtner et al. 1998), (**b**) Anatolian ground squirrel, female (Gür et al. 2009), (**c**) European ground squirrel, female; (**d**) idem., male (Hut et al. 2002). Only torpor bouts lasting at least 4 days are shown. *Insets* The corresponding body temperature recordings. In all cases, TBL expressed in subjective time is remarkably constant compared with the wide dispersion of TBL in astronomical time. This in turn results essentially from the variations in hibernaculum temperature. Modified from Malan (2010)

squirrel (Hut et al. 2002) TBL ranged 4–16 days against 16.9 ± 1.1 h in subjective time (Fig. 19.1c); in a male of the same species, TBL also ranged 4–16 days against 16.4 ± 1.0 h in subjective time (Fig. 19.1d). *F*-tests for nonlinear regression gave *P*-values lower or equal to 10^{-6}, most of them lower than 10^{-9}. Similar results were obtained on 13 other animals from the same studies (Malan 2010) and in two hamster species (Malan et al. in preparation). Whatever the length of astronomical time which has actually elapsed, the hibernator arouses from torpor at a constant subjective time given by the proposed endogenous torpor-arousal clock.

19.4 Discussion

19.4.1 Factors Contributing to Torpor Bout Length

The control of TBL by a temperature-dependent circadian clock provides a much more accurate prediction of TBL than any of the other explanations that have been proposed so far (Malan 2010). A circadian system similar to that which controls the sleep-activity cycle in euthermia gives the animal a safe mechanism to elicit arousal in due course. It also presents another selective advantage. By a passive thermodynamic effect, it will expand TBL when the ambient temperature T_a (and T_b) decreases. Increasing TBL reduces the number of arousals in a hibernation season, in which they represent up to 86% of the total energy expenditure (Heldmaier et al. 2004; Karpovich et al. 2009). The total energy expenditure will thus reach a minimum when the hibernation T_b is just above the threshold of cold-induced thermogenesis. The cost of regulatory thermogenesis is avoided, while the summated cost of arousals over the hibernation season is minimized. This would explain why in cold-temperate climates hibernating mammals most often select their hibernaculum temperature close to this optimum.

The control by a special circadian clock also explains several relationships that had been identified by previous authors, beyond the effect of T_b on TBL: (1) The effect of T_a: during torpor, when T_b is close to T_a, a TBL–T_b relationship necessarily results in a TBL–T_a relationship, although the latter will be less tight (Refs. in Malan 2010). (2) Seasonal effects. First, in field studies a seasonal evolution of hibernaculum temperature (T_a) is often present. The resulting time course of T_b and its effects on TBL may be confounded with a seasonal effect. Such indirect effects differ from the transients observed at the beginning ("test drops") or end of the hibernation season (Strumwasser 1959). These may result from the progressive substitution of the torpor-arousal clock to the SCN clock (or back) for the overall circadian control, and/or of an epigenetic control of the temperature dependence of the torpor-arousal clock itself. (3) The effect of diet lipid composition may be related to the effects of lipid diets on the circadian system (Refs. in Malan 2010).

In some cases, additional constraints may interfere with a simple clock control. (1) When T_a becomes lower than the threshold for eliciting regulatory thermogenesis TBL decreases as T_a decreases further (Buck and Barnes 2000; Geiser 2004; Ortmann and Heldmaier 2000; further references in Malan 2010). In this case a compromise has to be reached between the cold-induced thermogenesis and the requirements of the next arousal for drawing on energy resources. The resulting TBL–T_b relationship is still not understood. (2) Small-sized hibernators with little food stores such as bats may need to shorten their torpor bouts for feeding or drinking (Turbill et al. 2008). (3) In hibernators which live in milder climates and rely on external heat sources for arousal, temperature- or light-driven controls may interact with metabolism-driven circadian controls in a way that still needs to be elucidated (El Ouezzani et al. 2010; Kobbe et al. 2010; Nowack et al. 2010). In the near future, these questions as well as that of the interplay between the torpor-arousal and the SCN clocks during interbout arousals will benefit from the new techniques currently under development for circadian studies, such as in vivo luciferase reporter gene assays (Geusz et al. 2009).

19.4.2 Time Signal or Control?

Does the torpor-arousal clock act only as a time signal? The question holds for circadian clocks in general. Human major depression, which might share some mechanisms with hibernation, is correlated with disturbances of the circadian rhythm. Several treatments that correct the circadian functioning (phototherapy,

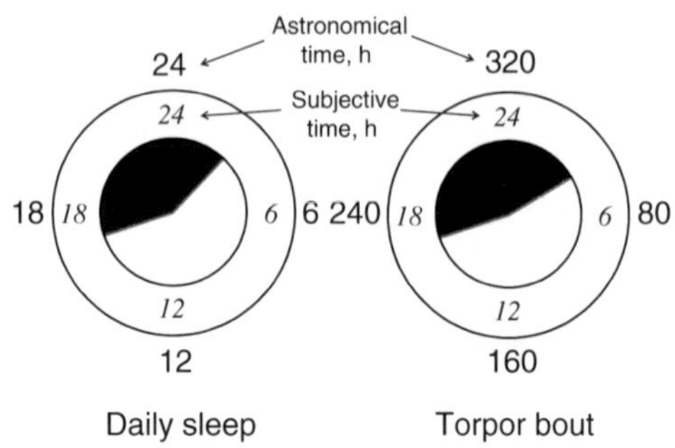

Fig. 19.2 Schematic representation of the clock control of daily sleep by the SCN clock in euthermia (*left*) and of the control of TBL by the torpor-arousal clock in hibernation (*right*). In euthermia, the SCN clock is synchronized to the photoperiod and the subjective time is identical to the astronomical time. In torpor, the torpor-arousal clock takes over. It is temperature-dependent and subjective time diverges from astronomical time. In this example (golden hamster, unpubl.) the torpor-arousal clock shows 24 h when 320 h have elapsed in the outside (astronomical time), but torpor stops at a fixed subjective time read on the torpor-arousal clock. *Black sectors* in the center represent, respectively the typical durations of daily sleep (*left*) and of a torpor bout (*right*) as recorded from the same individual (Malan et al. in preparation)

sleep deprivation, drugs) also result in a marked improvement of the depression (Kronfeld-Schor and Einat 2012; Monteleone et al. 2011). This suggests that the circadian system plays a direct role in its control. Similar conclusions have been reached for Huntington's disease (Pallier et al. 2007). More generally, reciprocal interactions between circadian system and metabolism are a major cause of obesity (Delezie and Challet 2011). The circadian system is now viewed as a very complex network of interacting central and local clocks, driven by photoperiod, temperature, and numerous metabolic signals and controlling a wide range of processes (Asher and Schibler 2011; Epperson et al. 2011; Huang et al. 2011). It is thus highly suited to control the multiple aspects of the torpor-arousal cycle.

19.5 Conclusion

In conclusion, the hypothesis of a temperature-dependent circadian clock accurately predicts the timing of arousal over a wide range of body temperatures and actual durations in astronomical time in the species studied. In all cases, the animal would simply arouse at a constant subjective time given by this clock. Originating from a simple question of timing, this study is leading us toward a much broader conception of the role of the circadian system in the control of the torpor-arousal cycle. Biochemical and physiological adaptations to torpor may be very specific, but the underlying control system would simply derive from the circadian system that operates in euthermic life. In particular, a striking similarity now emerges between the circadian control of daily sleep and the control of TBL by the torpor-arousal clock (Fig. 19.2).

References

Asher G, Gatfield D, Stratmann M, Reinke H, Dibner C, Kreppel F, Mostoslavsky R, Alt FW, Schibler U (2008) SIRT1 regulates circadian clock gene expression through PER2 deacetylation. Cell 134:317–328

Asher G, Schibler U (2011) Crosstalk between components of circadian and metabolic cycles in mammals. Cell Metab 13:125–137

Buck CL, Barnes BM (2000) Effects of ambient temperature on metabolic rate, respiratory quotient, and torpor in an arctic hibernator. Am J Physiol Regul Integr Comp Physiol 279:R255–R262

Challet E, Mendoza J (2010) Metabolic and reward feeding synchronises the rhythmic brain. Cell Tissue Res 341:1–11

Delezie J, Challet E (2011) Interactions between metabolism and circadian clocks: reciprocal disturbances. Ann NY Acad Sci 1243:30–46

El Ouezzani S, Janati IA, Magoul R, Pevet P, Saboureau M (2010) Overwinter body temperature patterns in captive jerboas (*Jaculus orientalis*): influence of sex and group. J Comp Physiol B Bioch Syst Environ Physiol 181:299–309

Epperson LE, Karimpour-Fard A, Hunter LE, Martin SL (2011) Metabolic cycles in a circannual hibernator. Physiol Genomics 43:799–807

Feillet CA, Mendoza J, Pevet P, Challet E (2008) Restricted feeding restores rhythmicity in the pineal gland of arrhythmic suprachiasmatic-lesioned rats. Eur J Neurosci 28:2451–2458

Geiser F (2004) Metabolic rate and body temperature reduction during hibernation and daily torpor. Annu Rev Physiol 66:239–274

Geusz ME, Blakely KT, Hiler DJ, Jamasbi RJ (2009) Elevated mPer1 gene expression in tumor stroma imaged through bioluminescence. Int J Cancer 126:620–630

Gür MK, Refinetti R, Gür H (2009) Daily rhythmicity and hibernation in the Anatolian ground squirrel under natural and laboratory conditions. J Comp Physiol B Bioch Syst Environ Physiol 179:155–164

Heldmaier G, Ortmann S, Elvert R (2004) Natural hypometabolism during hibernation and daily torpor in mammals. Respir Physiol Neurobiol 141:317–329

Huang W, Ramsey KM, Marcheva B, Bass J (2011) Circadian rhythms, sleep, and metabolism. J Clin Invest 121:2133–2141

Hut RA, Barnes BM, Daan S (2002) Body temperature patterns before, during, and after semi natural hibernation in the European ground squirrel. J Comp Physiol B Bioch Syst Environ Physiol 172:47–58

Ishida N (2009) Role of PPARalpha in the control of torpor through FGF21-NPY pathway: from circadian clock to seasonal change in mammals. PPAR Res 2009:412949

Karpovich SA, Tøien Ø, Buck CL, Barnes BM (2009) Energetics of arousal episodes in hibernating arctic ground squirrels. J Comp Physiol B Bioch Syst Environ Physiol 176: 691–700

Kobbe S, Ganzhorn JU, Dausmann KH (2010) Extreme individual flexibility of heterothermy in free-ranging Malagasy mouse lemurs (*Microcebus griseorufus*). J Comp Physiol B Bioch Syst Environ Physiol 181:165–173

Körtner G, Song X, Geiser F (1998) Rhythmicity of torpor in a marsupial hibernator, the mountain pygmy-possum (*Burramys parvus*), under natural and laboratory conditions. J Comp Physiol B Bioch Syst Environ Physiol 168:631–638

Kronfeld-Schor N, Einat H (2012) Circadian rhythms and depression: human psychopathology and animal models. Neuropharmacology 62:101–114

Malan A (2008) Metabolism and rythmicity: the torpor arousal cycle. In: Lovegrove BG, McKechnie AE (eds) Hypometabolism in animals. Interpak Books, Pietermaritzburg, pp 215–222

Malan A (2010) Is the torpor-arousal cycle of hibernation controlled by a non-temperature-compensated circadian clock? J Biol Rhythms 25:166–175

Melvin RG, Andrews MT (2009) Torpor induction in mammals: recent discoveries fueling new ideas. Trends Endocrinol Metab 20:490–498

Mendoza J, Challet E (2009) Brain clocks: from the suprachiasmatic nuclei to a cerebral network. Neuroscientist 15:477–488

Mistlberger RE (2011) Neurobiology of food anticipatory circadian rhythms. Physiol Behav 104:535–545

Monteleone P, Martiadis V, Maj M (2011) Circadian rhythms and treatment implications in depression. Prog Neuropsychopharmacol Biol Psychiatry 35:1569–1574

Morin PJ, Storey KB (2009) Mammalian hibernation: differential gene expression and novel application of epigenetic controls. Int J Dev Biol 53:433–442

Nowack J, Mzilikazi N, Dausmann KH (2010) Torpor on demand: heterothermy in the non-lemur primate *Galago moholi*. PLoS ONE 5:e10797

Oklejewicz M, Daan S, Strijkstra AM (2001) Temporal organisation of hibernation in wild-type and tau mutant Syrian hamsters. J Comp Physiol B Bioch Syst Environ Physiol 171:431–439

Ortmann S, Heldmaier G (2000) Regulation of body temperature and energy requirements of hibernating Alpine marmots (*Marmota marmota*). Am J Physiol Regul Integr Comp Physiol 278:R698–R704

Pallier PN, Maywood ES, Zheng Z, Chesham JE, Inyushkin AN, Dyball R, Hastings MH, Morton AJ (2007) Pharmacological imposition of sleep slows cognitive decline and reverses dysregulation of circadian gene expression in a transgenic mouse model of Huntington's disease. J Neurosci 27:7869–7878

Pezuk P, Mohawk JA, Yoshikawa T, Sellix MT, Menaker M (2010) Circadian organization is governed by extra-SCN pacemakers. J Biol Rhythms 25:432–441

Reddy AB, Karp NA, Maywood ES, Sage EA, Deery M, O'Neill JS, Wong GK, Chesham J, Odell M, Lilley KS, Kyriacou CP, Hastings MH (2006) Circadian orchestration of the hepatic proteome. Curr Biol 16:1107–1115

Ruby NF (2003) Hibernation: when good clocks go cold. J Biol Rhythms 18:275–286

Strumwasser F (1959) Factors in the pattern, timing and predictability of hibernation in the squirrel, *Citellus beecheyi*. Am J Physiol 196:8–14

Turbill C, Körtner G, Geiser F (2008) Timing of the daily temperature cycle affects the critical arousal temperature and energy expenditure of lesser long-eared bats. J Exp Biol 211:3871–3878

Twente JW, Twente J, Moy RM (1977) Regulation of arousal from hibernation by temperature in three species of *Citellus*. J Appl Physiol 42:191–195

Zhang EE, Liu AC, Hirota T, Miraglia LJ, Welch G, Pongsawakul PY, Liu X, Atwood A, Huss JW,, Janes J, Su AI, Hogemesch JB, Kay A (2009) A genome-wide RNAi screen for modifiers of the circadian clock in human cells. Cell 139:199–210

Chapter 20
Ultradian Episodes of Thermogenesis in Mammals: Implications for the Timing of Torpor Entry and Arousal

Carola W. Meyer, William Blessing and Gerhard Heldmaier

Abstract Mammals forage and feed in a periodic manner, and facultative thermogenesis in brown adipose tissue (BAT) contributes to the increases in heat production associated with feeding. Even in the absence of food, BAT thermogenesis occurs in an episodic manner. In the torpid or hibernating state, mammals do not feed and thermogenesis is substantially suppressed. However, periodic peaks and troughs of metabolic rate, ventilation and heart rate continue to occur at reduced amplitude. We hypothesize that the episodic pattern of thermogenesis in the normothermic and torpid state is part of a centrally mediated ultradian rhythm that serves maintenance of neuronal integrity and alertness. Periodic thermogenesis may also provide the neuronal and temporal framework for torpor entry and arousal. Therefore, identifying the central pathways that underlie ultradian fluctuations in energy metabolism may lead towards understanding and predicting torpor behaviour.

C. W. Meyer (✉) · G. Heldmaier
Animal Physiology, Philipps-Universität Marburg, 35032 Marburg, Germany

Present Address:
C. W. Meyer
Helmholtz Zentrum München, Deutsches Forschungszentrum für Gesundheit und Umwelt (GmbH), Ingolstädter Landstr. 1, 85764 Neuherberg, Germany
e-mail: carola.meyer@helmholtz-muenchen.de

W. Blessing
Human Physiology, Flinders Medical Center, Adelaide, SA, Australia

T. Ruf et al. (eds.), *Living in a Seasonal World*, 219
DOI: 10.1007/978-3-642-28678-0_20, © Springer-Verlag Berlin Heidelberg 2012

20.1 Body Temperature and Energy Metabolism

Body temperature depends on both heat production and on heat exchange with the environment. In endotherms, heat produced endogenously from metabolic processes contributes substantially to maintenance of a stable 'euthermic' body temperature (T_b) of about 35–37°C (mammals) (Scholander et al. 1950). Thus, in endotherms a change in T_b is often preceded by a change in energy metabolism. This sequence of events is prominent during entry into deep torpor or hibernation, when animals suppress their metabolic rate to save energy, thermoregulating at T_bs as low as −3°C (Heldmaier et al. 2004).

Variations in metabolic rate and T_b are also observed during the non-hibernating state. The most obvious fluctuation in 'euthermic' T_b is seen in association with the diurnal sleep-wake cycle, when T_b decreases during sleep phases and increases during waking in a circadian manner (Aschoff and Pohl 1970). In addition, changes in various physiological parameters, including T_b, are also evident at a periodicity of less than 24 h, and they are thus referred to as ultradian rhythms (Daan and Aschoff 1981; Refinetti and Menaker 1992; Stupfel et al. 1990b; Robinson and Fuller 1999). Ultradian fluctuations in T_b occur approximately every 1–3 h, and the amplitude of the variations can be substantial, as much as 2°C (abdominal) in rats, mice and hamsters. Figure 20.1 shows examples of the daily oxygen consumption and abdominal T_b traces of three different species from two mammalian clades (*Placentalia* and *Marsupialia*).

20.2 Thermogenesis in Brown Adipose Tissue Contributes to Metabolism and Body Temperature

The vegetative nervous system is a major determinant in the regulation of energy homeostasis, and thus likely to be a key player in the appearance of ultradian rhythmic thermogenic events. In mammals, plasma norepinephrine (NE) peaks every ~90 min in conjunction with an increase in overall sympathetic tone (Levin et al. 1978; Tapp et al. 1981). In some mammals, including humans, a major thermogenic effector targeted by sympathetic nerves is the brown adipose tissue (BAT). Upon cold stimulation, adrenergic signalling via β3-adrenoreceptors to BAT activates the intracellular release of free fatty acids and leads to UCP1-mediated dissipation of heat by uncoupling of mitochondrial respiration from ATP synthesis (Cannon and Nedergaard 2004). Thereby, BAT is a major site of heat production facilitating maintenance of T_b in the cold. In addition, BAT plays an important role for arousal from hypometabolic or hibernating states (Hart 1971). In rats, thermogenesis in BAT is associated with ultradian variations in T_b (Ootsuka et al. 2009), and SR59230A, a β3-adrenoceptor antagonist, inhibits ultradian BAT thermogenesis and interrupts associated episodic brain and body heating (Ootsuka et al. 2011).

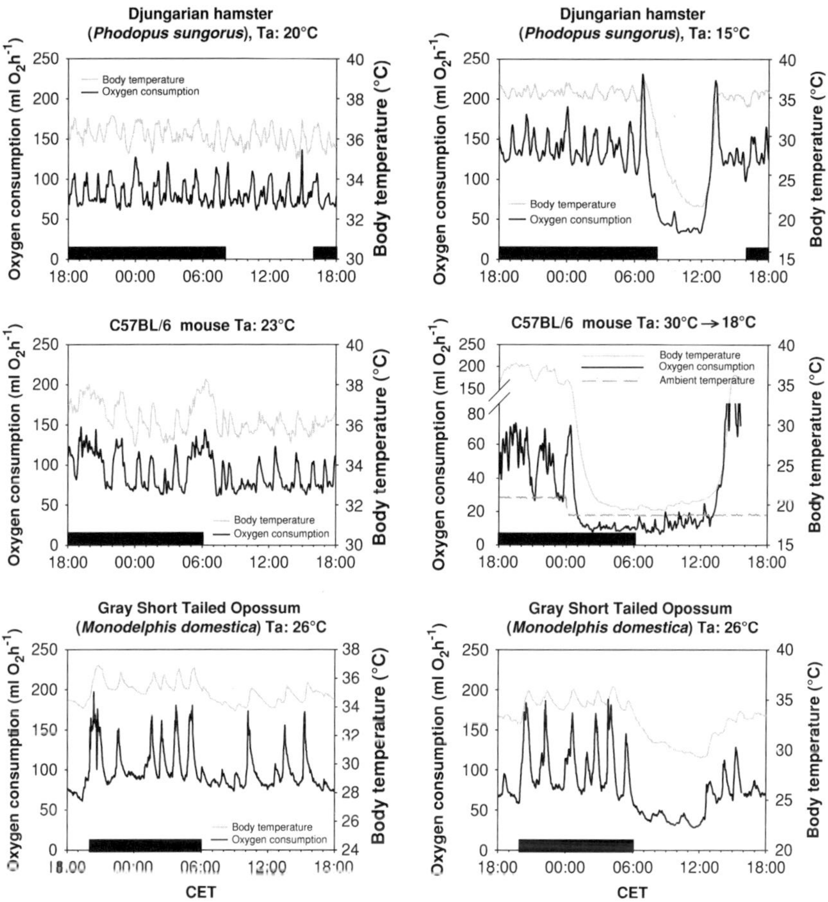

Fig. 20.1 24-h oxygen consumption and abdominal T_b time courses in three different mammalian species with (*right column*) or without (*left column*) a torpor episode. The resolution of readings for each variable was 2–5 min. Note the different ambient temperatures (T_a), as indicated on the top of each graph. In the C57BL/6 mouse (*M. musculus*), torpor was induced by previous food restriction for 24 h at 30°C and a lowering of ambient temperature from 30 to 18°C at midnight (*hatched line*) (Oelkrug et al. 2011). The *black bars* represent the periods of *lights off*. The body mass of the individuals displayed was: *P. sungorus:* 20–24 g (short-day acclimated hamsters), mouse: 22 g, *M. domestica*: 80 g

20.3 The Ultradian Pattern of Mammalian Thermogenesis is Regulated by Brain Central Command

Kleitman, 50 years ago, proposed that ultradian rhythms of physiological variables during the waking period of mammals corresponded to the REM sleep-non-REM sleep transitions observed during the circadian resting phase. He introduced the term Basic Rest-Activity-Cycle (BRAC) to describe the approximately 100 min periodicity in alertness and behavioural activity in mammals (reviewed by Kleitman 1982; Lavie and Kripke 1981). In distinction to circadian rhythms, ultradian rhythms are episodic, but not truly periodic (Aschoff and Pohl 1970). It is now established that the ultradian BRAC patterning, including patterning of BAT thermogenesis, occurs in the absence of any classical environmental *Zeitgeber,* e.g. light (Honma and Hiroshige 1978; Gerkema et al. 1993; Stupfel et al. 1990a; Rensing et al. 2001; Schibler and Naef 2005; Ootsuka et al. 2009). The BRAC organisation is therefore flexible, 'on call' (Aschoff and Gerkema 1985), and salient environmental events, such as the detection of a predator, or internal physiologic needs, may trigger particular BRAC episodes. This interaction between intrinsic brain circuitry and environmental events may presumably contribute to the large variability and irregularity in the periodicity of ultradian events. Thus, in mammals, the overall BRAC organisation is likely to result from brain central command. Hypothalamic, brainstem and spinal pathways regulating sympathetic outflow to BAT are now well characterised (Morrison et al. 2008) but the circuitry responsible for the patterning of the BRAC is not yet understood.

20.4 Food Intake Occurs in Association with Ultradian Thermogenesis and the Basic Rest-Activity-Cycle

Richter (1927) noted that in rats eating commences a short time after the onset of a behavioural activity, and that activity occurs episodically, at intervals varying from 1 to 2 h. Subsequent investigators further established that T_b increases substantially in association with eating and the processing of food (Abrams and Hammel 1964; Rampone and Reynolds 1991; De Vries et al. 1993; Himms-Hagen 1995; Secor 2009). Brobeck (1948) emphasised the concept of thermoregulatory feeding control, suggesting that eating commences as a thermoregulatory response to a fall in T_b, terminating when eating-induced heat production restores T_b. However, when food is available ad libitum the temperature starts to increase approximately 12 min before eating commences (De Vries et al. 1993), and at present the physiological significance of this increase is not understood.

Blessing et al. (2011) demonstrated that eating is integrated into the BRAC in a temporally precise manner (Fig. 20.2). The rise in T_b reflects the fact that behavioural activity and BAT thermogenesis occurs early in the active phase of the BRAC, approximately 15 min before eating commences. There is no correlation between meal size and the timing of the next feeding episode, and strikingly, even when food is not available rats still become active and touch the food container approximately 15 min after onset of an episode of BAT thermogenesis. Previously, Gerkema et al. (1993) observed that the ultradian activity rhythm of common voles (*Microtus arvalis*) is maintained when food and water is withdrawn or resting is prevented. We conclude that a programmed intrinsic brain activity acts

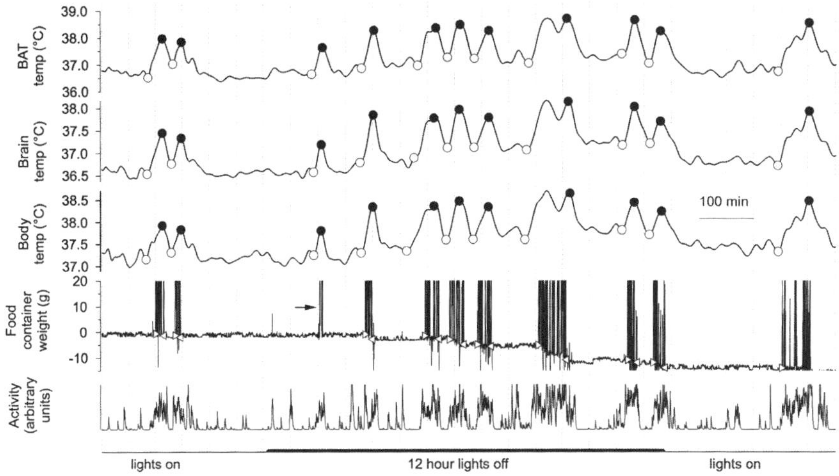

Fig. 20.2 Records of continuously recorded parameters in an individual Sprague–Dawley rat during both *light* and *dark* periods (resolution of readings: 1 Hz for each variable). *Open* and *filled black circles* indicate onset and peak times of the ultradian increases in temperature. The transient large amplitude changes in the weight of the food container indicate mechanical disturbance of the container by the rat. The open triangle markers indicate the onset and end of eating. The progressive decrease in weight documents the amount of food eaten. The horizontal *black arrow* on the food weight trace indicates disturbance of the food container without pellet removal (modified from Blessing et al. 2011)

as the pacemaker for periodic arousal and alertness of the normothermic and normometabolic animal, manifesting itself in an increase in thermogenesis and a drive to search and ingest food episodically.

20.5 Ultradian Thermogenic Episodes and Daily Torpor

Torpor is a strategy by which animals adjust their daily energy requirements, often in association with unfavourable seasons or environments. All species displayed in Fig. 20.1 have the ability to enter torpor either spontaneously (*P.s.*: Heldmaier and Steinlechner 1981, *M.d.*: S. Busse and C. Meyer, unpublished results), or induced by food restriction and cold (*M.m.*: Oelkrug et al. 2011). In each case, ultradian thermogenic bursts are seen prior to a torpor bout and they reoccur with similar frequency after termination of torpor. Even in torpor, metabolic rate and T_b are not stable, and periodic peaks and troughs in thermogenesis remain visible. In some species, including *Phodopus sungorus*, *Mus musculus* and *Monodelphis domestica* (Oelkrug et al. 2011, other unpublished data) repetitive torpor entry has been observed, indicating that the internal signalling for the onset of torpor is not only capable for one induction per day but can be activated several times or is maintained for a prolonged period of time. This indicates that ultradian thermogenic episodes persist in daily torpor, but their amplitude is dramatically suppressed.

During active period of the light–dark cycle, interaction with the external environment, including the search for food, requires complex perception, complex decision making and complex behaviours. This complexity requires extensive activation of cortical neural circuitry, involving extensive synaptic processing. Synaptic events are especially temperature-sensitive (Janssen 1992), and ultradian thermogenic episodes may support the maintenance and performance of cognitive functions. A generally similar explanation was proposed by Wehr (1992) for the increase in brain temperature during the occurrence of paradoxical 'brain awake' REM sleep. In contrast, during torpor episodes, animals do not search for food, feed or sleep, and there is often limited interaction with the external environment, obviating the full behavioural repertoire. This does not exclude the possibility that even in torpid endotherms, with often low baseline temperatures, a temporary further increase in temperature facilitates the neuronal processes mediating preparation for full arousal. Therefore, understanding the sequence of events that generate ultradian rhythmic thermogenic episodes may also lead towards understanding how torpor entry and spontaneous arousal is timed.

Hypometabolism, bradycardia, bradyventilation and reduced agility, during torpor suggest a parasympathetic involvement in torpor entry (Milsom et al. 1993; Milsom et al. 1999), while increased sympathetic discharge to BAT and to other thermogenic target tissues may trigger arousal. Based on the latter, one might therefore expect that mice with non-functional BAT or mice with deficits in adrenergic signalling should be reluctant to enter torpor, or they should be incapable to rewarm from torpor. We have shown that UCP1-KO mice can enter torpor and they spontaneously arise from torpor, but the rate of rewarming and the amount of energy required for arousal is higher compared to wildtype mice (Oelkrug et al. 2011). In contrast, mice lacking dopamine hydroxylase (dbh-KO), a key enzyme in NE and epinephrine synthesis, fail to enter torpor unless a synthetic β3-adrenergic receptor agonist or a centrally acting NE precursor substitutive is given (Swoap et al. 2006). Therefore, expression of fasting induced torpor in mice requires functional sympathetic transmission acting via β3-adrenoreceptor signalling (Swoap et al. 2006; Swoap and Weinshenker 2008).

More recently, it was shown that administration of 6-hydroxydopamine (6-OHDA), which destroys transmitter vesicles in peripheral sympathetic nerve endings, abolishes torpor behaviour in short-day acclimated Djungarian hamsters (*P. sungorus*) for up to 7 days (Braulke and Heldmaier 2010). The time required for re-establishment of torpor matches the time frame required for recovering vesicles in sympathetic varicosities. Interestingly, 6-OHDA treated hamsters also lost their ultradian rhythm of T_b for 2 days. These results not just imply that an intact adrenergic system is necessary to establish T_b fluctuations but the sequence of events also indicated that restoration of (daily) torpor may be linked to the re-establishment of ultradian thermogenesis. To experimentally test this hypothesis in future studies it will be important to investigate torpor behaviour and ultradian thermogenic rhythms in different transgenic mouse lines with deficits in sympathetic transmission.

Djungarian hamsters receiving atropine implants failed to enter torpor on the first day after implantation, and they also exhibited disturbed ultradian T_b rhythms during this period (Braulke and Heldmaier 2010). Therefore, the full expression of torpor and ultradian rhythms may require both intact parasympathetic (cholinergic) and sympathetic pathways. Contrasting the more long lasting effect of a single 6-OHDA injection, episodic T_b patterns and torpor with atropine were resumed after 24 h only, even though the implants continued to release atropine for up to 3 weeks. It therefore appears that blocking the cholinergic varicosities is compensated for.

Recently, we were able to provoke torpor-like metabolic suppression and reduced T_b in fasted mice by exposing individuals to a rapid drop in ambient temperature from 30 to 18°C (Oelkrug et al. 2011, and Fig. 20.1). The metabolic response is remarkable as mice would normally recruit thermoregulatory heat production in order to maintain normothermic T_b in the cold. Acute hypometabolism in response to cold, however, did not occur when the switch in ambient temperature coincided with the onset of the circadian activity phase in mice (at *lights off*). In contrast, 6 h after *lights off* cold stimulus was effective again. We speculate that the ability to enter torpor may only be permissive during certain times of the circadian day, i.e. in an ultradian manner. Further experiments will be required to reveal the periodicity of these permissive phases.

20.6 Ultradian Thermogenic Episodes and Hibernation

Hibernation is the most intensive form of torpor in endotherms. A hibernation bout includes a torpor bout and the adjacent euthermic period, and the entire hibernation season consists out of a continuous sequence of such hibernation bouts. The duration of hibernation bouts increases during the hibernation season, is often found in the order of 10 through 30 days in midwinter, and may become shorter towards the end of the hibernation season (Ortmann and Heldmaier 2000; Bieber and Ruf 2009).

There is good evidence that ultradian processes continue during hibernation. Many species of hibernators show discontinuous ventilation during deep hibernation, i.e. periods of apnoea are interrupted by short bursts of ventilation. In squirrels and in dormice the periods of apnoea last up to 1 h followed by a few minutes taking 10–15 breaths (Zimmer and Milsom 2002; Elvert and Heldmaier 2005). During ventilation heart rate increases probably as a consequence of vagal control of cardiorespiratory functions. This suggests that cardiorespiratory functions in deep hibernation not only depend upon the peripheral needs for oxygen but may also underlie central control for its periodicity (Milsom et al. 1997; Milsom et al. 2001).

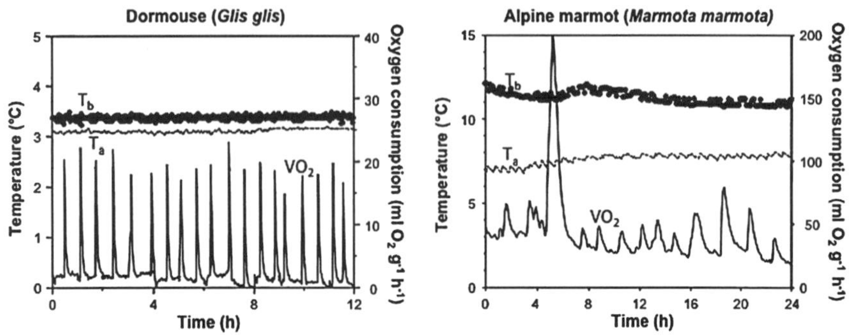

Fig. 20.3 Variations of oxygen consumption (VO_2) during deep hibernation in the Dormouse (*Glis glis*, 120 g body mass) and an Alpine marmot (*Marmota marmota,* 3,950 g body mass). In dormice oxygen consumption was recorded while they were hibernating in their sleeping box (volume 2 l, air flow 25 l min^{-1}, Elvert and Heldmaier 2005). Marmots were placed in perspex boxes for metabolic rate measurements during hibernation (volume 20 l, air flow 50 l min^{-1}, Ortmann and Heldmaier 2000). Both species showed more or less regular variations of oxygen consumption during hibernation. In dormice this is largely the result of discontinuous ventilation, as can be seen from almost zero oxygen consumption between bursts of ventilation (average peak interval 36 min). In marmots no clear pattern of discontinuous ventilation could be observed but they frequently showed bursts of heat generation to maintain a relatively high T_b in the torpid state (average peak interval 102 min). The magnitude of these bursts increases at lower ambient temperature (T_a) and the peaks may even reach the level of basal metabolic rate (Ortmann and Heldmaier 2000)

In hibernating marmots and dormice, repeated bursts of metabolic activity can be observed when metabolic rate is continuously recorded during deep hibernation (Fig. 20.3). These episodic bursts in metabolic activity are likely based on different regulatory systems. Dormice show discontinuous ventilation in deep hibernation and the bursts of gas exchange reflect ventilation following periods of apnoea (Elvert and Heldmaier 2005). In contrast, marmots maintain rather high T_bs during hibernation and the episodic metabolic activity probably reflects heat production controlled by thermoregulatory neural pathways. The underlying control mechanisms thus differ between the two species but this does not exclude the possibility that central command links ultradian ventilatory rhythms to metabolic (thermogenic) rhythms. It should be noted that in both cases these bursts of metabolic activity did not cause corresponding changes in T_b.

Short-term bursts of metabolic activity under central neuronal control are probably more common in deep hibernation than previously thought but the analysis of these bursts requires long-term records of metabolic rate, heart rate or ventilation rate, and may not be visible in long-term records of T_b. In the sequence of events that characterises hibernation, a periodic arousal to normometabolism and normothermia may represent the most extreme form of an ultradian thermogenic burst by which neuronal restorative processes are facilitated. Thereby neuronal integrity is maintained even after a long hibernation season (von der Ohe et al. 2007; Clemens et al. 2009).

20.7 Perspectives

We here introduce the hypothesis that the behavioural and physiological mechanisms driving alternations between resting and activity may be generally similar to corresponding states that allow for torpor entry and arousal. The brain pathways regulating occurrence of the ultradian physiological patterns that constitute the BRAC are, however, not understood. In contrast to circadian rhythms, functional integrity of the suprachiasmatic nuclei is not required (Gerkema et al. 1993). Orexin-synthesising neurons in the hypothalamus regulate many of the cognitive, behavioural and physiological functions engaged during the BRAC (Sakurai et al. 1998; Peyron et al. 1998; Saper et al. 2005; Zhang et al. 2010). Loss of these neurons in mice reduces the amplitude of the REM sleep rhythm (Kantor et al. 2009), and global loss of orexin is associated with BAT-hypoactivity and reduced energy metabolism (Sellayah et al. 2011). Therefore, orexinergic pathways may well be important for establishing ultradian thermogenic episodes as part of the BRAC output during waking, sleeping and torpor.

References

Abrams R, Hammel HT (1964) Hypothalamic temperature in unanesthetized albino rats during feeding and sleeping. Am J Physiol 206:641–646

Aschoff J, Gerkema M (1985) On diversity and uniformity of ultradian rhythms. Exp Brain Res 12(Suppl):321–334

Aschoff J, Pohl H (1970) Rhythmic variations in energy metabolism. Fed Proc 29:1541–1552

Bieber C, Ruf T (2009) Summer dormancy in edible dormice (*Glis glis*) without energetic constraints. Naturwissenschaften 96:165–171

Blessing W, Mohammed M, Ootsuka Y (2011) Heating and eating: brown adipose tissue thermogenesis precedes food ingestion as part of the ultradian basic rest-activity cycle in rats. Physiol Behav 105:966–974

Braulke LJ, Heldmaier G (2010) Torpor and ultradian rhythms require an intact signalling of the sympathetic nervous system. Cryobiology 60:198–203

Brobeck JR (1948) Food intake as a mechanism of temperature regulation. Yale J Biol Med 20:545–552

Cannon B, Nedergaard J (2004) Brown adipose tissue: function and physiological significance. Physiol Rev 84:277–359

Clemens LE, Heldmaier G, Exner C (2009) Keep cool: memory is retained during hibernation in Alpine marmots. Physiol Behav 98:78–84

Daan S, Aschoff J (1981) Short term rhythms in activity. In: Aschoff J, Daan S, Groos GA (eds) Vertebrate circadian systems: structure and physiology. Springer, Berlin, pp 305–331

De Vries J, Strubbe JH, Wildering WC, Gorter JA, Prins AJ (1993) Patterns of body temperature during feeding in rats under varying ambient temperatures. Physiol Behav 53:229–235

von der Ohe CG, Garner CC, Darian-Smith C, Heller HC (2007) Synaptic protein dynamics in hibernation. J Neurosci 27:84–92

Elvert R, Heldmaier G (2005) Cardiorespiratory and metabolic reactions during entrance into torpor in dormice, *Glis glis*. J Exp Biol 208:1373–1383

Gerkema MP, Daan S, Wilbrink M, Van der LF Hop MW (1993) Phase control of ultradian feeding rhythms in the common vole (*Microtus arvalis*): the roles of light and the circadian system. J Biol Rhythms 8:151–171

Hart JS (1971) Rodents. In: Whittow GC (ed) Comparative physiology of thermoregulation. Academic Press, New York, pp 1–149

Heldmaier G, Ortmann S, Elvert R (2004) Natural hypometabolism during hibernation and daily torpor in mammals. Respir Physiol Neurobiol 141:317–329

Heldmaier G, Steinlechner S (1981) Seasonal control of energy requirements for thermoregulation in the Djungarian hamster (*Phodopus sungorus*), living in natural photoperiod. J Comp Physiol 142:429–437

Himms-Hagen J (1995) Role of brown adipose tissue thermogenesis in control of thermoregulatory feeding in rats: a new hypothesis that links thermostatic and glucostatic hypotheses for control of food intake. Proc Soc Exp Biol Med 208:159–169

Honma KI, Hiroshige T (1978) Endogenous ultradian rhythms in rats exposed to prolonged continuous light. Am J Physiol 235:R250–R256

Janssen R (1992) Thermal influences on nervous system function. Neurosci Biobehav Rev 16:399–413

Kantor S, Mochizuki T, Janisiewicz AM, Clark E, Nishino S, Scammell TE (2009) Orexin neurons are necessary for the circadian control of REM sleep. Sleep 32:1127–1134

Kleitman N (1982) Basic rest-activity cycle—22 years later. Sleep 5:311–317

Lavie P, Kripke DF (1981) Ultradian circa 11/2 hour rhythms: a multioscillatory system. Life Sci 29:2445–2450

Levin BE, Goldstein A, Natelson BH (1978) Ultradian rhythm of plasma noradrenaline in rhesus monkeys. Nature 272:164–166

Milsom WK, Burlington RF, Burleson ML (1993) Vagal influence on heart rate in hibernating ground squirrels. J Exp Biol 185:25–32

Milsom WK, Harris MB, Reid SG (1997) Do descending influences alternate to produce episodic breathing? Respir Physiol 110:307–317

Milsom WK, Zimmer MB, Harris MB (1999) Regulation of cardiac rhythm in hibernating mammals. Comp Biochem Physiol A Mol Integr Physiol 124:383–391

Milsom WK, Zimmer MB, Harris MB (2001) Vagal control of cardiorespiratory function in hibernation. Exp Physiol 86:791–796

Morrison SF, Nakamura K, Madden CJ (2008) Central control of thermogenesis in mammals. Exp Physiol 93:773–797

Oelkrug R, Heldmaier G, Meyer CW (2011) Torpor patterns, arousal rates, and temporal organization of torpor entry in wildtype and UCP1-ablated mice. J Comp Physiol B 181:137–145

Ootsuka Y, de Menezes RC, Zaretsky DV, Alimoradian A, Hunt J, Stefanidis A, Oldfield BJ, Blessing WW (2009) Brown adipose tissue thermogenesis heats brain and body as part of the brain-coordinated ultradian basic rest-activity cycle. Neuroscience 164:849–861

Ootsuka Y, Kulasekara K, de Menezes RC, Blessing WW (2011) SR59230A, a beta-3 adrenoceptor antagonist, inhibits ultradian brown adipose tissue thermogenesis and interrupts associated episodic brain and body heating. Am J Physiol Regul Integr Comp Physiol 301:R987–R994

Ortmann S, Heldmaier G (2000) Regulation of body temperature and energy requirements of hibernating alpine marmots (*Marmota marmota*). Am J Physiol Regul Integr Comp Physiol 278:R698–R704

Peyron C, Tighe DK, van den Pol AN, De Lecea L, Heller HC, Sutcliffe JG, Kilduff TS (1998) Neurons containing hypocretin (orexin) project to multiple neuronal systems. J Neurosci 18:9996–10015

Rampone AJ, Reynolds PJ (1991) Food intake regulation by diet-induced thermogenesis. Med Hypotheses 34:7–12

Refinetti R, Menaker M (1992) The circadian rhythm of body temperature. Physiol Behav 51:613–637

Rensing L, Meyer-Grahle U, Ruoff P (2001) Biological timing and the clock metaphor: oscillatory and hourglass mechanisms. Chronobiol Int 18:329–369

Richter CP (1927) Animal behaviour and internal drives. Quart Rev Biol 2:307–343

Robinson EL, Fuller CA (1999) Endogenous thermoregulatory rhythms of squirrel monkeys in thermoneutrality and cold. Am J Physiol 276:R1397–R1407

Sakurai T, Amemiya A, Ishii M, Matsuzaki I, Chemelli RM, Tanaka H, Williams SC, Richarson JA, Kozlowski GP, Wilson S, Arch JR, Buckingham RE, Haynes AC, Carr SA, Annan RS, McNulty DE, Liu WS, Terrett JA, Elshourbagy NA, Bergsma DJ, Yanagisawa M (1998) Orexins and orexin receptors: a family of hypothalamic neuropeptides and G protein-coupled receptors that regulate feeding behavior. Cell 92:573–585

Saper CB, Cano G, Scammell TE (2005) Homeostatic, circadian, and emotional regulation of sleep. J Comp Neurol 493:92–98

Schibler U, Naef F (2005) Cellular oscillators: rhythmic gene expression and metabolism. Curr Opin Cell Biol 17:223–229

Scholander PF, Hock R, Walters V, Irving L (1950) Adaptation to cold in arctic and tropical mammals and birds in relation to body temperature insulation, and basal metabolic rate. Biol Bull 99:259–271

Secor SM (2009) Specific dynamic action: a review of the postprandial metabolic response. J Comp Physiol B 179:1–56

Sellayah D, Bharaj P, Sikder D (2011) Orexin is required for brown adipose tissue development, differentiation, and function. Cell Metab 14:478–490

Stupfel M, Gourlet V, Court L, Perramon A, Merat P, Lemercerre C (1990a) There are basic rest-activity ultradian rhythms of carbon dioxide emission in small laboratory vertebrates characteristic of each species. Prog Clin Biol Res 341A:179–184

Stupfel M, Gourlet V, Perramon A, Merat P, Court L (1990b) Ultradian and circadian compartmentalization of respiratory and metabolic exchanges in small laboratory vertebrates. Chronobiologia 17:275–304

Swoap SJ, Gutilla MJ, Liles LC, Smith RO, Weinshenker D (2006) The full expression of fasting-induced torpor requires beta 3-adrenergic receptor signaling. J Neurosci 26:241–245

Swoap SJ, Weinshenker D (2008) Norepinephrine controls both torpor initiation and emergence via distinct mechanisms in the mouse. PLoS One 3:e4038

Tapp WN, Levin BE, Natelson BH (1981) Ultradian rhythm of plasma norepinephrine in rats. Endocrinology 109:1781–1783

Wehr TA (1992) A brain-warming function for REM sleep. Neurosci Biobehav Rev 16:379–397

Zhang W, Sunanaga J, Takahashi Y, Mori T, Skurai T, Kanmura Y, Kuwaki T (2010) Orexin neurons are indispensable for stress-induced thermogenesis in mice. J Physiol 588(Pt 21): 4117–4125

Zimmer MB, Milsom WK (2002) Ventilatory pattern and chemosensitivity in unanesthetized, hypothermic ground squirrels (*Spermophilus lateralis*). Respir Physiol Neurobiol 133:49–63

Chapter 21
Spontaneous Daily Torpor Versus Fasting-Induced Torpor in the Djungarian Hamster (*Phodopus sungorus*): Two Sides of a Medal or Distinct Phenomena?

Victoria Diedrich and Stephan Steinlechner

Abstract During the winter months the Djungarian hamster, *Phodopus sungorus*, can enter spontaneous daily torpor (SDT) even in the presence of abundant food and at thermoneutral ambient temperature. This indicates that torpor is not only a response to energy shortage and that it has not only the well-known energy saving function. Here we summarise the hallmarks of SDT and compare it to the characteristics of fasting-induced torpor (FIT) in *P. sungorus*. Some obvious differences let us conclude that these are two forms of hypothermia and appear to be regulated by different control mechanisms. Additionally, recent evidences suggest that SDT, at least in the Djungarian hamsters, has benefits beyond the energetic advantages.

Abbreviations

2-DG	2-deoxy-D-glucose
i.p.	Intraperitoneal
MA	Mercaptoacetate
RQ	Respiratory quotient
SDT	Spontaneous daily torpor
FIT	Fasting-induced torpor

V. Diedrich (✉) · S. Steinlechner
Department of Zoology, University of Veterinary Medicine, Hannover, Germany
e-mail: Victoria.Diedrich@tiho-hannover.de

S. Steinlechner
e-mail: Stephan.Steinlechner@tiho-hanover.de

T. Ruf et al. (eds.), *Living in a Seasonal World*,
DOI: 10.1007/978-3-642-28678-0_21, © Springer-Verlag Berlin Heidelberg 2012

21.1 Introduction

Several means of inducing torpor-like states of hypothermia have been investigated in the past in order to uncover the factor(s) which is (are) responsible for the proximate induction of a spontaneously occurring torpor bout. Although many details could be elucidated in the last 30 years several basic questions remain unanswered. Especially the induction of daily torpor has remained a mystery, despite many attempts and initially promising reports (reviewed in Bouma et al. 2011).

The Djungarian hamster (*Phodopus sungorus*) uses spontaneous daily torpor (SDT) and we can safely assume that it serves as an important mechanism to survive the harsh winters of the Central Asian steppes. As the shortage of food resources and thus the supply of energy can be considered as one major challenge for endothermic mammals during the cold season, it seems reasonable to attribute the occurrence of SDT to reduced food availability (Weiner 1987). And indeed, the hamsters respond to prolonged food restriction with the expression of hypothermia (Ruby and Zucker 1992; Steinlechner et al. 1986). However, the hamsters become torpid even with food ad libitum and at thermoneutral ambient temperature (Heldmaier and Steinlechner 1981; Ruf et al. 1991). With regard to the huge amount of the literature dealing with different forms of hypometabolism and hypothermia, *P. sungorus* is one of the few species showing both SDT and FIT. Hence, the hamsters prove to be excellent model organisms to compare the two phenomena that share several similarities, but also certain differences characterising them as two distinct responses to an energetic challenge. We shall give a comprehensive overview and discuss these obvious differences in terms of seasonality, preconditions, circadian organisation and several other characteristics. In addition we raise a few questions about torpor that still await answers and review recent evidences that torpid states have benefits beyond energetic advantages, at least in the Djungarian hamster.

21.2 Seasonality

The general ability to express SDT in *P. sungorus* is exclusively restricted to the winter months (Heldmaier and Steinlechner 1981) or rather to short photoperiods, i.e. less than 13 h of light per day (Hoffmann 1982). SDT starts only after about 12 weeks of exposure to the winter-like photoperiodic conditions and then continues for about 3–4 months (Elliott et al. 1987; Ruf et al. 1993). SDT emerges together with several other seasonal adaptations such as reduction of body mass, involution of gonads, change from a brown summer to a white winter fur and activation of non-shivering thermogenesis (Heldmaier et al. 1982; Hoffmann 1973). These preparations for winter form an 'adaptive syndrome' (Heldmaier and Lynch 1986) and primarily constitute additional energy saving mechanisms, but may also serve as preconditions for the expression of SDT. Furthermore, external cues representing energetic challenges, like food restriction or low ambient temperatures, may facilitate the onset of the torpor season during the exposure to short photoperiods, or may increase the incidence of SDT, but they are not required to ultimately induce SDT. Several studies suggested the existence of a short photoperiod-induced seasonal cycle of endogenous readiness for SDT (Bartness et al. 1989; Elliott et al. 1987; Ruf et al. 1993). During this 'torpor season' the frequency of SDT is low during November, reaches a maximum during January and declines again thereafter as the animals change back to summer conditions.

In contrast, FIT in *P. sungorus* does not require any photoperiodic preconditions, but can be induced at any time of the year after a prolonged period of reduced food availability, which forces body mass to the low level normally occurring in winter.

21.3 Preconditions for Torpor Expression

21.3.1 Body Constitution

One major precondition that both SDT and FIT have in common in *P. sungorus* is the reduction of body mass prior to the first expression of hypothermia. The low body mass (~25 g) of hamsters showing FIT is the result of a forced reduction of food and thus of lower energy intake (Ruby et al. 1993), whereas hamsters, exposed to short photoperiods, spontaneously down-regulate their energy balance and body mass to a new, winter-adapted set point (Steinlechner et al. 1983). Experiments by Steinlechner and colleagues (later confirmed by Morgan et al. 2003) showed that a temporal food restriction during the short photoperiod-induced weight reduction leads to an additional drastic fall of body mass which is compensated for after restoration of ad libitum food availability as body mass returns to the original seasonal time course. In contrast, the body mass reduction during food restriction does not seem to be regulated, but to be a direct and faster (10–15 days) result of a decrease in energy availability which is supposed to be a much more challenging alteration of body constitution.

During acclimation to short photoperiods, it is primarily the loss of body fat that substantially contributes to the reduction of body mass in *P. sungorus* (Bartness 1996; Wade and Bartness 1984). Therefore, it has been argued that it is not the body mass per se, but the low amount of body fat together with the correspondingly low concentrations of leptin that serve as permissive SDT-inducing factors. According to their study, reduced leptin concentrations appear necessary for the occurrence of SDT, but simply reducing leptin concentrations below a threshold value is not sufficient to induce SDT (Freeman et al. 2004). On the other hand, increasing the concentration by daily injections of leptin during the torpor season does not influence torpor frequency or duration (Schmidt et al. 1997). As food restriction also leads to a decline in body fat reserves, leptin might also be involved in the expression of FIT. Although there is no evidence for this mechanism in Djungarian hamsters, Döring et al. (1998) revealed an inhibiting effect of exogenous leptin administration in fasting mice. Despite the fact that the body weight reduction is achieved in a different way, its general outcome offers a first, significant similarity between the occurrence of SDT and FIT.

These meaningful similarities encouraged scientists to conduct further studies in order to examine in detail, which energetic pathways and which metabolic fuels might be necessary for *P. sungorus* to enter torpor. Heldmaier et al. showed that during the first hours of SDT blood glucose, serum lipids and RQ were on a high level and only declined gradually later during the torpor bout. This suggests that glucose is the primary fuel during induction of SDT whereas lipids are only used during prolonged torpor. It also demonstrates that the induction of hypometabolism is not necessarily due to exogenous or endogenous substrate limitations (Heldmaier et al. 1999). In contrast, FIT is characterised by lipid utilisation already before the onset of hypothermia, combined with a metabolic/respiratory acidosis (Malan et al. 1988; Nestler 1990). Food restriction increases hypothalamic gene expression of orexigenic peptides while the expression of anorexigenic peptides is suppressed (Mercer et al. 2000; Rousseau et al. 2002). Hypothalamic gene expression of hamsters during adaptation to short photoperiods, however, shows a different pattern and it has been concluded that despite the reduction in body mass, the short day hamsters are not starving, but are in energetic balance (Morgan et al. 2003). By approaching this question from a completely different angle, Stamper et al. (1999) came to a similar conclusion: "The energetic state of short photoperiod hamsters apparently is not equivalent to that of food-restricted hamsters in long photoperiod…". These authors used a treatment with 2-deoxy-D-glucose (2-DG), an inhibitor of cellular glucose oxidation, which previously had been shown to induce a torpor-like state in *P. sungorus* (Dark et al. 1994), and compared it with the inhibition of fatty acid metabolism by mercaptoacetate (MA) in long photoperiod and during adaptation to short photoperiod. Torpor-like states could be induced by 2-DG in both long and short photoperiod, but contrary to expectation, 2-DG decreased the induction of torpor-like states in short photoperiod rather than facilitating it. Treatment with MA did not induce torpor-like states in both photoperiods, but caused a slight hypothermia only. Hence, the authors concluded that SDT in Djungarian hamsters exposed to a short photoperiod was not induced by increased responsiveness to glucose availability; instead, responsiveness to glucoprivation appeared to be blunted by these photoperiodic conditions. Thus, the occurrence of SDT may be unrelated to concurrent metabolic fuel availability (Stamper et al. 1999). The same laboratory showed in other studies that blood glucose concentration is significantly reduced during SDT. However, even a sustained, pronounced hypoglycaemia induced by insulin treatment did not trigger torpor bouts in *P. sungorus*. The authors concluded that a reduced blood glucose concentration is rather a consequence than the cause of torpor and this questions whether induction of torpor-like states by 2-DG is due to its glucoprivic actions (Dark et al. 1999).

21.3.2 Reproductive State

One striking difference when comparing SDT and FIT is the reproductive and hormonal state of *P. sungorus*. In contrast to other seasonal breeding mammals (Geiser et al. 2008; Geiser and Masters 1994; Morrow and Nicol 2009), Djungarian hamsters employ SDT exclusively during their non-reproductive season and even rely on the physiological and behavioural inhibition of reproduction as a prerequisite for SDT expression. The first appearance of SDT is preceded by a reduction of pituitary gonadotropin release (Duncan and Goldman 1984), followed by a complete regression of all sexual and accessory organs and, consequently, a drastic decline in sexual hormone concentrations (Elliott et al. 1987; Vitale et al. 1985). In order to document the strong relationship between the hamsters' reproductive axis and their onset of SDT, several experiments have been conducted by artificially manipulating the gonadal and hormonal state of *Phodopus*. In summary, castration of short photoperiod-exposed hamsters prolongs the duration of the individual torpor season (Bartness et al. 1989; Elliott et al. 1987), whereas SDT is effectively inhibited by exogenous testosterone administration (Vitale et al. 1985). Although testicular regression and thus low sexual hormone concentrations seem to be an important prerequisite for SDT, it is no prerequisite for the induction of FIT under long photoperiods (Vitale et al. 1985). Djungarian hamsters are capable of FIT during summer-like photoperiods in their maximal reproductive state (Ruby et al. 1993). Interestingly, exogenous testosterone remains a potent inhibitor of the occurrence of FIT in gonadectomized hamsters during a long photoperiod (Ruby et al. 1993), which is quite counterintuitive considering the lack of effect of high endogenous testosterone levels in intact animals. Similarly, in reproductively competent males and in those bearing testosterone-filled capsules producing supraphysiological blood testosterone concentrations 2-DG injection induced torpor-like states within 1 h of treatment (Dark et al. 1996).

21.4 Circadian Organization

FIT is not only expressed independently of the photoperiodic regime and can therefore occur at any time of the year, but it also lacks circadian rhythmicity (Fig. 21.2) (Ruby and Zucker 1992; Steinlechner et al. 1986). In contrast, SDT in *P. sungorus* is an inherent component of the circadian rhythm of body temperature and is consequently synchronised to the daily light–dark-cycle (Ruf et al. 1989). More precisely, a torpor episode is restricted to the photophase (resting phase) of the nocturnal hamsters (Fig. 21.1) and does not interfere with their nightly foraging activity. An animal can profit energetically most if torpor occurs at a time of day when the probability for successful foraging is highest compared to the risk of becoming prey. Due to this strong circadian control of SDT the animal can use the night time for foraging under the cover of darkness. Several studies showed the

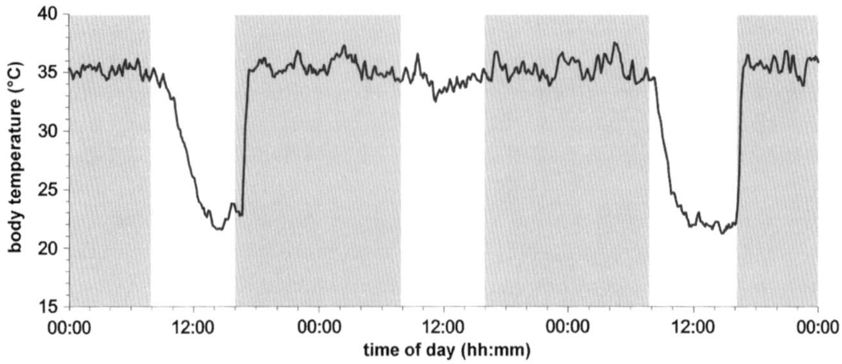

Fig. 21.1 Continuous measurement of body temperature in 6-min-intervals via i.p.-implanted telemetric device (Model X, Minimitter Co., Sunriver, Oregon) of an ad libitum fed Djungarian hamster during exposure to a winter-like short photoperiod (LD 8:16) and an ambient temperature of 23°C. The *grey shading* indicates the daily scotophase. During the photophase of day 1 and day 3, the hamster shows SDT bouts. The plateau phase of hypothermia is characterised by brief alternations of body temperature resulting form thermoregulatory bursts of metabolic rate

reliance of a torpor bout on the light–dark-cycle by systematically manipulating defined parts of the hamsters' circadian system, including the suprachiasmatic nuclei and the pineal gland, which play an essential role in regulating the timing of circadian and seasonal rhythms in all seasonal mammals (Goldman and Darrow 1983; Körtner and Geiser 2000). While an ablation of the suprachiasmatic nuclei merely resulted in a temporal disruption of the torpor expression pattern (Ruby and Zucker 1992), a pinealectomy substantially decreased (Lynch et al. 1980) or even completely prevented SDT (Steinlechner et al. 1986; Vitale et al. 1985). In contrast, pinealectomy turned out to have no effect on the expression of FIT, an observation that again corroborates the independence of this type of hypothermia from any circadian rhythmicity and additionally represents the most obvious difference to SDT (Steinlechner et al. 1986).

21.5 Characteristics of Torpor Bouts

A more detailed examination of individual bouts of SDT reveals a mean duration of 4–8 h, which perfectly fits into the winter-like short photoperiod. The variability of torpor duration seems to be a lot greater in FIT with individual bout lengths of 2–12 h (Steinlechner et al. 1986). Additionally, several studies described multi-bout events during FIT (Figs. 21.2, 21.3), whereby hamsters entered a second phase of hypothermia immediately after the termination of arousal from their first phase of hypothermia (Ruby and Zucker 1992; Steinlechner et al. 1986). Interestingly, the expression of these multiple bouts of hypothermia was enhanced after ablation of the suprachiasmatic nuclei (Ruby and Zucker 1992). Beside the

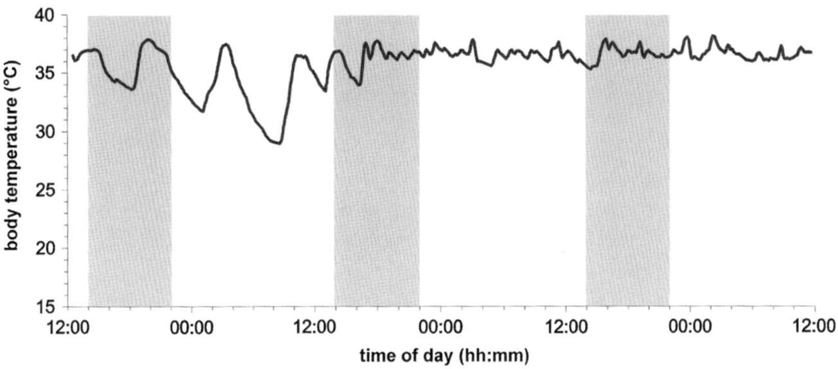

Fig. 21.2 Continuous measurement of body temperature in 6-min-intervals via i.p.-implanted telemetric device (Model X, Minimitter Co., Sunriver, Oregon) of a food-restricted (70% of ad libitum food intake) Djungarian hamster in a summer-like long photoperiod (LD 16:8) and an ambient temperature of 20 ± 1°C. The *grey shading* indicates the daily scotophase. The hamster shows 3 consecutive fasting-induced bouts of torpor (FIT). This multi-bout torpor covers the photophase, as well as the scotophase of day 1. There is no indication for a plateau phase of hypothermia, but a direct arousal after a considerably long entrance phase

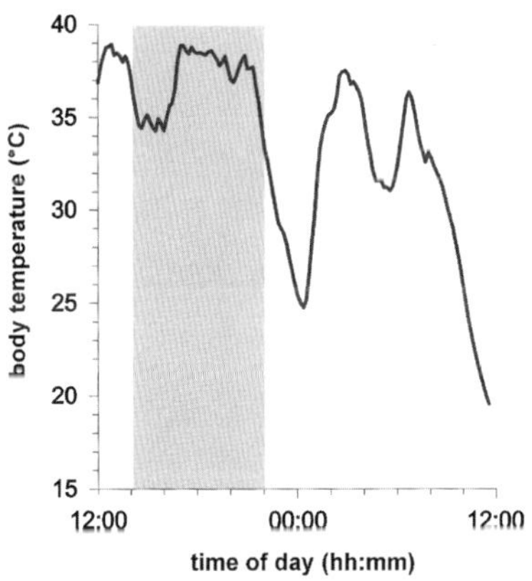

Fig. 21.3 Continuous measurement of body temperature in 10-min-intervals via i.p.-implanted telemetric device (Ecologger 2.4, Novikov, Novosibirsk) of a food-restricted (70% of ad libitum food intake) Djungarian hamster in a summer-like long photoperiod (LD 16:8) and an ambient temperature of 15 ± 1°C. The *grey shading* indicates the daily scotophase. The hamster failed to rewarm after the expression of multi-bout FIT

multi-bout FIT, (Steinlechner et al. 1986) observed a substantially lower frequency of expression of FIT compared to SDT, with the latter occurring two to four times per week in Djungarian hamsters (Ruf et al. 1993), but again with a high individual variability. When a colony of hamsters is kept under identical conditions throughout the winter, a few hamsters show torpor almost every day while others never enter a torpor bout during the whole winter season. The origin and nature of this high individual variability remains an enigma. One could speculate a genetic background with a range of phenotypes whereby hamsters with a high tendency to enter SDT might have a clear advantage to survive a severe Siberian winter. During relatively mild winters, hamsters with a low tendency to exhibit SDT might have an advantage because becoming torpid also imposes risks. In addition, hamsters not entering torpor might be able to maintain their territory or become reproductively competent earlier in the next spring. With a random distribution of relatively mild and harsh winters these different traits could become stable phenotypes.

There is also certain individual variability in the onset and occurrence of FIT, but in this case the main reason is considered to be an individual difference in condition, amount of fat stores and probably also in the efficiency for assimilation of food. In contrast to the ad libitum fed hamsters the fasting or even starving animals have no real choice—they either have to save energy or die!

21.6 Conclusion

The striking differences between SDT and FIT discussed above and summarised in Table 21.1 lead to the conclusion that these two types of hypothermia in *P. sungorus* appear to be mediated by different control mechanisms. The SDT demonstrates an inherent, precisely regulated, but indirect response to reduced food and energy availability as a long-term adaptation to the short photoperiods of the energetically challenging winter. In contrast, FIT resembles a more acute and direct reaction towards energetic limitations. It is an emergency measure and some authors even propose a possible failure of control mechanisms induced by a forced reduction of food and energy availability (Ruf et al. 1991). And in fact, we can provide several arguments suggesting that FIT shows characteristics of a struggle against breakdown of thermoregulation:

(1) In contrast to the quite regular circadian pattern of body temperature in winter-acclimatised hamsters, body temperature of hamsters after prolonged fasting is very irregular with 20–40% higher fluctuations, and barely shows a circadian rhythm (Fig. 21.2, see also Ruby and Zucker 1992). (2) Fasted animals are more restless and eager to search for food, resulting in a higher locomotor activity in fasted animals as compared to ad libitum fed animals. In contrast, in fed hamsters the animals with the highest torpor frequency show the lowest activity level at night (Ruf et al. 1991). (3) Bouts of FIT are not restricted to the resting phase, but can occur at any time of day. Hence, arousal takes place often at a time of day when successful foraging (i.e. replenishing of energy stores) would be unlikely. (4) Furthermore, FIT bouts are often very short and therefore they cost additional energy rather than save it. Torpor can only save energy if its total costs (incl. the high costs for rewarming) are lower than the costs of remaining normothermic during the same time (Wojciechowski and Jefimow 2006). (5) Compared to SDT bouts which are usually characterised by a distinct plateau phase of hypometabolism and hypothermia (Fig. 21.1), interspersed by thermoregulatory bursts of metabolic rate, bouts of FIT do not show this pattern.

Table 21.1 Summary of the major differences and similarities between spontaneous daily torpor (SDT) and fasting-induced torpor (FIT) in Djungarian hamsters

Dependent on …	SDT	FIT
Photoperiod	+	–
	(Heldmaier and Steinlechner 1981; Ruf et al. 1989)	(Steinlechner et al. 1986)
Ambient temperature	±	?
	(Elliott et al. 1987; Ruf et al. 1993)	
Circadian control	+	–
	(Ruf et al. 1989)	(Ruby and Zucker 1992; Steinlechner et al. 1986)
Preparatory time period	Long	Short
	(Elliott et al. 1987; Ruf et al. 1993)	(Ruby et al. 1993; Ruby and Zucker 1992)
Low body mass/fat mass	+	+
	(Bartness 1996; Wade and Bartness 1984)	(Ruby et al. 1993; Ruby and Zucker 1992)
Glucose metabolism	+	–
	(Heldmaier et al. 1999)	(Dark et al. 1999; Stamper et al. 1999)
Fat metabolism	±	+
	(Heldmaier et al. 1999)	(Malan et al. 1988; Nestler 1990)
Reproductive state	+	–
	(Bartness et al. 1989; Elliott et al. 1987; Vitale et al. 1985)	(Dark et al. 1996; Ruby et al. 1993; Yellon and Goldman 1987)
Plateau phase during torpor	+	?/–
	(Heldmaier and Steinlechner 1981)	(Steinlechner et al. 1986)

Instead, the entrance phase of a FIT bout often seems to be followed by an immediate arousal phase (Figs. 21.2, 21.3). (6) Finally, in the few cases where a hamster died during the long-term fasting, it always occurred during an attempt to rewarm (Fig. 21.3).

Although superficially the two phenomena appear very similar and they certainly share common features, a closer look reveals characteristic differences that point to different controlling mechanisms and even warrant the conclusion that (at least) in the Djungarian hamster they are not just two sides of a medal. If that stimulates colleagues to think about it and have a closer look at other species and at what happens physiologically during induced torpor versus during voluntary (unforced) torpor, than the purpose of our short review is fully served. After all, there are still many questions unanswered with respect to what triggers individual torpor bouts and what are the costs and benefits of torpor.

21.7 Open Questions

There is no doubt that torpor is ultimately an energy saving strategy. In *P. sungorus* the metabolic savings during SDT by itself were estimated to approach only 30% (Heldmaier 1989), but other, indirect savings due to torpor may increase this percentage considerably. Locomotor activity and food consumption, for example, are negatively correlated with torpor frequency. As a consequence a hamster that entered torpor almost every day had 75% lower food consumption in comparison to a hamster that remained normothermic during the same 15-day period (Ruf et al. 1991). Under certain circumstances, e.g. in desert species, the high energetic costs of arousal can even be reduced by basking behaviour (Warnecke et al. 2008). However, by focussing on the energetically beneficial aspects of torpor, other advantages an organism can gain from exhibiting torpor were probably overlooked. Today it becomes more and more apparent that a voluntary hypometabolism, followed by a reduction in body temperature is not only, and not always, a response to energy shortage. Why do Djungarian hamsters enter torpor when there is no acute energy deficiency? We hesitate to impute foresight to the hamsters, but SDT may be compared to a hoarding behaviour, which probably serves the same end. However, what makes an individual hamster decide to enter torpor? Might there be other advantages of torpor besides energy savings? Bieber and Ruf (2009) reported that adult dormice returned to dormancy during summer although they did not suffer from energy constraints for their own survival. This happened when a major purpose of life, namely reproduction, was unlikely to be successful due to a "crop failure", i.e. a seeding failure of beech and oak that allows the rapid fattening of the juveniles in fall. The authors suggest that this almost year-round torpor was evolved as a predator avoidance strategy helping to survive. This is further supported by a recent study (Turbill et al. 2011a) showing that hibernation in dormice is associated with increased longevity. Furthermore the same group reported recently that in *P. sungorus* the relative telomere length increases during the torpor season, suggesting an increased somatic maintenance and retarded ageing (Turbill et al. 2011b). This clearly goes well beyond energy saving strategies and shows that our knowledge is still very deficient in many aspects of this multifaceted phenomenon torpor.

References

Bartness TJ (1996) Photoperiod, sex, gonadal steroids, and housing density affect body fat in hamsters. Physiol Behav 60:517–529

Bartness TJ, Elliott JA, Goldman BD (1989) Control of torpor and body weight patterns by a seasonal timer in Siberian hamsters. Am J Physiol 257:R142–R149

Bieber C, Ruf T (2009) Summer dormancy in edible dormice (*Glis glis*) without energetic constraints. Naturwissenschaften 96:165–171

Bouma HR, Verhaag EM, Otis JP, Heldmaier G, Swoap SJ, Strijkstra AM, Henning RH, Carey HV (2011) Induction of torpor: mimicking natural metabolic suppression for biomedical applications. J Cell Physiol. doi:10.1002/jcp.22850

Dark J, Lewis DA, Zucker I (1999) Hypoglycemia and torpor in Siberian hamsters. Am J Physiol 276:R776–R781

Dark J, Miller DR, Licht P, Zucker I (1996) Glucoprivation counteracts effects of testosterone on daily torpor in Siberian hamsters. Am J Physiol 270:R398–R403

Dark J, Miller DR, Zucker I (1994) Reduced glucose availability induces torpor in Siberian hamsters. Am J Physiol 267:R496–R501

Döring H, Schwarzer K, Nuesslein-Hildesheim B, Schmidt I (1998) Leptin selectively increases energy expenditure of food-restricted lean mice. Int J Obes Relat Metab Disord 22:83–88

Duncan MJ, Goldman BD (1984) Hormonal regulation of the annual pelage color cycle in the Djungarian hamster, *Phodopus sungorus*. I. Role of the gonads and pituitary. J Exp Zool 230:89–95

Elliott JA, Bartness TJ, Goldman BD (1987) Role of short photoperiod and cold exposure in regulating daily torpor in Djungarian hamsters. J Comp Physiol A 161:245–253

Freeman DA, Lewis DA, Kauffman AS, Blum RM, Dark J (2004) Reduced leptin concentrations are permissive for display of torpor in Siberian hamsters. Am J Physiol Regul Integr Comp Physiol 287:R97–R103

Geiser F, Christian N, Cooper CE, Körtner G, McAllan BM, Pavey CR, Turner JM, Warnecke L, Willis CKR, Brigham RM (2008) Torpor in marsupials: recent advances. In: Lovegrove BG, McKechnie AE (eds) Hypometabolism in animals: hibernation, torpor and cryobiology. University of KwaZulu-Natal, Pietermaritzburg, pp 297–306

Geiser F, Masters P (1994) Torpor in relation to reproduction in the mulgara, *Dasycercus cristicauda* (Dasyuridae: Marsupialia). J Therm Biol 1:30–40

Goldman BD, Darrow JM (1983) The pineal gland and mammalian photoperiodism. Neuroendocrinology 37:386–396

Heldmaier G (1989) Seasonal acclimatization of energy requirements in mammals: functional significance of body weight control, hypothermia and hibernation. In: Wieser W, Gnaiger E (eds) Energy transformation in cells and organisms. Fischer, Stuttgart, pp 130–139

Heldmaier G, Klingenspor M, Werneyer M, Lampi BJ, Brooks SP, Storey KB (1999) Metabolic adjustments during daily torpor in the Djungarian hamster. Am J Physiol 276:E896–E906

Heldmaier G, Lynch GR (1986) Pineal involvement in thermoregulation and acclimatization. Pineal Res Rev 4:97–139

Heldmaier G, Steinlechner S (1981) Seasonal pattern and energetics of short daily torpor in the Djungarian hamster, *Phodopus sungorus*. Oecologia 48:265–270

Heldmaier G, Steinlechner S, Rafael J (1982) Nonshivering thermogenesis and cold resistance during seasonal acclimatization in the Djungarian hamster. J Comp Physiol 149:1–9

Hoffmann K (1973) The influence of photoperiod and melatonin on testis size, body weight, and pelage colour in the Djungarian hamster (*Phodopus sungorus*). J Comp Physiol 85:267–282

Hoffmann K (1982) The critical photoperiod in the Djungarian hamster *Phodopus sungorus*. In: Aschoff J, Daan S, Groos G (eds) Vertebrate circadian systems. Springer, Berlin, pp 297–304

Körtner G, Geiser F (2000) The temporal organization of daily torpor and hibernation: circadian and circannual rhythms. Chronobiol Int 17:103–128

Lynch GR, Sullivan JK, Gendler SL (1980) Temperature regulation in the mouse, *Peromyscus leucopus*: effects of various photoperiods, pinealectomy and melatonin administration. Int J Biometeorol 24:49–55

Malan A, Mioskowski E, Calgari C (1988) Time-course of blood acid-base state during arousal from hibernation in the European hamster. J Comp Physiol B 158:495–500

Mercer JG, Moar KM, Ross AW, Hoggard N, Morgan PJ (2000) Photoperiod regulates arcuate nucleus POMC, AGRP, and leptin receptor mRNA in Siberian hamster hypothalamus. Am J Physiol Regul Integr Comp Physiol 278:R271–R281

Morgan PJ, Ross AW, Mercer JG, Barrett P (2003) Photoperiodic programming of body weight through the neuroendocrine hypothalamus. J Endocrinol 177:27–34

Morrow G, Nicol SC (2009) Cool sex? Hibernation and reproduction overlap in the echidna. PLoS ONE 4:e6070

Nestler JR (1990) Intracellular pH during daily torpor in *Peromyscus maniculatus*. J Comp Physiol B 159:661–666

Rousseau K, Atcha Z, Cagampang FR, Le Rouzic P, Stirland JA, Ivanov TR, Ebling FJ, Klingenspor M, Loudon AS (2002) Photoperiodic regulation of leptin resistance in the seasonally breeding Siberian hamster (*Phodopus sungorus*). Endocrinology 143:3083–3095

Ruby NF, Nelson RJ, Licht P, Zucker I (1993) Prolactin and testosterone inhibit torpor in Siberian hamsters. Am J Physiol 264:R123–R128

Ruby NF, Zucker I (1992) Daily torpor in the absence of the suprachiasmatic nucleus in Siberian hamsters. Am J Physiol 263:R353–R362

Ruf T, Klingenspor M, Preis H, Heldmaier G (1991) Daily torpor in the Djungarian hamster (*Phodopus sungorus*): interactions with food intake, activity, and social behaviour. J Comp Physiol B160:609–615

Ruf T, Steinlechner S, Heldmaier G (1989) Rhythmicity of body temperature and torpor in the Djungarian hamster, *Phodopus sungorus*. In: Malan A, Canguilhem B (eds) Living in the cold II. John Libbey Eurotext, London, pp 53–60

Ruf T, Stieglitz A, Steinlechner S, Blank JL, Heldmaier G (1993) Cold exposure and food restriction facilitate physiological responses to short photoperiod in Djungarian hamsters (*Phodopus sungorus*). J Exp Zool 267:104–112

Schmidt I, Döring H, Stehling O, Nuesslein-Hildesheim B, Steinlechner S, Schwarzer K (1997) Leptin disinhibits rather than stimulates sympathetically mediated energy expenditure. In: Blum WF, Kiess W, Rascher W (eds) Leptin—the voice of adipose tissue. J. A. Barth Verlag, Edition J & J, Heidelberg, pp 133–139

Stamper JL, Dark J, Zucker I (1999) Photoperiod modulates torpor and food intake in Siberian hamsters challenged with metabolic inhibitors. Physiol Behav 66:113–118

Steinlechner S, Heldmaier G, Becker H (1983) The seasonal cycle of body weight in the Djungarian hamster: photoperiodic control and the influence of starvation and melatonin. Oecologia 60:401–405

Steinlechner S, Heldmaier G, Weber C, Ruf T (1986) Role of photoperiod: pineal gland interaction in torpor control. In: Heller HC, Wang LCH, Musacchia X (eds) Living in the cold: physiological and biochemical adaptations. Elsevier Science Publishing Co., Inc., New York, pp 301–307

Turbill C, Bieber C, Ruf T (2011a) Hibernation is associated with increased survival and the evolution of slow life histories among mammals. Proc Biol Sci 278:3355–3363

Turbill C, Smith S, Deimel C, Ruf T (2011b) Daily torpor is associated with telomere length change over winter in Djungarian hamsters. Biol Lett. doi:10.1098/rsbl.2011.0758

Vitale PM, Darrow JM, Duncan MJ, Shustak CA, Goldman BD (1985) Effects of photoperiod, pinealectomy and castration on body weight and daily torpor in Djungarian hamsters (*Phodopus sungorus*). J Endocrinol 106:367–375

Wade GN, Bartness TJ (1984) Effects of photoperiod and gonadectomy on food intake, body weight, and body composition in Siberian hamsters. Am J Physiol 246:R26–R30

Warnecke L, Turner JM, Geiser F (2008) Torpor and basking in a small arid zone marsupial. Naturwissenschaften 95:73–78

Weiner J (1987) Maximum energy assimilation rates in the Djungarian hamster (*Phodopus sungorus*). Oecologia 72:297–302

Wojciechowski MS, Jefimow M (2006) Is torpor only an advantage? Effect of thermal environment on torpor use in the Siberian hamsters (*Phodopus sungorus*). J Physiol Pharmacol 57(Suppl 8):83–92

Yellon SM, Goldman BD (1987) Influence of short days on diurnal patterns of serum gonadotrophins and prolactin concentrations in the male Djungarian hamster (*Phodopus sungorus*). J Reprod Fertil 80:167–174

Chapter 22
Does the Road Traveled Matter? Natural Versus Prematurely Induced Arousal from Torpor

Jenifer C. Utz and Frank van Breukelen

Abstract Although hibernating animals spontaneously arouse from torpor at regular intervals, the practice of prematurely inducing arousal is common. Herein we review the many differences between natural and prematurely induced arousal to address the question of whether these two paths to euthermy are truly synonymous events. We present data demonstrating that the duration of the interbout arousal (IBA) is significantly reduced following a prematurely induced arousal and that the time required to respond to the induction stimulus is influenced by the duration of time spent in torpor. There are numerous alterations in intracellular and whole animal physiology when arousal is prematurely induced; thus we recommend that careful consideration be given to experiments utilizing this type of arousal mechanism.

22.1 Introduction

Although hibernation physiology has been scrutinized for over a century, we are just beginning to appreciate that we must consider more than a bipartite physiology. Hibernating animals are frequently regarded as functioning in one of two dissimilar states, torpor or euthermy. Torpor is characterized by low body temperatures (T_b) and severely depressed metabolic rates, whereas the interbout arousal (IBA) is characterized by euthermic T_b and high metabolic rates (Barnes 1989; Carey et al. 2003).

J. C. Utz (✉) · F. van Breukelen
School of Life Sciences, University of Nevada, Las Vegas, NV, USA
e-mail: utzj@unlv.nevada.edu

T. Ruf et al. (eds.), *Living in a Seasonal World*,
DOI: 10.1007/978-3-642-28678-0_22, © Springer-Verlag Berlin Heidelberg 2012

Recently, we have begun to understand that not all torpid animals are the same. The internal workings of an animal in the first days of torpor are different from an animal approaching the end of a torpor bout. For example, the distribution of the ribosome pool and the types of transcripts associated with ribosomes differ markedly between early and late torpor (van Breukelen and Martin 2001; Pan and van Breukelen 2011). In addition to changes in cellular physiology, important differences are also evident at the level of systemic function. Cardiovascular and respiratory system functions vary across a torpor bout (Steffen and Riedesel 1982; Milsom et al. 1999, 2001). The episodic ventilation characteristic of torpor becomes continuous during arousal (Steffen and Riedesel 1982). Changes in heart rate during entrance and early torpor are primarily mediated via the parasympathetic nervous system, whereas arousal from torpor is primarily mediated via sympathetic tone (Milsom et al. 1999).

Increased awareness of the spectrum of physiological activity represented throughout a torpor bout necessitates a more careful consideration of the hibernation literature. In particular, greater consideration should be given to arousal, the transition from torpor to the IBA state. Although animals arouse spontaneously with highly predictable timing, researchers often induce a premature arousal from torpor. Natural and prematurely induced arousals have often been regarded as synonymous events; however, we must now consider the question, “Does the road traveled matter?” In other words, is the path of a natural arousal equivalent to the path of a prematurely induced arousal? In both cases, heat production and metabolic activity increase, ultimately driving the animal into the euthermic state of the interbout arousal. Although direct comparisons between natural and prematurely induced arousal are rare, meaningful differences in both cellular and systemic functions have been documented. When the details of the rewarming process are evaluated, it seems that the road traveled does indeed matter.

22.2 Physiological Transitions of Natural Arousal

Natural arousals occur spontaneously with highly predictable timing, presumably in response to endogenous cues. The duration of consecutive torpor bouts during mid-winter often varies by less than 10% (Twente and Twente 1965a, b; Twente et al. 1977). When compared to the process of entering torpor, arousal from torpor is a rapid event. Golden-mantled ground squirrels (*Callospermophilus lateralis*) can rewarm from near freezing temperatures to euthermy in a mere 2–3 h (Twente and Twente 1965a; Utz et al. 2007). Maximum rewarming rates for individual ground squirrels (*Callospermophilus lateralis*) can approach 1.5°C per minute (Utz et al. 2007). This rewarming from torpor is driven by dynamic changes in the activity of many organ systems.

The process of arousal involves incongruent changes in various physiological systems; in other words, not all processes are re-engaged concurrently. Oxygen consumption, blood flow, and cellular metabolism all resume at different rates and peak at different times. For example in arousing 13-lined ground squirrels (*Ictidomys tridecemlineatus*), blood flow posterior to the diaphragm is restricted until the thoracic temperature is ~25°C (Bullard and Funkhouser 1962). In arousing arctic ground squirrels (*Urocitellus parryii*), oxygen consumption peaks when abdominal T_b is ~5°C (Tøien et al. 2001). During arousal, animals must successfully reconcile the incongruence in oxygen consumption, blood flow, and changing body temperature in the midst of rapidly changing metabolic demand as cellular processes (transcription, protein metabolism, mitochondrial respiration) and systemic activities (heart rate, respiration) are re-engaged (van Breukelen and Martin 2001, 2002a, 2002b; van Breukelen et al. 2004; Velickovska et al. 2005; Velickovska and van Breukelen 2007; Martin et al. 1999; Steffen and Riedesel 1982). As an animal arouses, the increasing body temperature provides a foundation for the escalation of both cellular and systemic physiological activity.

At a fundamental level, heat is generated as a by-product of biochemical reactions. As temperature increases, the rate of most biochemical reactions also increases (Withers 1992). Therefore, the rate at which animals rewarm has been expected to progressively increase as body temperature increases until a euthermic set point is approached. Indeed, models have been presented for animals to rewarm as rapidly as possible (Stone and Purvis 1992). Rewarming quickly minimizes heat loss to the environment and conserves an animal's fuel stores. Figure 22.1 presents an idealized arousal according to the aforementioned models as well as actual measurements from an arousing animal. Although models for rewarming as quickly as possible are sensible, this depiction of arousal is somewhat inaccurate (Fig. 22.1). In actuality, the rate of rewarming decreases long before a euthermic set point is approached (Fig. 22.1). When golden-mantled ground squirrels (*Callospermophilus lateralis*) are arousing from a torpid body temperature of approximately 5°C, peak warming rates are consistently achieved around a body temperature of 16°C (Utz et al. 2007). In other words, animals start slowing down despite the need to increase body temperature by another 20°C. Perhaps there are detrimental consequences associated with warming too quickly that mitigate the additional energy expenditure of warming more slowly.

We previously investigated the effect of ambient temperature (a modulator of torpid body temperature) on the process of natural arousal. Ambient temperature was found to impact the rate of rewarming (RRW), the time required to achieve maximum RRW, and the body temperature associated with the maximum RRW in golden-mantled ground squirrels (*Callospermophilus lateralis*; Utz et al. 2007). One parameter was found to be independent of ambient temperature; all squirrels reached their maximum RRW when they had generated 30–40% of the heat required to reach a euthermic body temperature (Utz et al. 2007). These data suggest that arousal is more regulated than previously expected. Rather than simply allowing temperature to dictate metabolic reactions and thus rewarming rates, the data indicate complex regulation of heat gain and heat loss that is influenced both by body temperature and time.

Brown adipose tissue (BAT) and skeletal muscle are the major sites of heat production during arousal (Carey et al. 2003; Fons et al. 1997; Genin et al. 2003).

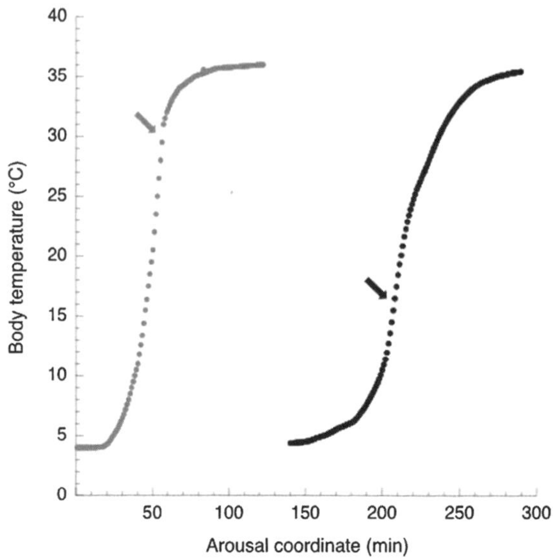

Fig. 22.1 Elevation in body temperature according to an energetically favorable model of rewarming as quickly as possible (*gray*) and measurements from an arousing golden-mantled ground squirrel (*black*) housed at an ambient temperature of 4°C. *Arrows* indicate the occurrence of peak rewarming rates. The energy saving model places peak rates at a higher body temperature than the real occurrence. The *x*-axis represents time in minutes with two distinct arousals plotted. The traces are offset rather than overlaid on this coordinate to facilitate comparison. The beginning and end of each trace was determined based on the rate of change in body temperature (Utz et al. 2007)

For a typical arousal, initial increases in T_b appear to be due to BAT activity with shivering thermogenesis beginning to contribute heat midway through the arousal (Fons et al. 1997). In two species of ground squirrels, the onset of shivering thermogenesis varies from when T_b >12°C (*Ictidomys tridecemlineatus* housed at 5°C) to T_b >15°C (*Urocitellus parryii* housed at 2°C; taken respectively from Phillips and Heath 2004; Tøien et al. 2001). Appropriate cardiovascular system function provides a foundation for the heat generation and dissipation that occur during arousal. BAT and skeletal muscle must be supplied with adequate metabolic fuels in order to facilitate the chemical reactions responsible for heat production (Cannon and Nedergaard 2004; Horwitz et al. 1968; Fons et al. 1997; Genin et al. 2003). This heat must then be conducted into other tissues via the cardiovascular system.

Regional blood flow is regulated during arousal. Blood flow posterior to the diaphragm is restricted in arousing 13-lined ground squirrels (*Ictidomys tridecemlineatus*) until the thoracic temperature is ~25°C (Bullard and Funkhouser 1962). Such restrictions of peripheral blood flow are common in other species of hibernators. Blood flow to the hind foot is markedly reduced during early arousal in hamsters (*Mesocricetus auratus*; Osborne et al. 2005). In the big brown bat (*Eptesicus fuscus*) anterior organs receive increasing fractions of cardiac output while posterior organs receive decreasing fractions of blood after the initiation of arousal (Rauch and Beatty 1975).

Certainly arousal from torpor is an intricate and multifaceted process; re-engaging the physiological activity of an entire organism is no small feat. Although many investigations have utilized arousing animals, limited attention has been directed toward how those animals aroused (Table 22.1). Given the dynamic nature of arousal, it seems reasonable to expect that a prematurely induced arousal from torpor might be difficult for an animal to regulate due to mismatches between external stimuli and endogenous timing cues.

22.3 Direct Comparisons Between Natural and Prematurely Induced Arousal

Recognizing arousal as a multifaceted, regulated process necessitates consideration of how an animal arouses from torpor. Although the occurrence of a natural arousal from torpor can be reliably predicted, much hibernation research has utilized animals that were induced to arouse prematurely (Table 22.1; Twente and Twente 1965a, b; Twente et al. 1977). Increasing awareness of the differences between natural and prematurely induced arousals is important since these two physiological states are often utilized rather interchangeably. Premature arousals occur in response to rapid changes in body temperature, loud noises, or physical disturbances (Twente and Twente 1965a, 1968: Pengelley and Fisher 1968). Table 22.1 presents a random sample of 50 publications investigating arousal; 9 records utilized natural arousals, 35 records utilized induced arousals, and the nature of the arousal was unclear for the remaining 6 records. Of the 35 studies utilizing induced arousals, only 8 represent *intentionally* induced arousals, i.e., prematurely inducing arousal was part of the experimental design. Experimental procedures, lack of consideration of the mechanism of arousal, or convenience for the investigator are the primary sources of the remaining prematurely induced arousals.

Though direct comparisons between natural and prematurely induced arousal are limited, meaningful differences in the path to euthermy have been documented both at an intracellular as well as a systemic level. Protein synthesis is a vital aspect of cellular physiology; both the quantity as well as the qualities of protein synthesis may be altered by prematurely inducing arousal (Derij and Shtark 1985; Pan and van Breukelen 2011). A study investigating rates of protein synthesis in various brain regions of the red-cheeked ground squirrel (*Spermophilus erythrogenys*) noted that incorporation of [^{3}H] leucine was much lower following a forced arousal than following a natural arousal (Derij and Shtark 1985). A more recent investigation demonstrated alterations in the types of proteins synthesized during natural and prematurely induced arousals (*Callospermophilus lateralis*; Pan and van Breukelen 2011).

Table 22.1 Categorization of 50 publications that utilized arousing animals

Year	Arousal type	Title
1950	Induced	Circulatory changes during the process of arousal in the hibernating hamster
1959	Induced	Notes on hibernation and awakening in arctic ground squirrels
1962	Induced	Estimated regional blood flow by rubidium 86 distribution during arousal from hibernation
1965	Unclear	Oxidation of glucose-U-C^{14} and palmitate-1-C^{14} by liver, kidney, and diaphragm from hamsters in cold exposure and hibernation
1965	Natural	Effects of core temperature upon duration of hibernation of *Citellus lateralis*
1968	Induced	Estimated heat contribution of brown fat in arousing ground squirrels (*Citellus lateralis*)
1968	Induced	Carbon dioxide fixation during hibernation and arousal from hibernation
1968	Induced	Plasma free amino acids in hibernation and arousal
1968	Intentionally induced	Ability of the ground squirrel, *Citellus lateralis*, to be habituated to stimuli while in hibernation
1968	Natural	Simultaneous recording of core temperature and energy expenditure during the hibernation cycle of *Mesocricetus auratus*
1968	Intentionally induced	Progressive irritability of hibernating *Citellus lateralis*
1969	Intentionally induced	Alteration of activity rhythm after induced arousal from hibernation
1970	Intentionally induced	Sliding set points for body weight in ground squirrels during the hibernation season
1970	Induced	Regional distribution of blood flow in the bat (*Myotis lucifugus*) during arousal from hibernation
1971	Unclear	Circulatory patterns of hibernators
1974	Induced	Temporal changes in AA catabolism during arousal from hibernation in the golden-mantled ground squirrel
1975	Induced	Comparison of regional blood distribution in big brown bat during torpor (summer), hibernation (winter), and arousal
1977	Intentionally induced	Respiratory and circulatory differences between induced and spontaneous arousals in hibernating hedgehogs
1977	Natural	Regulation of arousal from hibernation by temperature in three species of *Citellus*
1978	Intentionally induced	Comparison of induced and spontaneous arousals in hibernating hedgehogs
1979	Induced	Effect of cerebral injection of biogenic amines during arousal from hibernation
1982	Induced	Concentrations of lactate and pyruvate and temperature effects on lactate dehydrogenase in the tissues of the big brown bat during arousal from hibernation
1982	Intentionally induced	Effect of temperature on the duration of arousal episodes during hibernation
1982	Natural	Urine concentration by an undisturbed, naturally arousing hibernator (*S. lateralis*): water balance implications
1982	Natural	Pulmonary ventilation and cardiac activity in hibernating and arousing golden-mantled ground squirrels

(continued)

Table 22.1 (continued)

Year	Arousal type	Title
1983	Unclear	Body temperature, heart rate and oxygen consumption of normothermic and heterothermic western jumping mice (*Zapus principus*)
1987	Unclear	Glucose oxidation by adipose tissue of the edible dormouse during hibernation and arousal: effect of insulin
1988	Induced	Cardiac arrhythmias during arousal from hibernation in three species of rodents
1988	Induced	The influence of ambient temperature on the rate of arousal and behavioral changes during arousal from hibernation in the 13-lined ground squirrel
1988	Induced	Time-course of blood acid–base state during arousal from hibernation in the European hamster
1993	Intentionally induced	Induction of arousal in hibernating European hamsters by vasopressin infusion on the lateral septum
1997	Induced	Rates of rewarming, heart, and respiratory rates and their significance for oxygen transport during arousal from torpor in the smallest mammal, the Etruscan shrew (*Suncus etruscus*)
1999	Induced	Arousal from torpor in the Chilean mouse-opposum (*Thylamys legans*): does non-shivering thermogenesis play a role?
1999	Induced	Reversible depression of oxygen consumption in isolated liver mitochondria during hibernation
2000	Unclear	Hibernation induces oxidative stress and activation of NF-KB in ground squirrel intestine
2001	Induced	Ascorbate dynamics and oxygen consumption during arousal from hibernation in Arctic ground squirrels
2002	Induced	Activation of stress signaling molecules in bat brain during arousal from hibernation
2002	Unclear	Ubiquitin conjugate dynamics in the gut and liver of hibernating ground squirrels
2003	Induced	Brown fat and nonshivering thermogenesis in the gray mouse lemur (*Microcebus murinus*)
2003	Induced	State-dependent regulation of cortical blood flow and respiration in hamsters: response to hypercapnia during arousal from hibernation
2004	Induced	Comparison of surface temperature in 13-lined ground squirrel and yellow-bellied marmot during arousal from hibernation
2004	Natural	Increased heat loss affects hibernation in golden-mantled ground squirrels
2004	Natural	Patterns of body temperature variation and torpor in the numbat, *Myrmecobius fasciatus* (Marsupialia: Myrmecobiidae)
2005	Induced	Sympathetic alpha-adrenergic regulation of blood flow and volume in hamsters arousing from hibernation
2005	Induced	Arousal from hibernation alters contextual learning and memory
2007	Natural	Temporal and temperature effects on the maximum rate of rewarming from hibernation
2008	Induced	Simultaneous measurement of brain tissue oxygen partial pressure, temperature, and global oxygen consumption during hibernation, arousal and euthermy in non-sedated non-anesthitized Arctic ground squirrels

(continued)

Table 22.1 (continued)

Year	Arousal type	Title
2008	Natural	Rewarming rates and thermogenesis in hibernating echnidnas
2009	Induced	Energetics of arousal episodes in hibernating Arctic ground squirrels
2009	Induced	Physiological oxidative stress after arousal from hibernation in Arctic ground squirrels

Table 22.1 references are listed in order. Chatfield and Lyman (1950), Muscacchia and Hamilton (1959), Bullard and Funkhouser (1962), Baumber and Denyes (1965), Twente and Twente (1965a, b), Horwitz et al. (1968), Klain and Whitten (1968a, b), Pengelley and Fisher (1968), Robertson et al. (1968), Twente and Twente (1968), Popovic and Kent (1969), Mrosovsky and Fisher (1970), Rauch and Hayward (1970), Wells (1971), Whitten et al. (1974), Rauch and Beatty (1975), Tähti and Soivio (1977), Twente et al. (1977), Tähti and Soivio (1978), Glass and Wang (1979), Cuddihee and Fonda (1982), French (1982), Muchlinski and Carlisle (1982), Steffen and Riedesel (1982), Cranford (1983), Castex et al. (1987), Eagles et al. (1988), Hintz and Hall (1988), Malan et al. (1988), Hermes et al. (1993), Fons et al. (1997), Opazo et al. (1999), Martin et al. (1999), Carey et al. (2000), Tøien et al. (2001), Lee et al. (2002), van Breukelen and Carey (2002), Genin et al. (2003), Osborne and Hashimoto (2003), Phillips and Heath (2004), Kauffman et al. (2004), Cooper and Withers (2004), Osborne et al. (2005), Weltzin et al. (2006), Utz et al. (2007), Ma and Wu (2008), Nicol and Andersen (2008), Karpovich et al. (2009), Orr et al. (2009)

The metabolic depression of torpor necessitates drastic decreases in the cellular processes of transcription and translation. In addition to the canonical method of initiating translation via the 5′ G-cap, additional mechanisms of initiating protein synthesis are also utilized. One example is translation initiation via Internal Ribosome Entry Sites (IRES) wherein certain initiation factors recognize a secondary structure within the mRNA strand and facilitate the initial assembly of the ribosome downstream from the cap (Merrick 2004). The mechanism of translation initiation is meaningful when one considers that many stress proteins are synthesized via the IRES mechanism. IRES-mediated initiation has been implicated in responses to hypoxia, oxidative stress, and mild hypothermia (Conte et al. 2008; Lang et al. 2002; Holcik et al. 2000; Holcik and Sonenberg 2005; Chappell et al. 2001). As a golden-mantled ground squirrel (*Callospermophilus lateralis*) progresses through a torpor bout, there is a shift in the types of mRNA transcripts associated with the protein synthesis machinery, such that more IRES-containing transcripts are loaded on ribosomes during late torpor (Pan and van Breukelen 2011). IRES transcripts reach a peak level during early arousal (Pan and van Breukelen 2011). As the animal naturally arouses, the protein synthesis machinery is maximally loaded with transcripts that encode stress response proteins. The activity of these stress response proteins may be an important contributor to the successful transition between the cold temperatures and metabolic depression of torpor to the warm temperatures and high metabolic activity of the interbout arousal. When animals are induced to arouse prematurely, the ribosomes have not yet been maximally loaded with IRES transcripts (Pan and van Breukelen 2011).

When an animal is induced to arouse prematurely, the types of proteins that will initially be synthesized are functionally distinct from a natural arousal. Given the association of the IRES mechanism with stress response proteins, prematurely induced arousals may be more damaging than natural arousals.

Studies comparing spontaneous and induced arousals in hibernating hedgehogs (*Erinaceus europaeus*) documented alterations in respiratory and circulatory system functions (Tähti and Soivio 1977, 1978). Oxygen consumption, blood pressure, and heart rate all increased more rapidly during induced arousal (Tähti and Soivio 1977). During induced arousal, peak oxygen consumption occurred 30 min earlier and was 36% greater than for spontaneous arousal (Tähti and Soivio 1977). In addition, the relative timing of increases in respiratory and cardiovascular system function was also altered. During induced arousal cessation of apneas, acceleration of heart rate, and increases in blood pressure all occurred together, whereas during spontaneous arousal, increases in cardiovascular activity preceded increases in respiratory system activity (Tähti and Soivio 1977). Indeed, these alterations in cardiovascular and respiratory system function may be physiologically meaningful when one considers how an animal rewarms. The control of heat generation as well as heat dissipation is intricately linked to cardiovascular and respiratory system functions. The altered rate of oxygen consumption associated with prematurely inducing arousal may substantially impact studies on metabolism.

22.4 Influence of Arousal Type on Other Phases of a Torpor Bout

Prematurely inducing arousal from torpor not only alters aspects of the arousal process itself, but also changes aspects of the following period of IBA. The homeothermal period, which includes both arousal and IBA, was shorter than expected when Belding's ground squirrels (*Urocitellus beldingi*) were induced to arouse prematurely (French 1982). The reduction in homeothermal period was most pronounced when animals were disturbed early (26–50% of expected duration) in the torpor bout (French 1982).

Records from 10 golden-mantled ground squirrels (*Callospermophilus lateralis*) housed at 4°C were analyzed to determine the effect of prematurely induced arousal on the duration of the IBA; additional husbandry information can be found in the Materials and Methods section at the end of the chapter. Briefly, animals were induced to arouse mid-bout (approximately 40–60% of the expected bout duration) by handling for 30sec. The IBA following an induced arousal was ~25% shorter than the corresponding IBA following a natural arousal (Fig. 22.2). The average IBA duration following natural arousal was ~12.5 h, whereas it was ~9 h following induced arousal. This reduction in IBA duration was significant (paired t test, $p < 0.05$).

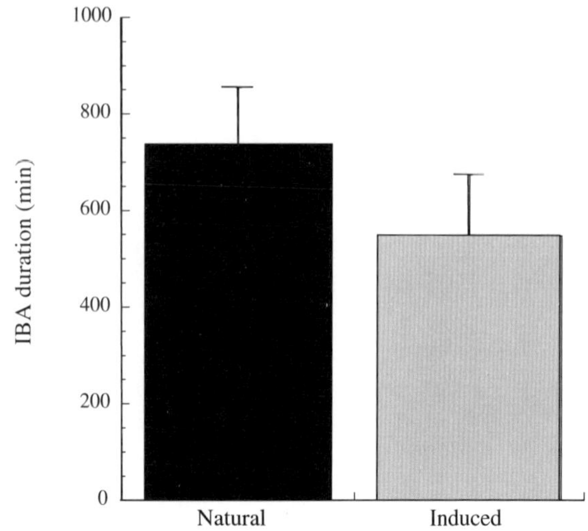

Fig. 22.2 Effect of induced arousal on IBA duration. *Bars* represent mean ± SE for natural (*black*) and induced (*gray*) arousal; $n = 10$. All animals were housed at 4°C. Type of arousal has a significant effect on IBA duration, paired *t* test $p < 0.05$

22.5 Neurological Aspects of Prematurely Induced Arousal

Although direct comparison with natural arousals are lacking, many interesting observations related to neurological function have been documented for prematurely induced arousal. For example, the circadian rhythm of running activity was measured in 13-lined ground squirrels (*Ictidomys tridecimlineatus*) prior to hibernation (Popovic and Kent 1969). Following a prematurely induced arousal, the activity patterns were measured in two experimental groups; one group was exposed to light cues while the second was maintained in constant darkness. In the presence of 11 h of light and 13 h of darkness, induced arousal initiated an activity rhythm that was out of phase with the pre-hibernation pattern (Popovic and Kent 1969). However, during the 3–4 days following arousal, the onset of running activity drifted and ultimately became consistent with the pre-hibernation timing (Popovic and Kent 1969). In complete darkness, a new rhythm was set and maintained. These results were interpreted to indicate that the circadian rhythm of a hibernator is entirely reset after an induced arousal (Popovic and Kent 1969).

In the 1960s, Pengelley and Fisher conducted experiments investigating the function of the nervous system at the low body temperatures of torpor. Their procedure involved physical manipulation of hibernating ground squirrels (*Callospermophilus lateralis*); specifically animals were tossed 2–3 feet in the air on a daily basis (Pengelley and Fisher 1968). The animals initially responded by arousing from torpor prematurely, however they ultimately became unresponsive to this stimulus and could be tossed 100 times without arousing (Pengelley and Fisher 1968). These results were interpreted as indicating that the nervous system was functional, given the ability of the animals to establish a habituated response to the stimulus (Pengelley and Fisher 1968).

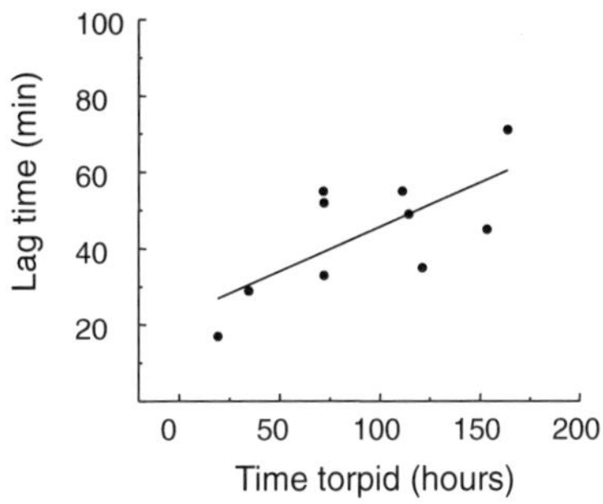

Fig. 22.3 Effect of torpor duration on responsiveness to the arousal stimulus. 10 animals housed at 4°C were subjected to 30 s of mild shaking, following differing amounts of time in torpor. The period of time between the application of the arousal stimulus and the initiation of arousal was measured and is referred to as the lag time. Time torpid had a significant effect on lag time, $p < 0.05$; $r^2 = 0.49$

Alternatively, in a more recent experiment arctic ground squirrels (*Urocitellus parryii*) were trained in a fear conditioning paradigm, and the hippocampal-dependent response of freezing behavior (maintaining a crouched position with no overt motion) was assessed 3 h and 24 h following a prematurely induced arousal from torpor (Weltzin et al. 2006). Evidence of cognitive enhancement, akin to better learning, was found in animals trained 24 h after arousal but not in animals trained 3 h after arousal (Weltzin et al. 2006). These results were interpreted as potentially indicating that functional synapses had yet to be reconstructed at the 3-h time point (Weltzin et al. 2006).

We recently investigated whether an animal's responsiveness to an induced arousal stimulus was influenced by the amount of time spent in torpor. The period of time spanning the application of the induced arousal stimulus to the initiation of arousal was measured (Fig. 22.3). This "lag time" was influenced by the amount of time the animal had spent in torpor. Lag times were relatively shorter when an animal was induced to arouse early in the torpor bout. As ground squirrels (*Callospermophilus lateralis*) progressed throughout the torpor bout, the lag time increased. The amount of time spent in torpor had a significant effect on the lag time (linear regression, $p < 0.05$).

22.6 Concluding Remarks

At first glance, a hibernation season seems to comprise two distinct and dissimilar physiological states: torpor and euthermy. Upon closer inspection, one gains an appreciation for the intricacies of the transition phases, entrance and arousal. Arousal is an intricate and rapid process. Animals may increase their body temperature by more than 35°C in a mere 2–3 h. During arousal, animals must successfully reconcile differing rates of oxygen consumption, blood flow, and changing T_b in the midst of rapidly changing metabolic demand. The physiological processes that are depressed during torpor are re-engaged during arousal. Within a cell transcription, protein metabolism, mitochondrial respiration, and mitosis are restored (van Breukelen and Martin 2001, 2002a, b; van Breukelen et al. 2004; Velickovska et al. 2005; Velickovska and van Breukelen 2007; Martin et al. 1999; Kruman et al. 1988). At the level of the entire organism heart rate, blood pressure, and respiration resume and ultimately attain standard euthermic values (Steffen and Riedesel 1982; Carey et al. 2003).

When animals are induced to arouse prematurely, numerous aspects of both intracellular as well as systemic physiology are altered. The types and the quantity of proteins produced are altered during prematurely induced arousal (Derij and Shtark 1985; Pan and van Breukelen 2011). Heart rate, blood pressure, respiratory rate, as well as oxygen consumption are altered during prematurely induced arousal (Tähti and Soivio 1977, 1978). Prematurely inducing arousal shortens the following period of interbout arousal (Fig. 22.2; French 1982). Additionally, the capacity for learning and memory as well as the ability to maintain a circadian cycle may be inhibited by prematurely inducing arousal (Weltzin et al. 2006; Popovic and Kent 1969).

Considering this substantial list of differences, we recommend that careful consideration be given to experimental design and data interpretation when arousing animals are utilized. When the mechanistic details and functional consequences of how an animal arouses are considered, it seems that, yes, the road traveled to euthermy does matter.

22.7 Materials and Methods

Golden-mantled ground squirrels, *Callospermophilus lateralis*, were live trapped from local populations in Southern Nevada, Southern California, or Southern Utah. Prior to the beginning of the hibernation season, temperature sensitive radiotelemeters (model VM-FH disk; Mini Mitter, Sun River, OR) were surgically implanted into the abdominal cavity. These radiotelemeters allow for precise measurement of T_b throughout the hibernation season. Animals were housed in an environmental chamber at an ambient temperature of 4°C in constant darkness during the hibernation season. The University of Nevada, Las Vegas Institutional Animal Care and Use Committee approved all procedures.

Body temperature (T_b) was recorded every minute to the hundredth of a degree. The first derivative (FD) and its associated standard deviation (SD) for temperature as a function of time were calculated. The initiation of arousal was defined as the point where the instantaneous RRW exceeded the FD + 3 SD (threshold rate; TR) for a 10-min period of consistent T_b. The first T_b measurement with an FD > TR was defined as the onset of arousal. The onset of arousal typically corresponded to a rate of $0.02°C \cdot min^{-1}$. The termination of arousal was defined as the first point at which the instantaneous rewarming rate was $\leq$ TR for the following 5 min. IBA duration is defined as the period of time spanning the termination of arousal to the initiation of entrance. Initiation of entrance is defined as the first point after which there are only decreases in body temperature. Time torpid is defined as the period of time spanning the start of torpor to the application of the arousal stimulus. Start of torpor is the first point following entrance where there is a 5-min period with no decrease in body temperature. Lag time is defined as the period of time spanning the application of the arousal stimulus to the initiation of arousal.

Following an undisturbed torpor bout, animals were induced to arouse prematurely by handling for 30 s. Records from 10 animals were subjected to a paired *t* test to determine the effect of arousal type on the duration of the IBA (Statview 4.1). A regression analysis was performed on a separate data set to determine whether the amount of time spent in torpor affected responsiveness to the arousal stimulus (Statview 4.1). A p-value < 0.05 was considered significant for both analyses.

Acknowledgments This work was supported by NSF grant IOB 0448396, awarded to Frank van Breukelen. Jenifer C. Utz was supported by an NSF Graduate Research Fellowship. We thank other members of the laboratory for their diligent efforts.

References

Barnes BM (1989) Freeze avoidance in a mammal: body temperatures below 0 degree C in an Arctic hibernator. Science 244:1593–1595

Baumber J, Denyes A (1965) Oxidation of glucose-U-C14 and palmitate-1-C14 by liver, kidney, and diaphragm from hamsters in cold exposure and hibernation. Can J Biochem 43:747–753

Bullard RW, Funkhouser GE (1962) Estimated regional blood flow by rubidium 86 distribution during arousal from hibernation. Am J Physiol 203:266–270

Cannon B, Nedergaard J (2004) Brown adipose tissue: function and physiological significance. Physiol Rev 84:277–359

Carey HV, Frank CL, Seifert JP (2000) Hibernation induces oxidative stress and activation of NF-kappa B in ground squirrel intestine. J Comp Physiol B 170:551–559

Carey HV, Andrews MT, Martin SL (2003) Mammalian hibernation: cellular and molecular responses to depressed metabolism and low temperature. Physiol Rev 83:1153–1181

Castex C, Tahri A, Hoo-Paris R, Sutter BC (1987) Glucose oxidation by adipose tissue of the edible dormouse (*Glis glis*) during hibernation and arousal: effect of insulin. Comp Biochem Physiol A 88:33–36

Chappell SA, Owens GC, Mauro VP (2001) A 5′ leader of Rbm3, a cold stress-induced mRNA, mediates internal initiation of translation with increased efficiency under conditions of mild hypothermia. J Biol Chem 276:36917–36922

Chatfield PO, Lyman CP (1950) Circulatory changes during process of arousal in the hibernating hamster. Am J Physiol 163:566–574

Conte C, Riant E, Toutain C, Pujol F, Arnal JF, Lenfant F, Prats AC (2008) FGF2 translationally induced by hypoxia is involved in negative and positive feedback loops with HIF-1alpha. PLoS ONE 3:e3078

Cooper CE, Withers PC (2004) Patterns of body temperature variation and torpor in the numbat, *Myrmecobius fasciatus* (Marsupialia: Myrmecobiidae). J Thermal Bio 29:277–284

Cranford JA (1983) Body temperature, heart rate and oxygen consumption of normothermic and heterothermic western jumping mice (*Zapus princeps*). Comp Biochem Physiol A 74:595–599

Cuddihee RW, Fonda ML (1982) Concentrations of lactate and pyruvate and temperature effects on lactate dehydrogenase activity in the tissues of the big brown bat (*Eptesicus fuscus*) during arousal from hibernation. Comp Biochem Physiol B 73:1001–1009

Derij LV, Shtark MB (1985) Hibernators' brain: protein synthesis in the neocortex and the hippocampus. Comp Biochem Physiol B 80:927–934

Eagles DA, Jacques LB, Taboada J, Wagner CW, Diakun TA (1988) Cardiac arrhythmias during arousal from hibernation in three species of rodents. Am J Physiol 254:R102–R108

Fons R, Sender S, Peters T, Jurgens KD (1997) Rates of rewarming, heart and respiratory rates and their significance for oxygen transport during arousal from torpor in the smallest mammal, the Etruscan shrew *Suncus etruscus*. J Exp Biol 200:1451–1458

French AR (1982) Effects of temperature on the duration of arousal episodes during hibernation. J Appl Physiol 52:216–220

Genin F, Nibbelink M, Galand M, Perret M, Ambid L (2003) Brown fat and nonshivering thermogenesis in the gray mouse lemur (*Microcebus murinus*). Am J Physiol Regul Integr Comp Physiol 284:R811–R818

Glass JD, Wang LC (1979) Effects of central injection of biogenic amines during arousal from hibernation. Am J Physiol 236:R162–R167

Hermes ML, Kalsbeek A, Kirsch R, Buijs RM, Pevet P (1993) Induction of arousal in hibernating European hamsters (*Cricetus cricetus* L.) by vasopressin infusion in the lateral septum. Brain Res 631:313–316

Hintz DR, Hall KD (1988) The influence of ambient temperature on rate of arousal and behavioral changes during arousal from hibernation in the 13-lined ground squirrel, *Spermophilus tridecimlineatus*. Bios 57:109–120

Holcik M, Sonenberg N, Korneluk RG (2000) Internal ribosome initiation of translation and the control of cell death. Trends Genet 16:469–473

Holcik M, Sonenberg N (2005) Translational control in stress and apoptosis. Nature Rev Mol Cell Biol 6:318–327

Horwitz BA, Smith RE, Pengelley ET (1968) Estimated heat contribution of brown fat in arousing ground squirrels (*Citellus lateralis*). Am J Physiol 214:115–121

Karpovich SA, Tøien Ø, Buck CL, Barnes BM (2009) Energetics of arousal episodes in hibernating arctic ground squirrels. J Comp Physiol B 179:691–700

Kauffman AS, Paul MJ, Zucker I (2004) Increased heat loss affects hibernation in golden-mantled ground squirrels. Am J Physiol Regul Integr Comp Physiol 287:R167–R173

Klain GJ, Whitten BK (1968a) Carbon dioxide fixation during hibernation and arousal from hibernation. Comp Biochem Physiol 25:363–366

Klain GJ, Whitten BK (1968b) Plasma free amino acids in hibernation and arousal. Comp Biochem Physiol 27:617–619

Kruman II, Ilyasova EN, Rudchenko SA, Khurkhulu ZS (1988) The intestinal epithelial cells of ground squirrel (*Citellus undulatus*) accumulate at G2 phase of the cell cycle throughout a bout of hibernation. Comp Biochem Physiol A Physiol 90:233–236

Lang KJ, Kappel A, Goodall GJ (2002) Hypoxia-inducible factor-1 alpha mRNA contains an internal ribosome entry site that allows efficiency translation during normoxia and hypoxia. Mol Biol Cell 13:1792–1801

Lee M, Choi I, Park K (2002) Activation of stress signaling molecules in bat brain during arousal from hibernation. J Neurochem 82:867–873

Ma YL, Wu SF (2008) Simultaneous measurement of brain tissue oxygen partial pressure, temperature, and global oxygen consumption during hibernation, arousal, and euthermy in non-sedated and non-anesthetized Arctic ground squirrels. J Neurosci Methods 174:237–244

Malan A, Mioskowski E, Calgari C (1988) Time-course of blood acid-base state during arousal from hibernation in the European hamster. J Comp Physiol B 158:495–500

Martin SL, Maniero GD, Carey C, Hand SC (1999) Reversible depression of oxygen consumption in isolated liver mitochondria during hibernation. Physiol Biochem Zool 72:255–264

Merrick WC (2004) Cap-dependent and cap-independent translation in eukaryotic systems. Gene 332:1–11

Milsom WK, Zimmer MB, Harris MB (1999) Regulation of cardiac rhythm in hibernating mammals. Comp Biochem Physiol A Mol Integr Physiol 124:383–391

Milsom WK, Zimmer MB, Harris MB (2001) Vagal control of cardiorespiratory function in hibernation. Exp Physiol 86:791–796

Mrosovsky N, Fisher KC (1970) Sliding set points for body weight in ground squirrels during the hibernation season. Can J Zool 48:241–247

Muchlinski AE, Carlisle AL (1982) Urine concentration by an undisturbed, naturally arousing hibernator (*Spermophilus lateralis*): water-balance implications. J Mammal 63:510–512

Musaccahia XJ, Hamilton RD (1959) Notes on hibernation and awakening in Arctic ground squirrels. J Mammal 40:201–204

Nicol SC, Andersen NA (2008) Rewarming rates and thermogenesis in hibernating echidnas. Comp Biochem Physiol A Mol Integr Physiol 150:189–195

Opazo JC, Nespolo RF, Bozinovic F (1999) Arousal from torpor in the Chilean mouse-opposum (*Thylamys elegans*): does non-shivering thermogenesis play a role? Comp Biochem Physiol A Mol Integr Physiol 123:393–397

Orr AL, Lohse LA, Drew KL, Hermes-Lima M (2009) Physiological oxidative stress after arousal from hibernation in Arctic ground squirrel. Comp Biochem Physiol A Mol Integr Physiol 153:213–221

Osborne PG, Hashimoto M (2003) State-dependent regulation of cortical blood flow and respiration in hamsters: response to hypercapnia during arousal from hibernation. J Physiol 547:963–970

Osborne PG, Sato J, Shuke N, Hashimoto M (2005) Sympathetic alpha-adrenergic regulation of blood flow and volume in hamsters arousing from hibernation. Am J Physiol Regul Integr Comp Physiol 289:R554–R562

Pan P, van Breukelen F (2011) Preference of IRES-mediated initiation of translation during hibernation in golden-mantled ground squirrels, *Spermophilus lateralis*. Am J Physiol Regul Integr Comp Physiol 301:R370–R377

Pengelley ET, Fisher KC (1968) Ability of ground squirrel *Citellus lateralis* to be habituated to stimuli while in hibernation. J Mammal 49:561–562

Phillips PK, Heath JE (2004) Comparison of surface temperature in 13-lined ground squirrel (*Spermophilus tridecimlineatus*) and yellow-bellied marmot (*Marmota flaviventris*) during arousal from hibernation. Comp Biochem Physiol A Mol Integr Physiol 138:451–457

Popovic VP, Kent BB (1969) Alteration of activity rhythm after induced arousal from hibernation. Rassegna di neurologia vegetativa 23:27–36

Rauch JC, Hayward JS (1970) Regional distribution of blood flow in the bat (*Myotis lucifugus*) during arousal from hibernation. Can J Physiol Pharm 48:269–273

Rauch JC, Beatty DD (1975) Comparison of regional blood distribution in *Eptesicus fuscus* (big brown bat) during torpor (summer), hibernation (winter), and arousal. Can J Zool 53:207–214

Robertson WD, Yousef MK, Johnson HD (1968) Simultaneous recording of core temperature and energy expenditure during the hibernation cycle of *Mesocricetus auratus*. Nature 219: 742–743

Steffen JM, Riedesel ML (1982) Pulmonary ventilation and cardiac activity in hibernating and arousing golden-mantled ground squirrels (*Spermophilus lateralis*). Cryobiology 19:83–91

Stone GN, Purvis A (1992) Warm-up rates during arousal from torpor in heterothermic mammals: physiological correlates and a comparison with heterothermic insects. J Comp Physiol B 162:284–295

Tähti H, Soivio A (1977) Respiratory and circulatory differences between induced and spontaneous arousals in hibernating hedgehogs (*Erinaceus europaeus L*). Ann Zool Fenn 14:198–203

Tähti H, Soivio A (1978) Comparison of induced and spontaneous arousals in hibernating hedgehogs. Experientia Suppl 22:321–325

Tøien Ø, Drew KL, Chao ML, Rice ME (2001) Ascorbate dynamics and oxygen consumption during arousal from hibernation in Arctic ground squirrels. Am J Physiol Regul Integr Comp Physiol 281:R572–R583

Twente JW, Twente JA (1965a) Effects of core temperature upon duration of hibernation of *Citellus lateralis*. J Appl Physiol 20:411–416

Twente JW, Twente JA (1965b) Regulation of hibernating periods by temperature. PNAS 54:1058–1061

Twente JW, Twente JA (1968) Progressive irritability of hibernating *Citellus lateralis*. Comp Biochem Physiol 25:467–474

Twente JW, Twente J, Moy RM (1977) Regulation of arousal from hibernation by temperature in three species of *Citellus*. J Appl Physiol 42:191–195

Utz JC, Velickovska V, Shmereva A, van Breukelen F (2007) Temporal and temperature effects on the maximum rate of rewarming from hibernation. J Thermal Bio 32:276–281

van Breukelen F, Martin SL (2001) Translational initiation is uncoupled from elongation at 18 degrees C during mammalian hibernation. Am J Physiol Regul Integr Comp Physiol 281:R1374–R1379

van Breukelen F, Carey HV (2002) Ubiquitin conjugate dynamics in the gut and liver of hibernating ground squirrels. J Comp Physiol B 172:269–273

van Breukelen F, Martin SL (2002a) Invited review: molecular adaptations in mammalian hibernators: unique adaptations or generalized responses? J Appl Physiol 92:2640–2647

van Breukelen F, Martin SL (2002b) Reversible depression of transcription during hibernation. J Comp Physiol B 172:355–361

van Breukelen F, Sonenberg N, Martin SL (2004) Seasonal and state-dependent changes of eIF4E and 4E-BP1 during mammalian hibernation: implications for the control of translation during torpor. Am J Physiol Regul Integr Comp Physiol 287:R349–R353

Velickovska V, Lloyd BP, Qureshi S, van Breukelen F (2005) Proteolysis is depressed during torpor in hibernators at the level of the 20S core protease. J Comp Physiol B 175:329–335

Velickovska V, van Breukelen F (2007) Ubiquitylation of proteins in livers of hibernating golden-mantled ground squirrels, *Spermophilus lateralis*. Cryobiol 55:230–235

Wells LA (1971) Circulatory patterns of hibernators. Am J Physiol 221:1517–1520

Weltzin MM, Zhao HW, Drew KL, Bucci DJ (2006) Arousal from hibernation alters contextual learning and memory. Behav Brain Res 167:128–133

Whitten BK, Burlington RF, Posiviata MA (1974) Temporal changes in amino acid catabolism during arousal from hibernation in the golden-mantled ground squirrel. Comp Biochem Physiol A 47:541–546

Withers PC (1992) Comparative animal physiology. Saunders College Publishing, Fort Worth

Chapter 23
The Hibernating Immune System

Hjalmar R. Bouma, Arjen M. Strijkstra, Fatimeh Talaei, Rob H. Henning, Hannah V. Carey and Frans G. M. Kroese

Abstract Several physiological changes take place during hibernation, which are thought to allow animals to conserve energy and limit organ injury as might otherwise occur due to the physiological extremes of torpor and arousal. Significant changes occur in the immune system during torpor. The number of circulating leukocytes drops by $\sim$90% during entrance into torpor and seems to be driven by low body temperature. Normal cell counts restore upon arousal. Recently, we demonstrated that clearance of circulating lymphocytes is due to retention in lymphoid organs caused by a reduced plasma level of sphingosine-1-phosphate (S1P). Besides its effects on leukocyte migration, hibernation affects complement function, phagocytosis capacity, cytokine production, lymphocyte proliferation, and antibody production. The reduced immune function might play a major role in the etiology of White Nose Syndrome (WNS) in hibernating bats. Further, the ability to induce a fully reversible state of immune suppression in humans might aid the treatment of several inflammatory and immune-mediated diseases. Unraveling the

H. R. Bouma (✉) · A. M. Strijkstra · F. Talaei · R. H. Henning
Department of Clinical Pharmacology, University Medical Center Groningen, Antonius Deusinglaan 1/FB20, 9713 AV Groningen, Groningen, The Netherlands
e-mail: h.r.bouma@med.umcg.nl

H. R. Bouma · A. M. Strijkstra
Department of Animal Behavior/Chronobiology, University of Groningen, Groningen, The Netherlands

H. V. Carey
Department of Comparative Biosciences, School of Veterinary Medicine, University of Wisconsin, Madison, WI, USA

F. G. M. Kroese
Department of Rheumatology and Clinical Immunology, University Medical Center Groningen, University of Groningen, Groningen, The Netherlands

T. Ruf et al. (eds.), *Living in a Seasonal World*,
DOI: 10.1007/978-3-642-28678-0_23,

mechanisms underlying the reduced immune function during torpor will not only enhance fundamental knowledge about the immune system, but might also lead to the development of a strategy to limit mortality due to WNS.

23.1 Introduction

Profound changes in physiology are observed during torpor. In deep torpor, ground squirrels reduce their heart rate from 350–400 to 5–10 beats per minute, respiration is depressed from approximately 40 breaths per minute, to less than one breath per minute and renal function is severely reduced (Carey et al. 2003; Storey 2010). Due to the profoundly reduced metabolism, typical body temperatures during deep torpor are ~0–4°C (Kenagy et al. 1989). Periods of deep torpor are interspersed by (shorter) euthermic periods called arousal. Although hibernators regularly go through periods with an extremely reduced body temperature followed by rapid rewarming and normalization of physiological parameters upon arousal, no signs of massive cell death or organ injury can be found (Fleck and Carey 2005; Sandovici et al. 2004; Zancanaro et al. 1999; Arendt et al. 2003; Talaei et al. 2011; van Breukelen et al. 2010). Specific adaptations of the cells in hibernating animals are thought crucial in limiting cellular stress during torpor and arousal. For example, levels of the pro-apoptotic proteins p53 and Caspase-3-activity are decreased, while levels of the anti-apoptotic proteins Bcl-X_L and Akt are increased in enterocytes during hibernation in the 13-lined ground squirrels (*Ictidomys tridecemlineatus*) as compared to summer euthermic animals (Fleck and Carey 2005). In addition, the production of Nrf2 and superoxide dismutase (SOD) in cardiac tissue (Morin et al. 2008) and the level of ascorbate (vitamin C) in plasma (Drew et al. 1999) are increased during torpor as compared to euthermic animals. These adaptations might lead to increased resistance against ischemia/reperfusion damage, as demonstrated experimentally in hibernating 13-lined ground squirrels in comparison with summer euthermic animals and rats (*Rattus norvegicus*) (Kurtz et al. 2006; Lindell et al. 2005).

In addition to the adaptations described above, important aspects of the immune system are also significantly affected by hibernation. Hibernation alters aspects of both the innate and adaptive arms of the immune system, including, e.g., lower complement levels, reduced phagocytotic capacity, decreased cytokine production, and diminished lymphocyte proliferation and antibody production (reviewed by Bouma et al. 2010a, b). The suppressed innate immune function, as measured by the ability to induce a febrile response, is illustrated by the fact that injection of lipopolysaccharide (LPS) in golden-mantled ground squirrels (*Callospermophilis lateralis*) does not induce a febrile response (or an arousal) when injected during torpor, while it does induce a febrile response when injected in aroused animals (Prendergast et al. 2002). Suppression of the adaptive immune function is demonstrated by the fact that skin allografts transplanted onto torpid 13-lined ground squirrels are not rejected until after the hibernation season (in spring), while rejection of allografts transplanted onto summer-active squirrels occurs within several weeks (Shivatcheva 1988). Although the underlying mechanism for the suppression of the immune function during torpor is largely unknown, part of it might be explained by the reduction of the number of circulating leukocytes during torpor. In this chapter we summarize the available literature on the immune system during mammalian hibernation.

23.2 Skin and Mucosal Barrier Function

The first line of defense against invading pathogens consists of the skin and mucosal barriers. Although no specific changes in skin of hibernating mammals have been reported, mucosal barrier integrity of the gut has been studied in more detail. During hibernation of the 13-lined ground squirrel, the mucosal surface area is reduced due to decrease in villus height (Carey 1990). Low body temperature impairs epithelial ion transport during torpor across the mucosal barrier, which increases during arousal to values above those measured in summer euthermic animals (Carey 1990). Although increased ionic epithelial permeability might suggest leakiness of the mucosal barrier, absorptive and secretory functions of the gut are maintained throughout hibernation (Carey 1990; Carey and Martin 1996; Carey and Sills 1992, 1996; Galluser et al. 1988). It is not clear whether the observed changes in the gut are induced by a regulated process or whether these changes are secondary to the anorexic state of the animals during hibernation. Taken together, important alterations not only in the epithelial barrier (as described above), but also in gut microbial composition (as described elsewhere in this book by us) take place during hibernation, which both play an important role in the defense against invading pathogens.

23.3 Innate Immunity

An important line of defense against microbial invasion is formed by the innate immune system. Opsonization of pathogens with complement factors leads to a significant enhancement of phagocytosis by monocytes, macrophages, and granulocytes. During torpor, the expression of C3 mRNA in the liver, an important protein of the complement system, is reduced (Maniero 2002). Binding of microbes and microbial products to pattern recognition receptors (PRR's) on leukocytes from the innate immune system (i.e. neutrophils and monocytes) stimulates the production and secretion of pro-inflammatory cytokines such as TNF-α and IL-1β. Toll-like receptors (TLR's) on (or in) cells of the innate immune system are an important class of PRR's. LPS is a well-known bacterial product that can activate leukocytes by interaction with TLR4. LPS forms a complex with the

LPS binding protein (LBP), which subsequently binds to CD14. Upon interaction with LPS-LBP, CD14 becomes activated and forms a dimer with TLR4, leading to the production of TNF-α, IL-1β, and IFN-α (Ulevitch and Tobias 1999). Spontaneous production of TNF-α by peritoneal macrophages (cultured at 37°C) from Arctic ground squirrels is diminished during the pre-hibernation period (October; in euthermic animals) and during torpor (Novoselova et al. 2000). Prendergast et al demonstrated that injection of LPS during torpor does not induce a febrile response or arousal in golden-mantled ground squirrels, while it does induce a febrile response when injected in aroused animals (Prendergast et al. 2002). Although a febrile response might follow the induction of an immune response, the absence of a febrile response does not necessarily imply immune suppression as an immune response can also occur in the absence of fever. Furthermore, intracerebroventricular injection of prostaglandin E1, an important signaling molecule inducing fever during inflammation, is able to induce arousal and fever in torpid animals (Prendergast et al. 2002). It has been suggested that these observations might well be explained by effects of torpor on the absorption and circulation of LPS after intraperitoneal injection. However, it is more likely to be due to specific suppression of the innate immune system during torpor, as the peritoneal cavity contains macrophages, which under euthermic circumstances mount an immune response by binding and reacting to microbes. Additional in vitro studies provided more information about the effects of torpor on the function of macrophages. Although the binding of LPS by splenic macrophages from golden-mantled ground squirrels is not affected by torpor or temperature (Maniero 2005), the spontaneous production of TNF-α by peritoneal macrophages from Arctic ground squirrels is significantly reduced during torpor and was not restored by rewarming of cells (Novoselova et al. 2000). Taken together, at least part of the reduced innate immune function during torpor is likely to be due to altered signal transduction downstream of PRR's (Fig. 23.1).

23.4 Neutrophil and Monocyte Dynamics

In all hibernating mammals studied so far, numbers of circulating leukocytes drop by ~90% during torpor. These species include the European hamster (*Cricetus cricetus*) (Reznik et al. 1975), the Syrian hamster (*Mesocricetus auratus)* (Bouma et al. 2011), the hedgehog (*Erinaceus europeaus* L.) (Suomalainen and Rosokivi 1973), the European ground squirrel (*Spermophilus citellus*) (Bouma et al. 2010b), the Arctic ground squirrel (*Urocitellus parryii*) (Toien et al. 2001), and the 13-lined ground squirrel (Frerichs et al. 1994; Spurrier and Dawe 1973). This reduction in the number of total leukocytes is due to a decrease in number of granulocytes (neutrophils, eosinophils, and basophils), lymphocytes, and monocytes. The ~10% remaining leukocytes in the blood during torpor are mainly neutrophils (90%) and lymphocytes (9%) (Szilagyi and Senturia 1972). Not only during deep torpor, but also during shallow daily torpor in the Djungarian hamster (*Phodophus sungorus*) a profound reduction in the number of circulating lymphocytes (Bouma et al. 2011), neutrophils, and

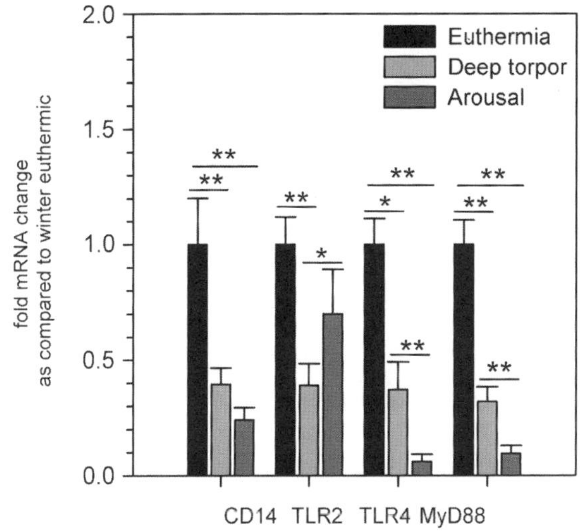

Fig. 23.1 Deep torpor affects the expression of CD14, toll-like receptor 2 and 4 (TLR2; TLR4) and the adaptor protein MyD88. In the Syrian hamster (*Mesocricetus auratus*), the expression of CD14, TLR2, TLR4 and MyD88 in the spleen is lower during deep torpor. The expression of TLR2 increases upon arousal, while the expression of TLR4 and MyD88 is further reduced upon arousal. Gene expression was determined by real-time quantitative PCR in snap-frozen spleens and normalized to GAPDH as housekeeping gene of $n = 5$–9 animals per group. Bars represent average fold change ± standard error of the mean (S.E.M.) as compared to winter euthermic values.

monocytes (unpublished data) has been observed. Further, we demonstrated in the Syrian hamster that the decrease in the number of circulating neutrophils (Fig. 23.2a; closed circles) and monocytes (Fig. 23.2b; closed circles) show a similar pattern of decline during entrance into torpor as the body temperature drops (as do lymphocytes, vide infra). These dynamics may point at an important role of body temperature in the induction and restoration of neutropenia and monocytopenia during hibernation. Indeed, during forced cooling of anesthetized summer euthermic hamsters, a similar reduction of circulating neutrophils (Fig. 23.2a; open circles) and monocytes (Fig. 23.2b; open circles) was found as in animals entering torpor. Numbers of circulating leukocytes are restored to normal levels upon arousal. Although we showed that temperature plays a major role by governing clearance of circulating neutrophils and monocytes, the underlying mechanism remains to be unraveled. A reduction in the number of free circulating neutrophils might be either explained by apoptosis of cells or retention at certain sites. If apoptosis represents the underlying mechanism of removal of neutrophils and monocytes from the circulation during torpor, then restoration of normal cell counts upon arousal depends on generation and release of cells from the bone marrow. Bone marrow from hibernating (i.e. torpid and aroused) squirrels contains relatively more mature granulocytes than bone marrow from non-hibernating squirrels (Szilagyi and Senturia 1972). This might suggest that release of mature granulocytes is relatively decreased in torpid animals as compared to maturation rates and could reflect an adaptation to rapidly restore the pool of circulating neutrophils

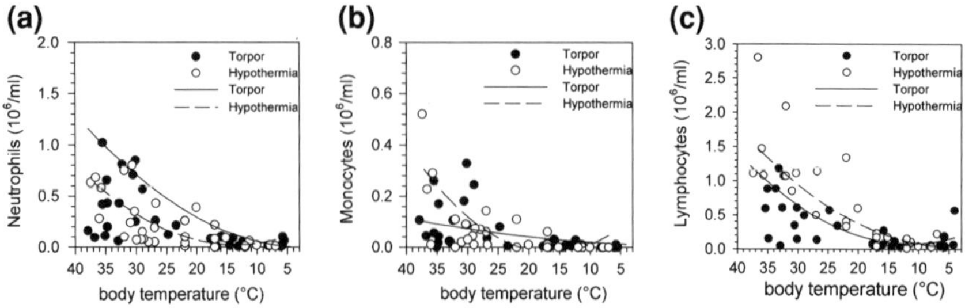

Fig. 23.2 Low body temperature governs clearance of circulating leukocytes during deep torpor and forced hypothermia in the Syrian hamster. Both during entrance into deep torpor (*closed circles* and *uninterrupted curves*) and forced cooling of anesthetized summer euthermic Syrian hamsters (*Mesocricetus auratus*) (*open circles* and *dashed curve*), the numbers of circulating neutrophils (**a**), monocytes (**b**), and lymphocytes (**c**); reprinted with permission from PNAS (Bouma et al. 2011) decrease as body temperature drops. Blood (~200 μl) was drawn by cardiac puncture upon euthanization in the case of hibernating animals or through the canulated jugular vein in the case of anesthetized hypothermic animals into EDTA-coated cups (minicollect EDTA-K3; Greiner Bio-One, Alphen a/d Rijn, The Netherlands) for automated hematological analysis using a Sysmex XE-2100 (Sysmex, Etten-Leur, The Netherlands) (Briggs et al. 2000; Ruzicka et al. 2001). Body temperature was measured rectally. Curves were fitted (*quadratic*) using Sigmaplot 11.0 for Windows

upon arousal. However, the observation that the number of circulating *mature* neutrophils rises upon arousal (Suomalainen and Rosokivi 1973) favors a storage-and-release mechanism to explain rapid restoration of circulating monocytes upon arousal. Retention of neutrophils in the lungs has been suggested by Inkovaara and Suomalainen (1973). Neutrophil retention might be due to margination (reversible adherence) of the vessel wall, caused by an altered balance between adhering factors as integrins, selectins, and other adhesion molecules and detaching forces, such as blood flow velocity (Beekhuizen and van Furth 1993; von Vietinghoff and Ley 2008). Margination is a process that occurs at sites of inflammation, but can also occur in the absence of inflammation. For example, studies in rabbits have shown that under normal circumstances ~61% of the blood granulocytes are not "free circulating", but adhering to the vessel wall and can be mobilized by injecting dexamethasone (Nakagawa et al. 1998). Retention of cells by means of margination to the wall of small vessels either locally at specific sites or systemically, might well explain the rapid reduction and restoration of the number of circulating neutrophils and monocytes during hibernation.

23.5 Adaptive Immunity

Hibernation affects both humoral and cellular immune responses. The reduced adaptive immune function is shown by the fact that rejection of transplanted skin allografts onto torpid 13-lined ground squirrels is delayed by 3–4 times as compared to allografts transplanted onto summer euthermic animals (Shivatcheva 1988). It is

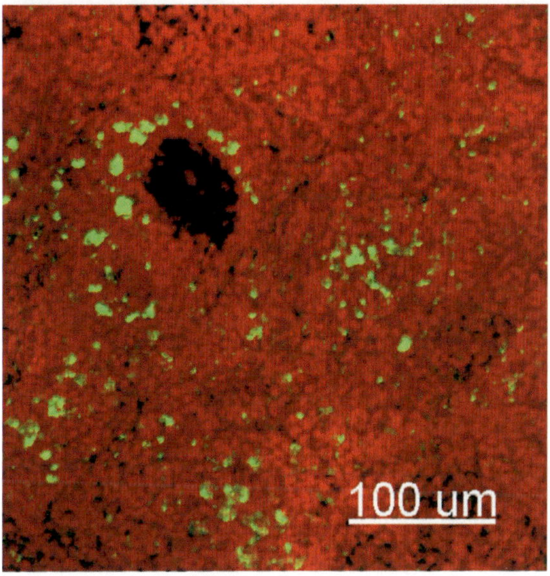

Fig. 23.3 Lymphocytes are retained in cervical lymph node during deep torpor in the Syrian hamster. Figure shows representative fluorescent microscopic image of CFDA-SE-labeled lymphocytes (*green*) and TPO-3-TO (*red*) from a torpid Syrian hamster (*Mesocricetus auratus*). Animals were splenectomized and a canula was placed into the jugular vein connected to a subcutaneous access port (Soloport, Instech Solomon, USA) implanted between the scapulas for infusion of cells during torpor. Lymphocytes were isolated from the spleen and labeled by incubation in 25 μM CFDA-SE in saline (15 min, 37°C) (Invitrogen, the Netherlands) and injected through the canula that into the same animal during torpor. After infusion of the labeled lymphocytes during torpor, an arousal was induced by gently handling the animals. The animals were euthanized 2 days after entrance into the subsequent torpor bout. Snap-frozen lymphoid tissues were embedded in Tissue Tek (Sakura, The Netherlands), sectioned into 200 μm thick slices, and counter-stained using TPO-3-TO (*red*) (Invitrogen, the Netherlands). Images were taken at 200x magnification (Leica SP2 AOBS) and processed using Imaris 6.4

not clear whether the delayed rejection of skin allografts is due to a reduced adaptive immune function as might coincide with the reduced metabolism during torpor or whether additional immune suppression during arousal might also be involved. Further, a reduced humoral immune function is demonstrated by a reduced and delayed antibody production after injection of sheep or rabbit red blood cells into torpid ground squirrels (Andjus et al. 1964; Jaroslow and Serrell 1972).

23.6 Lymphocyte Migration

As described above, leukopenia during torpor affects all major types of leukocytes, including lymphocytes. We demonstrated that clearance of circulating lymphocytes is due to their retention in peripheral lymphoid organs. In the Syrian hamster, low body temperature governs the loss of lymphocytes from the circulation. A near complete absence of lymphocytes was not only observed during deep torpor (Fig. 23.2c; closed circles), but could also be induced to a similar extent by forced hypothermia of anesthetized summer euthermic animals (Fig. 23.2c; open circles) (Bouma et al. 2011). By injecting fluorochrome-labeled lymphocytes we provided evidence that these cells are retained in peripheral lymphoid organs during torpor (Fig. 23.3; Bouma et al. 2011). The number of circulating lymphocytes correlates strongly with body temperature during deep torpor and forced hypothermia (Bouma et al. 2011). The correlation between the number of circulating lymphocytes and the plasma level of sphingosine-1-phosphate (S1P) suggests a potential role for this bioactive lipid in regulation of lymphocyte numbers during hibernation, as S1P known to regulate lymphocyte egress from lymph nodes in other models as well (Mandala et al. 2002; Matloubian et al. 2004; Pappu et al. 2007). Indeed, by antagonizing the S1P-receptor upon arousal, we precluded restoration of normal lymphocyte counts (Bouma et al. 2011). The reduced S1P plasma level is likely the result of a temperature-dependent reduction of S1P release from erythrocytes, which are regarded as the main source of plasma S1P (Bouma et al. 2011; Pappu et al. 2007). Thus, low body temperature drives lymphopenia during torpor through S1P-dependent retention of cells in secondary lymphoid organs.

23.7 Lymphocyte Proliferation

An important functional aspect of lymphocytes that allows them to induce an adaptive immune response is their clonal expansion after activation. The capacity of T-lymphocytes to proliferate spontaneously and upon activation by Concanavalin A (ConA)-induced T-lymphocyte proliferation in Arctic ground squirrels is diminished in the prehibernation period (October) and during torpor as compared to summer euthermic and aroused squirrels (Novoselova et al. 2000). To our knowledge, no study investigated proliferative capacity of B-lymphocytes in hibernating animals. Taken together, not only is cytokine production by T-lymphocytes reduced during torpor, but proliferation of these cells is also reduced. In addition to reduced proliferation of mature T-lymphocytes, proliferation of immature lymphocytes might also be reduced during hibernation, as the thymus of ground squirrels contains almost no lymphocytes during the hibernation period (Galletti and Cavallari 1972; Novoselova et al. 2004). It has been suggested that 5′-AMP released from brown adipose tissue surrounding the thymus inhibits proliferation of lymphocytes in the thymus (Atanassov et al. 1995).

23.8 Antibody production

The function of B-lymphocytes during torpor has been investigated in vivo by measuring the induction of a humoral immune response following injection of antigens into the European ground squirrel (Andjus et al. 1964; Jaroslow and Serrell 1972).

Either rabbit red blood cells (RRBCs) or sheep red blood cells (SRBCs) were used as antigens in these experiments. In these animals the formation of antibody-producing (plaque-forming) cells in the spleen and immune hemaglutinins was delayed for the duration of torpor (up to 40 days) (Andjus et al. 1964; Jaroslow and Serrell 1972). However, hemaglutinin titers rose rapidly during arousal from torpor in response to the antigen that was injected (Andjus et al. 1964). Not only natural deep torpor, but also forced hypothermia of summer euthermic squirrels delays the formation of immune hemaglutinins until rewarming (Andjus et al. 1964). The reduced capacity to generate antibody-producing cells and produce antibodies is likely due to the low body temperature. Secretion of IgA into the lumen is an important effector mechanism in the defense against infections by counteracting the attachment of bacteria to the mucosal epithelial cells (Kraehenbuhl and Neutra 1992; Woof and Mestecky 2005). Spontaneous production of IgA in the gut is increased during hibernation as demonstrated in the 13-lined ground squirrel (Kurtz and Carey 2007). The relative numbers of cells that are capable of producing IgA, being B-lymphocytes (i.e. CD45RA^{+}) and specifically IgA^{+}-cells, are increased among lamina propria lymphocytes both during torpor and arousal (Kurtz and Carey 2007). Increased production of IgA might be stimulated by the increased mucosal levels of IL-4 and IL-10 in hibernating 13-lined ground squirrels (Kurtz and Carey 2007). Also, mucosal levels of TNF-α and IFN-γ are increased in the small intestine of 13-lined ground squirrels during hibernation (Kurtz and Carey 2007). Production of IFN-γ induced by activation of mononuclear cells (i.e. monocytes and lymphocytes) with the T-lymphocyte-specific mitogen phytohemagglutinin (PHA) in vitro at 37°C, however, is lower in cells derived from torpid animals as compared to euthermic ground squirrels (Kandefer-Szerszen 1988). IFN-γ strongly activates macrophages and stimulates proliferation of cytotoxic cells, which are essential responses against intracellular microorganisms. Thus, although the induction of a systemic humoral immune response appears to be reduced by torpor, mucosal humoral responses, as reflected by secretion of IgA in the gut lumen are higher in hibernating animals. These observations demonstrate organ-specific alterations in the immune system during torpor, and may reflect the key role of the intestinal immune system in maintaining an effective barrier to resident gut microbes during the winter fast.

23.9 Implications

Hibernating mammals display major alterations in their physiology, including increased resistance to ischemia/reperfusion damage and a reduced immune function. Such changes might be crucial in allowing the animal to limit organ injury due to the repetitive cooling and rewarming throughout hibernation. In humans, therapeutic hypothermia is frequently used during major surgery of brain and heart to limit neuronal injury (Arrich et al. 2009) and during preservation in transplantation medicine (Belzer et al. 1967; Terasaki et al. 1996). However, hypothermia during cardiac surgery is associated with increased renal injury (Kourliouros et al. 2010) and prolonged cold storage of organ grafts may lead to inferior long-term graft survival following transplantation (Salahudeen 2004). In addition to its applicability in human medicine, understanding the immune function in hibernators is also of growing importance as large populations of hibernating bats are currently threatened by the White Nose Syndrome (WNS). This condition has a mortality rate of 75–100% and is caused by a psychrophilic (cold-loving) fungus which thrives on bats in torpor (Barlow et al. 2009; Blehert et al. 2009; Buchen 2010; Puechmaille et al. 2010; Zimmerman 2009) and is thought to be related to the suppressed immune function during hibernation (Bouma et al. 2010a; Wibbelt et al. 2010). Therefore, understanding the hibernating immune system might be of substantial clinical and ecological relevance.

References

Andjus RK et al (1964) Influence of hibernation and of intermittent hypothermia on the formation of immune hemagglutinins in the ground squirrel. Ann Acad Sci Fenn Biol 71:26–36

Arendt T et al (2003) Reversible paired helical filament-like phosphorylation of tau is an adaptive process associated with neuronal plasticity in hibernating animals. J Neurosci 23:6972–6981

Arrich J et al (2009) Hypothermia for neuroprotection in adults after cardiopulmonary resuscitation. Cochrane Database Syst Rev 4:CD004128

Atanassov CL et al (1995) Anti-lymphoproliferative activity of brown adipose tissue of hibernating ground squirrels is mainly caused by AMP. Comp Biochem Physiol C Pharmacol Toxicol Endocrinol 112:93–100

Barlow A et al (2009) Investigations into suspected white-nose syndrome in two bat species in Somerset. Vet Rec 165:481–482

Beekhuizen H, van Furth R (1993) Monocyte adherence to human vascular endothelium. J Leukoc Biol 54:363–378

Belzer FO et al (1967) 24-hour and 72-hour preservation of canine kidneys. Lancet 2:536–538

Blehert DS et al (2009) Bat white-nose syndrome: an emerging fungal pathogen? Science 323:227

Bouma HR et al (2010a) Hibernation: the immune system at rest? J Leuk Biol 88:619–624

Bouma HR et al (2010b) Blood cell dynamics during hibernation in the European ground squirrel. Vet Immunol Immunopathol 136:319–323

Bouma HR et al (2011) Low body temperature governs the decline of circulating lymphocytes during hibernation through sphingosine-1-phosphate. Proc Natl Acad Sci USA 108: 2052–2057

Briggs C et al (2000) New quantitative parameters on a recently introduced automated blood cell counter—the XE 2100. Clin Lab Haematol 22:345–350

Buchen L (2010) Disease epidemic killing only US bats. Nature 463:144–145

Carey HV (1990) Seasonal changes in mucosal structure and function in ground squirrel intestine. Am J Physiol 259:R385–R392

Carey HV et al (2003) Mammalian hibernation: cellular and molecular responses to depressed metabolism and low temperature. Physiol Rev 83:1153–1181

Carey HV, Martin SL (1996) Preservation of intestinal gene expression during hibernation. Am J Physiol 271:G804–G813

Carey HV, Sills NS (1992) Maintenance of intestinal nutrient transport during hibernation. Am J Physiol 263:R517–R523

Carey HV, Sills NS (1996) Hibernation enhances D-glucose uptake by intestinal brush border membrane vesicles in ground squirrels. J Comp Physiol B 166:254–261

Drew KL et al (1999) Ascorbate and glutathione regulation in hibernating ground squirrels. Brain Res 851:1–8

Fleck CC, Carey HV (2005) Modulation of apoptotic pathways in intestinal mucosa during hibernation. Am J Physiol Regul Integr Comp Physiol 289:R586–R595

Frerichs KU et al (1994) Local cerebral blood flow during hibernation, a model of natural tolerance to "cerebral ischemia". J Cereb Blood Flow Metab 14:193–205

Galletti G, Cavallari A (1972) The thymus of marmots: spontaneous, natural seasonal thymectomy? Acta Anat (Basel) 83:593–605

Galluser M et al (1988) Adaptation of intestinal enzymes to seasonal and dietary changes in a hibernator: the European hamster (*Cricetus cricetus*). J Comp Physiol B 158:143–149

Inkovaara P, Suomalainen P (1973) Studies on the physiology of the hibernating hedgehog. 18. On the leukocyte counts in the hedgehog's intestine and lungs. Ann Acad Sci Fenn Biol 200:1–21

Jaroslow BN, Serrell BA (1972) Differential sensitivity to hibernation of early and late events in development of the immune response. J Exp Zool 181:111–116

Kandefer-Szerszen M (1988) Interferon production in leukocytes of spotted sousliks—effect of hibernation on the interferon response in vitro. J Interf Res 8:95–103

Kenagy GJ et al (1989) Annual cycle of energy and time expenditure in a golden-mantled ground squirrel population. Oecologia 78:269

Kourliouros A et al (2010) Low cardiopulmonary bypass perfusion temperatures are associated with acute kidney injury following coronary artery bypass surgery. Eur J Cardiothorac Surg 37:704–709

Kraehenbuhl JP, Neutra MR (1992) Molecular and cellular basis of immune protection of mucosal surfaces. Physiol Rev 72:853–879

Kurtz CC, Carey HV (2007) Seasonal changes in the intestinal immune system of hibernating ground squirrels. Dev Comp Immunol 31:415–428

Kurtz CC et al (2006) Hibernation confers resistance to intestinal ischemia-reperfusion injury. Am J Physiol Gastrointest Liver Physiol 291:G895–G901

Lindell SL et al (2005) Natural resistance to liver cold ischemia reperfusion injury associated with the hibernation phenotype. Am J Physiol Gastrointest Liver Physiol 288:G473–G480

Mandala S et al (2002) Alteration of lymphocyte trafficking by sphingosine-1-phosphatesphingosine-1-phosphate receptor agonists. Science 296:346–349

Maniero GD (2002) Classical pathway serum complement activity throughout various stages of the annual cycle of a mammalian hibernator, the golden-mantled ground squirrel, *Spermophilus lateralis*. Dev Comp Immunol 26:563–574

Maniero GD (2005) Ground squirrel splenic macrophagesmacrophages bind lipopolysaccharide over a wide range of temperatures at all phases of their annual hibernation cycle. Comp Immunol Microbiol Infect Dis 28:297–309

Matloubian M et al (2004) Lymphocyte egress from thymus and peripheral lymphoid organs is dependent on S1P receptor 1. Nature 427:355–360

Morin P Jr et al (2008) Expression of Nrf2 and its downstream gene targets in hibernating 13-lined ground squirrels, *Spermophilus tridecemlineatus*. Mol Cell Biochem 312:121–129

Nakagawa M, Terashima T, D'yachkova Y, Bondy GP, Hogg JC, van Eeden SF (1998) Glucocorticoid-induced granulocytosis: contribution of marrow release and demargination of intravascular granulocytes. Circulation 98:2307–2313

Novoselova EG et al (2000) Production of tumor necrosis factor in cells of hibernating ground squirrels *Citellus undulatus* during annual cycle. Life Sci 67:1073–1080

Novoselova EG et al (2004) Effect of the transplanted thymus of hibernating ground squirrels on the age-related thymus involution in rats. Dokl Biol Sci 397:272–273

Pappu R et al (2007) Promotion of lymphocyte egress into blood and lymph by distinct sources of sphingosine-1-phosphatesphingosine-1-phosphate. Science 316:295–298

Prendergast BJ et al (2002) Periodic arousal from hibernation is necessary for initiation of immune responses in ground squirrels. Am J Physiol Regul Integr Comp Physiol 282: R1054–R1062

Puechmaille SJ et al (2010) White-nose syndrome fungus (Geomyces destructans) in bat, France. Emerg Infect Dis 16:290–293

Reznik G et al (1975) Comparative studies of blood from hibernating and nonhibernating European hamsters (*Cricetus cricetus* L). Lab Anim Sci 25:210–215
Ruzicka K et al (2001) The new hematology analyzer Sysmex XE-2100: performance evaluation of a novel white blood cell differential technology. Arch Pathol Lab Med 125:391–396
Salahudeen AK (2004) Consequences of cold ischemic injury of kidneys in clinical transplantation. J Investig Med 52:296–298
Sandovici M et al (2004) Differential regulation of glomerular and interstitial endothelial nitric oxide synthase expression in the kidney of hibernating ground squirrel. Nitric Oxide 11:194–200
Shivatcheva TM (1988) Survival of skin allografts in European ground squirrels, *Spermophilus citellus* L., during hibernation. Folia Biol (Krakow) 36:213–221
Spurrier WA, Dawe AR (1973) Several blood and circulatory changes in the hibernation of the 13-lined ground squirrel, *Citellus tridecemlineatus*. Comp Biochem Physiol A Comp Physiol 44:267–282
Storey KB (2010) Out cold: biochemical regulation of mammalian hibernation—a mini-review. Gerontology 56:220–230
Suomalainen P, Rosokivi V (1973) Studies on the physiology of the hibernating hedgehog. 17. The blood cell count of the hedgehog at different times of the year and in different phases of the hibernating cycle. Ann Acad Sci Fenn Biol 198:1–8
Szilagyi JE, Senturia JB (1972) A comparison of bone marrow leukocytes in hibernating and nonhibernating woodchucks and ground squirrels. Cryobiology 9:257–261
Talaei F et al (2011) Reversible remodelling of lung tissue during hibernation in the Syrian hamster. J Exp Biol 214(pt8):1276–1282
Terasaki PI et al (1996) Fit and match hypothesis for kidney transplantation. Transplantation 62:441–445
Toien O et al (2001) Ascorbate dynamics and oxygen consumption during arousal from hibernation in Arctic ground squirrels. Am J Physiol Regul Integr Comp Physiol 281:R572–R583
Ulevitch RJ, Tobias PS (1999) Recognition of gram-negative bacteria and endotoxin by the innate immune system. Curr Opin Immunol 11:19–22
van Breukelen F et al (2010) Vertebrate cell death in energy-limited conditions and how to avoid it: what we might learn from mammalian hibernators and other stress-tolerant vertebrates. Apoptosis 15:386–399
von Vietinghoff S, Ley K (2008) Homeostatic regulation of blood neutrophil counts. J Immunol 181:5183–5188
Wibbelt G et al (2010) Emerging diseases in Chiroptera: why bats? Biol Lett 6:438–440
Woof JM, Mestecky J (2005) Mucosal immunoglobulins. Immunol Rev 206:64–82
Zancanaro C et al (1999) The kidney during hibernation and arousal from hibernation. A natural model of organ preservation during cold ischemia and reperfusion. Neprol Dial Transplant 14:1982–1990
Zimmerman R (2009) Ecology. Biologists struggle to solve bat deaths. Science 324:1134–1135

Chapter 24
The Relationship Between White Nose Syndrome and Dietary PUFA Levels in Bats

Craig L. Frank, Patricia M. Diaz and Thomas H. Kunz

Abstract Six bat species in northeastern North America spend winter hibernating in caves/mines. Hibernation is characterized by multi-day periods of torpor, when metabolic rates are reduced. Body fat levels increase from 7 to 27% in the fall, just prior to the onset of hibernation by bats. Some groups of hibernating bats in northeastern North America display abnormal behaviors and increased overwinter mortality called white nose syndrome (WNS). Large numbers of dead bats are found in hibernation sites affected by WNS. Previous research has demonstrated that the increased over-mortality is due to the depletion of depot fat reserves by February, which is in turn caused by greatly shortened torpor bouts. Published studies on hibernating mammals have revealed that moderately high levels of polyunsaturated fatty acids (PUFAs) are required in the diet during the fall for torpor. We thus predicted that bats suffering from WNS have lower levels of PUFAs in their fall diets than those that do not suffer from WNS. We tested this hypothesis by analyzing white adipose tissue (WAT) samples from: (a) little brown bats (*Myotis lucifugus*) collected from different hibernation sites and (b) big brown bats (*Eptesicus fuscus*), a species not greatly affected by WNS. Our analyses reveal that *M. lucifugus* populations prone to WNS have significantly less of the PUFA α-linolenic acid in their fall diets than those where WNS does not occur. The fall diets of *E. fuscus* contain significantly more of the PUFA linoleic acid than those of *M. lucifugus* from the same site. Our findings thus support the hypothesis that bats suffering from WNS have lower levels of PUFAs in their fall diets.

C. L. Frank (✉) · P. M. Diaz
Department of Biological Sciences, Fordham University, Louis Calder Center, P.O. Box 887, Armonk, NY 10504, USA
e-mail: frank@fordham.edu

T. H. Kunz
Department of Biology, Boston University, 5 Cummington Street, Boston, MA 02215, USA

T. Ruf et al. (eds.), *Living in a Seasonal World*,
DOI: 10.1007/978-3-642-28678-0_24,

24.1 Introduction

White Nose Syndrome (WNS) is a recently discovered pathological state that occurs during hibernation, which results in greatly increased overwinter mortality rates. This syndrome is known to greatly affect five species in the northeastern portion of North America (Turner et al. 2011): little brown (*Myotis lucifugus*), Indiana (*Myotis sodalis*), northern long-eared (*Myotis septentrionalis*), small-footed (*Myotis leibii*), and tricolored (*Perimyotis subflavus*) bats. When affected by WNS, bats cluster near the entrance of caves/mines during hibernation, areas where torpid bats are not normally found. Bats affected by WNS also fly outside the caves/mines during the daylight hours of January/February, a period when they are not normally active outside. Large numbers of dead bats are also found during late January–early February in the areas where WNS occurs. A white fungus identified as *Geomyces destructans* grows on the muzzle, wings, and ears of bats suffering from WNS starting in late January/early February (Blehert et al. 2009; Gargas et al. 2009; Meteyer et al. 2009). Recent laboratory experiments have demonstrated that cutaneous infection with this fungus is the cause of WNS (Lorch et al. 2011). Although first discovered in central New York during the winter of 2006–2007, WNS is now found in 16 states and four Canadian provinces (Turner et al. 2011).

Six of the nine bat species found in northeastern North America spend the winter hibernating in caves/mines: *M. lucifugus*, *M. sodalis*, *M. septentrionalis*, *M. leibii*, *P. subflavus*, and the big brown bat (*Eptesicus fuscus*). These bat species usually congregate (swarm) near caves/mines starting in mid-September, with hibernation beginning by late October to mid-November. Emergence from hibernation and departure to their spring feeding/maternity areas typically peaks by mid-April. All six of these hibernating species feed almost exclusively on insects, with other arthropods comprising a small portion of the diet as well (Brack and LaVal 1985; Kunz 1974; Kellner and Harestad 2005). During the late summer/early fall swarming period, the body fat content of little brown bats (*M. lucifugus*) increases from 7 to 27% body mass (Kunz et al. 1998; Reynolds and Kunz 2000). Depot fat is the primary energy source utilized during mammalian hibernation (Kayser 1965; Dark 2005). Recent field studies on the torpor patterns of free-ranging *M. lucifugus* revealed that the cause of mortality associated with *G. destructans* infection during hibernation is a profound increase in the frequency of arousal episodes from torpor. Populations of *M. lucifugus* with WNS during hibernation have 50–60% shorter torpor bouts, and sometimes maintain significantly higher body temperatures during torpor, than those previously reported for *M. lucifugus* prior to the occurrence of WNS in North America (Reeder et al., *in review*). Body fat levels consequently decreased to minimum levels ($\sim$7% body mass) about 2 months earlier than normal in bats afflicted with WNS (Frank et al., *in review*). Field observations indicate that the mortality rates of hibernating *E. fuscus* are not substantially increased in the same caves/mines where other bat species have been decimated by WNS (Turner et al. 2011), and histological analyses of *E. fuscus* collected from WNS affected areas revealed no signs of cutaneous fungal infections (Blehert et al. 2009). Furthermore, some populations of *M. lucifugus* have not been affected by WNS even though *G. destructans* has been found in these areas for years. The factors that enable some bat species and populations to resist the mortality associated with cutaneous infection by *G. destructans* and thus WNS are unknown.

One possible factor affecting the overwinter mortality associated with WNS may be the concentration of polyunsaturated fatty acids found in the natural fall diet. A polyunsaturated fatty acid (PUFA) has more than one carbon–carbon double bond per molecule, as opposed to either a saturated fatty acid containing no carbon–carbon double bonds, or a monounsaturated fatty acid containing only one such bond per molecule (Gunstone 1996). Mammals can synthesize saturated and monounsaturated fatty acids, but they are incapable of producing PUFAs. Most plant species, however, produce 2 PUFAs: linoleic acid (18 carbon atoms, two double bonds) and α-linolenic acid (18 carbon atoms, three double bonds). When mammals consume PUFAs they are incorporated into cell membranes and storage lipids (Gunstone 1996). Laboratory experiments with chipmunks, three ground squirrel species, two species of prairie dogs, marmots, mice, and two marsupials have all revealed that the level of linoleic acid in the diet during fattening influences their ability to hibernate (see Ruf and Arnold 2008 for review). Previous laboratory studies with ground squirrels have shown that hibernation ability is greatest when the linoleic acid (18:2) content of the diet is moderately high (33–74 mg linoleic acid/g). Squirrels fed a 33–74 mg linoleic acid/g diet: (1) were more likely to hibernate, (2) spent less time fasting prior to the onset of torpor, (3) had lower T_b during torpor, and (4) had longer torpor bouts than those maintained on diets containing either less or more linoleic acid (Frank 1992, 2002; Frank and Storey 1995, 1996; Frank et al. 1998). It has also been demonstrated that the amount of α-linolenic acid (18:3) in the diet influences the torpor patterns in a manner identical to that for linoleic acid (Frank et al. 2004). Natural variations in dietary PUFA contents have been shown to have similar effects on the torpor patterns of free-ranging arctic ground squirrels (*Spermophilus parryii*) as well (Frank et al. 2008).

The diets of *Myotis spp.* consist almost entirely of insects found in the orders Coleoptera, Diptera, Hymenoptera, Lepidoptera, Neuroptera, and Trichoptera (Brack and LaVal 1985; Kellner and Harestad 2005). Insect species vary in their ability to synthesize PUFAs. Insects in the orders Diptera and Lepidoptera cannot synthesize either linoleic or α-linolenic acids, therefore the PUFA level of these tissues depends entirely upon the diet (Urich 1990; Buckner 1993). Insects of the orders Neuroptera, Orthoptera, Homoptera, and Hymenoptera, in contrast, can synthesize linoleic acid. Insect PUFA contents thus vary with both species as well as their diet (Urich 1990; Buckner 1993). Any factor that increases torpor bout length should mitigate the effects of WNS, since cutaneous infection with *G. destructans* causes mortality by reducing torpor bout length. Higher dietary PUFA levels increase torpor bout length, thus we predict that: (1) the fall diets of bat species that are not susceptible to WNS have relatively higher levels of PUFAs than those of species that are susceptible to WNS found in the same habitat and (2) the fall diets of bat populations where WNS does not occur have relatively higher PUFA contents than the diets of populations where WNS occurs.

A high ratio of *n*-6 PUFAs to *n*-3 PUFAs incorporated into the membranes of cardiac muscle cells may offset the effects of low temperature on the activity of the sarcoplasmic reticulum Ca^{+2}–Mg^{+2} pump, thereby preserving cardiac function during torpor (Ruf and Arnold 2008). The *n*-6/*n*-3 PUFA ratio of both cardiac and liver cellular membranes isolated from the tissues of free-ranging alpine marmots (*Marmota marmota*) increase rapidly before the onset of hibernation (Arnold et al. 2011). The *n*-6/*n*-3 PUFA ratios for the diets of bat species/populations resistant to WNS may thus be significantly greater as well. These hypotheses were tested by analyzing the PUFA contents of WAT samples from free-ranging little brown (*M. lucifugus*) and big brown (*E. fuscus*) bats collected at multiple hibernation sites.

24.2 Materials and Methods

The average amount of polyunsaturated fatty acids in natural bat diets during the entire fall feeding/fattening period was estimated by measuring the amount (% of all fatty acids) of PUFAs in subcutaneous WAT sampled from the carcasses of bats collected either just prior to hibernation (October), or during hibernation (February). This method was previously developed and used successfully to determine the fatty acid compositions of the fall diets of free-ranging ground squirrels in two areas (Frank et al. 1998, 2008). The concentrations of both linoleic and α-linolenic acids in mammalian WAT depend upon the amount of these fatty acids in the diet during fattening (Gunstone 1996), generally increasing with dietary PUFA content.

Population differences in WAT (diet) PUFA contents. A total of ten adult *M. lucifugus* with no obvious signs of WNS were collected from two different caves/mines, one located in Kentucky, the other in Ohio, where WNS does not occur. 11 more adult *M. lucifugus* were collected from three different mines/caves, one each located at Massachusetts, New York, and Vermont, all where WNS occurs. All bats were collected while torpid during October 2008, at the onset of hibernation. They were immediately sacrificed upon capture by an Isoflurane overdose and frozen at −20°C for later analysis. A 100 mg biopsy of WAT was taken from each bat carcass from the caudal area. All WAT samples were then analyzed for total fatty acid composition using the gas chromatography procedures described in Frank et al. (2008).

Species differences in WAT (diet) PUFA contents. The NY State Department of Environmental Conservation collected a total of 14 adult *M. lucifugus* from four mines in New York State during the February period of 2008 and 2009. All *M. lucifugus* in these mines were displaying the behaviors and fungal growth associated with WNS. A total of seven adult *E. fuscus* were also collected from the same mines during this period. All bats were immediately sacrificed upon capture by an Isoflurane overdose and frozen at −20°C for later analysis. Bat carcasses were analyzed for total body fat content to further determine whether or not big brown bats are greatly affected by WNS. Adult *M. lucifugus* and *E. fuscus* are frequently infected with rabies (Mondul et al. 2003) which can also lead to rapid body fat loss during hibernation, thus all bat carcasses were first tested for rabies following the methods summarized in Blanton et al. (2008). None of the bats collected tested positive for rabies, thus all carcasses were included in the analyses. Total body fat contents were determined using petroleum ether extractions with a Soxhlet apparatus following the techniques of Dobush et al. (1985). A 100 mg biopsy of WAT was taken from each bat carcass from the caudal area prior to total body fat analyses. All WAT samples were then analyzed for total fatty acid composition using the gas chromatography procedures described in Frank et al. (2008).

24.3 Results

Population differences in WAT (diet) PUFA contents. Linoleic and α-linolenic acids were found in the WAT, all little brown bats collected. The WAT of little brown bats (*M. lucifugus*) collected from sites where WNS occurs and that from *M. lucifugus,* where WNS did not occur had statistically equivalent mean levels of linoleic acid ($t = 0.64$, d.f. = 19, $P = 0.95$; Fig. 24.1a). The mean α-linolenic acid content of WAT from bats collected at sites free of WNS (Fig. 24.1a), in contrast, was 1.4 times greater than that from *M. lucifugus* collected at WNS sites ($t = 2.769$, d.f. = 19, $P = 0.012$). The mean (± SE) *n*-6/*n*-3 PUFA ratio of the WAT from the WNS sites was thus 2.71 ± 0.32, whereas that for WAT from the sites where WNS does not occur was 1.87 ± 0.19, and significantly lower ($t = -2.164$, d.f. = 19, $P = 0.043$).

Species differences in WAT (diet) PUFA contents. Linoleic and α-linolenic acids were found in the WAT of both bat species. The mean linoleic acid content of WAT from big brown bats (*E. fuscus*) was 1.3 times greater (Fig. 24.1b) than that of the WAT from little brown bats (*M. lucifugus*) collected at the same sites ($t = 2.526$, d.f. = 19, $P = 0.021$). The mean α-linolenic acid contents of WAT from *E. fuscus* and *M. lucifugus* were statistically equivalent (Fig 24.1b), however, ($t = -1.868$, d.f. = 13, $P = 0.09$). The mean total (%) body fat content of *E. fuscus* during February was 1.5-fold greater (Table 24.1) than that of *M. lucifugus* collected during the same period ($t = 2.561$, d.f. = 17, $P = 0.02$). The mean (±SE) *n*-6/*n*-3 PUFA ratio for WAT from *E. fuscus* was thus 9.53 ± 2.13, whereas that for the WAT from *M. lucifugus* was 3.59 ± 0.43, and significantly less ($t = 3.854$, d.f. = 17, $P = 0.001$).

24.4 Discussion

The analyses of WAT collected from little brown bats (*M. lucifugus*) from the sites where WNS did and did not occur indicate that the fall diets of little brown bats in areas where WNS does not occur have a greater α-linolenic acid content than the

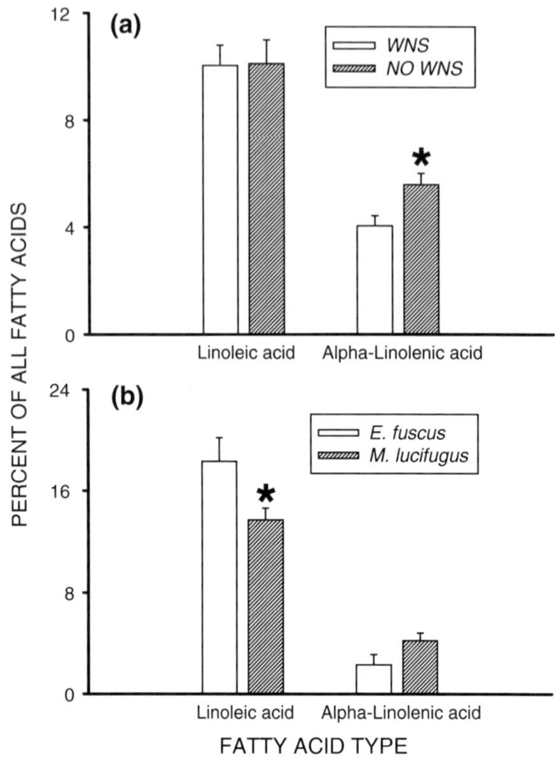

Fig. 24.1 Histograms indicating the mean (+SE) contents (% all fatty acids) of linoleic and α-linolenic acids in white adipose tissues from: **a** different populations of *M. lucifugus* in October and **b** both *M. lucifugus* and *E. fuscus* collected from the same sites in February. *Significantly different from the corresponding mean at the $P < 0.05$ level

Table 24.1 Mean (±SE) body compositions of free-ranging bats during February

Bat species	Mean % body fat	Mean lean mass (g)
M. lucifugus	6.62 ± 0.63	5.34 ± 0.15
E. fuscus	10.35 ± 1.57[a]	12.05 ± 0.30[a]

[a] Significantly different from *M. lucifugus* at the $P < 0.05$ level

fall diets of those found in hibernation sites where WNS occurs. Our findings thus support the hypothesis that the fall diets of bat populations, where WNS does not occur, have relatively higher PUFA contents than those of the same species where WNS occurs. These differences resulted in a 31% lower average n-6/n-3 PUFA ratio for the WAT (and presumably diet) of bat populations where WNS does not occur. It is thus unlikely that the small differences in α-linolenic acid content observed can account for population differences in WNS mortality.

The analyses of WAT collected from both big brown bats (*E. fuscus*) and little brown bats (*M. lucifugus*) at the same hibernation sites indicate that the fall diets of big brown bats have a higher linoleic acid content than those of little brown bats feeding in the same area. Our findings thus support the hypothesis that the fall diets of bat species that are not susceptible to WNS have relatively greater PUFA contents than the diets of species that are susceptible to WNS. The differences in linoleic acid levels produced a 2.7-fold greater *n*-6/*n*-3 PUFA ratio, on average, for WAT from *E. fuscus*. There is substantial evidence that a high *n*-6/*n*-3 PUFA ratio in cellular membranes enhances torpor (Ruf and Arnold 2008). Consequently, differences in dietary PUFA contents, resulting from contrasting insect prey specializations between these two bat species, may be one factor that enables *E. fuscus* to better resist WNS than *M. lucifugus*. Our total body fat analyses also further support the observation that *E. fuscus* is less susceptible to WNS. The proportion of body mass that was total body fat was 1.5-fold greater for *E. fuscus* than for *M. lucifugus* during February, which is about 2 months prior to the end of hibernation in New York.

The petroleum ether extraction methods of Dobush et al. (1985) were used to measure body fat contents in our study assay, both depot and structural lipids, and do not distinguish between them. The structural lipids in mammalian tissues are phospholipids, cholesterol, ceramides, and sphingolipids, none of which are either stored or mobilized as metabolic energy sources. Mammalian depot lipids, in contrast, are triacylglycerols which are stored in adipocytes, and they are both mobilized and consumed during fasting (Gunstone 1996). Total body fat content (% live mass) thus represents the sum of both depot lipids, which can be mobilized to support metabolism during fasting, and structural lipids, which cannot. Consequently, mammals that have depleted all of their depot lipid reserves will still have total body fat content of 4 8% live mass (Pond 1998), composed of structural lipids. The *M. lucifugus* in our study therefore probably depleted all of their depot fat (energy) reserves by February, since by this time their total body fat content was around 6.6%. In contrast, it does not appear that *E. fuscus* did not deplete their depot fat reserves by February, since their mean total body fat contents were >10% during this period. Healthy adult *M. lucifugus* have a mean lean body mass of about 7.7 g (Kunz et al. 1998; Reynolds and Kunz 2000). The mean lean body mass of the *M. lucifugus* afflicted with WNS that were collected during February in our study was only 69% of this normal healthy level (Table 24.1); however, which indicates that they have already depleted all of their depot fat reserves and are now metabolizing lean tissue to support their metabolism.

It thus appears that relatively higher levels of linoleic acid in the diet during the fall fattening period may enable some bat species to resist WNS. Laboratory hibernation experiments are required to fully test this prediction. Field studies should also be conducted to determine whether or not the torpor bouts of *E. fuscus* are affected by the presence of *G. destructans* in the cave/mine during hibernation. Further investigation of the role of dietary PUFAs in bat torpor will not only lead to new insights into the biology of hibernation, but to new discoveries regarding the pathology of WNS in North America as well.

Acknowledgments We thank Alan Hicks and Carl Hertzog for their generous assistance. This study was supported by NSF grants IOS-0818222 awarded to C.L.F., and IOS- 0840762 awarded to T.H.K. and C.L.F.

References

Arnold W, Ruf T, Frey-Roos F, Bruns U (2011) Diet-independent remodeling of cellular membranes precedes seasonally changing body temperature in a hibernator. PloS ONE 6(4): e18641

Blanton JD, Robertson K, Palmer D, Rupprecht C (2008) Rabies surveillance in the United States during 2008. JAVMA 235:676–689

Blehert DS, Hicks A, Behr M, Meteyer C, Berlowski-Zier B, Buckles E, Coleman J, Darling S, Gargas A, Niver R, Okoniewski J, Rudd R, Stone W (2009) Bat white nose syndrome: an emerging fungal pathogen? Science 323(5911):227

Brack V, LaVal R (1985) Food habits of the Indiana bat in Missouri. J Mammal 66:308–315

Buckner JS (1993) Cuticular polar lipids of insects. In: Stanley-Samuelson DW, Nelson DR (eds) Insect lipids: chemistry, biochemistry, and biology. University of Nebraska Press, Lincoln

Dark J (2005) Annual lipid cycles in hibernators: integration of physiology and behavior. Annu Rev Nutr 25:469–497

Dobush GR, Ankney CD, Krementa DG (1985) The effect of apparatus, extraction time, and solvent type on lipid extraction of snow geese. Can J Zool 63:1917–1920

Frank CL (1992) The influence of dietary fatty acids on hibernation by golden-mantled ground squirrels (*Spermophilus lateralis*). Physiol Zool 65:906–920

Frank CL, Storey KB (1995) The optimal depot fat composition for hibernation by golden-mantled ground squirrels (*Spermophilus lateralis*). J Comp Physiol B 164:536–542

Frank CL, Storey KB (1996) The effect of total unsaturate content on hibernation. In: Geiser F, Hulbert A, Nicol SJ (eds) Adaptations to the cold. University of New England Press, Armidale

Frank CL, Dierenfeld ES, Storey KB (1998) The relationship between lipid peroxidation, hibernation, and food selection in mammals. Amer Zool 38:341–349

Frank CL (2002) The effects of short-term variations in diet fatty acid composition on mammalian torpor. J Mammal 83:1019–1031

Frank CL, Hood WR, Donnelly MC (2004) The role of α-Linolenic acid (18:3) in mammalian torpor. In: Barnes M, Carey H (eds) Life in the cold: evolution, mechanisms, adaptation and application. Institute of Arctic Biology Press, Fairbanks

Frank CL, Karpovich S, Barnes B (2008) The relationship between natural variations in dietary fatty acid composition and the torpor patterns of a free-ranging arctic hibernator. Physiol Biochem Zool 81:486–495

Gargas A, Trest MT, Christensen M, Volk TJ, Blehert DS (2009) *Geomyces destructans* sp. Nov. associated with bat white nose syndrome. Mycotaxon 108:147–154

Gunstone FD (1996) Fatty acid and lipid chemistry. Blackie Academic and Professional, Glasgow

Kayser C (1965) Hibernation. In: Mayer W, VanGelder R (eds) Physiological mammalogy, volume II. Academic Press, New York

Kellner AME, Harestad AS (2005) Diets of bats in coastal rainforests on Vancouver Island, British Columbia. Northwest Nat 86:45–48

Kunz TH (1974) Feeding ecology of a temperate insectivorous bat (*Myotis velifer*). Ecology 55:693–711

Kunz TH, Wrazen JH, Burnett CD (1998) Changes in body mass and body composition in pre-hibernating little brown bats (*M. lucifugus*). Ecoscience 5:8–17

Lorch JM, Meteyer, C, Behr M, Boyles J, Cryan P, Hicks A, Ballmann A, Coleman J, Redell D, Reeder, D, Belhert D (2011) Experimental infection of bats with *Geomyces destructans* causes white-nose syndrome. Nature 480:376–378. doi:10.1038/nature10590

Meteyer CU, Buckles E, Blehert D, Hick A, Green D, Shearn-Bochsler V, Thomas N, Gargas A, Behr M (2009) Histopathologic criteria to confirm white-nose syndrome in bats. J Vet Diagn Invest 21:411–414

Mondul A, Krebs J, Childs J (2003) Trends in national surveillance for rabies among bats in the United States (1993–2000). JAVMA 222:633–639

Pond C (1998) The fats of life. Cambridge University Press, New York

Reynolds RS, Kunz T (2000) Changes in body composition during reproduction and postnatal growth in the little brown bat, *Myotis lucifugus* (Chiroptera: Vespertilionidae). Ecoscience 7:10–17

Ruf T, Arnold W (2008) Effects of polyunsaturated fatty acids on hibernation and torpor: a review and hypothesis. Amer J Physiol 294:R1044–R1052

Turner GG, Reeder D, Coleman J (2011) A five-year assessment of mortality and geographic spread of white-nose syndrome in North American bats and a look to the future. Bat Res News 52:13–27

Urich K (1990) Comparative animal biochemistry. Springer, New York

Chapter 25
Impact of Hibernation on Gut Microbiota and Intestinal Barrier Function in Ground Squirrels

Hannah V. Carey, Amanda C. Pike, Christopher R. Weber, Jerrold R. Turner, Annie Visser, Silvia C. Beijer-Liefers, Hjalmar R. Bouma and Frans G. M. Kroese

Abstract Like many hibernators, the 13-lined ground squirrel fasts during winter. The loss of food intake not only affects the hibernator's gut structure and function but also has the potential to modify the resident gut microbiota. Here we examined the effect of hibernation on numbers of cecal microbes, microbial production of short chain fatty acids, intestinal macromolecular permeability, and expression of the tight junction protein occludin. Compared with the active season, bacteria detected by probes specific for *Lactobacillus–Enterococcus* and *Eubacteria–Clostridium* groups were reduced during hibernation but the *Bacteroides–Prevotella* group was less affected. Cecal concentrations of short chain fatty acids were similar in active season squirrels, aroused hibernators, and hibernators in early torpor, but lower in late torpor. Gut permeability to a macromolecular marker increased during hibernation, as did tight junction localization and phosphorylation of occludin. The results suggest that hibernation selectively alters the gut microbiota, possibly reflecting

---

Present Address
H. V. Carey (✉) · A. C. Pike
Department of Comparative Biosciences, School of Veterinary Medicine, University of Wisconsin–Madison, 2015 Linden Dr, Madison, WI 53706, USA
e-mail: careyh@vetmed.wisc.edu

C. R. Weber · J. R. Turner
Department of Pathology, The University of Chicago, Chicago, IL, USA

A. Visser · S. C. Beijer-Liefers · F. G. M. Kroese
Department of Rheumatology and Clinical Immunology, University Medical Center Groningen, University of Groningen, Groningen, The Netherlands

H. R. Bouma
Department of Clinical Pharmacology, University Medical Center Groningen, Groningen, The Netherlands

H. R. Bouma
Department of Animal Behavior/Chronobiology, University of Groningen, Groningen, The Netherlands

T. Ruf et al. (eds.), *Living in a Seasonal World*,
DOI: 10.1007/978-3-642-28678-0_25,

differential sensitivity to loss of dietary polysaccharides. Hibernation impairs intestinal barrier function, which may require compensatory mechanisms to enhance tight junctional integrity and thus maintain a healthy relationship of the hibernator host with its resident gut microbiota until food intake resumes in the spring.

25.1 Introduction

Mammals have evolved in close association with their resident gut microbes (Ley et al. 2008). In nonruminants, the density and diversity of the microbiota increases distally along the gastrointestinal tract, with microbial numbers reaching as high as 10^{12}–10^{14} in the hindgut (cecum and colon). Estimates of community richness range in the thousands of species, representing more than 50–100 times the number of genes in the host genome (Vaughan et al. 2000). Resident gut microbes have a complex, symbiotic relationship with their hosts, in which the host reaps multiple benefits from its symbionts and in turn provides a nurturing environment for growth of the microbial community. The microbiota can influence a diverse array of host characteristics including nutrition and metabolism, fat storage, innate and adaptive immunity, enterocyte proliferation and differentiation, mucus composition, and resistance to pathogen colonization (Backhed et al. 2004; Backhed et al. 2005; Crawford et al. 2009; Hooper et al. 2001; Velagapudi et al. 2010). In turn, host biology, environment, and diet have significant effects on microbiota composition and production of metabolic by-products (Bailey et al. 2010; Ley et al. 2006).

Gut microbes degrade dietary polysaccharides as well as host-derived substrates such as mucus glycans, and nutrients in sloughed epithelial cells (Martens et al. 2008; Sonnenburg et al. 2005). Some of the by-products of microbial fermentation, including short-chain fatty acids (SCFA), are then absorbed by the host to meet energetic and nutritional needs. Studies based largely on germ-free animal models, imposed fasts, and dietary manipulations have provided strong support that host diet is a major determinant of the composition of the microbiota (Backhed et al. 2005; Leser et al. 2000; Muegge et al. 2011; Sonnenburg et al. 2005). Many seasonal hibernators fast completely for several months, making them useful models to examine the effect of extreme dietary changes on the resident microbial community. However, little is known about how hibernation affects gut microbiota composition or the production of metabolites such as SCFA, which could be a potential source of recycled energy to the hibernator during the winter fast. It is also unclear how hibernation affects intestinal barrier function, which is crucial in limiting access of commensal and pathogenic bacteria to the host's systemic environment.

Here we addressed some of these gaps in knowledge by using the 13-lined ground squirrel (*Ictidomys tridecemlineatus*) to determine how hibernation affects cecal microbiota, cecal SCFA levels, intestinal macromolecular permeability, and expression of the tight junction protein, occludin.

25.2 Methods

Animals and tissue collection. Ground squirrel use for these studies was approved by the UW-Madison School of Veterinary Medicine Animal Care and Use Committee. Squirrels were either collected directly from the wild in the vicinity of Madison, WI or were offspring of pregnant females in captivity. After collection or weaning squirrels were fed rodent chow (Teklad #7001) supplemented with sunflower seeds. In fall, some squirrels were moved to a 4°C dark room, and food and water were removed 2 weeks after the start of torpor cycles. Squirrels were studied in the active season (May through August) or during hibernation (November through March). For SCFA studies active squirrels were fasted overnight. Food was removed from hibernator cages ~1 week after start of torpor cycles. Hibernators were sampled in early Winter (<2.5 months after start of hibernation) or late winter (>3 months after start of hibernation). Early and late winter hibernators were sampled either in the interbout arousal (IBA) state, when T_b is ~36°C; or during torpor when T_b is ~5°C; when values in those states were similar within a winter group they were combined. In some studies, Torpor was differentiated as early torpor (ET, <24 h after entering torpor) or late torpor (LT, at least 7 *d* of continuous torpor); when those groups were similar they were combined into one "Torpor" group.

Squirrels were euthanized by isofluorane anesthesia followed by decapitation. Ceca were harvested and contents were weighed and divided into two portions: one was frozen for DNA extraction, and the other was sonicated, centrifuged, and supernatant collected. An aliquot was acidified with sulfuric acid and frozen for later SCFA analysis. Cecal tissues were frozen for later immunoblotting. Other squirrels were anesthetized and perfused with 4% paraformaldehyde.

Microbiota analysis. DNA was extracted from cecal samples and used in *q*PCR with four group-specific primers based on the 16S rDNA. The primers were: Universal probe for all bacteria (Bact338), UnivF340-ACTCCTACGGGAGGCAGCAGT and UnivR 514-ATTACCGCGGCTGCGGC; *Bacteroides* group (Bac303), BactF285-GGTT CTGAGAGGAAGGTCCC and UnivR338-GCTGCCTCCCGTAGGAGT; Lactobacilli/cocci (Lab158), LabF362-AGCAGTAGGGAATCTTCCA and LabR677-CAC-CGCTACACATGGAG; *C. coccoides/Eub. rectalis/Ros. rectalis* (Erec482), UnivF 338-ACTCCTACGGGAGGCAGC and C.cocR491-GCTTCTTAGTCAGGTACCGT CAT. The reference strains for these four groups were, respectively: *E. coli*, *Bacteroides fragilis*, *Lactobacillus acidophilus,* and *Ruminococcus productus.* Internal standard curves were constructed from serial dilutions of reference bacterial strain genomic DNA to convert *q*PCR values into number of bacteria per gram cecal contents.

Cecal SCFA levels. Cecal fluid supernatants were thawed and subject to gas chromatography for measurement of total SCFA concentrations and molar proportions of propionate, acetate, and butyrate. SCFA analyses were carried out by Dr. Glen Broderick (USDA Dairy Forage Research Center, Madison, WI) using conventional techniques (Broderick et al. 2004).

Gut permeability. Food was removed from active season squirrels 4 h prior to experiments and hibernators were used during IBA periods. The paracellular permeability marker FITC-labeled dextran (4 kD) (0.6 mg/g body mass) was administered under light anesthesia by gavage. Squirrels were returned to cages and 3.5 h later were euthanized and trunk blood collected. Plasma was read at absorbances of 490 nm (excitation) and 520 nm (emission) and a standard curve was used to calculate flux of FITC-dextran from gut lumen to blood. Data were expressed in arbitrary absorbance units (a.u.).

Occludin expression. Cecal tissue lysates were used for immunoblotting using anti-occludin antibodies (Invitrogen, Camarillo, CA). Relative proportions of total occludin and the hyperphosphorylated form (slower migrating band) were quantitated by densitometry and normalized to actin band intensity. Immunohistochemistry was performed on perfusion-fixed sections of small intestine and cecum using conventional techniques.

25.3 Statistics

Data were analyzed by unpaired *t*-tests or ANOVA as appropriate, with significance set at $P \leq 0.05$. If ANOVA was significant, post hoc comparisons among groups were assessed by Fisher's LSD test.

25.4 Results

Total numbers of bacteria per gram cecal contents as represented by the universal probe Bact338 did not differ among active season samples and those from hibernators, despite a trend for lower abundance in late winter (Fig. 25.1a). Bacterial numbers expressed as a percentage of Bact338 and normalized to active season values are shown in Fig. 25.1b. Relative numbers of *Bacteroides–Prevotella* (Bact303) tended to be lowest in late hibernation, although there were no significant differences among groups. *Lactobacillus–Enterococcus* (Lab158) was particularly affected by hibernation, falling to nearly undetectable levels, and *Eubacteria–Clostridium* (Erec 482) was also significantly lower in the two hibernation groups.

Cecal concentrations of acetate, propionate, and butyrate were similar in summer, early torpor, and IBA squirrels but were lower in late torpor squirrels (Fig. 25.2a). The molar concentration of acetate increased during hibernation, whereas that for propionate and butyrate fell (Fig. 25.2b). The length of time in the hibernation season did not affect cecal SCFA concentrations.

An in vivo permeability assay was used to determine if hibernation altered gut macromolecular permeability. Appearance of FITC-Dextran in plasma was five-fold higher in aroused hibernators (1033 ± 326 a.u., $n = 7$) than in active season squirrels (196 ± 75 a.u., $n = 7$) ($P < 0.05$). Altered intestinal permeability could result from changes in expression of tight junction proteins that regulate

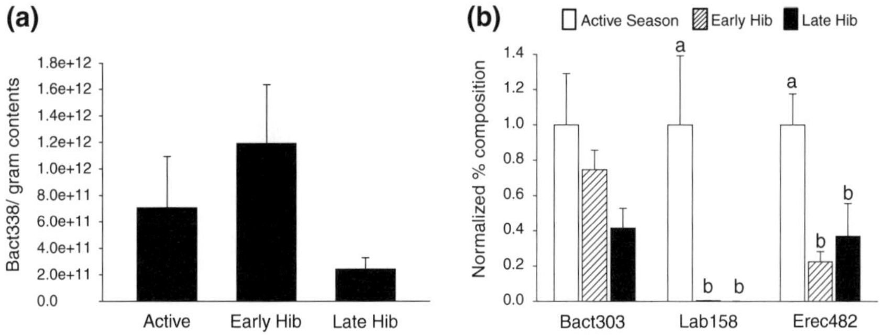

Fig. 25.1 Overall numbers of gut bacteria per gram cecal contents were similar in active season squirrels and early hibernators, but were reduced by late hibernation in *Lactobacillus–Enterococcus* and *Eubacteria–Clostridium* species. **a** Abundance of bacteria that hybridized to Bact338, a universal probe for all bacteria. **b** Normalized % composition. Bacterial numbers were expressed as percentage of total bacteria (Bact338), normalized to active season value. Bact303: *Bacteroides–Prevotella*; Lab158: *Lactobacillus–Enterococcus*; Erec482: *Eubacteria–Clostridium*. Shown are means ± SE. $N = 13–14$ for active season, 14 for early winter, and seven for late winter. Letters indicate significant differences among groups, $P < 0.05$

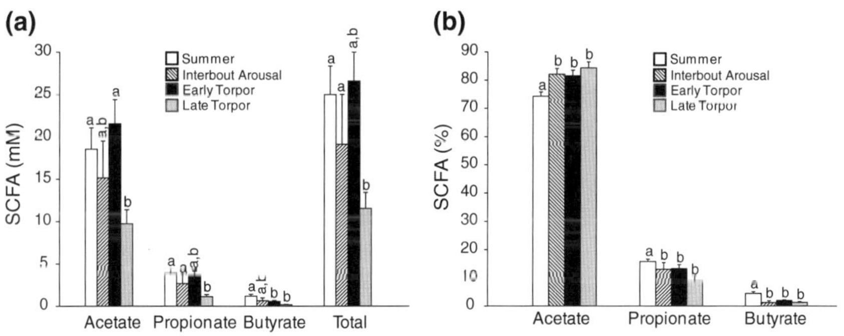

Fig. 25.2 Cecal SCFA concentrations were altered during torpor. **a** Individual and total SCFA concentrations in summer and winter ground squirrels. **b** Percent composition of each SCFA. Means with different letters are significantly different, $P < 0.05$. Shown are means ± SE. $N = 7$ summer, 6 IBA, nine early torpor, nine late torpor

paracellular passage of ions and macromolecules. To address this we examined occludin expression in ground squirrel small intestine and cecum. Occludin staining was particularly strong in hibernators in the apical intercellular spaces, where tight junction protein complexes are located (Fig. 25.3). Immunoblots of cecal lysates supported these results (Fig. 25.4), and indicated that the proportion of the total occludin signal represented by the hyperphosphorylated form (slower migrating band) increased about threefold in hibernators compared with active season squirrels.

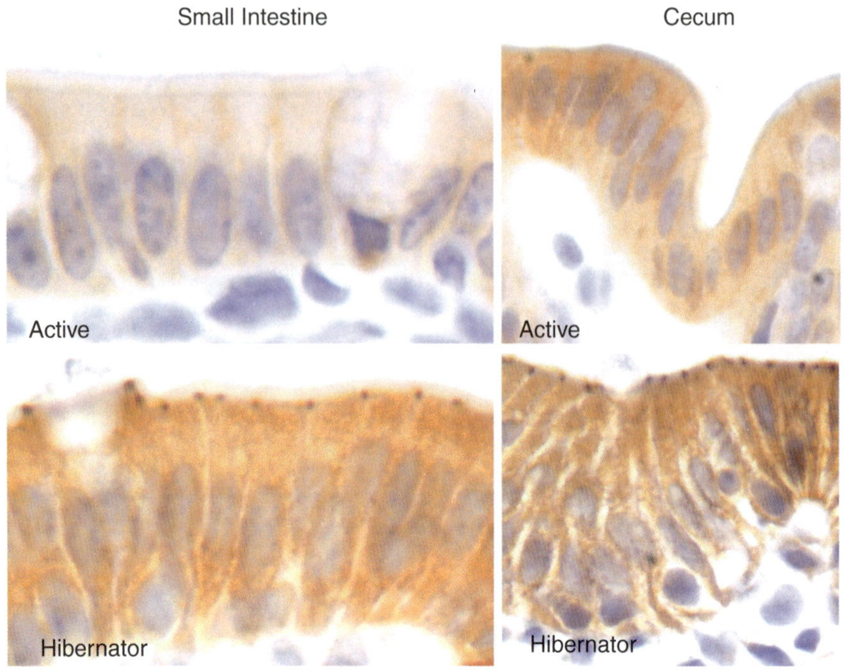

Fig. 25.3 Occludin tight junction localization is increased in the small intestine and cecum (*brown stain*) in hibernating ground squirrels. Occludin localization to the apical intercellular spaces was pronounced in hibernators compared to active ground squirrels

25.5 Discussion

Resident gut microbes exist in a symbiotic relationship with their hosts, with provision of substrates for microbial metabolism being a major benefit for the microbes. Although gut microbes can metabolize host-derived substrates including mucus glycans, and nutrients in sloughed epithelial cells, the host diet provides the major source of substrates to support microbial growth (Backhed et al. 2005; Leser et al. 2000; Muegge et al. 2011; Sonnenburg et al. 2005). The absence of dietary substrates and reduced gut mass in hibernators make the nutritional environment for microbes very different compared with the active season, and this has the potential to significantly alter numbers, composition, and metabolic capabilities of the resident microbes. Here we used quantitative PCR to assess the effect of the winter fast on bacterial numbers in active and hibernating ground squirrel guts. Total numbers of bacteria per gram as estimated by the universal probe for Domain Bacteria (Bact338) were not significantly affected by hibernation, possibly due to high degrees of variability in some active squirrels and early winter hibernators. However, the strong trend for lower numbers in late hibernation suggests that the extended winter fast likely reduces abundance of at least some bacterial taxa. Reduction in the *Lactobacillus–Enterococcus* group as a percentage of all

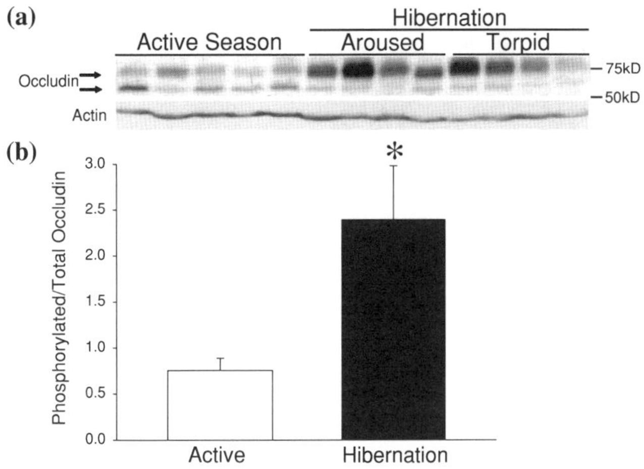

Fig. 25.4 Occludin phosphorylation is increased during hibernation. **a** Representative immunoblots of occludin and actin signals from cecal lysates. *Upper arrow* indicates slower migrating (hyperphosphorylated) occludin; *lower arrow* is slower migrating band. Molecular masses are shown on right. **b** Densitometric analysis of occludin bands (means ± SE) from 10 active season and 12 hibernating squirrels (torpor and IBA combined). Values represent arbitrary absorbance units of *upper* (phosphorylated) band divided by total signal (sum of *upper* and *lower bands*), normalized to density of actin bands for each lane. *, $P < 0.05$

bacteria was particularly pronounced in hibernators, and relative numbers of the *Eubacteria–Clostridium* group also fell during hibernation. In contrast, hibernation had less of an effect on percent composition of *Bacteroides–Prevotella*, which may reflect the ability of many members of this group to forage on complex glycans found in mucus and epithelial cell surfaces when dietary polysaccharides are not available (Martens et al. 2008; Sonnenburg et al. 2005). Thus, hibernation appears to selectively alter the gut microbiota, likely reflecting differential sensitivity to loss of dietary substrates among bacterial groups. It is also possible that reduced bacterial numbers in late versus early season hibernators may be due to the effect of repeated fluctuations in squirrel T_b on microbial metabolism and proliferation rates.

Despite the absence of dietary substrates during winter, microbial fermentation clearly continues in the hibernator gut as indicated by persistence of SCFA in cecal contents. It should be cautioned that interpretation of these data is limited because luminal SCFA levels reflect the dynamics of both microbial production and absorption across the gut epithelium, which vary depending on the metabolic and thermal state of the animal. Nevertheless, the data support the idea that microbes are metabolizing host-derived substrates such as mucus glycans during hibernation and thereby continue to provide a fuel source to their hosts (Martens et al. 2008; Sonnenburg et al. 2005). More detailed analysis with an expanded set of bacterial probes as well as unbiased pyrosequencing of the microbiota may reveal changes in community diversity that reflects a shift toward those taxa that are adept at utilizing host-derived substrates. The extent to which microbial SCFA contribute to a fed squirrel's energy balance is unknown, but based on other small mammals it may range from 5 to 20% (Bergman 1990). Studies in germ-free mice and those with a conventional microbiota suggest that microbial SCFAs contribute to energy balance under fasted conditions, in part by enhancing ketone production by the liver (Crawford et al. 2009). This suggests that recycling of host-derived substrates by gut microbes may contribute to energy balance in hibernators, particularly in late winter when fat stores are depleted.

Only a few studies have examined the impact of hibernation on gut microbiota in mammals. A study using culture-based techniques, which underestimate the numbers and complexity of the microbiota, concluded that the cecal microbiota in ground squirrels was reduced in number and composition but many bacteria were still cultured (Barnes 1970; Barnes and Burton 1970). In a previous study we cultured microbes from ground squirrel cecal tissue and detected more community richness than previously reported including additional species of *Enterobacteriaceae*, *Firmicutes*, *Pseudomonads*, and *Cytophaga–Flavobacterium* group bacteria in both hibernating and summer squirrels (Cloud-Hansen et al. 2007). In marmots that had recently emerged from hibernation but not yet refed, SCFA levels were reduced but still detectable in hindgut contents relative to summer animals (Hume et al. 2002), suggesting the presence of a microbial community during the winter fast. In Syrian hamsters numbers of cecal microbes and SCFA concentrations in hibernating (torpid) hamsters were closer to those in fed active hamsters than in active hamsters fasted for 4 days (Sonoyama et al. 2009). This may be because hibernating hamsters feed during periodic arousals, which would provide dietary substrates for microbial growth and fermentation. Culture-independent analysis of hamster cecal contents also suggested that fasting had more effect than hibernation on taxonomic distribution of 16S rRNA gene sequences, relative to fed animals (Sonoyama et al. 2009).

Intestinal barrier function is crucial to organismal survival as it restricts passage of luminal molecules that have the potential to induce inflammation, such as bacterial lipopolysaccharide, small proteins, and other antigens (Turner 2009). The tight junctions between epithelial cells are the primary regulators of barrier function, and, along with the intestinal immune system, contribute to the maintenance of a healthy relationship between hosts and their resident microbiota. Maintenance of barrier function is particularly important for hibernators that fast during winter because nutrient deprivation in mammals can increase intestinal epithelial permeability (Ferraris and Carey 2000). We showed previously that ionic tissue conductance increases in hibernator intestine studied in vitro (Carey 1990, 1992), and the results presented here suggest that permeability to macromolecules is also elevated in vivo. In animal models and humans increased tight junctional permeability is often associated with alterations in mucosal immune regulation (Turner 2009). Thus, intestinal permeability changes induced by hibernation may correlate with alterations that occur in the hibernator's intestinal immune system (Kurtz and Carey 2007).

Occludin is one of a number of intestinal tight junction proteins that regulates size and selectively of the intestinal epithelial barrier (Shen et al. 2011). The immunohistochemical results suggest greater occludin localization to tight junctions in small intestine and cecum in hibernators than in active season squirrels, and this is supported by the greater amount of occludin in the hyperphosphorylated form on western blots. Hyperphosphorylation of occludin appears to concentrate the protein at tight junctions and is associated with increased tight junctional integrity (Farshori and Kachar 1999). Although these results may seem counterintuitive given the increased permeability during hibernation, there is support for upregulation of occludin when gut permeability is elevated, such as under stress conditions (Dokladny et al. 2006, 2008). New studies also suggest that occludin is important in the setting of cytokine-induced barrier dysfunction, as occludin overexpression protects mice against barrier loss and diarrhea induced by tumor necrosis factor (TNF) (Marchiando et al. 2010). Levels of some pro-inflammatory cytokines, including TNF, increase in intestinal mucosa during IBAs in hibernating squirrels (Kurtz and Carey 2007). Thus, increased occludin expression during hibernation may be a compensatory response that limits the extent of permeability increases and thereby maintains homeostasis during the extended winter fast.

In summary, the data presented here together with our previous results (Kurtz and Carey 2007) suggest that despite the lack of food intake during hibernation the gut maintains an active role in physiological homeostasis, due in part to the host-gut microbe symbiosis that continues during winter. Hibernation selectively alters the gut microbiota, possibly reflecting differential sensitivity of microbes to loss of dietary polysaccharides. The winter fast is associated with altered intestinal barrier function, which may require compensatory responses that bolster immune defenses and enhance tight junctional integrity, thus maintaining a healthy relationship of the hibernator host with its resident gut microbiota until food intake resumes in the spring.

Acknowledgments We thank Michael Grahn for assistance with animal care, and Dr. Glen Broderick (USDA Dairy Forage Research Center, Madison, WI) for SCFA analyses. This work was supported in part by funds from the University of Wisconsin School of Veterinary Medicine.

References

Backhed F, Ding H, Wang T, Hooper LV, Koh GY, Nagy A, Semenkovich CF, Gordon JI (2004) The gut microbiota as an environmental factor that regulates fat storage. Proc Natl Acad Sci U S A 101:15718–15723

Backhed F, Ley RE, Sonnenburg JL, Peterson DA, Gordon JI (2005) Host-bacterial mutualism in the human intestine. Science 307:1915–1920

Bailey MT, Walton JC, Dowd SE, Weil ZM, Nelson RJ (2010) Photoperiod modulates gut bacteriabacteria composition in male Siberian hamsters (*Phodopus sungorus*). Brain Behav Immun 24:577–584

Barnes EM (1970) Effect of hibernation on the intestinal flora. Am J Clin Nutr 23:1519–1524

Barnes EM, Burton GC (1970) The effect of hibernation on the caecal flora of the thirteen-lined ground squirrel (*Citellus tridecemlineatus*). J Appl Bacteriol 33:505–514

Bergman EN (1990) Energy contributions of volatile fatty acids from the gastrointestinal tract in various species. Physiol Rev 70:567–590

Broderick GA, Uden P, Murphy ML, Lapins A (2004) Sources of variation in rates of in vitro ruminal protein degradation. J Dairy Sci 87:1345–1359

Carey HV (1990) Seasonal changes in mucosal structure and function in ground squirrel intestine. Am J Physiol Regul Integr Comp Physiol 259:R385–R392

Carey HV (1992) Effects of fastingfasting and hibernation on ion secretion in ground squirrel intestine. Am J Physiol Regul Integr Comp Physiol 263:R1202–R1208

Cloud-Hansen KA, Villiard KM, Handelsman J, Carey HV (2007) Thirteen-lined ground squirrels (*Spermophilus tridecemlineatus*) harbor multi-antibiotic resistant bacteria. J Am Assoc Lab Animal Sci 46:17–20

Crawford PA, Crowley JR, Sambandam N, Muegge BD, Costello EK, Hamady M, Knight R, Gordon JI (2009) Regulation of myocardial ketone body metabolism by the gut microbiota during nutrient deprivation. Proc Natl Acad Sci U S A 106:11276–11281

Dokladny K, Moseley PL, Ma TY (2006) Physiologically relevant increase in temperature causes an increase in intestinal epithelial tight junction permeability. Am J Physiol Gastrointest Liver Physiol 290:G204–G212

Dokladny K, Ye D, Kennedy JC, Moseley PL, Ma TY (2008) Cellular and molecular mechanisms of heat stress-induced up-regulation of occludin protein expression: regulatory role of heat shock factor-1. Am J Pathol 172:659–670

Farshori P, Kachar B (1999) Redistribution and phosphorylation of occludin during opening and resealing of tight junctions bacteria in cultured epithelial cells. J Membr Biol 170:147–156

Ferraris RP, Carey HV (2000) Intestinal transport during fasting and malnutrition. Ann Rev Nutr 20:195–219

Hooper LV, Wong MH, Thelin A, Hansson L, Falk PG, Gordon JI (2001) Molecular analysis of commensal host-microbial relationships in the intestine. Science 291:881–884

Hume ID, Beiglbock C, Ruf T, Frey-Roos F, Bruns U, Arnold W (2002) Seasonal changes in morphology and function of the gastrointestinal tract of free-living alpine marmots (*Marmota marmota*). J Comp Physiol B 172:197–207

Kurtz CC, Carey HV (2007) Seasonal changes in the intestinal immune system of hibernating ground squirrels. Dev Comp Immunol 31:415–428

Leser TD, Lindecrona RH, Jensen TK, Jensen BB, Moller K (2000) Changes in bacterial community structure in the colon of pigs fed different experimental diets and after infection with *Brachyspira hyodysenteriae*. Appl Environ Microbiol 66:3290–3296

Ley RE, Hamady M, Lozupone C, Turnbaugh PJ, Ramey RR, Bircher JS, Schlegel ML, Tucker TA, Schrenzel MD, Knight R, Gordon JI (2008) Evolution of mammals and their gut microbes. Science 320:1647–1651

Ley RE, Peterson DA, Gordon JI (2006) Ecological and evolutionary forces shaping microbial diversity in the human intestine. Cell 124:837–848

Marchiando AM, Shen L, Graham WV, Weber CR, Schwarz BT, Austin JR 2nd, Raleigh DR, Guan Y, Watson AJ, Montrose MH, Turner JR (2010) Caveolin-1-dependent occludin endocytosis is required for TNF-induced tight junction regulation in vivo. J Cell Biol 189: 111–126

Martens EC, Chiang HC, Gordon JI (2008) Mucosal glycan foraging enhances fitness and transmission of a saccharolytic human gut bacterial symbiont. Cell Host Microbe 4:447–457

Muegge BD, Kuczynski J, Knights D, Clemente JC, Gonzalez A, Fontana L, Henrissat B, Knight R, Gordon JI (2011) Diet drives convergence in gut microbiome functions across mammalian phylogeny and within humans. Science 332:970–974

Shen L, Weber CR, Raleigh DR, Yu D, Turner JR (2011) Tight junction pore and leak pathways: a dynamic duo. Annu Rev Physiol 73:283–309

Sonnenburg JL, Xu J, Leip DD, Chen CH, Westover BP, Weatherford J, Buhler JD, Gordon JI (2005) Glycan foraging in vivo by an intestine-adapted bacterial symbiont. Science 307:1955–1959

Sonoyama K, Fujiwara R, Takemura N, Ogasawara T, Watanabe J, Ito H, Morita T (2009) Response of gut microbiota to fasting and hibernation in Syrian hamsters. Appl Environ Microbiol 75:6451–6456

Turner JR (2009) Intestinal mucosal barrier function in health and disease. Nat Rev Immunol 9:799–809

Vaughan EE, Schut F, Heilig HG, Zoetendal EG, de Vos WM, Akkermans AD (2000) A molecular view of the intestinal ecosystem. Curr Issues Intest Microbiol 1:1–12

Velagapudi VR, Hezaveh R, Reigstad CS, Gopalacharyulu PV, Yetukuri L, Islam S, Felin J, Perkins R, Boren J, Oresic M, Backhed F (2010) The gut microbiota modulates host energy and lipid metabolism in mice. J Lipid Res 51:1101–1112

Chapter 26
Cardiac Electrical Alternans and Ventricular Fibrillation During Hypothermia in Non-Hibernating Versus Hibernating Animals: Role of Propagation Velocity and Dispersion of Repolarization

Alexey V. Glukhov, Yuriy V. Egorov, Igor R. Efimov and Leonid V. Rosenshtraukh

Abstract The heart of hibernating species is resistant to lethal ventricular fibrillation (VF) induced by hypothermia. Spatially discordant cardiac alternans (SDA) is a promising predictor of VF, yet its role in the mechanism of hypothermic arrhythmogenesis in both non-hibernating and hibernating mammals remains unclear. We optically mapped the posterior epicardial surface of Langendorff-perfused hearts of summer active (SA, n = 6), winter hibernating (WH, n = 7), ground squirrels (GS) *Spermophilus undulatus*, and rabbits (n = 5). Action potential duration (APD) and conduction velocity (CV) dynamic restitution as well as APD/CV alternans were measured at baseline (37°C) and during hypothermia (27–17°C). In rabbits, hypothermia significantly increased the magnitude of SDA which enhanced the ventricular repolarization gradient, caused conduction delays, conduction block, and the onset of VF (0% at 37°C vs. 60% at 17°C, $p < 0.01$). In contrast, no arrhythmia was observed in both SA and WH hearts at any temperature. No difference in APD dispersion was observed between animals. In rabbits, hypothermia caused a heterogeneous CV slowing which significantly exaggerated conduction anisotropy (from 2.0 ± 0.1 at 37°C to 2.8 ± 0.1 at 17°C, $p < 0.01$). A steeper slope in the CV restitution curve and bigger amplitude in CV alternans was observed in rabbit hearts compared with GS. The amplitude of SDA well correlated with the amplitude of CV alternans. In conclusion, the mechanism underlying SDA formation during hypothermia is shown to be associated with anisotropic CV alternans conditioned by an enhanced dispersion of repolarization. Increased conduction anisotropy formed

---

A. V. Glukhov (✉) · Y. V. Egorov · L. V. Rosenshtraukh
Laboratory of Heart Electrophysiology, Cardiology Research Center,
3rd Cherepkovsaya St., 15A, Moscow 121552, Russia
e-mail: glookhov@gmail.com

A. V. Glukhov · I. R. Efimov
Department of Biomedical Engineering, Washington University,
St. Louis, MO, USA

T. Ruf et al. (eds.), *Living in a Seasonal World*,
DOI: 10.1007/978-3-642-28678-0_26, © Springer-Verlag Berlin Heidelberg 2012

functional barriers which facilitated development of SDA between cells with different ionic properties through the electrotonic uncoupling of neighboring regions of myocardium. The factors of hibernating species responsible for their resistance to SDA and VF can be accredited to their safe and dynamically stable conduction anisotropy along with a low dispersion of repolarization.

Abbreviations

VF	Ventricular fibrillation
GS	Ground squirrel
SA	Summer active
WH	Winter hibernating
SCA and SDA	Spatially concordant and discordant alternans
APD	Action potential duration
CV	Conduction velocity
DI	Diastolic interval
SR	Sarcoplasmic reticulum

26.1 Introduction

The heart of non-hibernating mammals, including humans, demonstrates a limited tolerance to cold which results in life-threatening conditions associated with cardiac arrest primarily due to ventricular fibrillation (VF) (Badeer 1958; Mouritzen and Andersen 1966; Mattu et al. 2002; Fedorov et al. 2005, 2008; Glukhov et al. 2007; Hsieh et al. 2009). In contrast, the heart of hibernating mammals continues to beat at near freezing temperatures during the hibernation season without any cardiac dysfunction and/or malignant arrhythmias (Johansson 1996; Wang et al. 2002).

Electrophysiological parameters determining a predisposition to ventricular tachyarrhythmia during hypothermia include: enhanced dispersion of repolarization (Salama et al. 1998; Fedorov et al. 2008), slowed conduction (Duker et al. 1983; Fedorov et al. 2005), heterogeneity in conduction velocity (CV) (Fedorov et al. 2008), shortened wavelength (Glukhov et al. 2007; Fedorov et al. 2008), and a steep restitution slope (Hsieh et al. 2009). Recently, we have demonstrated that the mechanism of natural protection of the hibernator heart against hypothermic arrhythmias is associated with the homogeneous preexisting [i.e., opposite to dynamic (Weiss et al. 2000)] activation and repolarization pattern (Fedorov et al. 2008). At the same time, both experimental and cardiac modeling studies have shown that preexisting heterogeneity is not the only cause of VF (Weiss et al. 2000). Dynamic changes of Action potential duration (APD)/CV could be another mechanism of VF induction that requires no preexisting heterogeneity of any kind (Weiss et al. 2000; Watanabe et al. 2001). Although a variety of dynamic factors influence activation and repolarization properties, one that has received particular interest recently is cardiac alternans.

Electromechanical cardiac alternans refers to the beat-to-beat alternation of APD and intracellular Ca^{2+} transients in a repeating pattern of long-short–long-short or large-small–large-small, respectively. Cardiac alternans, first observed in the form of pulse alternans (Traube 1872) and later described as electrocardiographic *T* wave alternans (Lewis 1910) has been recently linked to the genesis of reentrant arrhythmias (Pastore et al. 1999) and shown to be a good marker of risk for sudden cardiac death in patients (Rosenbaum et al. 1994). Cardiac alternans has been observed in a surprisingly wide variety of clinical and experimental arrhythmogenic conditions including hypothermia (Hirayama et al. 1993; Hsieh et al. 2009).

Cardiac alternans may be spatially concordant (SCA) or discordant (SDA). SCA is less arrhythmogenic than SDA (Pastore et al. 1999; Qu et al. 2000). Although APD and hence the refractory period alternate, for any given beat, the refractory period is either long or short everywhere, and hence the dispersion of refractoriness is not greatly amplified. However, once APD alternans become SDA, the dispersion of refractoriness is greatly amplified, producing a favorable substrate for the initiation of re-entry. During SDA, some regions of tissue alternate in a long-short-long pattern, whereas other regions simultaneously alternate in a short-long-short pattern. These out-of-phase regions are separated by a nodal line in which no alternans is present. At a nodal line, the spatial APD gradient is the steepest, predisposing this region to localized conduction block and facilitating VF induction (Pastore et al. 1999; Qu et al. 2000; Watanabe et al. 2001). However, a systematic analysis of rate-dependent changes has never been applied to electrophysiological properties of hibernators' myocardium in the settings of both hypothermia-associated cardiac alternans and arrhythmia induction.

26.2 Methods

26.2.1 Animals

All experiments were conducted in accordance with the NIH Guide for the Care and Use of Laboratory Animals (NIH publication no. 80–23). We studied the wild GS *Spermophilus undulates* during summer active (SA, $T_{\text{body}} = 37.5 \pm 0.5$°C; $n = 6$; 594 ± 49 g) and winter hibernating (WH, torpid state, $T_{\text{body}} = 2.5 \pm 0.5$°C; $n = 7$, 587 ± 56 g) seasonal states. During hibernation period (October–March), WH animals were kept in a darkened cold room (~2°C) and had been in more than two successive bouts of hibernation before experiments. We used rabbits ($n = 5$, 2.62 ± 0.45 kg) as a non-hibernating control. Rabbits and SA were anesthetized with urethane 1.5 g/kg for in vitro experiments. Euthanasia was performed by decapitation. The isolated heart preparation was performed as described previously (Fedorov et al. 2005, 2008).

26.2.2 Optical Mapping Experiments

The optical mapping system has been previously described in detail (Fedorov et al. 2005). The entire anterior ventricular epicardial surface of the left ventricle was mapped with a 127 × 128 pixel CCD camera (Dalsa, Waterloo, Canada) with a field-of-view of 16 × 16 to 18 × 18 mm. To reduce noise, we applied a moving average filter from three to seven frames in time, or a median filtration of 8 × 8 × 3; that is, three by three pixels in space and three frames in time. The excitation–contraction uncoupler BDM (15 mmol/l; Sigma, Saint Louis, MO) was added to the perfusate to suppress motion artifacts.

Dependences of both APD and CV on the preceding diastolic interval (DI) were determined using the dynamic restitution S1S1 protocol (Koller et al. 1998; Fedorov et al. 2008). To characterize the spatiotemporal pattern of repolarization, we measured the mean APD70 throughout the mapped 256 epicardial sites, and maximum dispersion (D_{max}, in ms). D_{max} was calculated as the difference between the minimum and the maximum values of APD70 along the field of mapping.

26.2.3 Statistical Analysis

Continuous variables were expressed as mean ± SD (or ±SEM for Fig. 26.1d, e). Continuous variables were evaluated with an unpaired Student *t*-test for successive measurements using Bonferroni correction. Comparisons between different animal groups were performed using *t*-test for samples with different sizes with Bonferroni correction. Differences were considered significant at $p < 0.05$.

26.3 Results

26.3.1 Arrhythmogenesis and Spatial Organization of APD Alternans During Hypothermia

The restitution theory used for the description of dynamical changes in electrophysiological heterogeneity refers to the fact that APD and CV both depend on the preceding DI, which is the rest period between repolarization and the next excitation. When DI is too small, both APD and CV start to oscillate between neighboring beats. Indeed, in both GS and rabbits, the gradual S1S1 shortening led to APD alternans (Fig. 26.1a). Hypothermia resulted in APD prolongation, shifted the APD alternans curves to the longer pacing intervals (Fig. 26.1a) and increased the amplitude of APD alternans (Fig. 26.1b). In WH, the maximum amplitude of APD alternans was smaller compared with that in rabbits and SA at 37 and 27°C. However, during severe hypothermia (17°C), the amplitude of APD alternans in SA and WH significantly exceeded that measured in rabbits which was

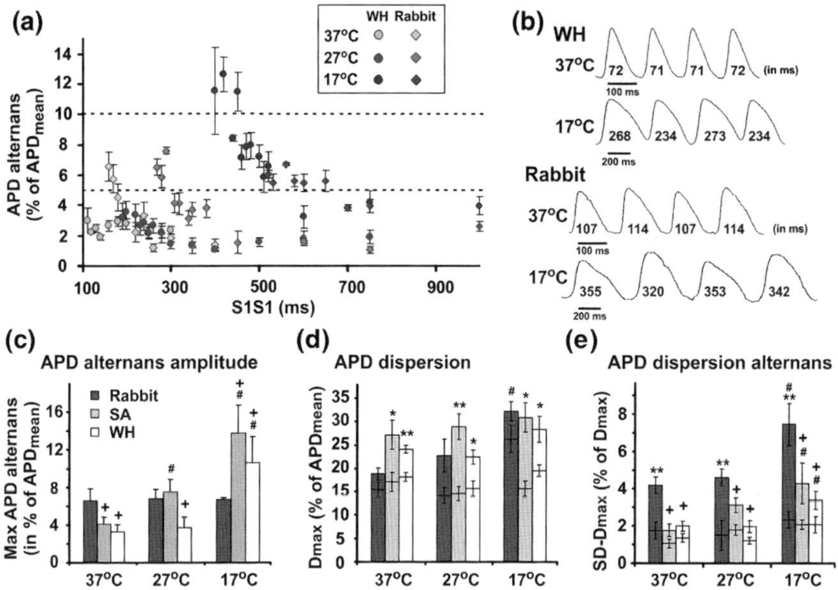

Fig. 26.1 APD alternans during hypothermia. **a** Relative beat-to-beat APD alternans. **b** Typical examples of optical traces with APD alternans recorded at 37 and 17°C in WH and rabbit hearts. **c** Maximum values of the APD alternans amplitude. **d** Maximum APD dispersion (D_{max}) throughout the mapped area. Each column has two compartments: one (*lower*) represents D_{max} measured at slow pacing (300 ms at 37°C, 750 ms at 27°C, and 2,000 ms at 17°C), and another (*upper*) represents D_{max} measured at the shortest S1S1. **e** Standard deviation of D_{max}. Two compartments in each column represent the same values as described for **d**. * and **—$p < 0.05$ and $p < 0.01$ versus lower frequency within each temperature; $^{\#}p < 0.05$ versus 37°C; $^{+}p < 0.01$ versus rabbits

relatively stable at all temperatures (Fig. 26.1c). Hypothermia also increased the relative APD dispersion in all animal groups (Fig. 26.1d).

As mentioned above, cardiac alternans may be SCA or SDA. Once APD alternans become SDA, APD dispersion starts to oscillate in a beat-to-beat manner mainly due to the unstable APD distribution pattern. To characterize SDA quantitatively, we used a standard deviation of the APD dispersion (SD-D_{max}) calculated for consecutive beats. As Fig. 26.1e indicates, SD-D_{max} measured at basic pacing does not differ between groups. However, when paced faster, SD-D_{max} measured in rabbit hearts was significantly exaggerated compared with GS indicating an occurrence of SDA in rabbits.

Figure 26.2 illustrates the spatial organization of APD alternans observed in rabbits at 17°C. During slow pacing, the APD distribution pattern had a stable base-to-apex gradient (Fig. 26.2a) which reached about 30% of the apical APD and was not changed between neighboring beats. The gradual shortening of the pacing interval to S1S1 = 600 ms (Fig. 26.2b) led to the occurrence of SCA when apical and basal regions alternate in the same beat-to-beat manner. During SCA, the pattern and dispersion of APD distribution were not significantly different from that at slow

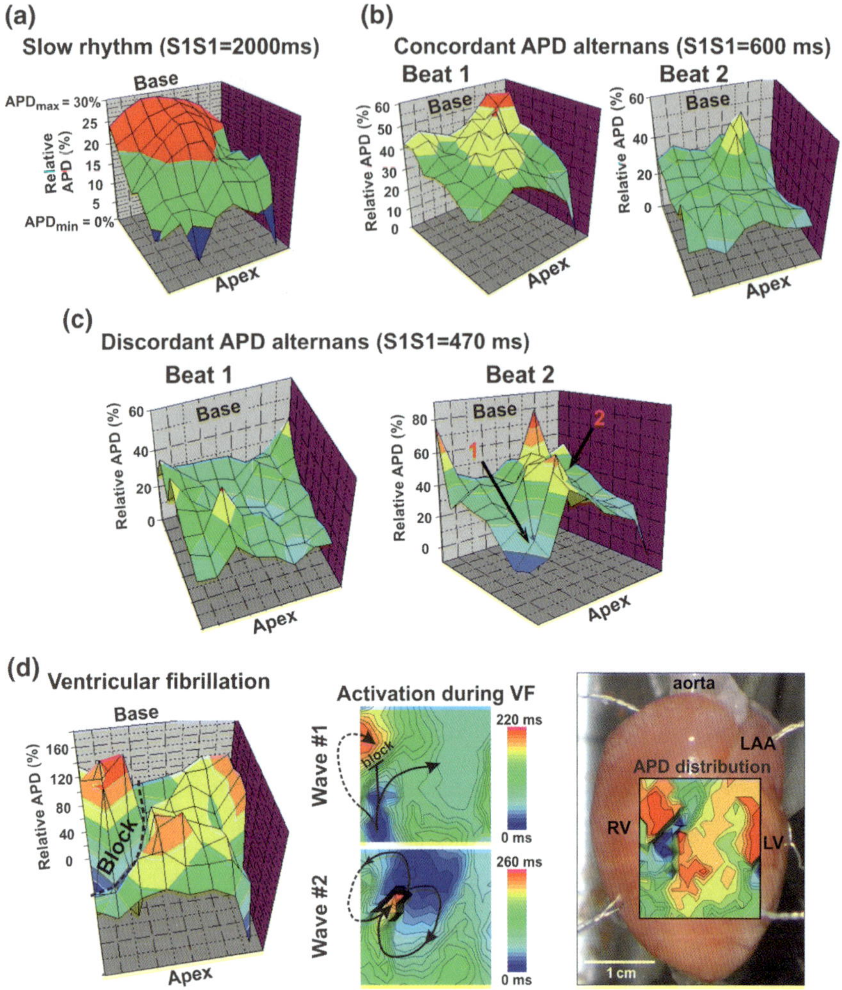

Fig. 26.2 Arrhythmogenesis and spatial organization of APD alternans. Representative examples of pseudo-3D APD distribution contour maps measured at slow pacing (**a**), during SCA (**b**), SDA (**c**), and fast pacing-induced VF (**d**) in the rabbit heart at 17°C. *X*–*Y* axes show the mapped heart area, *Z*-axis represents the relative distribution of the APD throughout the mapped area as a percent of minimal APD at each S1S1. For VF, two consequent activation maps are shown. On the photograph of an isolated rabbit heart, APD distribution map is correlated with anatomical features of the mapped area

pacing (base-to-apex APD gradient reached about 40% of the apical APD) and were also similar from beat #1 to beat #2. SCA caused minimal beat-to-beat oscillations of APD. At a somewhat faster rate (CL = 470 ms), SDA occurred, and, most importantly, the pattern of repolarization alternans changed importantly (Fig. 26.2c). Beat #1 resulted in a homogeneous APD distribution with two prominent peaks located about 15 mm apart. At beat #2, the repolarization pattern varied substantially, exhibiting a significantly heterogeneous APD distribution pattern characterized by three prominent peaks unrelated to that at beat #1.

Furthermore, during SDA the repolarization was half of that during SCA (80% versus 30% of the apical APD). The SDA-induced repolarization gradient was sufficiently large to cause secondary conduction delays (areas #1 and #2 labeled for beat #2 in Fig. 26.2c) resulting in conduction block (dotted line in Fig. 26.2d) and the induction of VF (Fig. 26.2d). Two activation maps (middle panel) reconstructed for the current (left panel) and subsequent beats demonstrate that the impulse was blocked in the area with prolonged APD (around the interventricular septum), and the following retrograde activation initiated a re-entry circuit which induced VF.

In rabbit hearts, hypothermia resulted in more than a fourfold increase in the amplitude of SDA (SD-D_{max} measured at fast pacing, upper part of column on Fig. 26.1e) compared with the amplitude of SCA (SD-D_{max} measured at slow pacing, lower part of column). It correlates to an increased probability of pacing-induced VF (0% at 37°C vs. 60% at 17°C, $p < 0.01$). Unlike the rabbit hearts, no arrhythmia was observed in either the SA or WH hearts at any temperature. During all recordings of VF, there was both an increase in APD dispersion ($D_{max} \geq 30\%$) and an enhanced amplitude of SDA (SD-$D_{max} \geq 7\%$). Increased APD dispersion alone was not enough for VF induction. Also, it should be noted that profound SDA ($SD - D_{max} = 8.5\%$) observed in one rabbit heart at moderate APD dispersion ($D_{max} = 19.5\%$) did not result in VF.

26.3.2 Role of CV in the Mechanism of SDA Formation

To estimate the role of APD propagation in the mechanism of SDA formation, we characterized CV properties. Hypothermia induced a more substantial and heterogeneous CV slowing in rabbits compared with SA and WH (Fig. 26.3b). The preferential suppression of propagation in the transverse direction increased conduction anisotropy in rabbits and SA but not in WH (2.12 ± 0.04, 2.07 ± 0.06, and 2.01 ± 0.04 vs. 2.77 ± 0.10, 2.35 ± 0.13, and 2.15 ± 0.10 in rabbits, SA, and WH at 37 and 17°C; $p < 0.01$). Moreover, rabbits had a steeper slope of anisotropy restitution compared with SA and WH. Hypothermia exaggerated the slope of CV anisotropy curve in rabbit hearts, but not in WH hearts (Fig. 26.3a). SA animals were located between rabbits and WH. At 17°C, the activation pattern in rabbit hearts was disorganized and demonstrated multiple local conduction blocks (Fig. 26.3b). Moreover, CV restitution measured in rabbit hearts had a steeper dependence on the preceding DI as characterized by a higher ($p < 0.05$) CV restitution curve slope (27 ± 2° and 23 ± 3° in rabbits vs. 15 ± 2°/14 ± 4° and 14 ± 3°/13 ± 6° in SA/WH, at 37 and 27°C, respectively). Surprisingly, at 17°C, CV restitution curves were almost flat in all animal groups and slopes were not statistically different (3 ± 1° in rabbits vs. 3 ± 1° and 5 ± 1° in SA and WH, *NS*).

The gradual S1S1 shortening led to CV oscillations. Figure 26.3c shows rate dependence curves for beat-to-beat CV alternans. Beat-to-beat oscillations

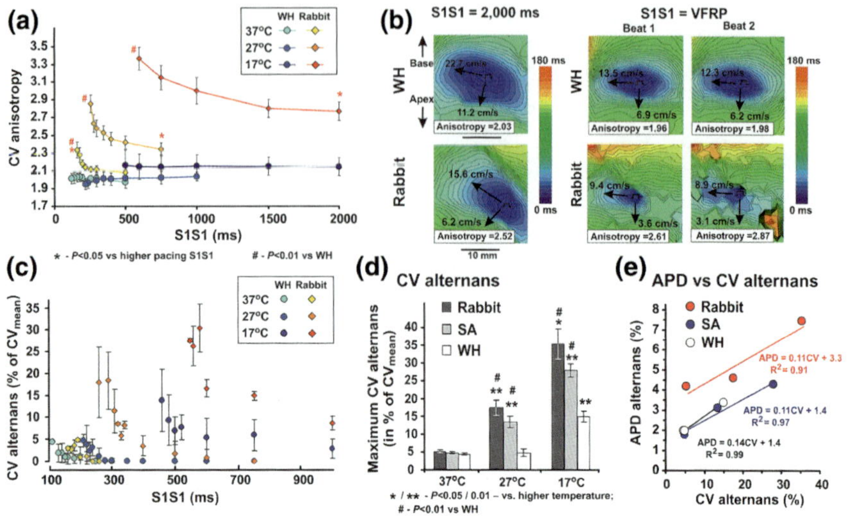

Fig. 26.3 Role of CV in the mechanism of APD alternans formation. **a** Conduction anisotropy versus pacing interval at different temperatures. The anisotropy index was calculated as a ratio of longitudinal and transversal components of CV. **b** Typical examples of activation maps measured in WH and rabbit hearts at 17°C during slow (S1S1 = 2,000 ms) and fast (S1S1 = 750 ms) pacing. All maps were reconstructed using the same color time scale. **c** Relative (as a percent of the mean CV) beat-to-beat alternans of transversal CV versus S1S1. **d** Maximum values of CV alternans. **e** Correlation between the amplitude of APD alternans and CV alternans

(Fig. 26.3b, right panel) in transversal CV were calculated. Hypothermia increased the amplitude of CV alternans in all animals (Fig. 26.3d); however, these changes were more pronounced in rabbits than in WH: at 27°C rabbits exhibited more than a threefold increase in the amplitude of CV alternans. SA animals were located between rabbits and WH. In all animals, the amplitude of SDA well correlated with the amplitude of CV alternans (Fig. 26.3e).

26.4 Discussion

Although it is unclear why neighboring cells under apparently identical conditions would respond differently with respect to their alternans phase, two possible and not necessarily mutually exclusive, mechanisms have been proposed to explain the appearance of SDA. The first is based on a steep CV restitution mechanism, i.e., when CV of a propagating wave has a steep dependence on the preceding diastolic interval (Qu et al. 2000; Watanabe et al. 2001; Mironov et al. 2008).

In the case of CV restitution, a sufficiently short DI causes the wavefront to slow. As it slows, its distance from the wave ahead increases, resulting in a longer DI.

As its DI increases, the APD also lengthens. Meanwhile, the wave's changing APD affects the DI of the wave behind it, so that the next wave's APD will also oscillate, and so forth for each successive wave. The important consequence is that the APD of the same wave changes while propagating through the tissue, becoming short in some areas and long in others. S1S1 at which CV restitution is engaged becomes a major determinant of SCA to SDA conversion.

In the present study, we demonstrated that hypothermia slowed CV, preferentially suppressing propagation in the transverse direction which increased conduction anisotropy (Fig. 26.3a). CV restitution measured in rabbit hearts had a steeper dependence on the preceding DI as characterized by a higher CV restitution curve slope. Hypothermia also increased the amplitude of CV alternans (Fig. 26.3c, d); however, these changes were more pronounced in rabbits than in GS. Moreover, the amplitude of SDA was well correlated with the amplitude of CV alternans in all animal groups (Fig. 26.3e). It also should be noted that CV alternans started earlier than APD alternans (Figs. 26.1a vs. 26.3c).

Impaired Na^+ channel function and decreased gap junctional conductance both predict a decrease in CV at low temperatures. Liu et al. (1991) showed that in hedgehog cardiac preparations, the peak of Ca^{2+} current did not significantly change at low temperatures and the Na^+ current was less suppressed by cooling in hibernator than in rat. These ionic differences may contribute to hypothermia tolerance and resistance to conduction block during hypothermia in hibernators.

The role of gap junctions could be particularly important in the setting of hypothermia and hibernation. It has been suggested that cellular uncoupling caused by hypothermia, which results from downregulation (Fedorov et al. 2005, 2008), lateralization, and de-phosphorylation (Hsieh et al. 2011) of the principal gap junction protein in the ventricles, connexin 43 (Cx43), may unmask the intrinsic differences in APD, enhance APD dispersion (Salama et al. 1998; Fedorov et al. 2008; Hsieh et al. 2009) and allow neighboring myocytes to alternate with opposite phase because of differences in calcium cycling properties and APD restitution. Immunohistochemical analysis revealed a significant up-regulation of Cx43 during hibernation (Saitongdee et al. 2000; Fedorov et al. 2005) as well as the overexpression of Cx45 (Fedorov et al. 2005), which is not normally expressed in the ventricle working myocardium. It could represent a compensatory response to maintain sufficient intercellular gap junction communication during physiologic conditions that would otherwise reduce conductance.

26.4.1 The Role of Regional Ionic and Calcium Cycling Heterogeneity in the Mechanism of SDA Formation

An alternative mechanism requires the presence of preexisting heterogeneities, and SDA can be formed via an appropriately timed stimulus or a change in pacing rate (Pastore et al. 1999; Chinushi et al. 2003). In heterogeneous tissue the amplitude of alternans varies spatially, making it possible to time a stimulus such that it reverses the alternans phase in only one part of the tissue. This mechanism is associated with pacing history, or the short-term memory phenomenon which is based on the spatial heterogeneities of calcium cycling and the sarcolemmal ionic currents which govern repolarization (Pruvot et al. 2004). According to this mechanism, repolarization alternans arises when the heart rate exceeds the capacity of myocytes to cycle intracellular Ca^{2+} (Spear and Moore 1971; Pruvot et al. 2004). A primary role of this mechanism is believed to be attributed to sarcoplasmic reticulum (SR) calcium cycling. In contrast to intracellular Ca^{2+} dysregulation observed in non-hibernating animals and humans during hypothermia, hibernating animals demonstrate an enhanced capability to maintain intracellular Ca^{2+} homeostasis (Liu et al. 1991; Johansson 1996; Wang et al. 2002). Measurements of intracellular Ca^{2+} indicated that the Ca^{2+} transient decays faster in GS cells than in rat cells owing to a faster rate of Ca^{2+} uptake and a greater level of Ca^{2+} accumulation in the SR (Wang et al. 2000). However, although these data suggest that the enhanced capability of hibernators' myocardium to maintain intracellular Ca^{2+} homeostasis could be responsible for resistance to SDA associated with short-term memory, further experiments are needed to elucidate the mechanism precisely.

Though our data clearly demonstrates that activation time is one of the factors modifying electrical alternans, enhanced APD dispersion also requires SDA formation. Spatial tissue heterogeneity seems to be necessary but is not sufficient for SDA formation. Therefore, based on our data and the literature, we conclude that the mechanism underlying SDA formation during hypothermia is primarily associated with SDA conduction velocity alternans conditioned by an enhanced dispersion of repolarization. The factors of hibernating species contributing to their resistance to the formation of SDA and VF seem to be the establishment of safe and dynamically stable conduction and the low dispersion of repolarization.

Acknowledgments This work was supported by the Stanley and Lucy Lopata Endowment (for IRE), Russian Foundation for Basic Research Grant 05-04-48311, and Russian President Foundation for Scientific School Grant SS-6211.2006.7 (for LVR).

References

Badeer H (1958) Ventricular fibrillation in hypothermia; a review of factors favoring fibrillation in hypothermia with and without cardiac surgery. J Thorac Surg 35(2):265–273

Chinushi M, Kozhevnikov D et al (2003) Mechanism of discordant T wave alternans in the in vivo heart. J Cardiovasc Electrophysiol 14(6):632–638

Duker GD, Olsson SO et al (1983) Ventricular fibrillation in hibernators and nonhibernators. Cryobiology 20(4):407–420

Fedorov VV, Glukhov AV et al (2008) Electrophysiological mechanisms of antiarrhythmic protection during hypothermia in winter hibernating versus nonhibernating mammals. Heart Rhythm 5(11):1587–1596

Fedorov VV, Li L et al (2005) Hibernator *Citellus undulatus* maintains safe cardiac conduction and is protected against tachyarrhythmias during extreme hypothermia: possible role of Cx43 and Cx45 up-regulation. Heart Rhythm 2(9):966–975

Glukhov AV, Egorov IV et al (2007) The effect of hypothermia on the wavelength and vulnarability to ventricular arrhythmias in mammals. Ross Fiziol Zh Im I M Sechenova 93(3):289–299

Hirayama Y, Saitoh H et al (1993) Electrical and mechanical alternans in canine myocardium in vivo. Dependence on intracellular calcium cycling. Circulation 88(6):2894–2902

Hsieh YC, Lin SF et al (2009) Therapeutic hypothermia (30 degrees C) enhances arrhythmogenic substrates, including spatially discordant alternans, and facilitates pacing-induced ventricular fibrillation in isolated rabbit hearts. Circ J 73(12):2214–2222

Hsieh YC, Yeh HI et al (2011) Short-duration therapeutic hypothermia causes prompt connexin43 gap junction remodeling in isolated rabbit hearts. Circ J 75(7):1706–1716

Johansson BW (1996) The hibernator heart—nature's model of resistance to ventricular fibrillation. Cardiovasc Res 31(5):826–832

Koller ML, Riccio ML et al (1998) Dynamic restitution of action potential duration during electrical alternans and ventricular fibrillation. Am J Physiol 275(5 Pt 2):H1635–H1642

Lewis T (1910) Notes upon alternation of the heart. Q J Med 4:141–144

Liu B, Arlock P et al (1991a) Temperature effects on the Na and Ca currents in rat and hedgehog ventricular muscle. Cryobiology 28(1):96–104

Liu B, Wang LC et al (1991b) Effect of low temperature on the cytosolic free Ca^{2+} in rat ventricular myocytes. Cell Calcium 12(1):11–18

Mattu A, Brady WJ et al (2002) Electrocardiographic manifestations of hypothermia. Am J Emerg Med 20(4):314–326

Mironov S, Jalife J et al (2008) Role of conduction velocity restitution and short-term memory in the development of action potential duration alternans in isolated rabbit hearts. Circulation 118(1):17–25

Mouritzen CV, Andersen MN (1966) Mechanisms of ventricular fibrillation during hypothermia. Relative changes in myocardial refractory period and conduction velocity. J Thorac Cardiovasc Surg 51(4):579–584

Pastore JM, Girouard SD et al (1999) Mechanism linking T-wave alternans to the genesis of cardiac fibrillation. Circulation 99(10):1385–1394

Pruvot EJ, Katra RP et al (2004) Role of calcium cycling versus restitution in the mechanism of repolarization alternans. Circ Res 94(8):1083–1090

Qu Z, Garfinkel A et al (2000) Mechanisms of discordant alternans and induction of reentry in simulated cardiac tissue. Circulation 102(14):1664–1670

Rosenbaum DS, Jackson LE et al (1994) Electrical alternans and vulnerability to ventricular arrhythmias. N Engl J Med 330(4):235–241

Saitongdee P, Milner P et al (2000) Increased connexin43 gap junction protein in hamster cardiomyocytes during cold acclimatization and hibernation. Cardiovasc Res 47(1):108–115

Salama G, Kanai AJ et al (1998) Hypoxia and hypothermia enhance spatial heterogeneities of repolarization in guinea pig hearts: analysis of spatial autocorrelation of optically recorded action potential durations. J Cardiovasc Electrophysiol 9(2):164–183

Spear JF, Moore EN (1971) A comparison of alternation in myocardial action potentials and contractility. Am J Physiol 220(6):1708–1716

Traube L (1872) Ein fall von pulsus bigeminus nebst bemerkungen tiber die lebershwellungen bei klappenfehlern und uber acute leberatrophic. Ber Klin Wschr 9:185

Wang SQ, Hong Q et al (2000) Recording of calcium transient and analysis of calcium removal mechanisms in cardiac myocytes from rats and ground squirrels. Sci China C Life Sci 43:191–199

Wang SQ, Lakatta EG et al (2002) Adaptive mechanisms of intracellular calcium homeostasis in mammalian hibernators. J Exp Biol 205(Pt 19):2957–2962

Watanabe MA, Fenton FH et al (2001) Mechanisms for discordant alternans. J Cardiovasc Electrophysiol 12(2):196–206

Weiss JN, Chen PS et al (2000) Ventricular fibrillation: how do we stop the waves from breaking? Circ Res 87(12):1103–1107

Chapter 27
Neonatal Deep Hypothermia: Heart Function and Metabolism

Richard W. Hill and Jacob J. Manteuffel

Abstract We present evidence, based particularly on studies of *Peromyscus leucopus*, that neonatal rodents in deep hypothermia ($T_b \leq 7°C$)—although apneic—steadily take up O_2 across the lungs and distribute it via the circulatory system. We argue that the myocardium, respiratory rhythmogenic neurons, and possibly other vital tissues depend for their survival during deep hypothermia on this continuing O_2 supply. In regards their steady O_2 uptake and its significance, neonatal rodents resemble rodent hibernators during deep hypothermia. However, details differ strikingly. Neonates, having stopped breathing on entry to hypothermia, take up O_2 via the mouth and nares by apneic mechanisms. Moreover, neonates lose sinoatrial (SA) pacing of ventricular contraction on entry to hypothermia and depend on quasi-rhythmic ventricular escape contractions to maintain cardiac activity. During deep hypothermia, physiologically limited mechanisms of respiratory and circulatory O_2 transport combine to provide neonate tissues with a limited rate of O_2 supply that is vitally important.

27.1 Introduction

In the lives of placental mammals, there are two stages at which deep hypothermia may be well tolerated: hibernation and the period immediately following birth. "Deep" or "profound" hypothermia here refers to a core $T_b \leq 7°C$. Many species

R. W. Hill (✉)
Department of Zoology, Michigan State University,
East Lansing, MI 48824, USA
e-mail: hillr@msu.edu

J. J. Manteuffel
Department of Emergency Medicine, Henry Ford Hospital,
2799 West Grand Blvd, Detroit, MI 48202, USA

T. Ruf et al. (eds.), *Living in a Seasonal World*,
DOI: 10.1007/978-3-642-28678-0_27,

can tolerate it and later return unscathed (or relatively so) to normothermia during neonatal development, hibernation, or both.

Inaccurate claims regarding hibernation sometimes arise because of failure to recognize the neonatal period. For example, the ability of the heart in hibernators to keep beating at tissue temperatures near freezing is commonly touted as "unique" for mammals (e.g., Eagles et al. 1988), even though the neonatal heart is often capable of the same thing. Similarly, the heart of rodent hibernators has been recognized for its distinctive resistance to ventricular fibrillation in hypothermia (Johansson 1996), although fibrillation is also almost never observed in profoundly hypothermic rodent neonates (see later). Such cases highlight the need for wider recognition that there are two life stages—not one—in which mammals may tolerate deep hypothermia.

These considerations notwithstanding hibernation justly deserves the particular attention it has received. One reason is that the deep hypothermia of hibernation is orchestrated by the animal; hibernators not only enter hibernation under their own endogenous control but also return to normothermia through endogenous thermogenesis (Heldmaier and Ruf 1992). Neonatal deep hypothermia, by contrast, is more dependent on the environment. It often is induced by environmental conditions that overwhelm the animal's thermoregulatory capability, and termination of the hypothermia is often dependent on environmental warming.

Despite differences, hibernation and neonatal hypothermia have many similarities, and surely a full understanding of mammalian deep hypothermia will ultimately depend on studies of both. However, research on neonatal deep hypothermia has lagged. In both hibernation and neonatal deep hypothermia, all tissues function without permanent damage at cellular temperatures that are >30°C lower than the norm of 37°C that has been so notably conserved in the evolution of normothermic placental mammals. Metabolism is profoundly depressed in both states. Beyond these obvious similarities, there are others that are less well understood. For example, brown adipose tissue (BAT) and non-shivering thermogenesis (NST) are highly emphasized in both hibernators and neonates. Possibly BAT and NST have been favored in both groups because of their common experience of low T_bs. Also, the striking absence of ventricular fibrillation in both groups already mentioned may well reflect similar adaptations at the molecular level in membrane composition and Ca^{2+} processing (Ruf and Arnold 2008).

An important similarity between hibernation and neonatal deep hypothermia is that circulatory O_2 delivery continues in both states. This similarity was obscured until late in the twentieth century because of reports in the 1940s and 1950s that—based on electrocardiographic (EKG) studies of several species—the heart stops along with breathing during deep hypothermia in neonates that survive deep hypothermia (Adolph 1948, 1951; Fairfield 1948). With cardiac arrest "established," investigators concluded that the tissues of neonates are cut off from supply of O_2 during deep hypothermia, and Adolph (1956) and (1963) referred to them as anoxic (although perhaps using terms a bit differently than today). In 1969, a major review of "mechanisms of mammalian tolerance to *low body temperatures*" (Andjus 1969; italics ours) used data drawn entirely from studies of *asphyxiated* individuals to address neonatal hypothermia, on the assumption (Adolph 1963) that tolerance of anoxia by neonates is central to their tolerance of deep hypothermia. In contrast, Fitzgerald reported in 1955 that sustained O_2 consumption occurs in neonatal laboratory mice at T_b = 0–10°C, suggesting that circulatory O_2 delivery might occur at some temperatures at which the heart was reputedly in arrest. Fitzgerald's paper was largely ignored, however (10 citations to date). Hill (2000) finally reported that EKG evidence of heart action is typically observed throughout neonatal deep hypothermia at T_as that are survived in all of several rodent species studied when high-sensitivity EKG recording is employed. Moreover, Hill (2000) presented preliminary evidence that O_2 consumption continues in neonates in deep hypothermia. The earlier reports of cardiac arrest—implying tissue anoxia—almost certainly arose from use of methods too coarse to detect the EKG signals present (Hill 2000), and Fitzgerald (1955) was correct!

We know now, therefore, that hibernation and neonatal deep hypothermia are similar in the important respect that circulatory O_2 delivery continues. They are dramatically different in crucial details of cardiac and respiratory function, however. Our goal here is to elaborate the relationship between heart function and O_2 delivery in neonatal rodents during deep hypothermia—a state in which they do not breathe. We focus on the white-footed mouse (*Peromyscus leucopus*), a native North American rodent known to start breeding prior to the end of winter at cold latitudes (e.g., Michigan and Ontario). *P. leucopus* neonates survive at presumptive $T_b < 1$°C, meaning they are among the most hypothermia-tolerant rodent neonates.

27.2 Methods

We recorded EKGs throughout 2.9–3.3 h exposures to T_a = 0–3°C in 2- to 10-day-old neonates of *P. leucopus* exposed to ordinary air. A Grass 79C polygraph with 7P5 preamplifier and 7DAE driver amplifier was employed with Grass E2 platinum electrodes. Electrodes were attached subdermally to the two forelegs and left hindleg (plus a ground attached at midback) after ca. 20 min of cooling, when an animal was lethargic but still breathing. EKG recording was then continuous as the animal cooled to deep hypothermia and later was rewarmed (at T_a = 20°C) and recovered breathing. Because the amplitude of EKG waveforms often decreases profoundly during hypothermia, extensive electrical shielding was employed at all steps in signal transmission. Most recording was from a single EKG lead, typically Lead I (right and left forelegs), or Lead III (right foreleg and left hindleg). EKGs were recorded in the same manner in 2- to 10-day-old neonates of laboratory rats (*Rattus norvegicus*), prairie voles (*Microtus ochrogaster*), and house mice (*Mus domesticus*) of two types (C3H and feral).

We measured O_2 consumption in 3- to 10-day-old *P. leucopus* (without EKG electrodes) using a closed, recirculating respirometry system constructed entirely of glass and stainless-steel tubing except for short lengths of Tygon tubing. Recirculating air flow was driven by a peristaltic pump (Cole-Parmer), with Ascarite and Drierite columns used to keep levels of CO_2 and gaseous H_2O low. The O_2 partial pressure was recorded continuously with an Applied Electrochemistry S3A O_2 analyzer. Because the S3A is exquisitely sensitive to total pressure in the gas stream, the pressure inside the closed system was monitored continuously and maintained constant with a custom-built, all-glass pressure-compensating device. An aliquot of air held in a Brooks Volumeter was passed through the O_2 analyzer at start and end to ascertain system stability.

27.3 Results

Our EKG results point to the following conclusions in both *P. leucopus* and the other species studied: the heart beats during deep hypothermia in individuals that survive. This steady beating is dependent on ventricular escape.

Patterns of EKG response during entry and exit from hypothermia are similar in all species (although lowest tolerated T_a varies with species). These patterns are exemplified here using *P. leucopus*. About 1.1 h after a neonatal *P. leucopus* is placed at $T_a = 0$–3°C, *P* waves—indicative of sinoatrial (SA) pacing—disappear (Figs. 27.1, 27.2). This loss of *P* waves is followed by ventricular cardiac arrest, ranging from momentary to 7–8 min in duration (Figs. 27.1c, 27.2c). Thereafter, rhythmic ventricular contraction is restored—albeit at a reduced rate—in the form of escape beats paced by one or more ventricular initiation sites (Figs. 27.1c–d, 27.2d–e). We observed restoration of ventricular action by escape beats following loss of *P* waves in all individuals of all species. The waveform of escape beats in the first minutes after loss of *P* waves in *P. leucopus* often resembles the waveform of ventricular contraction prior to *P*-wave loss (Figs. 27.1c, 27.2d)—suggesting a pacing site in the conducting system. However, the waveform of escape beats typically soon becomes dramatically different from that of SA-paced beats (compare Figs. 27.1a, d and 27.2a, e), indicating that novel sites take over initiation of depolarization. After the EKG stabilizes—and continuing then for ca. 1.6–2.0 h until rewarming—the waveform in Figs. 27.1d and 27.2e is particularly common. In some cases, it is maintained throughout deep hypothermia. In other cases, the waveform in deep hypothermia changes from time to time, as in Fig. 27.3a, indicating a shift from one ventricular pacing site to a different one. In 55 *P. leucopus* studied, 26 individuals exhibited only one ventricular pacing site during the period of escape beats, whereas 29 exhibited 2–5 sites. Average heart rate in steady-state deep hypothermia ($T_a = 0$–3°C) was 2.2 beats/min.

When a deeply hypothermic neonate, regardless of species, is rewarmed at $T_a = 20$°C, the initial effect is that escape beats gradually accelerate. Later, a comparatively abrupt further acceleration of heart rate occurs with the return of (1) SA pacing—evidenced by *P* waves and a steady *P–R* interval—and (2) breathing. Exemplifying with *P. leucopus*, in some cases during rewarming, SA pacing returns after an animal has taken its first breath (Fig. 27.4). Prompt transition then occurs over just a few minutes from arrhythmic (Fig. 27.4c) to rhythmic

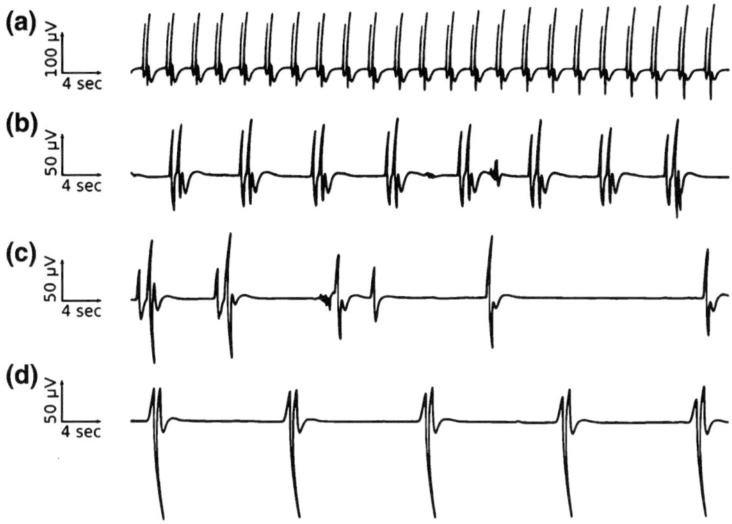

Fig. 27.1 EKG (Lead I) during entry into deep hypothermia of a 10-day-old *P. leucopus* at $T_a = 0–1°C$. EKG segment (**a**) started 46 min after exposure to low T_a began (although exposure had been interrupted for electrode attachment), and it shows regular sinus rhythm slowed by incipient hypothermia; (**b**) started 13 min after **a**, just prior to disappearance of *P* waves. Note the prolongation of electrical events in **b** because of deepening hypothermia; (**c**) started 2 min after **b** and shows the loss of *P* waves; the two ventricular beats to the right were escape beats, marked by a lower ventricular contraction rate than the SA-paced beats to the left; (**d**) started 7 min after **c**. The waveform of the ventricular escape beats had stabilized in **d**, indicating pacing by a consistent site. Most beats over the ensuing 30 min exhibited the same waveform as in **d**, although the ventricular contraction rate decreased during that period to half that seen in **d**

(Fig. 27.4d) SA-paced ventricular beats. Alternatively, *P* waves may appear prior to the first breath (Fig. 27.5). SA pacing is then relatively slow, and sometimes arrthymic, in the minutes (Fig. 27.5c) before the first breath. Thereafter, however, SA pacing becomes rhythmic, if it was not before, and accelerates dramatically (Fig. 27.5e).

Essentially every type of arrhythmia is observed, on occasion, during entry and exit from deep hypothermia in neonates of all species studied. Ventricular arrests occur, nearly always terminating spontaneously by return of beats; 16% of *P. leucopus*, for example, exhibited arrests lasting >5 min, averaging 26 min. Atrio-ventricular (A-V) conduction arrhythmias often observed include both Type I (Figs. 27.3b, 27.5d) and Type II (Figs. 27.3c, 27.5d) second degree A-V block (i.e., Mobitz I and II) and third degree A-V block (Fig. 27.3d). Premature ventricular contractions (PVCs) and ventricular bigeminy (a related phenomenon) are also often seen (Figs. 27.3e–g). Other arrhythmias include arrhythmic SA pacing (Fig. 27.4c), atrial flutter, and premature atrial complexes. EKGs indicative of ventricular fibrillation are seen only rarely and end in spontaneous restoration of rhythmic ventricular contraction, not arrest. Besides arrhythmias, another point of interest is that Osborn (J) waves occur occasionally during hypothermia induction (first beat in Fig. 27.1c; Fig. 27.2b).

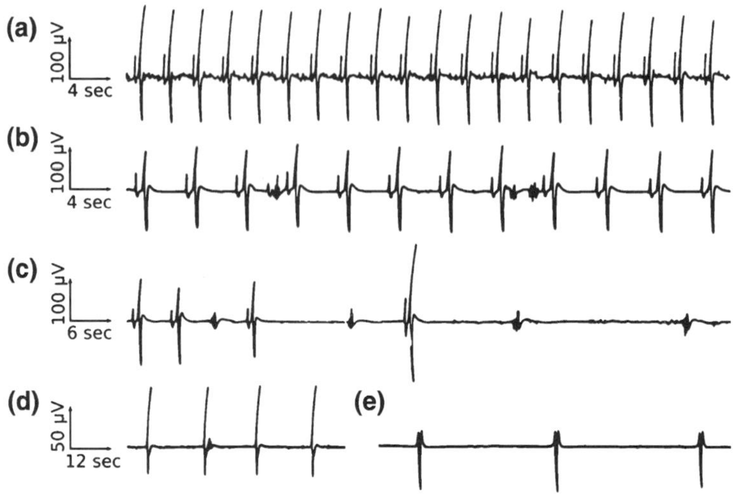

Fig. 27.2 EKG (Lead I) during entry into deep hypothermia of a 9-day-old *P. leucopus* at $T_a = 0\text{–}1°C$. EKG segment (**a**) started 58 min after exposure to low T_a began (although exposure had been interrupted for electrode attachment), and it shows regular sinus rhythm slowed by incipient hypothermia; (**b**) started 8 min after **a**, just prior to disappearance of *P* waves; (**c**) started 1 min after **b** and shows the final *P* waves. Following the ventricular beat in the middle of **c**, ventricular arrest lasted 1.5 min. The low-amplitude, multiphasic waveforms in **c** probably represent continuation of similar waveforms in **b** attributable to skeletal muscle, although non-conducting *P* waves cannot be ruled out; (**d**) started 1.65 min after the final ventricular beat in **c** and shows the 2nd–5th escape beats; (**e**) started 13 min after **d**. The waveform of the ventricular escape beats had stabilized, indicating that escape-beat pacing was originating at a consistent site. Most beats in the 30 min following **e** exhibited the same waveform as in **e**. Scaling is identical in **d** and **e**

Neonatal *P. leucopus* in deep hypothermia consume O_2 steadily at a T_a-dependent rate (Fig. 27.6). In experiments separate from those included in Fig. 27.6, we blocked the mouth and nares of neonatal *P. leucopus* in steady-state hypothermia ($T_a = 2°C$) with wet cotton. In all cases ($n = 9$), O_2 uptake dropped to zero but resumed when the block was removed.

27.4 Discussion

The steady uptake of O_2 from the atmosphere by neonatal *P. leucopus* (Fig. 27.6) occurs despite the fact that neonatal rodents undergo respiratory arrest as they are cooled (commencing after ca. 1.1 h in *P. leucopus* at $T_a = 2°C$) and are in continuous apnea during deep hypothermia (Adolph 1963; Hill 2000; Tattersall and Milsom 2003). Two types of evidence indicate that, despite apnea, the uptake of O_2 from the atmosphere by deeply hypothermic *P. leucopus* neonates is via the mouth and nares. First, as reported here, blockage of the mouth and nares reversibly stops O_2 consumption. Second, as described by Hill (2000), when *P. leucopus* in hypothermia are masked, O_2 uptake occurs from the air in the mask and not from air outside the mask.

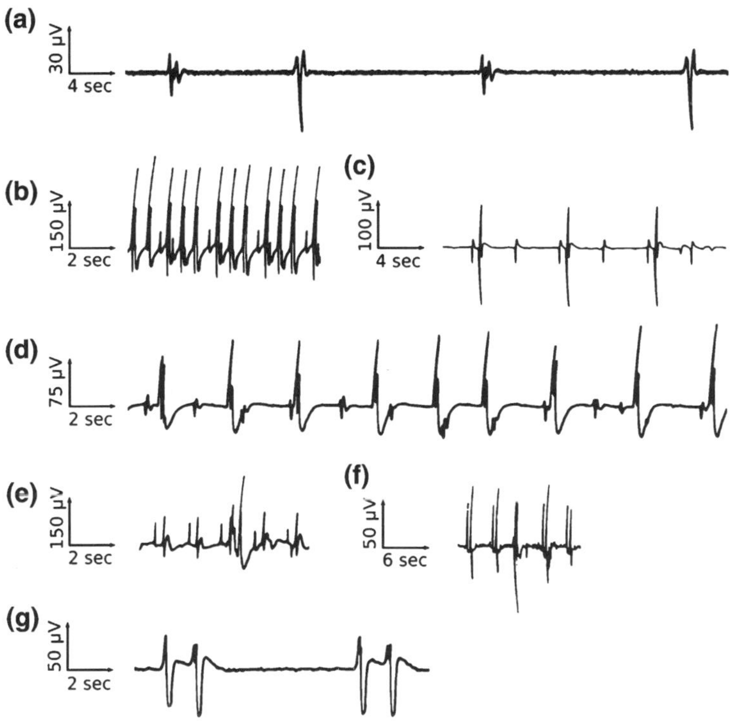

Fig. 27.3 Cardiac phenomena observed in *P. leucopus*. Interpretation is based on single-lead data, sometimes subject to ambiguity compared to multi-lead data. **a** Multifocal pacing of ventricular escape beats during deep hypothermia in an 8-day-old at T_a = 2–3°C (Lead III). Both waveforms were observed repeatedly over a 3-min period before the pacing site for the higher amplitude waveform took over and dominated pacing almost exclusively for >20 min. **b** Second degree A-V block, Type I, (Mobitz I) in a 6-day-old during rewarming at T_a = 20°C after hypothermia at T_a = 2–3°C (Lead III). **c** Second degree A-V block, Type II, (Mobitz II) in an 8-day-old during entry to hypothermia at T_a = 2–3°C (Lead III). **d** Third degree A-V block in a 7-day-old during rewarming at T_a = 20°C after hypothermia at T_a = 2–3°C (Lead III). **e** and **f** Premature ventricular contraction in (**e**) a 9-day-old during entry to hypothermia at T_a = 2–3°C (Lead III) and (**f**) another 9-day-old during entry to hypothermia at T_a = 0–1°C (Lead I). In both cases the premature waveform was observed several times although each trace reproduced here shows only one occurrence. **g** Ventricular bigeminy in a 6-day-old during rewarming at T_a = 20°C after hypothermia at T_a = 2–3°C (Lead III)

The rate of O_2 consumption by *P. leucopus* neonates at T_a = 0–4°C (Fig. 27.6) averages ca. 1% of the rate in same-age neonates at T_b = 35°C (Hill 1976). This extent of O_2 consumption strikingly resembles that in hibernators; for example, the four species of hibernating mice and hamsters tabulated by Heldmaier and Ruf (1992) exhibit O_2 consumption rates at low T_b averaging ca. 1% of their euthermic rates. Thus, despite apnea, neonates in deep hypothermia are similar to hibernators in O_2 delivery.

Apneic O_2 consumption is documented also in a number of other mammals in hypothermia (e.g., Sullivan and Szewczak 1998). Three mechanisms are believed to account for pulmonary O_2 uptake without breathing: diffusion of

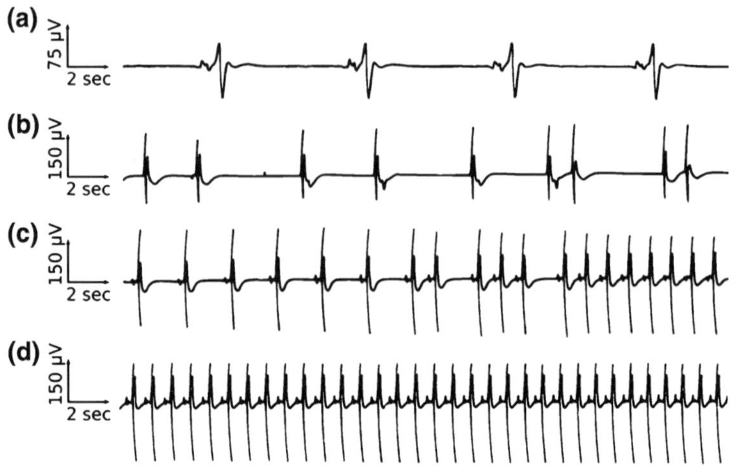

Fig. 27.4 EKG (Lead III) during exit from deep hypothermia in a 6-day-old *P. leucopus*. EKG segment (**a**) started immediately after the animal was returned to $T_a = 20°C$ after 3 h at $T_a = 2–3°C$; (**b**) started 9.5 min after **a** and only 50 s before the animal took its first breath. Note that *P* waves have not yet reappeared; (**c**) started 20 s after the first breath. The *P* waves seen in **c** were the first observed since *P* waves were lost during entry into hypothermia; (**d**) started 2 min after **c**, immediately following the animal's second breath. Note that scaling of time is the same in **a–d**. Progressive acceleration of heart action is evident from start to finish in the period of rewarming depicted in this figure

O_2 down the pulmonary airways following an O_2 partial-pressure gradient, bulk inflow of air via the airways caused by a metabolically produced relative vacuum in the lungs (less CO_2 production than O_2 consumption), and pulsatile movements of air in major airways caused by beating of the heart next to the airways, the so-called "cardiogenic mixing" (Malan 1982; Sullivan and Szewczak 1998). Although the third mechanism might have been discounted for deeply hypothermic neonates prior to discovery of sustained cardiac activity, the beating of the heart now known to occur makes cardiogenic mixing possible.

The EKGs we recorded from deeply hypothermic neonates of *P. leucopus* and three other species show that *P* waves cease as deep hypothermia becomes established—indicating that the SA node goes silent or the atria become unresponsive to SA stimulation. SA pacing of ventricular contraction is then lost. Nonetheless, in all four species, ventricular contraction typically continues in a quasi-regular rhythm during hypothermia, paced by ventricular sites (i.e., in escape rhythms) (Figs. 27.1, 27.2). The rate of ventricular contraction in deep hypothermia, ca. 2.2 beats/min in *P. leucopus* at $T_a = 0–3°C$, is low because not only is myocardial T_b low but also heart rate drops 40–50% when SA-paced beats are replaced by escape beats (Fig. 27.1c). Similarly low heart rates, 2–4 beats/min, are observed in ground squirrels of two species in hibernation at similar T_b (Dawe and Morrison 1955). During rewarming (Figs. 27.4, 27.5), neonates of all species studied first exhibit acceleration of escape

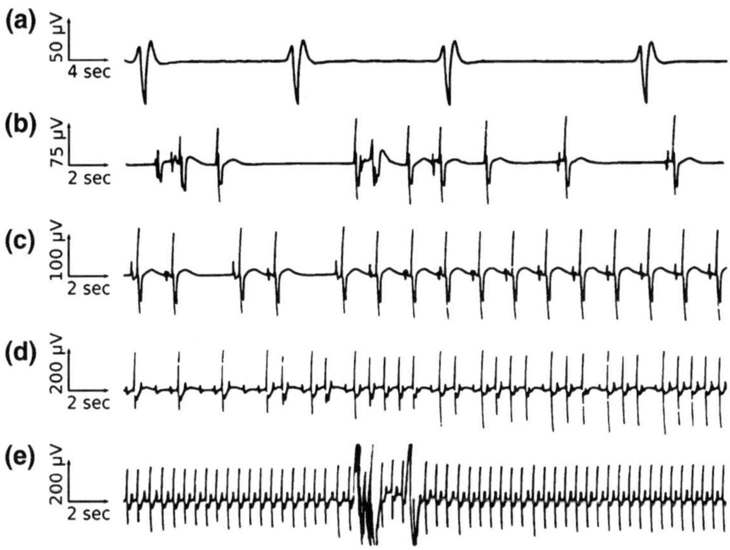

Fig. 27.5 EKG (Lead III) during exit from deep hypothermia in a 10-day-old *P. leucopus*. EKG segment (**a**) started immediately after the animal was returned to $T_a = 20°C$ after 3.2 h at $T_a = 2–3°C$; (**b**) started 16 min after **a**, when the first *P* waves appeared. In this neonate, *P* waves returned >5 min before breathing was restored; (**c**) started 2 min after **b**; (**d**) started 3.3 min after **c**, immediately after the animal took its first breath. Following the first breath, as **d** shows, a period of second degree A-V block occurred before a more regular, rapid rhythm was established; the block was initially Type II (Mobitz II) with a 3:1 A-V ratio, then transitioned to Type I (Mobitz I). After the rhythm became consistent at the end of **d**, it remained so (no further arrthymia) except for gradual acceleration—evident in (**e**), which started 1.5 min after **d**. The irregular signal near the middle of **e** (truncated at *top* and *bottom* in this reproduction) is the animal's fifth breath. Note that the time scale is the same in **b**–**e**

ventricular beats while remaining apneic. Later, *P* waves reappear and SA pacing is resumed, starting often with intermittent capture beats (Fig. 27.5b). Apnea is ended by spontaneous reappearance of breathing at roughly the same time as *P* waves reappear. Although individuals vary in detail, the restoration of breathing is inevitably followed by rapid acceleration of SA-paced ventricular contraction, suggesting that myocardial hypoxia develops during rewarming while apnea continues (and tissue metabolism increases, Fig. 27.6). Further research is needed to determine fully the extent to which these cardiac responses to deep hypothermia are responses to local tissue conditions (e.g., temperature) or mediated by neural controls (Milsom et al. 1999).

Based on our focal study on *P. leucopus*, continued O_2 uptake from the atmosphere and beating of the heart during deep hypothermia are vitally important. According to several sorts of observations, they promote tissue oxygenation during hypothermia and are required for a neonate to survive hours of deep hypothermia. Two preliminary points to make are (1) the rate of O_2 uptake in deep hypothermia (Fig. 27.6)—1% of euthermic rate—is similar to the rate in hibernation, during which the vital importance of steady O_2 uptake is well recognized and (2), based on separate studies, we have

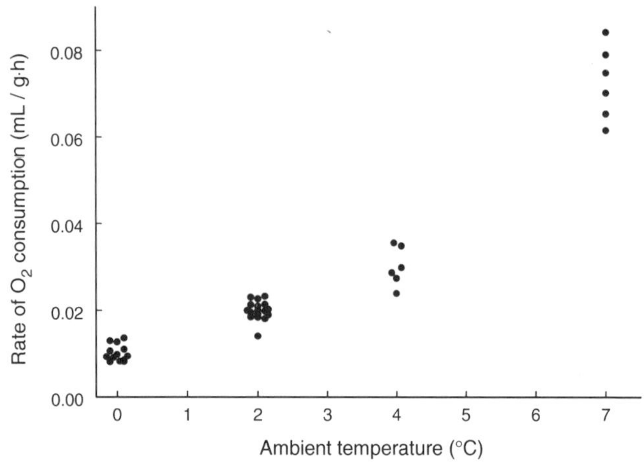

Fig. 27.6 Rate of O_2 consumption as a function of T_a in apneic 3- to 10-day-old *P. leucopus*. T_b approximated T_a. Data for all ages are combined because age (and size) had no consistent effect on weight-specific O_2 consumption. Mice were apneic during measurement except possibly, in unusual cases, at $T_a = 7°C$. When mice were rewarmed at $T_a = 20°C$, the time that passed before first breath was 13–33 (mean = 20.8) min after hypothermia at 0°C, 9–22 (mean = 13.8) min after 2°C, 4–14 (mean = 6.5) min after 4°C, and 1–4 (mean = 2.5) min after 7°C (at each T_a, young neonates tended to breathe sooner than old ones, e.g., means of 17.9 and 24.2 min for ages 3–4 versus 9–10 days after 0°C, and 10.1 versus 14.7 min after 2°C). Mice were studied in sets of 2–5 littermates (huddling prevented); each symbol represents one such set ($n = 14$ at 0°C, 18 at 2°C, 6 at 4°C, and 6 at 7°C). Some symbols at 0–4°C are displaced slightly along the *x*-axis for clarity (e.g., actual T_a never fell below 0°C). Exposure to T_a lasted 3.1–3.5 h. Data are from the final hour. Six sets of mice studied in the same manner at $T_a = 10°C$, a T_a at which they breathe, averaged 0.24 ml O_2/g·h

verified electro-mechanical coupling, meaning that escape ventricular EKG waveforms in hypothermia indicate ventricular contraction. Based on extensive experiments on *P. leucopus*, we have found that if an apneic neonate in deep hypothermia is placed in a N_2 atmosphere, its EKG starts to deteriorate within ca. 10 min—exhibiting decreasing amplitude and/or rate, ending ultimately in arrest. Return to air reverses this deterioration provided arrest has not yet developed. These results indicate that the myocardium is oxygenated during ordinary deep hypothermia by pulmonary uptake of O_2 from the atmosphere and circulatory O_2 transport. Another noteworthy observation is that, based on our EKG studies in air in both *P. leucopus* and the other species studied, prolonged cardiac arrest in deep hypothermia makes for a poor prognosis: when, on occasion, an arrest lasting >40 min occurs, the odds that the neonate will die are greatly increased, even at a T_a usually survived. This observation suggests that blood circulation is essential for survival despite the fact that tissue O_2 needs are slight (Fig. 27.6). Insight into the reason for death after cardiac arrest comes from the fact that during rewarming, although the heart always beats in all species studied (it invariably restarts if it had been in arrest), neonates destined to die never take a breath, whereas those that take a breath recover. Failure to take a first breath, according to Mellen et al. (2002), implies damage to the respiratory rhythm-generating circuit in the medullary pre-Bötzinger complex. Such damage would be a logical consequence of long cardiac arrest if the relevant neurons require oxygenation. We hypothesize, therefore, that we see evidence in two tissues of a vital need for respiratory and circulatory O_2 delivery during apneic deep hypothermia. Specifically, we hypothesize that (1) the EKG deteriorates in an N_2 atmosphere because the myocardium, ordinarily perfused with oxygenated blood despite apnea, is then perfused with relatively deoxygenated blood, and (2) neurons of the pre-Bötzinger rhythmogenic circuit are fatally damaged during prolonged cardiac arrest because of being deprived of O_2 supply.

To conclude, neonatal rodents are notably similar to rodent hibernators in that respiratory and circulatory O_2 delivery continue throughout deep hypothermia. However, hibernators and neonates differ strikingly in respiratory and circulatory detail. Hibernators continue to breathe throughout deep hypothermia despite, in some cases, periodicity in ventilatory frequency or episodic apnea (Zimmer and Milsom 2001). Neonates, by contrast, stop breathing as they enter deep hypothermia but continue O_2 uptake by apneic mechanisms via the mouth and nares (Fig. 27.6). Hibernators typically maintain SA pacing during hypothermia, evidenced by *P* waves and SA initiation of QRS complexes (Dawe and Morrison 1955; Nardone 1955). In neonates, SA pacing of ventricular contraction stops and stays stopped, but the neonatal heart exhibits a remarkable propensity for ventricular escape in which one or more ventricular sites pace ventricular contractions that are often strikingly rhythmic over tens of minutes. Physiologically limited mechanisms of respiratory and circulatory O_2 transport combine in deeply hypothermic neonates to provide neonate tissues with a limited rate of O_2 supply that nonetheless is similar to the rate in hibernators and just as vitally important.

Acknowledgments The success of this research depended on contributions from Michael Cook, Susan Hill, Brock Horsley, and Bradley White. Vincent Shaw and Daphne Swope helped with manuscript preparation. George Eyster and David Matisoff, both clinical cardiologists, provided invaluable assistance with interpretation of EKGs. Much gratitude to all.

References

Adolph EF (1948) Tolerance to cold and anoxia in infant rats. Am J Physiol 155:366–377

Adolph EF (1951) Responses to hypothermia in several species of infant mammals. Am J Physiol 166:75–91

Adolph EF (1956) Effects of low body temperature on tissue oxygen utilization. In: Dripps RD (ed) The physiology of induced hypothermia. National Academy of Sciences, Washington, DC

Adolph EF (1963) How do infant mammals tolerate deep hypothermia? In: Hardy JD (ed) Temperature, its measurement and control in science and industry: Part 3. Biology and medicine. Reinhold, New York

Andjus RK (1969) Some mechanisms of mammalian tolerance to low body temperatures. Symp Soc Exp Biol 23:351–394

Dawe AR, Morrison PR (1955) Characteristics of the hibernating heart. Am Heart J 49:367–384

Eagles DA, Jacques LB, Taboada J, Wagner CW, Diakun TA (1988) Cardiac arrhythmias during arousal from hibernation in three species of rodents. Am J Physiol 254:R102–R108

Fairfield J (1948) Effects of cold on infant rats: body temperatures, oxygen consumption, electrocardiograms. Am J Physiol 155:355–365

Fitzgerald LR (1955) Oxygen consumption of newborn mice at low temperatures. Am J Physiol 182:105–110

Heldmaier G, Ruf T (1992) Body temperature and metabolic rate during natural hypothermia in endotherms. J Comp Physiol B 162:696–706

Hill RW (1976) The ontogeny of homeothermy in neonatal *Peromyscus leucopus*. Physiol Zool 49:292–306

Hill RW (2000) Anoxia tolerance to oxygen necessity: paradigm shift in the physiology of survival of apneic deep hypothermia in neonatal rodents. In: Heldmaier G, Klingenspor M (eds) Life in the cold. Springer, Berlin

Johansson BW (1996) The hibernator heart—Nature's model of resistance to ventricular fibrillation. Cardiovasc Res 31:826–832

Malan A (1982) Respiration and acid-base state in hibernation. In: Lyman CP, Willis JS, Malan A, Wang LCH (eds) Hibernation and torpor in mammals and birds. Academic, New York

Mellen NM, Milsom WK, Feldman JL (2002) Hypothermia and recovery from respiratory arrest in a neonatal rat in vitro brain stem preparation. Am J Physiol 282:R484–R491

Milsom WK, Zimmer MB, Harris MB (1999) Regulation of cardiac rhythm in hibernating mammals. Comp Biochem Physiol A 124:383–391

Nardone RM (1955) Electrocardiogram of the arctic ground squirrel during hibernation and hypothermia. Am J Physiol 182:364–368

Ruf T, Arnold W (2008) Effects of polyunsaturated fatty acids on hibernation and torpor: a review and hypothesis. Am J Physiol 294:R1044–R1052

Sullivan SG, Szewczak JM (1998) Apneic oxygen uptake in the torpid pocket mouse *Perognathus parvus*. Physiol Zool 71:624–632

Tattersall GJ, Milsom WK (2003) Hypothermia-induced respiratory arrest and recovery in neonatal rats. Resp Physiol Neurobiol 137:29–40

Zimmer MB, Milsom WK (2001) Effects of changing ambient temperature on metabolic, heart, and ventilation rates during steady state hibernation in golden-mantled ground squirrels (*Spermophilus lateralis*). Physiol Biochem Zool 74:714–723

Chapter 28
Seasonal Changes in Thermoregulatory Strategies of Tegu Lizards

William K. Milsom, Colin Sanders, Cleo Leite, Augusto S. Abe, Denis V. Andrade and Glenn Tattersall

Abstract During the post-reproductive period (December–March) black and white tegus [*Tupinambis merianae* (Duméril and Bibron)] basked during the day and regulated maximum daily T_b around 33–37°C. Heart rate (HR) and breathing frequency (fR) increased during the day to relatively constant levels. Nighttime values of HR and fR, however, decreased progressively from December to March while nighttime T_b was fairly constant ($\sim$25°C). During March to mid-April the lizards slowly entered into hibernation. If HR is correlated to metabolic rate under steady state conditions, declines in nighttime HR at constant T_b would suggest there was a continuum from sleep to nightly torpor to multi-day bouts of torpor to hibernation. During hibernation, tegus in artificial outdoor burrows exhibited periodic bouts of emergence. This was not seen in tegus in natural burrows. During arousal from hibernation there was an increase in preferred day time T_b during periods of emergence and an increase in nighttime T_b that exceeded the increase in nighttime burrow temperature (T_{burrow}). This gave rise to nighttime differences between T_b and T_{burrow} that at their peak were as much as 6–8°C. Differences between T_b and $T_{ambient}$ of 2–4°C could be maintained in an environmental chamber with no bedding for extended periods suggesting that the temperature differential was due in part to either decreased heat loss, increased heat production, or both.

W. K. Milsom · C. Sanders
Department of Zoology, University of British Columbia, Vancouver, Canada

C. Leite · A. S. Abe · D. V. Andrade
Depto de Zoologia, Instituto Nacional de Ciência e Tecnologia em Fisiologia Comparada, Univ Estadual Paulista, Rio Claro, SP 13506-900, Brazil

G. Tattersall (✉)
Department of Biology, Brock University, St. Catharines, ON, Canada
e-mail: gtattersall@brocku.ca

T. Ruf et al. (eds.), *Living in a Seasonal World*,
DOI: 10.1007/978-3-642-28678-0_28,

Keywords Reptiles · Tegu lizards · Torpor · Dormancy · Hibernation · Cardiorespiratory control

28.1 Seasonality and the Tegu Lizard Life Cycle

The tegu lizard, *Tupinambis merianae*, depends largely on external heat sources to regulate body temperature. As a consequence, individuals living in seasonal habitats undergo seasonal changes in body temperature, metabolism and activity. Most terrestrial ectothermic vertebrates employ behavioural and physiological mechanisms to provide some degree of autonomy from environmental change but the extent of this independence is generally small. We have been conducting a long-term study of the black and white tegu lizard in Brazil with a focus on both the strategies employed during the cooler, dryer winter period to reduce metabolism as well as the spring reproductive period to enhance metabolism.

The tegu lizard is an active omnivore that can grow as an adult to over 5 kg body weight. The main habitat of this lizard in Brazil is characterised by savannah-like regions, called 'Cerrado' (Hueck 1972). The yearly cycle of the tegu can be divided into five main periods based upon the physiological-behaviour states: the reproductive (early active), post-reproductive (late active), entrance (into hibernation), hibernation and arousal (from hibernation) periods. Mid-September through to December (spring in the southern hemisphere) is a period of high activity associated with courtship and breeding (the reproductive period). December–February is a post-reproductive period of reduced activity. Throughout the reproductive and post-reproductive periods the lizards are foraging and replenishing energy stores. During March to mid April the lizards slowly enter into hibernation. This period is marked by a drastic decrease in activity and complete aphagia (the animals are no longer interested in eating) (Lopes and Abe 1999). Hibernation occurs from mid April through to the end of July during which time the tegus spend almost all their time inactive in their burrows (Abe 1983). In August to mid-September they arouse (Lopes and Abe 1999). During this period the tegus progressively emerge from the burrows with increasing frequency.

28.2 Winter Dormancy (Hibernation)

While many are reluctant to use the term hibernation to describe dormancy in reptiles, in the case of the tegu lizard this is a period of active metabolic suppression (Abe 1983, 1995; de Souza et al. 2004; Andrade and Abe 1999) that occurs seasonally during the winter (Abe 1983; Lopes and Abe 1999). By strictest definition it is a form of active winter sleep [hibernare (v) = to winter] and will be referred to as such here.

28.2.1 Anticipatory Decreases in Metabolic Rate: Programmed Nocturnal Torpor?

During the fall (February–April) tegus are still active in the day, warming to usual active temperatures and are still exposed to progressive changes in photoperiod and ambient temperature (Köhler and Langerwerf 2000). During this period, however, nighttime values of heart rate (HR) and breathing frequency (fR) decrease progressively (Fig. 28.1). Minimum nighttime body temperature (T_{bmin}) through December–March, on the other hand, are fairly constant (T_{bmin} hovered around 25°C) (Fig. 28.1) (Sanders et al. 2011). If HR is correlated to metabolic rate under steady state conditions, as suggested by several studies (Butler et al. 2000, 2002; Clark et al. 2004; Piercy et al. 2010), declines in nighttime HR at constant T_{b} would suggest that there is a nightly torpor that increases in magnitude over this period of time. This is consistent with previous studies that have shown that tegus depress metabolism in the autumn in anticipation of the hibernation period when confined in the dark for several days at constant temperature (Abe 1983, 1995; de Souza et al. 2004; Andrade and Abe 1999; Milsom et al. 2008).

As the year progresses, during the late active period and during entrance into hibernation (March and April), tegus emerge from their burrows later in the morning and retreat into their burrows earlier in the afternoon (Fig. 28.2). In most tegus, daytime maximum T_{b} also drops slightly (Fig. 28.1). Daytime maximum levels of HR and breathing show similar trends suggesting that a behavioural metabolic suppression (behavioural selection of lower temperatures or warming for shorter periods) is also occurring during the day late in the active period and during the entrance phase; however, the extent of this is small (Sanders et al. 2011).[1]

28.2.2 Daily Torpor/Multi-Day Bouts of Torpor/Hibernation: A Natural Progression?

At the onset of the active period, daytime quiescence is rare and associated with inclement weather. If rainfall only occurs for half a day, tegus still usually emerge once the rain stops. When full days are spent in the burrow due to unfavourable conditions, T_{b}, fR, HR and presumably metabolic rate remain low. T_{b}, however, remains elevated relative to burrow temperature suggesting that this is simple quiescence (Sanders et al. 2011). Throughout the late active period and the period leading up to hibernation the tegus begin to spend more time in the burrow, often for more than one day, even when weather is favourable. As the seasons progress, during

[1] T29F-7B implantable biopotential/temperature amplifier/encoders [5.7 × 2.8 × 0.92 cm; (60 g or roughly 2% of body weight)] (Konigsberg Instruments, Inc., Pasadena, CA, USA) inserted to lie between the fat bodies ventral to the digestive tract in the abdominal cavity were used to monitor T_{b}, HR and breathing frequency.

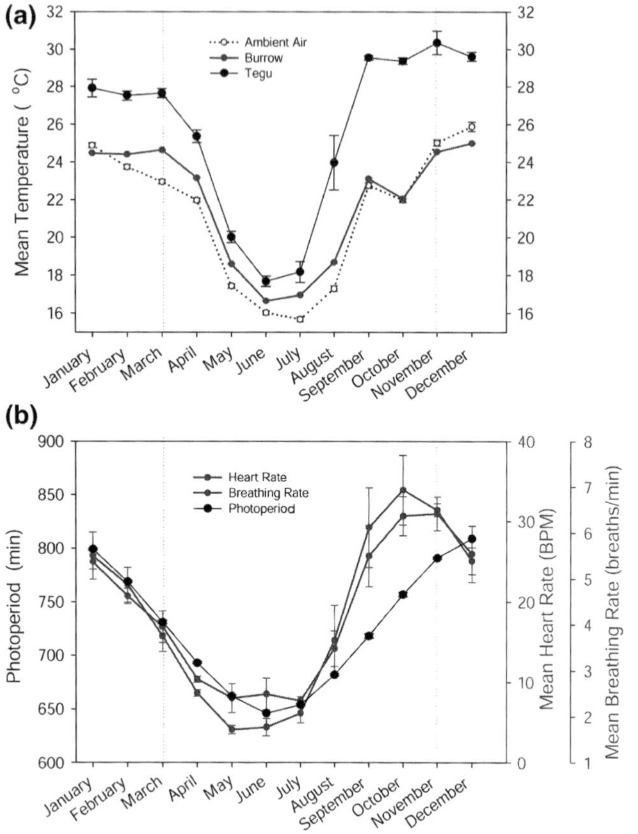

Fig. 28.1 Mean monthly values for **a** ambient and burrow temperatures, and core body temperature, and **b** photoperiod, heart rate and breathing rate for four tegus

Fig. 28.2 Times of sunrise and sunset at the study site (Rio Claro, SP, Brazil) for the study period. Time of emergence from the burrow and retreat into the burrow are also shown for 7 months in the late active season (January), during entrance into hibernation (March), for rare incidents of emergence in hibernation (June, July), during arousal from hibernation (August) and during the early active season (September and December)

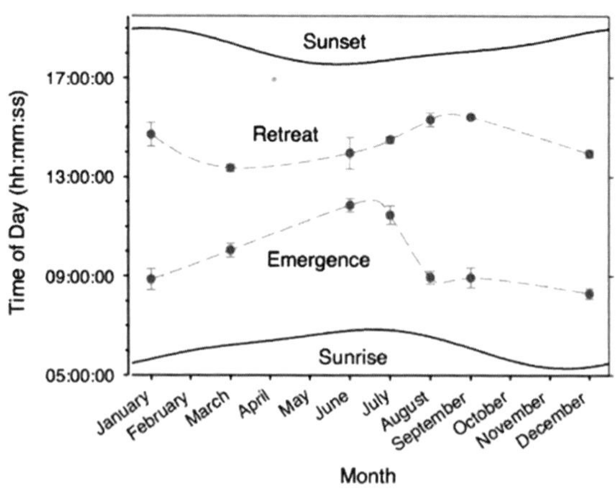

these periods T_b falls to the temperature of the burrow. This becomes particularly pronounced in April when, although rainfall is infrequent and light, the tegus spend roughly half the month in their burrows (Sanders et al. 2011). The hallmark of final entrance into hibernation is when the burrow entrances are blocked with detritus and litter, indicating a behavioural commitment to hibernate. This suggests there is a continuum from sleep to nightly torpor to multi-day bouts of torpor to hibernation. This is consistent with a recent report that tegu lizards can reduce their metabolism to the low rates seen in hibernation at all times of the year when given sufficient time in the cold and dark (Milsom et al. 2008). It is also consistent with the views of some (Wilz and Heldmaier 2000; Heldmaier et al. 2004) but not all (Geiser and Ruf 1995; Geiser 2003) studies of mammalian hibernation.

28.2.3 Do Tegus Arouse Periodically During Hibernation?

While there is still much debate over the causes and benefits of periodic arousal from hibernation (Barnes et al. 1993; Wang 1993; van Breukelen and Martin 2002; Carey et al. 2003), it is common to all mammalian hibernators except the bears. Their occurrence is normally very rhythmic with the lengths of the hibernation bouts increasing in the early season and then decreasing again in the spring (Twente and Twente 1967). Most mammals do not leave their burrows during these bouts of periodic arousal (Willis 1982). Whether this occurs in species of reptiles that hibernate in temperate zones is not known. Species of reptiles that hibernate in subtropical regions are less constrained to remain in their burrows during periods of arousal and in a recent study it has been shown that hibernating *Varanus rosenbergi* spontaneously arouse fairly frequently during their hibernating period (Rismiller and McKelvey 2000).

For *Tupinambis merianae*, the data are equivocal. Animals held in artificial burrows exhibit both short bouts of arousal and emergence throughout the hibernation season (Fig. 28.2). These arousals occur randomly, with no distinct pattern in any animal, suggesting that they are not the consequence of an underlying biological rhythm. They also are not tightly correlated to local factors such as temperature change, noise, or disturbance (Sanders et al. 2011). At present it is not clear what the underlying cause of the arousals is. Animals dormant in natural burrows do not emerge over the hibernation season. Whether they undergo periods of arousal without emerging from the burrow is not clear. There is some evidence to indicate that tegus in artificial burrows do arouse on occasion without leaving the burrow.

28.3 Reproductive (Spring) Thermogenesis?

Shortly after emerging from hibernation, the biological priority of the tegu is to reproduce. Activity from September through to November is elevated as tegus not only seek food to replenish stores depleted by hibernation, but also engage in courtship,

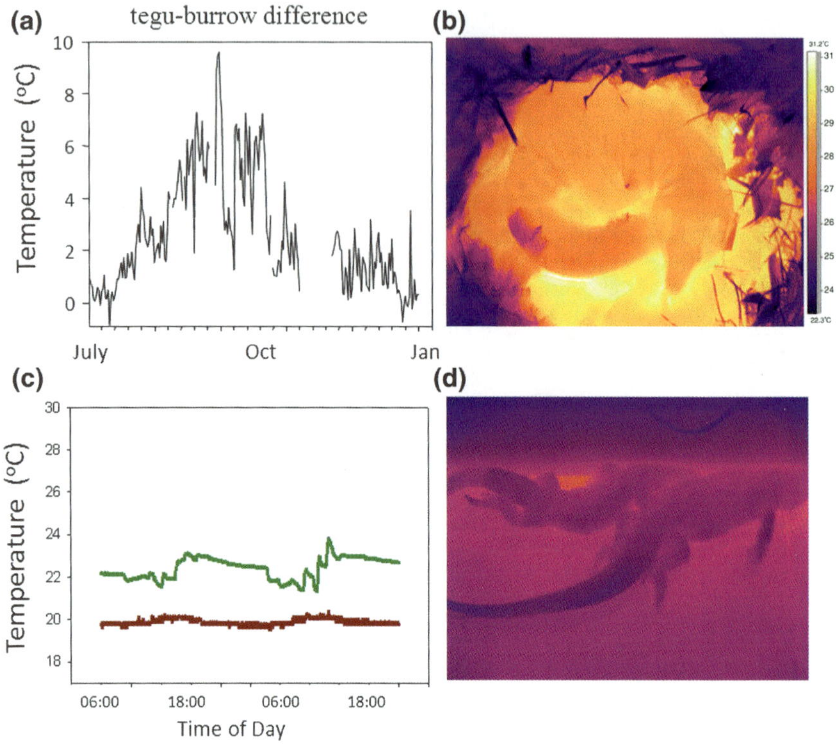

Fig. 28.3 **a** The mean differential between tegu body temperature and burrow temperature during arousal from hibernation and through the reproductive season. **b** Tegus in artificial burrows could be as much as 6°C warmer than the burrow at the end of the night. **c** Tegus inside an environmental chamber remained 2–3°C warmer than the chamber over extended periods. **d** Despite the fact their skin was at or below ambient temperature

mating, nesting and egg laying. Not surprisingly, concomitant with this elevated activity is an elevated metabolism as evidenced by high daytime and nighttime HR and fR (Fig. 28.1) (Sanders et al. 2011). This suite of changes is not uncommon in reptiles (Rismiller and Heldmaier 1991) and can be accompanied by an elevation in preferred T_b and is attributed to 'mating unrest' (Rismiller and Heldmaier 1991). Tegus also increase their preferred daytime T_b in August and most notably, increase their nighttime T_b to exceed the increase in nighttime burrow temperature. This gives rise to a nighttime difference between T_b and T_{burrow} that is maintained throughout this period. At its peak, the $T_b - T_{burrow}$ difference is as much as 6–8°C (Fig. 28.3a) (Sanders et al. 2011).

The underlying cause(s) of this $T_b - T_{burrow}$ differential is unclear. It has recently been shown that if tegus are maintained in an environmental chamber at constant temperature with normal photoperiod at this time of year, they are capable of maintaining this differential for at least 10 days (length of the study period) (Fig. 28.3) (Milsom et al. 2011). Their ability to do this could not be due to activity as it still occurs during the nighttime quiescent period. Certainly, basic thermal inertia due to body size will contribute but this too cannot explain their ability to sustain the differential for this period of time. This leaves two possibilities. The first is that it is due to changes in vascular tone and peripheral blood flow acting to decrease thermal conductance. Thermal conductance is inversely proportional to body mass in reptiles (McNab and Auffenberg 1976) and the size of the tegus will be of benefit in this regard. The second is that it reflects some degree of thermogenesis during this period. This remains to be determined. There is no doubt that this would also be enhanced by behavioural means, including the use of bedding/nesting material in the burrows as well as social huddling (Fig. 28.3b).

28.4 Perspective

The ongoing study of the ability of tegu lizards to adapt to seasonal change is providing new insight into the physiological capacity of ectothermic vertebrates to both suppress and enhance their body temperatures and metabolic rates. This will certainly help to conserve energy during periods of environmental challenge and to promote reproductive capacity during the mating season.

Acknowledgments This research was supported by grants from the Conselho Nacional de Desenvolvimento Científico e Tecnológico (CNPq) to ASA, from the Fundação de Amparo a Pesquisa do Estado de São Paulo (FAPESP) and Fundação para o Desenvolvimento da Unesp (FUNDUNESP) to DVA, and the Natural Sciences and Engineering Research Council of Canada to WKM and GJT. All experiments were performed with approval of the UBC animal care committee (A09-0232).

References

Abe AS (1983) Observations on dormancy in tegu lizard, *Tupinambis teguixin* (Reptilia, Teiidae). Naturalia 8:135–139

Abe AS (1995) Estivation in South American amphibians and reptiles. Brazil J Med Biol Res 28:1241–1247

Andrade DV, Abe AS (1999) Gas exchange and ventilation during dormancy in the tegu lizard, *Tupinambis merianae*. J Exp Biol 202:3677–3685

Barnes BM, Omtzigt C, Daan S (1993) Hibernators periodically arouse in order to sleep. In: Carey C, Florant GL, Wunder BA, Horwitz B (eds) Life in the cold: ecological, physiological, and molecular mechanisms. Westview Press, Boulder, pp 555–558

Butler PJ, Woakes AJ, Bevan RM, Stephenson R (2000) Heart rate and rate of oxygen consumption during flight of the barnacle goose, *Branta leucopsis*. Comp Biochem Physiol A 126:379–385

Butler PJ, Frappell PB, Wang T, Wikelski M (2002) The relationship between heart rate and rate of oxygen consumption in Galapagos marine iguanas (*Amblyrhynchus cristatus*) at two different temperatures. J Exp Biol 205:1917–1924

Carey HV, Andrews MT, Martin SL (2003) Mammalian hibernation: cellular and molecular responses to depressed metabolism and low temperature. Physiol Rev 83:1153–1181

Clark TD, Butler PJ, Frappell PB (2004) Digestive state influences the heart rate hysteresis and rates of heat exchange in the varanid lizard *Varanus rosenbergi*. J Exp Biol 208:2269–2276

Geiser F (2003) Hibernation and daily torpor in mammals: ecophysiological aspects. Comp Biochem Physiol A 134:S92

Geiser F, Ruf T (1995) Hibernation versus daily torpor in mammals and birds: physiological variables and classification of torpor patterns. Physiol Zool 68:935–966

Heldmaier G, Ortmann S, Elvert R (2004) Natural hypometabolism during hibernation and daily torpor in mammals. Respir Physiol Neurobiol 141:317–329

Hueck K (1972) As Florestas da América do Sul. Editora Universidade de Brasília, Polígono

Köhler G, Langerwerf B (2000) Tejus: Lebensweise, Plege, Zucht. Herpeton, Offenbach

Lopes HR, Abe AS (1999) Biologia reprodutiva e comportamento do teiú, *Tupinambis merianae*, em cativero (Reptilia, Teiidae). In: Fang TG, Montenegro OL, Bodmer RE (eds) Manejo y Conservación de Fauna Silvestre en América Latina. Editorial Instituto e Ecología, La Paz, pp 259–272

McNab BK, Auffenberg W (1976) The effect of large body size on the temperature regulation of the Komodo dragon, *Varanus komodoensis*. Comp Biochem Physiol 55A:345–350

Milsom WK, Andrade DV, Brito SP, Toledo LF, Wang T, Abe AS (2008) Seasonal changes in daily metabolic patterns of tegu lizards (*Tupinambis merianae*) placed in the cold (17°C) and dark. Physiol Biochem Zool 81:165–175

Milsom WK, Leite C, Sanders C, Abe AS, Andrade DV, Tattersall GJ (2011) Heterothermic behavior of the black and white tegu lizard during the reproductive season (in preparation)

Piercy J, Rogers K, Andrade D, Abe AS, Tattersall G, Milsom WK (2010) The relationship between body temperature, heart rate, breathing rate, and the rate of oxygen consumption in the tegu lizard (*Tupinambis merianae*) at various levels of activity. J Comp Physiol B (in revision)

Rismiller PD, Heldmaier G (1991) Seasonal changes in daily metabolic patterns of *Lacerta viridis*. J Comp Physiol B 161:482–488

Rismiller PD, McKelvey MW (2000) Spontaneous arousal in reptiles? Body temperature ecology of Rosenberg's Goanna, *Varanus rosenbergi*. In: Heldmaier G, Klingenspor M (eds) Life in the cold. Springer, New York, pp 57–64

Sanders CE, Tattersall GJ, Andrade DV, Abe AS, Milsom WK (2011) Daily and annual cycles in thermoregulatory behavior and cardio-respiratory physiology of black and white tegu lizards. J Comp Physiol B (in revision)

de Souza SCR, de Carvalho JE, Abe AS, Bicudo JEPW, Bianconcini MSC (2004) Seasonal metabolic depression, substrate utilisation and changes in scaling patterns during the first year cycle of tegu lizards (*Tupinambis merianae*). J Exp Biol 207:307–318

Twente JW, Twente JA (1967) Seasonal variation in the hibernating behaviour of *Citellus lateralis*. In: Fisher KC, Dawe AR, Lyman CP, Schöbaum E, South FE Jr (eds) Mammalian Hibernation III. Elsevier Inc, New York, pp 47–63

van Breukelen F, Martin SL (2002) Reversible depression of transcription during hibernation. J Comp Physiol B 172:355–361

Wang LCH (1993) Neurochemical regulation of arousal from hibernation. In: Carey C, Florant GL, Wunder BA, Horwitz B (eds) Life in the cold: ecological, physiological, and molecular mechanisms. Westview Press, Boulder, pp 559–561

Willis JS (1982) The mystery of the periodic arousal. In: Lyman CP, Willis JS, Malan A, Wang LCH (eds) Hibernation and torpor in mammals and birds. Academic Press, New York, pp 92–103

Wilz M, Heldmaier G (2000) Comparison of hibernation, estivation and daily torpor in the edible dormouse, *Glis glis*. J Comp Physiol B 170:511–521

Part III
Molecular Mechanisms and Metabolic Control

Chapter 29
Phylogenetic Background of Hibernation and Hibernation-Specific Proteins in Sciuridae

Tsuneo Sekijima, Hiroko Ishiniwa and Noriaki Kondo

Abstract In mammals, hibernation is expressed in only a limited number of species, and the molecular mechanisms underlying hibernation are not well understood. We have previously demonstrated a decrease in hibernation-specific protein (HP) in the blood of chipmunks during hibernation. Recently, we revealed that HP was regulated by an endogenous circannual rhythm and conveys a signal to the brain essential for hibernation control. In most mammalian hibernators other than some squirrel species, however, the relationship between HP and hibernation remains unknown. Here, we describe the phylogenetic relationships between HP and hibernation in Sciuridae. The expression of HP gene was observed only in hibernating species. Non-hibernating species did not express HP gene even though they possessed the HP-20, -25, and -27 genes. The hibernation-associated HP cycle found in the chipmunk also occurred in other hibernating species. The phylogenetic tree in Sciuridae demonstrated that hibernating species expressing HP like ground squirrel and chipmunk, which belong to Marmotini, diverged recently from non-hibernating species not expressing HP, such as Funambulini and Protoxerini. Thus, HP is closely related to the evolution of hibernation in Sciuridae.

T. Sekijima (✉) · H. Ishiniwa
Graduate School of Science and Technology, Niigata University, 2-8050 Ikarashi, Nishi-ku, Niigata 950-2181, Japan
e-mail: sekijima@gs.niigata-u.ac.jp

N. Kondo
Tamagawa University Research Institute, 6-1-1 Tamagawagakuen, Machida-shi, Tokyo 194-8610, Japan

T. Ruf et al. (eds.), *Living in a Seasonal World*,
DOI: 10.1007/978-3-642-28678-0_29,

29.1 Introduction

Hibernation in mammals is a physiological system in which the body adapts to seasonal environmental change. During hibernation, the body temperature of mammalian hibernators drops to below 10°C and the metabolic rate is reduced to only a few percent of the euthermic level. This remarkable lowering of body temperature, while lethal for humans, never causes cellular or tissue damage in hibernators. Mammalian hibernators maintain thermoregulatory control with a lowered set point during hibernation and normal physiological functions so they do not suffer damage due to hypothermia. The mechanism of this outstanding physiological phenomenon, however, is still not completely or properly understood.

Hibernation-specific protein (HP) found in the blood of the chipmunk *Tamias sibiricus* may provide important clues for elucidating the mechanism of mammalian hibernation. HP is a protein complex composed of four subunits, HP-20, -25, -27, and -55, and is expressed in the liver (Kondo and Kondo 1992; Takamatsu et al. 1993). The content of HP in the blood is regulated by a free-running circannual rhythm and decreases during hibernation (Kondo and Joh 1996). On the contrary, HP content in the cerebrospinal fluid (CSF) is increased during hibernation, though the content is remarkably lower in the CSF than in the blood. Recently, we tried to block CSF-HP activity using an antibody and succeeded at decreasing the duration of hibernation, which suggested that HP plays an important role in carrying hormonal signals essential for hibernation to the brain (Kondo et al. 2006).

In 4070 species of mammals, it is known that at least 183 species hibernate including the chipmunk (Kawamichi 2000). In almost all mammalian hibernators other than the chipmunk and some squirrels (Takamatsu et al. 1993), however, the relationship between HP and hibernation remains unclear. Clarifying whether HP is involved in the induction of hibernation in various mammalian species is essential to understand the need for HP in mammalian hibernation and the evolutionary process of hibernation regulated by HP. In this study, we first investigated the phylogenetic background in Sciuridae with respect to the existence of the HP gene and its gene expression, and then monitored quantitative changes in HP content to ascertain, whether hibernating species other than the chipmunk possess HP rhythm that is synchronized with the hibernation cycle.

29.2 Materials and Methods

29.2.1 Animals

Ten species belonging to Sciuridae, i.e., Northern palm squirrel (*Funambulus pennanti*), Smith's bush squirrel (*Paraxerus cepapi*), Siberian flying squirrel (*Pteromys volans*), Eurasian red squirrel (*Sciurus vulgaris*), Columbian ground squirrel (*Spermophilus columbianus*), Mexican ground squirrel (*Spermophilus mexicanus*), Richardson's ground squirrel (*Spermophilus richardsoni*), 13-lined ground squirrel (*Spermophilus tridecemlineatus*), Chipmunk (*Tamias sibiricus*), and Maritime striped squirrel (*Tamiops maritimus*), were used in this study. Among these species, 5 hibernating species (Columbian ground squirrel, Mexican ground squirrel, Richardson's ground squirrel, 13-lined ground squirrel, Chipmunk) were housed under constant darkness and cold conditions (24 h of dark, room temperature at 5°C) to inducing hibernation. The remaining 5 non-hibernating species (Northern palm squirrel, Smith's bush squirrel, Siberian flying squirrel, Eurasian red squirrel, and Maritime stripped squirrel) were also housed under the above-mentioned condition throughout a year for collecting plasma and tissue samples, but thereafter kept under constant warm laboratory condition (12 h photoperiod, room temperature at 23°C) to remove the physical stress by low room temperature. All animals were caged individually and were fed Rat Lab Chow (Charles River Laboratories) and water ad libitum.

29.2.2 Collection of Plasma

To confirm the existence of HP and to monitor the changes in HP content, the blood of all 10 squirrels species was collected from the femoral vein every month. For each blood sample, plasma was separated by centrifugation and then stored at −80°C until use.

29.2.3 Western Blotting and Quantitative Analysis of HP

The proteins in plasma were separated by SDS-PAGE (12.5% acrylamide gel) and transferred onto PVDF membrane using an electro-blotting apparatus. Rabbit antisera for HP-20, -25, -27, and -55 were used as primary antibodies. Detection was performed with an ECL Advance color reagent kit (Amersham Bioscience). Image was generated by scanning a photograph of the developed membrane with Adobe Photoshop software. The densities of signals for HP-20, -25, -27, and -55 were measured with Scion image software (http://www.scioncorp.com/frames/fr_download_now.htm). The contents of HP in all samples were normalized as the percentage of HP content in plasma to the maximum of a certain control individual.

29.2.4 Southern Blot Analysis

High-molecular weight chromosomal DNA was prepared from livers of the 10 squirrels species using a High Pure™ PCR Template Preparation Kit (Roche Diagnostics). A total of 10 μg of DNA was digested with *Eco*RI, subjected to 0.7% agarose gel electrophoresis, transferred to a nylon membrane (Roche), and then fixed by UV Crosslinker FS-800 (Funakoshi). The membrane was hybridized in 50% formamide, 5 × SSC (standard saline citrate), 1% SDS, and 1% blocking reagent with ^{32}P-random-prime-labeled chipmunk HP-20, -25, and -27 cDNA fragments at 42°C for 16 h. The membrane was washed with 2 × SSC-0.1% SDS at 37°C for 10 min, 0.1 × SSC-0.1% SDS at 37°C for 10 min, and 0.1 × SSC-0.1%SDS at 65°C for 5 min. The signals were detected using an Imaging Plate BAS-SR 2025 (FUJI PHOTO FILM), and measured by Bioimaging Analyzer BAS-5000 (FUJI PHOTO FILM).

29.2.5 Real-Time RT-PCR Analysis

Total RNA from livers of the 10 squirrels species was isolated using Sepasol-RNA I (Nacalai Tesque). Single-stranded cDNA was synthesized from the total RNA (1 μg) using a Transcriptor High Fidelity cDNA Synthesis Kit (Roche Diagnostics) with oligo $(dT)_{18}$ primer. The expression levels of HP-20, -25, -27, and β-actin mRNA were obtained by quantitative real-time PCR using Light Cycler R2.0 (Roche Diagnostics). Specific primers for these genes were as follows, HP20-f: 5′-ATGAC CGACG TGTGG AGA-3′; HP20-r: 5′-GCAGG GGCAT TTTAC ACT-3′; HP25-f: 5′-GTCTG AATGC CTGCA CAAAG AG-3′; HP25-r: 5′-GATGG ACAGG ACCAA AATCC A-3′; HP27-f: 5′-TGTAA TGTTC CAG-GA CCTCA-3′; HP27-r: 5′-CCGAT GTTCC TTTGC TGT-3′; β-actin-f: 5′-ACTGG GACGA CATGG AGAAA A-3′; β-actin-r: 5′-CCACA CGCAG CTCGT TGTA-3′. The real-time PCR amplifications of these genes were performed for 45 cycles of denaturing (95°C, 10 s), annealing (60°C, 10 s), and extension (72°C, 10 s) with a single fluorescence measurement following initial denaturation (95°C, 10 min). After amplification and quantification cycles, melting curve analysis (65–97°C with a heating rate 0.1°C/s and a continuous fluorescence measurement) was carried out to validate whether specific PCR products were generated. Negative control samples were processed in the same manner except for the template. A standard curve was constructed by plotting the cross point (Cp) against the dilution series of a particular sample. Here, the Cp represents the cycle number at which the fluorescence signal is greater than a defined threshold, one in which all the reactions are in the logarithmic phase of amplification.

29.2.6 DNA Sequence Determination

Total genomic DNA was extracted from muscle or tail of the 10 squirrel species using a High Pure PCR Template Preparation Kit (Roche Diagnostics). As the genetic marker for constructing phylogenetic tree, we chose the two genes, proto-oncogene c-myc and recombination activating gene 1 (RAG1), because the phylogenetic trees by these two genes had stronger Bayesian and bootstrapping support than 12S/16S/IRBP tree (Steppan et al. 2004). The sequenced genes were 1953 bp of c-myc including exon 2 and a part of intron 2, and 2199 bp of RAG1 (Steppan et al. 2004). The amplification products of the two genes were obtained by the method described in Steppan et al. (2004). Amplified samples were purified using a QIAquick Gel Extraction Kit (QIAGEN) and sequenced directly on an ABI 3100 automated sequencer, with Big-Dye Terminator chemistry (ABI). Chromatograms were edited and aligned by CLUSTAL W (Thompson et al. 1994) and alignments were then modified by eye.

29.2.7 Construction of Phylogenetic Tree

Sequence information of the species other than the squirrel species that we sequenced was acquired from GenBank Accession Nos. AY239477-AY239500, AY241468-AY241500, and AY241510-AY241544. *Aplodontia* was placed as an outgroup to Sciuridae. Phylogenetic analyses were conducted under equally weighted maximum-parsimony (MP) and maximum-likelihood (ML) criteria with PAUP* version 4.0b10 (Swofford 2002) for total gene (c-myc and RAG1). The initial tree was generated by MP. ML parameter values were estimated under a nested array of substitution models for the MP trees with Modeltest 3.04 (Posada and Crandall 1998). The best fit model for the gene regions was identified as GTR + I + Γ by both the likelihood ratio test and Akaike Information Criterion. Bootstrapping and decay indices were performed on all data partitions. Bayesian analyses were conducted on the same data with MrBayes (Huelsenbeck and Ronquist 2002) and used GTR + I + Γ model as in the ML analyses. The detailed procedure used was that described by Steppan et al. (2004).

29.3 Results and Discussion

29.3.1 Expression of HP Gene in Sciuridae

Western blot analyses revealed that the specific signals of HP-20, -25, -27, and -55 were recognized in hibernating species only, i.e., Columbian ground squirrel, Mexican ground squirrel, Richardson's ground squirrel, 13-lined ground squirrel, and Chipmunk, while there was no HP signal in any non-hibernating species (Fig. 29.1). To reinforce these results, the expressions of HP mRNA were analyzed by quantitative real-time PCR. Similar to the result for Western blot analysis, real-time PCR also showed that only hibernating species expressed HP-20, -25, and -27 mRNA while non-hibernating species did not (data not shown). To ascertain further whether the non-expression of HP observed in non-hibernating species is due to the absence of HP genes or the lack of HP gene expression despite possessing HP genes, we investigated the presence of the genes

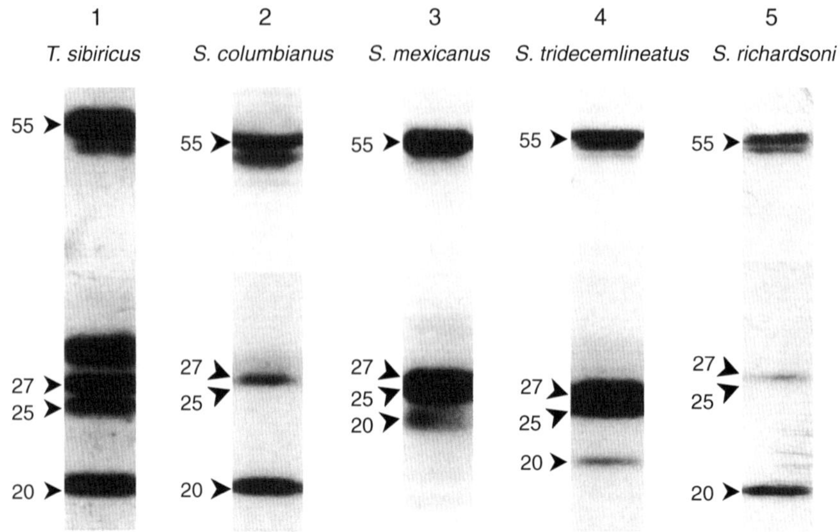

Fig. 29.1 Results of Western blot analysis on hibernating species in Sciuridae. Lanes: *1* Chipmunk (*T. sibiricus*), *2* Columbian ground squirrel (*S. columbianus*), *3* Mexican ground squirrel (*S. mexicanus*), *4* 13-lined ground squirrel (*S. tridecemlineatus*), *5* Richardson's ground squirrel (*S. richardsoni*). *Arrowheads* indicate the specific signals corresponding to chipmunk HP-20, -25, -27, and -55, respectively. The bands of HP-25 and HP-27 overlapped in ground squirrels

corresponding to chipmunk HP-20, -25, and -27 genes in the 10 squirrel species. Southern blot analyses revealed specific hybridization signals with all 10 squirrel species (data not shown). Based on these results, it appeared that the expression of HP was recognized only in hibernating species, and non-hibernating species did not express HP even though they possess HP-20, -25, and -27 genes. Kojima et al. (2001) compared the gene structure and promoter activity of HP-25 gene of a non-hibernating species, the tree squirrel *Callosciurus caniceps*, with that of the chipmunk, and concluded that the tree squirrel HP-25 gene was a pseudogene since it had almost no promoter activity and lost binding to HNF-4 which is involved in transcriptional regulation of the HP-25 gene. In this study, the lack of HP gene expression observed in non-hibernating species may indicate that HP genes are pseudogenes.

Next, we monitored the quantitative changes in HP content in the blood as well as hibernation cycles in all 5 hibernating species including chipmunk. The contents of HP-20, -25, and -27 of the Columbian ground squirrel, Mexican ground squirrel, Richardson's ground squirrel, and 13-lined ground squirrel were increased during arousal and decreased during hibernation, a tendency that was very similar to the change in HP content in the chipmunk (Fig. 29.2). Hibernation-associated HP rhythm occurred in all hibernating species. These results suggest that hibernation control in which HP intervened is not only limited to chipmunk but also prevails in other hibernating species, at least in Sciuridae.

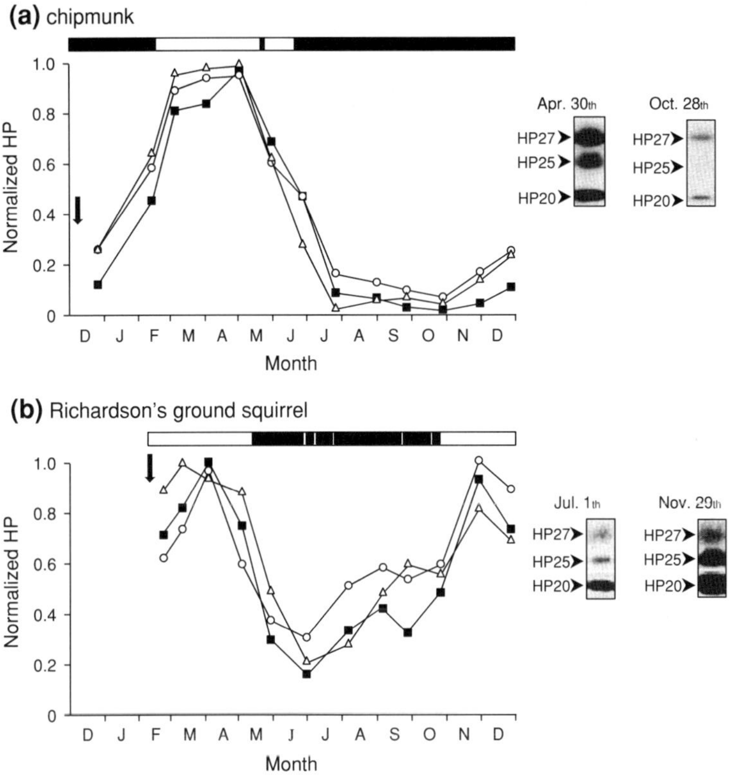

Fig. 29.2 Relationships between blood HP and hibernation period in chipmunk (**a**) and Richardson's ground squirrel (**b**). HP content in plasma collected monthly was determined by Western blot analysis, and then normalized to its maximum content of a control individual. *Open circles*, *filled quadrates*, and *open triangles* show the normalized values of HP-20, -25, and -27, respectively. In the upper column, the period of hibernation is shown as closed areas. *Arrowhead* indicates the date when cold exposure was started. *Inset* shows Western blot analysis of plasma collected during the non-hibernating and hibernating periods, and each HP is shown by an *arrowhead*

29.3.2 Phylogenetic Relationship Between Hibernating and Non-Hibernating Squirrels

Steppan et al. (2004) classified the Sciuridae into 5 major lineages, (A) the giant squirrel *Ratufa*, (B) the pygmy squirrel *Sciurillus*, (C) a clade consisting of the flying squirrels Pteromyinae, New World tree squirrels Sciurini, Microsciurini, and the red squirrel *Tamiasciurus*, (D) the Callosciurini, and (E) a clade consisting of the African ground squirrels Xerini, Holarctic ground squirrels Marmotini, African Protoxerini, and the African members of the Funambulini, based on ML analysis.

Here, a phylogenetic tree of Sciuridae was reconstructed by adding our squirrel sequence data to the data of Steppan et al. (2004) (Fig. 29.3). The squirrel species

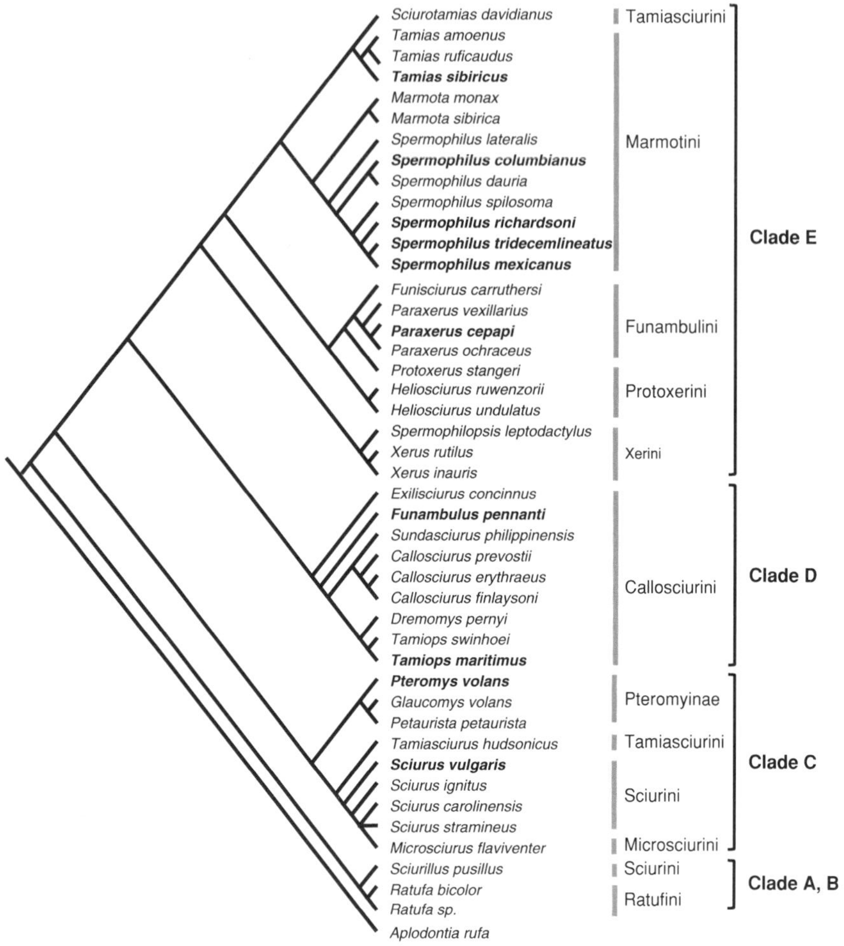

Fig. 29.3 Phylogenetic tree reconstructed by adding our squirrel sequence data to the data of Steppan et al. (2004). Species that we used are shown in *bold*

that we used covered clade C to clade E; the Eurasian red squirrel and Siberian flying squirrel in clade C, Maritime striped squirrel and Northern palm squirrel in clade D, Smith's bush squirrel, Mexican ground squirrel, 13-lined ground squirrel, Richardson's ground squirrel, Columbian ground squirrel, and Chipmunk in clade E. All of the hibernators in Sciuridae belonged to clade E (Fig. 29.3). The phylogenetic tree also demonstrated that hibernating species like ground squirrel and chipmunk, which belonging to Marmotini, diverged recently following emergence of non-hibernating species like Funambulini and Protoxerini. Considering that the HP-25 gene in the tree squirrel, one of the non-hibernating species, was a pseudogene (Kojima et al. 2001), it is likely that the common ancestor of both hibernating and non-hibernating species, i.e., progenitor of Sciuridae before emergence of non-hibernating species, was a hibernator, and that the expression and function of HP were lost in the process during evolvement from the common ancestor to the non-hibernating species.

Acknowledgments We are grateful to Norikazu Hara, Hiroki Nagase, Sayuri Maruyama, Haruka Maruyama, Taeko Ishima for their excellent technical assistance and useful suggestions throughout this study.

References

Huelsenbeck JP, Ronquist F (2002) MrBayes: a program for the Bayesian inference of phylogeny, 2.01. Rochester, NY

Kawamichi T (2000) Ecology for hibernation. In: Kawamichi T, Kondo N, Morita T (eds) Hibernating mammal. Tokyo University Press, Tokyo (In Japanese)

Kondo N, Kondo J (1992) Identification of novel blood proteins specific for mammalian hibernation. J Biol Chem 267:473–478

Kondo N, Joh T (1996) Endogenous circannual rhythm of hibernation-specific proteins (HP) associated with hibernation in chipmunks. In: Geiser F, Hulbert AJ, Nicol SC (eds) Adaptation to the cold. University of New England Press, Armidale

Kondo N, Sekijima T, Kondo N, Takamatsu N, Tohya K, Ohtsu T (2006) Circannual control of hibernation by HP complex in the brain. Cell 125:161–172

Kojima M, Shiba T, Kondo N, Takamatsu N (2001) The tree squirrel HP-25 gene is a pseudogene. Eur J Biochem 268:5997–6002

Posada D, Crandall KA (1998) Modeltest: testing the model of DNA substitution. Bioinformatics 14:817–818

Steppan SJ, Storz BL, Hoffmann RS (2004) Nuclear DNA phylogeny of the squirrels (Mammalia: Rodentia) and the evolution of arboreality from c-myc and RAG1. Mol Phylogenet Evol 30:703–719

Swofford DL (2002) PAUP*. Phylogenetic analysis using parsimony (*and other methods). Version 4. Sinauer Associates, Sunderland

Takamatsu N, Ohba K, Kondo J, Kondo N, Shiba T (1993) Hibernation associated gene regulation of plasma proteins with a collagen-like domain in mammalian hibernatiors. Mol Cell Biol 13:1516–1521

Thompson JD, Higgins DG, Gibson TJ (1994) CLUSTAL W: improving the sensitivity of progressive multiple sequence alignment through sequence weighting position specific gap penalties and weight matrix choice. Nucleic Acids Res 22:4673–4680

Chapter 30
Adenosine, AMP, and Daily Torpor

Steven J. Swoap, Benjamin Iliff and Son Le

Abstract When calorically restricted at cool ambient temperatures, mice conserve energy by entering torpor, during which metabolic rate (MR), core body temperature (T_b), heart rate (HR), and locomotor activity (LMA) decrease. Others have suggested that injection of adenosine monophosphate (AMP) mimics the torpor pathway. We have postulated that AMP is dephosphorylated to adenosine which drastically slows HR, and the reduced oxygen delivery forces metabolism to fall with an accompanying decrease in T_b. We would also predict an increase in anaerobic metabolism as oxygen delivery becomes limiting. To test this hypothesis, we measured circulating lactate, a marker for anaerobic metabolism, in C57Bl/6 mice, 30 min after treatment with AMP, ADP, adenosine, methacholine, and during fasting-induced torpor. HR slowed drastically with the treatments, from >600 bpm to <200 bpm within 1 min. Lactate levels rose from control values of 1.8 mM to >5 mM when treated with AMP or ADP, while lactate levels dropped during natural torpor (0.8 mM). Further, we show that adenosine signaling is necessary for the initiation and maintenance of torpor in fasted mice. Mice were subcutaneously (s.c.) infused with either saline or aminophylline, an adenosine receptor blocker, during fasting. Over the course of a 23 h fast, aminophylline significantly blunted the length of time in torpor fivefold. Also, when aminophylline was infused once torpor had begun, it reversed the hypometabolism, hypothermia, bradycardia, and hypoactivity of torpor, while saline did not. Finally, intracerebroventricular infusion of a polar adenosine receptor antagonist, which does not cross the blood–brain barrier, during fasting-induced torpor also induced emergence from torpor. These findings support the hypothesis that adenosine is necessary for the induction and maintenance of torpor in mice.

S. J. Swoap (✉) · B. Iliff · S. Le
Department of Biology, Williams College, Williamstown, MA 01267, USA
e-mail: sswoap@williams.edu

T. Ruf et al. (eds.), *Living in a Seasonal World*,
DOI: 10.1007/978-3-642-28678-0_30,

30.1 Introduction

When exposed to a cool ambient temperature during a period of restricted caloric intake, both wild and laboratory strains of the laboratory mouse *Mus musculus* enter bouts of daily torpor (Morton 1978; Hudson et al. 1979). During a torpor bout, MR decreases to a level below the basal MR, and T_b falls well below 37°C (Hudson et al. 1979; Williams et al. 2002; Swoap et al. 2009). Likewise, HR and blood pressure drop substantially and LMA drops to near zero (Williams et al. 2002; Swoap et al. 2009). In laboratory mice, torpor bouts are approximately 6 h in length, separated by short inter-bout arousal periods during which MR, T_b, HR, BP, and LMA return to levels typical during euthermia (Hudson et al. 1979; Geiser 2004; Swoap et al. 2009). As daily torpor invokes comprehensive and drastic physiological and biochemical changes, the state can be difficult to define. Daily torpor has been arbitrarily defined as a T_b below 30°C (Barclay et al. 2001). However, we have taken less of an arbitrary approach with a definition of daily torpor as $T_b < 34$°C based on an equation from a recent publication (Willis 2007). While this could be considered a shallow bout of hypothermia and hypometabolism, we will use this as a working definition of torpor in this study.

Recent research has shown that peripheral injection of the nucleotide AMP causes a bout of hypothermia that is similar in depth to torpor (Zhang et al. 2006; Swoap et al. 2007; Zhang et al. 2009), raising the question of whether AMP might be activating a natural pathway to torpor. It is possible that AMP induces hypothermia through its action on AMP-activated kinase (AMPK), or its allosteric activation/inhibition of metabolic enzymes like phosphorylase b, or some other unknown mechanism. By blunting the hypothermic effect of AMP using an adenosine receptor blocker, we hypothesized that AMP is first dephosphorylated to adenosine, and it is adenosine that induces the physiological/pathophysiological effects (Swoap et al. 2007). Indeed we and others have shown that adenosine injections do induce hypothermia in mice (Anderson et al. 1994; Swoap et al. 2007). While others have shown that AMP can have its effect in mice missing one of the three adenosine receptors (Daniels et al. 2010), the evidence suggesting that adenosine plays a key role in torpor regulation is compelling.

Administration of an adenosine receptor antagonist reverses the spontaneous entrance into torpor of arctic ground squirrels (Jinka et al. 2011). Also, administration of an adenosine receptor agonist could induce torpor in arctic ground squirrels, but only during the hibernation season (Jinka et al. 2011). Further supporting a role for neuronal adenosine signaling in the regulation of torpor is the finding that intracerebroventricular injection of an adenosine receptor antagonist during the initiation phase of hibernation causes a return to euthermia in Syrian hamsters (Tamura et al. 2005). Besides hypothermia, the other effects of adenosine are also consistent with torpor. Adenosine acts as a signal of metabolic stress, acting on the inhibitory A1 adenosine receptor during hypoxia to reduce MR (Blood et al. 2003). Additionally, adenosine's effects mimic the cardiovascular changes of torpor (Yang et al. 2009).

In addition to its torpor-mimicking metabolic and cardiovascular effects, adenosine acts on neural pathways that are involved in the regulation of torpor. Torpor is thought to be entered through non-rapid eye movement (NREM) sleep, while adenosine promotes sleep through A1-mediated inhibition of the cholinergic basal forebrain (Berger 1984; Deboer et al. 1995; Deboer et al. 1996; Alam et al. 1999; Heller et al. 2004). A1- and A2-specific adenosine receptor agonists also inhibit GABA release from cultured arcuate nucleus neurons; the presence of a functioning arcuate nucleus is necessary for torpor in mice, suggesting that adenosine may have a role in regulating torpor through modulation of neuropeptide Y (Chen et al. 1997; Paul et al. 2005; Gluck et al. 2006; Dark et al. 2008).

Because the effects of adenosine receptor activation mimic torpor and adenosine receptor blockade interrupts hibernation in Syrian hamsters and arctic ground squirrels, we sought to determine whether: (1) AMP injections induce a pathophysiological state due to the bradycardia associated with its injection; and (2) adenosine receptor signaling is necessary for daily torpor in mice.

30.2 Materials and Methods

30.2.1 Animals

C57BL/6 mice were either bred locally for peripheral drug infusion/injection experiments or purchased from The Jackson Laboratory (Bar Harbor, ME) with a cannula pre-implanted in the lateral ventricle for central drug infusion experiments. Mice were individually housed at 22°C with 12 h light/dark phases. Experiments were run between the beginning of the dark phase and 1 h before the start of the next dark phase, with the last hour of the light phase reserved for animal care. Mice were fed ad libitum except in cases where noted. All procedures were approved by the Williams College Animal Care and Use Committee.

30.2.2 Radiotelemeter and Cannula Implantation

Each mouse was implanted with a telemeter capable of measuring T_b, HR, and LMA (TA10ETA-F20; Data Sciences International, St. Paul, MN). Mice were anesthetized with 5% isoflurane in oxygen, followed by maintenance with 2.5% isoflurane. The telemeter was implanted in the abdominal cavity and EKG leads were placed subcutaneously (s.c.), approximating a Lead II configuration. A custom-made 25 gauge stainless steel subcutaneous cannula was implanted intrascapularly in mice used for experiments requiring peripheral infusions. Meloxicam (5 mg/kg every 24 h, s.c.) was administered for perioperative and postoperative analgesia. A heating pad was used as an external heating source during surgery, and each cage was placed on a heating pad for 48 h after surgery. Mice were allowed to recover for at least 10 days before experimental testing.

30.2.3 Drug Treatments

All drugs were purchased from Sigma-Aldrich Corporation. Drugs given intraperitoneally (i.p.) or s.c. (under the dorsal skin) were administered in sterile 0.9% NaCl solution, while drugs given intracerebroventricularly (i.c.v.) were administered in sterile phosphate-buffered saline (PBS). The efficacy of adenosine antagonist treatment was confirmed by evaluating the effect of each treatment on adenosine-induced hypothermia. Drug infusion through s.c. or i.c.v. cannulae was controlled using a calibrated syringe pump (PHD-22; Harvard Apparatus, Holliston, MA), connected via polyethylene tubing to a cagetop-mounted swivel (375/25PS; Instech Laboratories, Inc., Plymouth Meeting, PA).

30.2.4 Experiments

All experiments were conducted according to a randomized repeated measure, crossover design. In the first experiment, eight female mice were injected i.p. with either AMP (800 mg/kg), ADP (800 mg/kg), adenosine (100 mg/kg), methacholine (8 mg/kg), saline, or fasted to induce torpor. Mice were injected 10.5 h into the dark phase, which is a typical time for entry into torpor. Thirty minutes after the injection, or 30 min into the bout of torpor, tail vein blood was collected ($\sim$20 μl) for measurement of blood lactate and blood glucose. Blood lactate was measured with a lactate meter (www.lactate.com) that was calibrated with five standards ranging from 0.5 to 5 mM. Blood glucose was measured with a glucose meter, purchased from a local drugstore, and calibrated at five standards ranging from 50 to 300 mg/dl. The standard curves for glucose and lactate had an r^2 value of 0.95 and 0.98, respectively. The standard curve data points were measured in duplicate, while the samples were measured once. Mice were allowed to recover for 7 days between treatments. In the second experiment, mice ($n = 7$) were either fasted at the onset of the dark phase or fed ad libitum and continuously infused s.c. with aminophylline (6 mg/kg-hr) or saline. In the third experiment, mice were fasted at the onset of the dark phase. Once mice had reached a torpid state ($n = 4$), they were s.c. infused with aminophylline (100 mg/kg) or saline (100 μl over the course of 10 min) to evaluate whether adenosine signaling is necessary for the maintenance of torpor. Treatment with either aminophylline or saline was randomized, and 7 days were allowed for recovery between the fasts. The fourth and fifth experiments used the polar adenosine receptor antagonist 8-sulfophenyltheophylline (8-SPT), which does not cross the blood–brain barrier (Ngai et al. 1993;

Pekny et al. 1998). Mice were fasted at the onset of the dark phase. Once mice had reached a torpid state (experiment 4, n = 6; experiment 5, n = 5), they were infused with either saline or 8-SPT, either s.c. (50 mg/kg in 100 μl over 25 min—experiment 4) or i.c.v. (10 μg in 5 μl over 5 min—experiment 5). Mice were allowed at least 7 days before treatment with the other infusate.

30.2.5 Indirect Calorimetry

MR was measured by indirect calorimetry during experiment 3. Mice were housed in clear, air-tight polycarbonate cages. Room air was dried with calcium sulfate and pumped into the cage at a regulated flow rate of 350 ml/min. Outflow from the cage was pulled through an O_2 analyzer at a rate of 140 ml/min (S-3A/II; AEI Technologies, Naperville, IL). Up to three cages were serially sampled, with an empty cage serving as a reference. Ambient pressure and temperature were concurrently measured electronically (APR-1 and C10T, Data Sciences International). Data from gas sensors were digitized and recorded using LabVIEW 8 (National Instruments, Austin, TX), while pressure and temperature data were recorded using Dataquest A.R.T. 3.01 acquisition and analysis software (Data Sciences International).

30.2.6 Data Analysis

All results are reported at means ± SE. Statistical analyses were performed in SPSS 15.0 (SPSS Inc., Chicago, IL). In experiment 1, repeated-measures ANOVAs, followed by post hoc Newman–Keuls tests, were used to determine statistical significance in blood lactate and glucose levels. In experiment 2, a repeated-measures ANOVA was performed, with fed/fasted condition and aminophylline/saline treatment condition as within-subject variables. This test was followed by a planned comparison to evaluate the difference between aminophylline and saline treatment in fasted mice, using a paired Student's t-test. For all other experiments, paired t-tests were performed between control and experimental conditions. The α level for significance was set at 0.05.

30.3 Results and Discussion

30.3.1 Experiment 1: Blood Lactate and Glucose Measurements

We have previously shown that injection of AMP drastically slows HR (Swoap et al. 2007). We hypothesize that the decrease in blood/oxygen delivery associated with this injection induces the hypometabolism and would thus induce peripheral hypoxia, where oxygen demand outstrips oxygen supply. As a result, these mice might rely on anaerobic metabolism to fuel cellular activities. To test this hypothesis, we measured circulating lactate and glucose after injection of hypothermia-inducing compounds or during fasting-induced torpor (Fig. 30.1b). In fasted mice, blood was sampled 30 min after the onset of the steep decline in core T_b. In these mice, circulating lactate fell from control values of ~1.7 to 0.8 mM (see Fig. 30.1b), suggesting that mice entering torpor are not in a hypoxic situation. Circulating lactate was elevated to 5.3 mM 30 min after the injection of AMP (Fig. 30.1b). ADP and adenosine also elevated circulating lactate. Methacholine, a muscarinic acetylcholine (Ach) receptor agonist, was tested to determine whether simply slowing HR could induce hypothermia. Mice injected with methacholine did become hypothermic (~27°C) in response to a slowed HR (~190 bpm) and also began to rely on anaerobic metabolism as lactate levels were also elevated (Fig. 30.1b). As reported previously (Zhang et al. 2009), AMP also induced hyperglycemia as did ADP and adenosine (Fig. 30.1c). However, torpid mice did not display elevated circulating glucose (Fig. 30.1c). It should be noted, though, that measurement of glucose 11 h into the dark phase can be problematic because mice eat typically during the dark phase, which might influence circulating blood sugar. However, the repeated-measures, crossover design effectively removes any bias due to differences between individual animals. To the extent that food intake or other variables could alter results, one would expect those effects to be randomly distributed between the groups, increasing the noise in the results but not affecting the overall conclusion. These data support our hypothesis (Swoap 2008) that the effect of AMP is through deprivation of oxygen at peripheral tissues, with ensuing anaerobic metabolism. These data also suggest that the fasting-induced state of torpor is vastly different from the hypothermia that is induced with injection of AMP.

30.3.2 Experiment 2: Continuous Subcutaneous Aminophylline Infusion

Mice were either fed or fasted for 23 h while being infused s.c. with either saline or aminophylline. The dosage used was sufficient to blunt adenosine-induced hypothermia (data not shown). Figure 30.2a shows a typical T_b tracing of the same mouse in each of the four conditions. In the fasted condition, aminophylline treatment significantly decreased time in torpor relative to saline treatment (96 ± 45 vs. 543 ± 43 min, respectively—see Fig. 30.2b), minimum T_b (33.2 ± 0.4 vs. 30.9 ± 0.3°C), and minimum HR (311 ± 41 vs. 240 ± 26 bpm, respectively). These data all suggest that subcutaneous infusion of the non-specific adenosine receptor antagonist, aminophylline, can prevent a mouse from entering a bout of daily torpor.

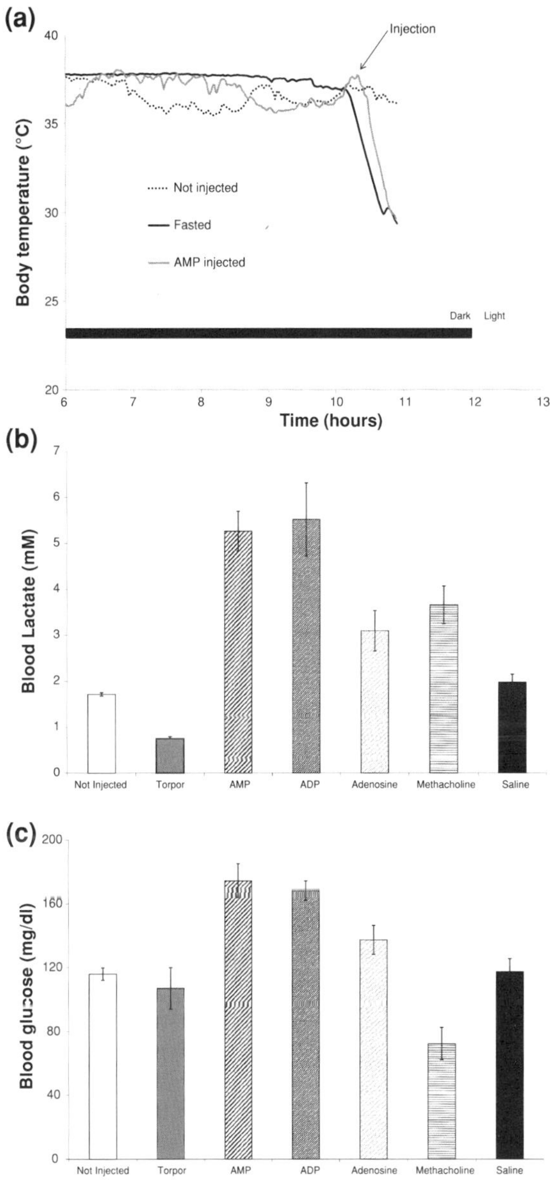

Fig. 30.1 **a** Typical tracings of core T_b of the same mouse under three different conditions (not injected, fasted, and peripherally injected with AMP at 800 mg/kg). In the fasting case, the fast was initiated at the beginning of the dark phase ($T = 0$ h). In each trial, the mouse was removed from the telemetry receiver for blood sampling (20 μl per trial) 30 min after the injection or after the visible steep decline in core T_b. The same mouse was also injected with saline, ADP, adenosine, and methacholine, but only three tracings are shown for clarity. Mice were allowed to recover for 7 days between each trial. Circulating lactate levels (**b**) and glucose (**c**) were measured in eight mice in each condition (30 min after an injection of one of five compounds, or 30 min into the start of a torpor bout). Lactate was significantly elevated in mice that received AMP, ADP, Adenosine, or methacholine (a muscarinic AchR antagonist). Torpid mice had reduced levels of lactate. AMP, ADP, and adenosine all had a hyperglycemic effect in mice

30.3.3 Experiment 3: Acute Subcutaneous Aminophylline Infusion

The previous experiment showed that we could prevent mice from entering torpor after a fast. We next wanted to test whether mice in a bout of torpor require adenosine receptor activation for the maintenance of torpor. In this experiment,

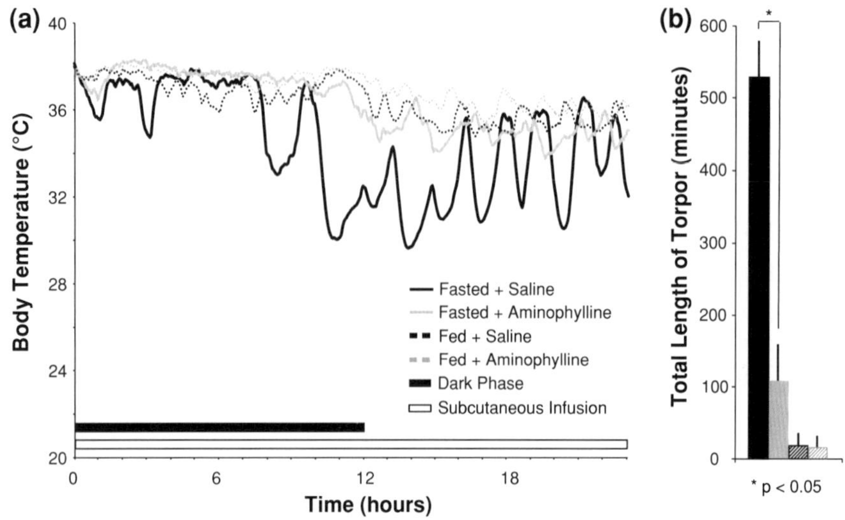

Fig. 30.2 **a** Typical T_b tracings from a mouse in each of four conditions (fed and infused with saline, fasted and infused with saline, fed and infused with aminophylline, fasted and infused with aminophylline). Mice were fasted at the onset of the dark phase, which is when the subcutaneous infusion began. **b** The length of time spent under 34°C. Subcutaneous administration of aminophylline significantly blunted the length of time spent in torpor

mice were fasted at the onset of the dark phase. During the first torpor bout which was $\sim$10 h after initiation of the fast, saline or aminophylline was infused through a previously implanted s.c. cannula. The MR at the time of infusion was $\sim$0.4 ml O_2/min, the $T_b \sim 28$°C, and HR was $\sim$200 bpm (see Fig. 30.3). Infusion of saline had no effect on any of the parameters measured. However, aminophylline caused significant increases in MR (to 1.4 ml O_2/min), T_b (36°C), HR (650 bpm) and LMA (Fig. 30.3), and it appears that these mice experience full arousal from torpor. These data suggest that adenosine signaling is required for maintenance of daily torpor in mice.

30.3.4 Experiment 4: Acute Subcutaneous 8-SPT Infusion

After finding that treatment with the adenosine receptor antagonist aminophylline was sufficient to both prevent the induction and maintenance of torpor bouts, further experiments were executed to determine the location and mechanism of action of adenosine receptor antagonism. We used the polar non-selective antagonist 8-sulfophenyltheophylline (8-SPT) as this does not cross the blood–brain barrier. Mice were fasted and infused s.c. with saline or 8-SPT during the first torpor bout of the light phase. As with aminophylline, infusion of 8-SPT prevented

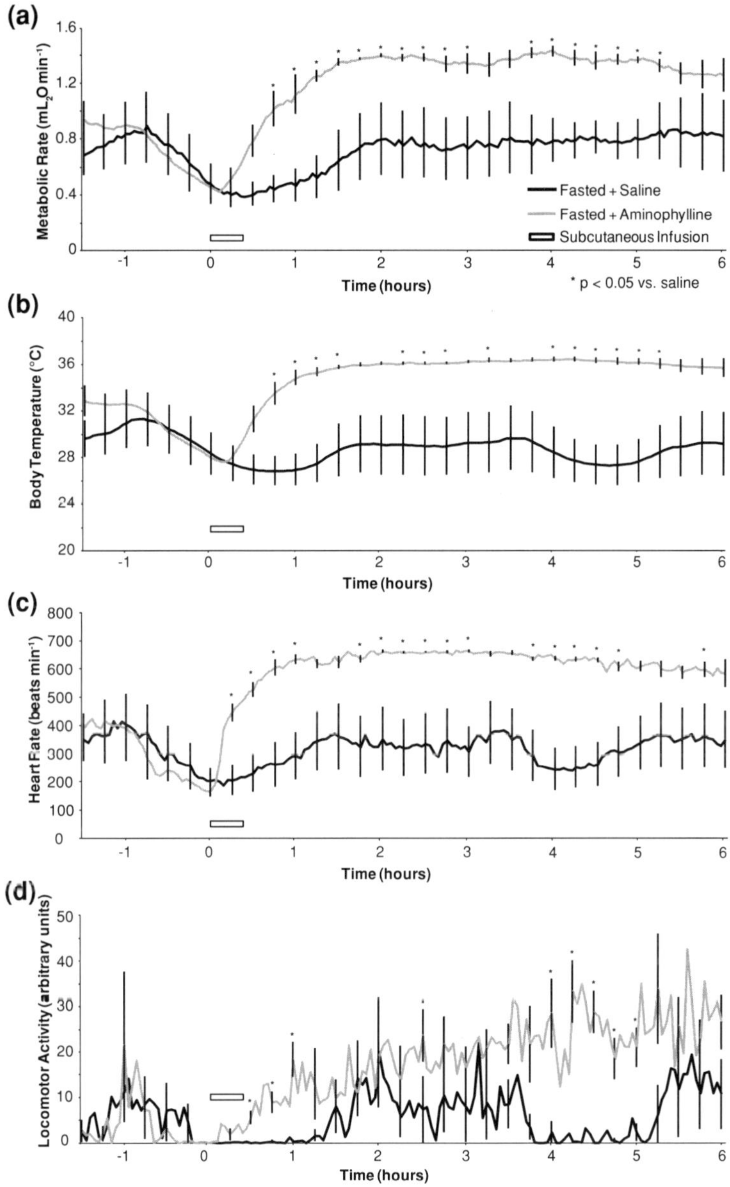

Fig. 30.3 Mice were fasted at the beginning of the dark phase, and allowed to enter into torpor. As T_b approached 28°C ($\sim$10 h after initiation of the fast), mice were infused s.c. with either saline or aminophylline. As assessed by MR, T_b, HR, and LMA, aminophylline caused an immediate arousal from torpor

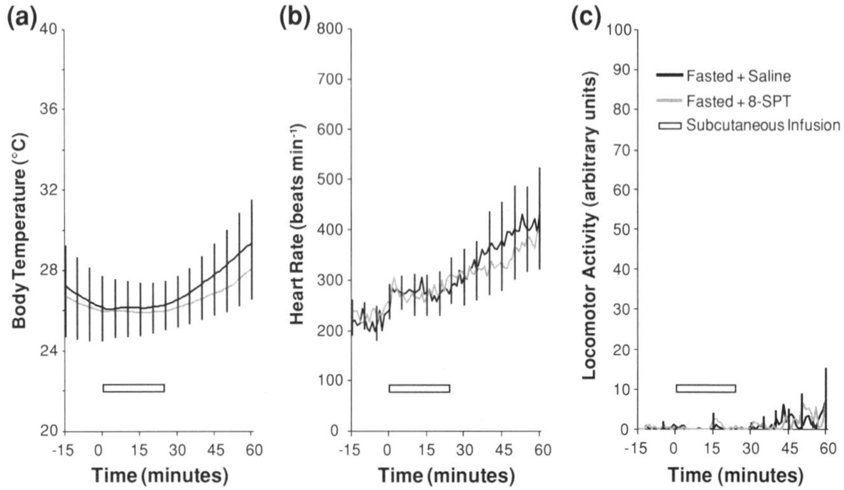

Fig. 30.4 Mice were fasted at the beginning of the dark phase, and allowed to enter into torpor. As core T_b approached 26°C, mice were infused s.c. for 25 min with either saline or 8-SPT, a polar adenosine receptor antagonist that does not cross the blood–brain barrier. No significant differences were found for T_b, HR, and LMA between saline and 8-SPT infusion

the hypothermia associated with adenosine injection (data not shown). Neither saline nor the peripheral administration of 8-SPT significantly affected T_b, HR, or LMA in torpid mice (Fig. 30.4).

30.3.4.1 Experiment 5: Acute Central 8-SPT Infusion

To determine whether the negative result produced by peripheral 8-SPT was due to its lack of central adenosine antagonism, we then examined whether central infusion of 8-SPT blunted fasting-induced torpor. Mice were fasted to induce torpor, and once torpid, were infused i.c.v. with PBS vehicle or 8-SPT to determine whether blockade of central adenosine receptors is sufficient to prevent the maintenance of torpor. Central 8-SPT treatment caused a rapid increase in T_b, HR, and LMA relative to treatment with PBS vehicle (Fig. 30.5).

Collectively, these data support the hypothesis that adenosine receptor activation centrally is required for the induction and maintenance of fasting-induced torpor in mice. The difference between AMP-induced hypothermia and torpor explains an interesting phenomenon observed in our dataset. While both central and peripheral adenosine treatment can induce hypothermia which can be blunted by pre-treatment with an adenosine antagonist, only central infusion of the antagonist 8-SPT is capable of reversing torpor in mice. This suggests that there are two different mechanisms by which adenosine induces hypothermia: one central route, which could involve inhibition of sympathetic efferent pathways, and one

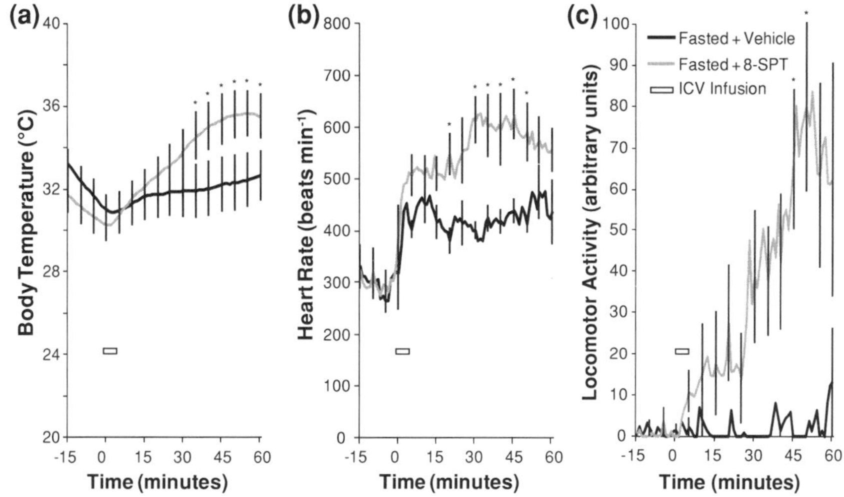

Fig. 30.5 Mice were fasted at the beginning of the *dark phase*, and allowed to enter into torpor. As core T_b approached 30°C, mice were infused i.c.v. for 5 min with either saline or 8-SPT. As assessed by T_b, HR, and LMA, central 8-SPT caused an immediate arousal from torpor

peripheral route involving direct negative inotropic and chronotropic effects on the heart (Belardinelli et al. 1989), resulting in tissue hypoxia and hypometabolism. Our results suggest that daily torpor is regulated by central adenosine signaling, and that the hypothermia of torpor is unlike the hypothermia caused by peripheral adenosine analogues. Adenosine freely crosses the blood–brain barrier, so it is not clear whether in torpor it is primarily produced in the brain and acting as an autocrine/paracrine signal, or produced ubiquitously, acting as an endocrine signal. Our findings are consistent with torpor bouts of other mammals shown to be dependent on adenosine receptor activation (Tamura et al. 2005; Jinka et al. 2011). These experiments allow us to create a model for adenosine's role in mediating torpor in mice. When temperatures become cool, mice compensate by increasing thermogenesis to maintain euthermia. However, when food is scarce, attempting to maintain a stable T_b causes an energy deficit, resulting in slightly reduced levels of ATP. The equilibrium between ATP, ADP, AMP, and adenosine shifts, greatly increasing adenosine levels, which participate in the signaling for the torpid state.

References

Alam MN, Szymusiak R, Gong H, King J, McGinty D (1999) Adenosinergic modulation of rat basal forebrain neurons during sleep and waking: neuronal recording with microdialysis. J Physiol Lond 521(3):679–690

Anderson R, Sheehan MJ, Strong P (1994) Characterization of the adenosine receptors mediating hypothermia in the concious mouse. Br J Pharmacol 113(4):1386–1390

Barclay RMR, Lausen CL, Hollis L (2001) What's hot and what's not: defining torpor in free-ranging birds and mammals. Can J Zool Revue Canadienne De Zoologie 79(10):1885–1890
Belardinelli L, Linden J, Berne RM (1989) The cardiac effects of adenosine. Prog Cardiovasc Dis 32(1):73–97
Berger RJ (1984) Slow-wave sleep, shallow torpor, and hibernation—homologous states of diminished metabolism and body temperature. Biol Psychol 19(3–4):305–326
Blood AB, Hunter CJ, Power GG (2003) Adenosine mediates decreased cerebral metabolic rate and increased cerebral blood flow during acute moderate hypoxia in the near-term fetal sheep. J Physiol 553(3):935–945
Chen G, vandenPol AN (1997) Adenosine modulation of calcium currents and presynaptic inhibition of gaba release in suprachiasmatic and arcuate nucleus neurons. J Neurophysiol 77(6):3035–3047
Daniels IS, Zhang J, O'Brien WG, Tao Z, Miki T, Zhao Z, Blackburn MR, Lee CC (2010) A role of erythrocytes in adenosine monophosphate initiation of hypometabolism in mammals. J Biol Chem 285(27):20716–20723
Dark J, Pelz KM (2008) NPY Y1 receptor antagonist prevents NPY-induced torporlike hypothermia in cold-acclimated siberian hamsters. Am J Physiol Regul Integr Comp Physiol 294(1):R236–R245
Deboer T, Tobler I (1995) Temperature-dependence of EEG frequencies during natural hypothermia. Brain Res 670(1):153–156
Deboer T, Tobler I (1996) Natural hypothermia and sleep deprivation: common effects on recovery sleep in the Djungarian hamster. Am J Physiol Regul Integr Comp Physiol 271(5):R1364–R1371
Geiser F (2004) Metabolic rate and body temperature reduction during hibernation and daily torpor. Annu Rev Physiol 66(1):239–274
Gluck EF, Stephens N, Swoap SJ (2006) Peripheral ghrelin deepens torpor bouts in mice through the arcuate nucleus neuropeptide Y signaling pathway. Am J Physiol Regul Integr Comp Physiol 291(5):R1303–R1309
Heller HC, Ruby NF (2004) Sleep and circadian rhythms in mammalian torpor. Annu Rev Physiol 66:275–289
Hudson JW, Scott IM (1979) Daily torpor in the laboratory mouse, mus-musculus var albino. Physiol Zool 52(2):205–218
Jinka TR, Tøien Ø, Drew KL (2011) Season primes the brain in an arctic hibernator to facilitate entrance into torpor mediated by adenosine A1 receptors. J Neurosci 31(30):10752–10758
Morton SR (1978) Torpor and nest-sharing in free-living sminthopsis casssicaudata (marsupialia) and mus musculus (rodentia). J Mammal 59(3):569–575
Ngai AC, Winn HR (1993) Effects of adenosine and its analogues on isolated intracerebral arterioles. Extraluminal and intraluminal application. Circ Res 73(3):448–457
Paul MJ, Freeman DA, Park JH, Dark J (2005) Neuropeptide y induces torpor-like hypothermia in siberian hamsters. Brain Res 1055(1–2):83–92
Pekny M, Stanness KA, Eliasson C, Betsholtz C, Janigro D (1998) Impaired induction of blood-brain barrier properties in aortic endothelial cells by astrocytes from GFAB-deficient mice. Glia 22(4):390–400
Swoap SJ (2008) The pharmacology and molecular mechanisms underlying temperature regulation and torpor. Biochem Pharmacol 76(7):817–824
Swoap SJ, Gutilla MJ (2009) Cardiovascular changes during daily torpor in the laboratory mouse. Am J Physiol Regul Integr Comp Physiol 297(3):R769–R774
Swoap SJ, Rathvon M, Gutilla M (2007) AMP does not induce torpor. Am J Physiol Regul Integr Comp Physiol 293(1):R468–R473
Tamura Y, Shintani M, Nakamura A, Monden M, Shiomi H (2005) Phase-specific central regulatory systems of hibernation in Syrian hamsters. Brain Res 1045(1–2):88–96
Williams TD, Chambers JB, Henderson RP, Rashotte ME, Overton JM (2002) Cardiovascular responses to caloric restriction and thermoneutrality in C57BL/6J mice. Am J Physiol Regul Integr Comp Physiol 282(5):R1459–R1467

Willis CK (2007) An energy-based body temperature threshold between torpor and normothermia for small mammals. Physiol Biochem Zool 80(6):643–651

Yang J-N, Chen J-F, Fredholm BB (2009) Physiological roles of A1 and A2A adenosine receptors in regulating heart rate, body temperature, and locomotion as revealed using knockout mice and caffeine. Am J Physiol Heart Circ Physiol 296(4):H1141–H1149

Zhang F, Wang S, Luo Y, Ji X, Nemoto EM, Chen J (2009) When hypothermia meets hypotension and hyperglycemia: the diverse effects of adenosine 5[prime]-monophosphate on cerebral ischemia in rats. J Cereb Blood Flow Metab 29(5):1022–1034

Zhang J, Kaasik K, Blackburn MR, Lee CC (2006) Constant darkness is a circadian metabolic signal in mammals. Nature 439(7074):340–343

Chapter 31
On the Dissimilarity of 5′-AMP Induced Hypothermia and Torpor in Mice

Arjen M. Strijkstra, Tim Koopmans, Hjalmar R. Bouma, Sietse F. de Boer, Roelof A. Hut and Ate S. Boerema

Abstract Administration of adenosine-5′-monophosphate (5′-AMP) can induce an artificial but endogenously reversible torpor-like state in mice. The dynamics of body temperature and the relation between body temperature and metabolic rate may indicate the (dis)similarity of this artificial torpor-like state to natural torpor in intact animals. We investigated these in C57BL/6J mice by (1) comparing cooling rates during 5′-AMP induced hypothermia to cooling rates during high workload induced torpor, and by (2) estimating the relative contributions of metabolic suppression and passive temperature (Q_{10}) effects in the 5′-AMP induced hypothermic state. We did the latter by back-extrapolating the relation between body temperature and metabolic rate in hypothermic conditions to the euthermic temperature level, using calculated Q_{10}-values. The data indicate that (1) cooling rate in 5′-AMP induced hypothermia is about 1.8 times faster than in natural torpor in workload conditions, and that (2) Q_{10} effects can entirely explain the metabolic reduction of 5′-AMP induced hypothermia, indicating that active metabolic suppression may be lacking. Together, this suggests fundamental differences between 5′-AMP induced

A. M. Strijkstra
Animal Management, University of Applied Sciences Van Hall Larenstein, Leeuwarden, The Netherlands

A. M. Strijkstra · T. Koopmans · R. A. Hut · A. S. Boerema
Departments of Chronobiology, University of Groningen, Groningen, The Netherlands

A. M. Strijkstra (✉) · H. R. Bouma
Department of Clinical Pharmacology, University Medical Center Groningen, Sector F FB20, A. Deusinglaan 1, 9713 AV Groningen, The Netherlands
e-mail: a.m.strijkstra@rug.nl

S. F. de Boer
Behavioural Physiology, University of Groningen, Groningen, The Netherlands

A. S. Boerema
Molecular Neurobiology, University of Groningen, Groningen, The Netherlands

T. Ruf et al. (eds.), *Living in a Seasonal World*,
DOI: 10.1007/978-3-642-28678-0_31,

hypothermia and natural torpor, limiting the validity of the paradigm to the study of effects of hypothermic conditions and temperature related metabolic effects.

31.1 Introduction

Hibernating mammals reduce energy expenditure by actively suppressing metabolism in a state of torpor (Heldmaier et al. 2004). Many small mammal species, including house mice, also use torpor to balance their energy budget in reaction to seasonal challenges, but do so on a daily basis (Geiser and Ruf 1995; Hut et al. 2011).

In mice, food restriction and acute fasting may lead to spontaneous torpor (Hudson and Scott 1979; Overton and Williams 2004; Swoap et al. 2007; Dikic et al. 2008). Also in simulated food shortage, mice may regularly enter torpor (Schubert et al. 2010; Hut et al. 2011). Torpor appears a natural behavior in mice, and house mice apparently handle low body temperature (T_b) well.

Torpor-like states can be induced with pharmacological substances, such as H_2S, 3-iodothyronamide, and adenosine 5′-monophosphate (5′-AMP) (reviewed by Bouma et al. 2012). 5′-AMP leads to reduced metabolic rate (MR) and low T_b (Zhang et al. 2006; Lee 2008). Arguably, 5′-AMP may be involved in natural torpor, since 5′-AMP is elevated in mice entering torpor upon fasting (Zhang et al. 2006; Lee 2008). However, the similarity is questionable, because of the acute artificial induction and absence of preparatory physiology and behavior associated with torpor. Additionally, 5′-AMP treated mice cool down faster than in fasting induced torpor (Swoap et al. 2007).

Metabolic suppression appears the main defining characteristic of natural torpor (Heldmaier 2011). Metabolic suppression occurs in all types of torpor: during continuous torpor in Bears (Tøjen et al. 2011), during deep torpor in hibernation (Heldmaier et al. 2004) and during daily torpor (Heldmaier et al. 2004; Song et al. 1995). There is much evidence to support whole animal metabolic suppression in torpor (Heldmaier et al. 2004; Song et al. 1995) and suppression of molecular processes, such as DNA read-off, protein production, cell proliferation and mitochondrial metabolism (Carey et al. 2003).

The MR reduction of torpor is based on the reset from the euthermic T_b set point to lower set points, on endogenous metabolic suppression and on passive thermodynamic effects (Heller 1979; Song et al. 1995; Heldmaier et al. 2004; Heldmaier 2011). The latter two effects can be disentangled in intact animals by estimating the contribution of passive effects by back-extrapolating its (Q_{10}-based) relation between MR and T_b under torpid conditions to euthermic T_b levels (Song et al. 1997; Heldmaier et al. 2004; Heldmaier and Elvert 2004). Such analyses may indicate if a hypothermic state has metabolic suppression.

Thus, 5′AMP may induce a hypothermic state by a change in thermoregulation (most probably shut-down), further actively regulated metabolic effects (metabolic suppression), and passive effects of low temperature itself (i.e. Q_{10}-effects).

Critical information on these effects may originate from experimentally induced variation of metabolic and physiological effects in the 5′-AMP induced hypothermic state. For thermoregulation related effects, a contrast of MR below the thermoneutral zone and at the lower critical temperature (LCT) is useful. At the LCT, animals control T_b by the lowest rate of metabolic heat production (basal metabolic rate, BMR) without the need for additional regulation for heat loss by peripheral vasodilation and evaporation. At the LCT, effects of 5′-AMP that are not related to shut-down of thermoregulation (i.e. metabolic suppression, Q_{10}-effects) should be most clearly indicated. Comparison between steady state MR at high T_b levels (at high T_a) with transient MR measured at the same T_b during cooling may indicate dissimilarities if metabolic suppression is important for the cooling effects of 5′-AMP.

We further investigated the similarity of 5′-AMP-induced hypothermia with torpor, using (1) comparison of the T_b pattern with spontaneous torpor, and (2) metabolic arguments. To elaborate on the fast cooling rate relative to fasting-induced torpor (Swoap et al. 2007), we compare the cooling rate with the cooling rate of spontaneous torpor in mice under high workload conditions (Schubert et al. 2010; Hut et al. 2011). To investigate metabolic suppression and passive temperature effects, we measured MR and T_b during 5′-AMP-induced hypothermia at three ambient temperatures between the normally used 20°C and 29°C, close to the LCT.

31.2 Methods

31.2.1 Animals

Male C57BL/6JOlaHsd mice ($n = 7$) were obtained from Harlan. Mice were housed individually in standard Macrolon cages on saw dust bedding with added nesting material as cage enrichment and with water and food (RMHB 2181, Arie Blok, Woerden) ad libitum. Animals were maintained in a climate room at 20 ± 1°C with a 12 h light/12 h dark cycle (lights on at 09:00 hours). During oxygen consumption recordings, animals were housed and recorded in a climate room at the targeted ambient temperature (T_a): 20 ± 1°C, 25 ± 1°C, or 29 ± 1°C. The experiments were approved by the Animal experimental ethics committee of the University of Groningen (DEC-5637A; DEC-5637B).

31.2.2 Body Temperature Telemetry

Body temperature (T_b) was measured by telemetry. Ta10ETA-F20 transmitters (Datasciences International, St. Paul, USA) were surgically implanted under isoflurane anesthesia (2.5%). Mice were allowed to recover for 2 weeks before recording. Recordings were made using Datascience receivers (RA-1010), signal interface (APR-1), and data-acquisition software (Datasciences ART). T_b telemetry was recorded on a 1 min time base.

31.2.3 Metabolic Rate Recording

For measurement, mice were housed for two consecutive days in airtight plastic measurement boxes of 10 × 10 × 15 cm with nesting material from the animal's home cage, and apple or gel as a water source. Measurements were started at 09:00.

MR was measured by respirometry in an open flow system. Outside air was compressed, dried, and led through the measurement box and a reference air channel at 30 l/h, and air samples of 6 l/h were taken from the overflow from measurement box or reference channel using flow controllers (Brooks, type 5850). The air sample was dried (Molecular sieve, 3 Å, Merck) and led through a paramagnetic oxygen analyzer (Servomex Xentra 4100) and an infra-red carbondioxide gas analyzer (Servomex 1440). Oxygen consumption (in ml O_2/h) was calculated by multiplying the difference between oxygen concentrations of inlet and overflow air with inlet flow, corrected for volumetric changes of the respired air using the measured respiratory quotient. A time resolution of 1 measurement per 3 min was usually reached.

31.2.4 5′-AMP Hypothermia Induction

5′-AMP salt was obtained from Sigma Aldrich and dissolved in sterile saline. 5′-AMP induced hypothermia was produced by *i.p.* injection (dose 7.5 μmol/g (2.6 mg/g) body weight,) 1.0–1.8 h in the light (resting) phase. After injection, animals were immediately returned to their measurement box.

31.2.5 Torpor Induction by Simulated Food Shortage: Workload Protocol

After the 5′-AMP trials, four mice were housed in standard Macrolon cages equipped with a 15 cm diameter running wheel. Running wheel activity was recorded in 2-min bins. Telemetric T_b recordings were done for 2–9 consecutive days in different phases of the ‘work for food’ protocol. A full description of the present experiment will be published elsewhere.

From April 21, animals were fed 45 mg food pellets (F0165; Bio-Serv, Frenchtown NJ, USA). Starting May 19, 2010, food pellets were delivered after a set number of wheel revolutions using computer controlled pellet dispensers (Med Associates Inc., St. Albans VT, USA). Mice started on 100 revolutions per pellet, which generated an ad libitum food condition. From May 20, workload was increased by adding ten revolutions per pellet to a maximum of 260 revolutions per pellet on June 14. Body mass was regularly measured and workload was individually titrated to keep the mice above 75% of their initial body mass. In the late high workload situation this resulted in an average reward ratio of 166 (SEM 26) revolutions per pellet, where body mass was 24.2 g (SEM 0.6 g), 78.3% (SEM 1.7%) of the initial body mass. T_b data of cooling during torpor bouts occurring in the high workload phase between June 14 and July 1 were used. In 7–9 days of recording, mice showed 3–5 long torpor bouts of 8.93 h (SEM 1.15 h) duration.

31.2.6 Data Handling

For the 5′-AMP experiments, the cooling phase was defined to start after injection. For the workload situation, torpor started when T_b continuously decreased towards a level below 26°C. Rewarming started when T_b increased towards full euthermia by >0.01°C/min. Statistical analysis and curve fitting was done with Sigmaplot/Sigmastat software.

31.3 Results and Discussion

31.3.1 Cooling Rate: Workload Induced Torpor Versus 5′-AMP Induced Hypothermia

Cooling curves of 5′-AMP induced hypothermia and workload induced torpor are shown in Fig. 31.1. Data are shown for the four individual animals that were subject to both challenges. Fitting an exponential decreasing function ($f(\text{time}) = T_a + C \times 10^{-\text{cooling rate} \times \text{time}}$) to the T_b data revealed a 1.75 (SEM 0.085) times higher cooling rate (average 0.0067/min, SEM 0.0006/min) than those of workload induced torpor (average 0.0038/min, SEM 0.0002/min; paired t test $t_3 = 6.97$, $p < 0.01$). The other parameters T_a and C ($= T_{b\ \text{at}\ t=0} - T_a$) were not significantly different ($p < 0.05$).

The difference in cooling rate was close to that reported for the difference with fasting induced torpor (Swoap et al. 2007): the fastest observed rate of change was 1.85 times faster for 5′-AMP induced hypothermia. 5′-AMP induced hypothermia appears associated with unnaturally fast cooling rates.

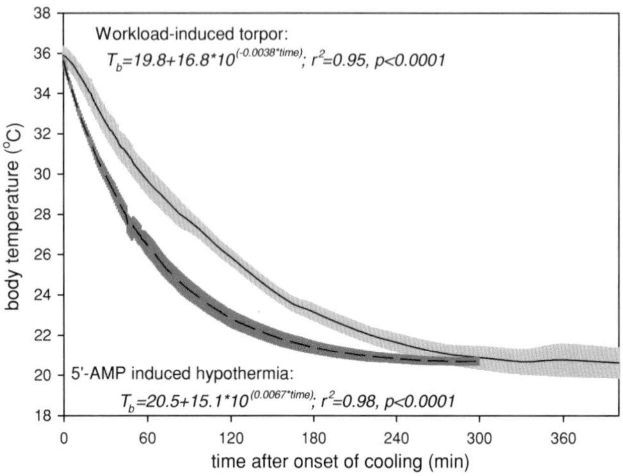

Fig. 31.1 Average body temperature of the same individual house mice ($n = 4$) during cooling in torpor bouts occurring in high workload conditions (*solid line*, *light grey* area denotes SEM, $n = 402$ points per individual) and during cooling in 5′-AMP induced hypothermia (*dashed line*, *dark grey* area denotes SEM, $n = 300$ points per individual). Equations indicate fit results of a Newtonian cooling equation to the data. r^2-values indicate goodness of fit, p-values indicate significance

31.3.2 5′-AMP Induced Hypothermia

An overview of average T_b and concurrent MR data of the 5′-AMP induced hypothermia trials is given in Fig. 31.2a, b, c. Data are shown as averages per 10 min intervals (20°C: $n = 7$; 25°C: $n = 6$, 29°C: $n = 7$) for the 12 h light phase. The 5′-AMP induced hypothermic state depended on T_a. Lowest T_b during cooling was (expectedly) different between T_a groups (repeated measures (rm)ANOVA: $F_{2,19} = 135.9$, $p < 0.001$; post hoc Holm-Sidak (H–S) tests: $p < 0.001$ between all groups), reaching T_b levels close to T_a. The duration between injection to start of rewarming was longer at lower T_a (rmANOVA: $F_{2,19} = 70.6$, $p < 0.001$; H–S tests: $p < 0.001$ between all groups). Although the minimal $T_b - T_a$ gradient was slightly higher at $T_a = 25$°C and at $T_a = 29$°C compared to $T_a = 20$°C, this was not significant (rmANOVA: $F_{2,19} = 2.06$, $p = 0.17$). Standard metabolic rate (SMR) was higher at lower T_a (rmANOVA: $F_{2,19} = 318.1$, $p < 0.001$; H–S tests: $p < 0.05$ between all groups), as may be expected for MR at temperatures around the LCT and below.

31.3.3 Metabolic Rate

The SMR value at $T_a = 29$°C (45.89 ml O_2/h, SE 1.33 ml O_2/h) was considered the lowest MR in euthermic conditions (BMR). This value was close to estimated

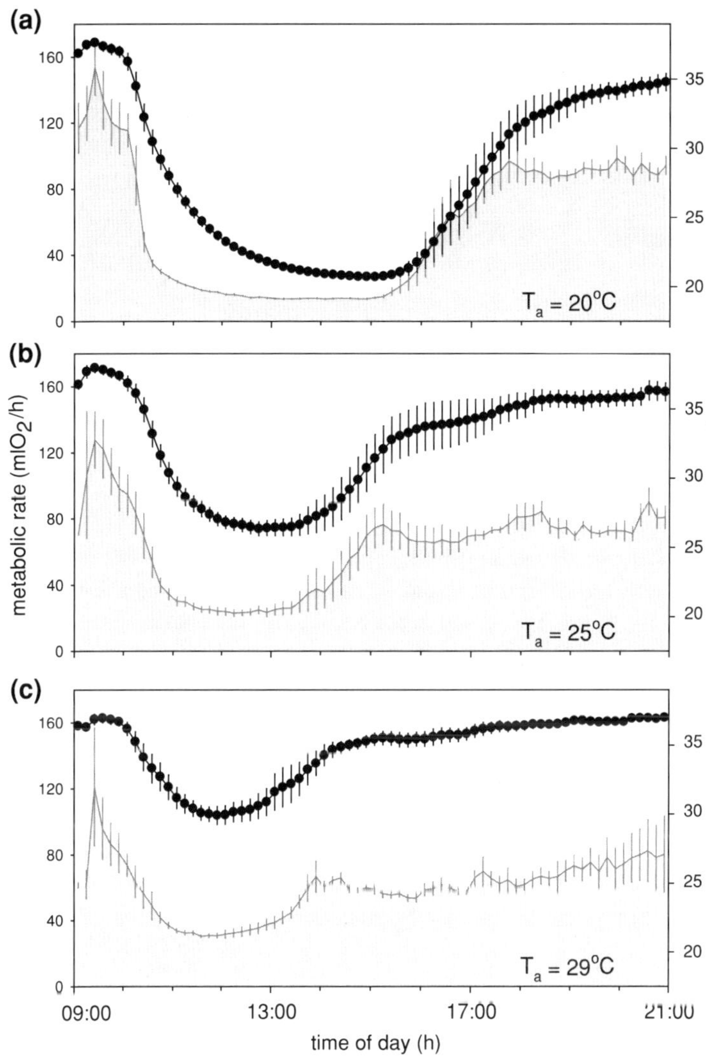

Fig. 31.2 Body temperature (*black dots* mean, SEM) and metabolic rate (ml O_2/h, *grey* area under mean, SEM) during the light phase (09:00–21:00) of house mice when challenged with an *i.p.* injection of 7.5 μmol/g 5′-AMP dissolved in saline. Timing of injection for the 20°C data (*panel A*) was at 10:11, at 25°C (*panel B*) at 10:20 and at 29°C (*panel C*) at 10:18. See Table 31.1 for further information

values of BMR for daily heterotherms: (111% of 1.26 ml O_2/g/h, derived from $^{10}\log$ (BMR) = 0.678-0.381 × $^{10}\log$ (BM); Geiser 2004).

MR during the last 30 min interval before rewarming was lower at lower T_a (rmANOVA: $F_{2,19} = 78.8$, $p < 0.001$; H–S tests: $p < 0.001$ between all groups). The MR value at $T_a = 20$°C appeared similar to relatively high compared to MR in fasting and cold induced torpid mice measured at 16°C: 16.5 ml O_2/h, derived from the equation $^{10}\log(MR) = 0.823 + 0.261 \times {}^{10}\log(BM)$ by Dikic et al. (2008), and 5.1–22.2 ml O_2/h measured at 18°C (Oelkrug et al. 2011). When comparing the MR values at $T_a = 20$°C and $T_a = 29$°C to expected values in torpid animals, these MR data appeared high at 0.43 ml O_2/g/h (146% of expected for daily torpor between 15 and 25°C; Geiser 2004) and 0.95 ml O_2/g/h (152% of expected for torpor between 25 and 33°C; Geiser 2004). Steady state 5′-AMP induced hypothermia appears to have a ~50% higher MR than torpid animals of other species, perhaps related to the differences in paradigms where torpor is measured (i.e. acutely induced versus natural behaviour).

31.3.4 Metabolic and Body Temperature

MR data during cooling was plotted against concurrent T_b (per 5 min interval) together with lines indicating MR when a $Q_{10} = 2.5$ is assumed (see Fig. 31.3). The data appear rather parallel to the $Q_{10} = 2.5$ lines at $T_b < 30$°C. It is also clear that at $T_b > 30$°C, the $T_a = 20$°C and $T_a = 25$°C MR data increase relatively fast compared to the $T_a = 29$°C MR data. The initially high SMR and high $T_b - T_a$ gradient probably leads to fast initial reduction of T_b and MR when cooling begins.

If metabolic suppression would have an important role in the MR reduction, transient MR (when T_b is still cooling towards T_a) may be expected to be lower than steady state MR (when T_b is constant and near to T_a), because it may be suppressed beyond temperature effects alone. MR at T_b between 29 and 30°C was 33.9 ml O_2/h at $T_a = 20$°C and 31.1 ml O_2/h $T_a = 25$°C (i.e. transient situations), similar to steady state MR at $T_a = 29$°C (rmANOVA: $F_{2,19} = 0.75$, $p = 0.50$). Transient MR at $T_b = 26$–27°C was 23.0 ml O_2/h (SEM 0.8 ml O_2/h) at $T_a = 20$°C, similar to steady state MR at $T_a = 25$°C of 23.7 ml O_2/h (paired t test: $t_5 = 0.42$, $p = 0.70$). This indicates that in 5′-AMP induced hypothermia, steady state MR and transient MR have similar temperature dependences, suggesting a limited role for metabolic suppression in the reduction of MR in both conditions.

31.3.5 Q_{10} Effects

MR, T_b and Q_{10} data are given in Table 31.1. Q_{10}-values were calculated using the equation:

$$Q_{10} = (MR_2/MR_1)^{10/(T_{b2}-T_{b1})}$$

This was done in three different ways: (1) between the 30 min interval before injection and the last 30 min interval during cooling, (2) between the first and last 30 min interval during cooling, and (3) between the MR level between $T_b = 29$–30°C on the cooling curve and the last 30 min interval during cooling.

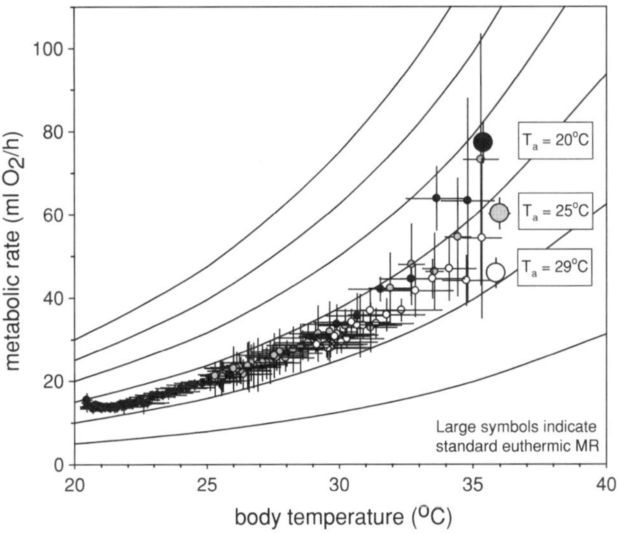

Fig. 31.3 Average data of metabolic rate (MR in ml O_2/h) as a function of body temperature (T_b in °C) on the cooling curves of 5′-AMP induced hypothermia (*small symbols*) and standard euthermic conditions (*large symbols*) at three ambient temperatures (*black symbols* $T_a = 20$°C, *grey symbols* $T_a = 25$°C, *open symbols* $T_a = 29$°C). Data were collapsed in 5 min intervals, averages and SEM values were made between individuals. The 6 *solid lines* indicate hypothetical MR changes when the Q_{10}-value for temperature related effects is 2.5 for 6 hypothetical MR values at 20°C (5, 10, 15, 20, 25 and 30 ml O_2/h)

Approach (1) included confounding effects of homeostatic regulation and activity before the 5′-AMP injection inactivated the animal and shut down thermoregulation. Approach (2) excluded confounding effects of homeostatic regulation and activity, but included a high SMR onset and a high $T_b - T_a$ gradient. Approach (3) excluded all confounding influences by only using the $T_b < 30$°C data of the cooling curves.

Q_{10} values using approach (1) were indistinguishable: 3.63 at $T_a = 20$°C, 3.53 at $T_a = 25$°C, and 4.40 at $T_a = 29$°C (rmANOVA: $F_{2,19} = 0.90$, $p > 0.05$). These values are above the usual physiological range of Q_{10} values (2–3), suggesting metabolic suppression of some kind. Q_{10} values using approach (2) were lower, but also indistinguishable between T_a groups: 2.87 at $T_a = 20$°C, 2.86 at $T_a = 25$°C, and 2.58 at $T_a = 29$°C (rmANOVA: $F_{2,19} = 0.95$, $p > 0.05$). Q_{10} values using approach (3) were even lower and also indistinguishable between T_a-groups: 2.68 at $T_a = 20$°C and 2.37 at $T_a = 25$°C, and 2.58 at $T_a = 29$°C (rmANOVA: $F_{2,19} = 0.28$, $p > 0.05$).

Q_{10} values of approach (3) were used for back-extrapolation of hypothermic T_b and MR levels to euthermic conditions. Average T_b and MR data of the last 30 min of the cooling curve and T_b data of the euthermic BMR measurements were used as input variables. The euthermic MR estimations were 65.4 ml O_2/h for $T_a = 20$°C data, 52.4 ml O_2/h for $T_a = 25$°C data, and 55.2 ml O_2/h for

Table 31.1 Body mass, metabolic rate (MR), body temperature (T_b) and timing data of conditions before and during 5′-AMP induced hypothermia at $T_a = 20°C$, $T_a = 25°C$ and $T_a = 29°C$, and their derived Q_{10} values and extrapolated MR estimates

		$T_a = 20°C$	$n = 7$	$T_a = 25°C$	$n = 6$	$T_a = 29°C$	$n = 7$	*p*-value#
		Mean	SEM	Mean	SEM	Mean	SEM	
Body mass	g	32.6	0.94	32.9	0.95	32.6	0.79	NS
Euthermic								
Standard MR	ml O_2/h	77.24	1.81	60.22	1.52	45.89	1.33	<0.0001
T_b (at SMR)	°C	35.39	0.12	35.98	0.16	35.85	0.10	NS
Pre-injection								
MR (last 30 min)	ml O_2/h	115.47	11.07	94.08	10.93	75.14	8.62	<0.0001
T_b (last 30 min)	°C	36.89	0.28	37.09	0.34	36.45	0.20	NS
Post-injection								
Time of injection	hh:mm	10:11	00:03	10:20	00:04	10:18	00:06	NS
MR (first 30 min)	ml O_2/h	44.55	3.16	46.98	2.71	43.16	1.14	NS
T_b (first 30 min)	°C	31.77	0.37	32.91	0.19	33.49	0.44	
Transient MR								
MR (T_b = 29–30°C)	ml O_2/h	33.39	3.65	31.11	1.00			NS
MR (T_b = 26–27°C)	ml O_2/h	23.03	0.82					NS#
Steady state MR								
MR (last 30 min)	ml O_2/h	14.03	0.40	23.72	2.20	31.03	1.04	<0.0001
T_b (last 30 min)	°C	20.69	0.15	26.25	0.50	29.86	0.50	<0.0001
$T_b - T_a$ gradient	°C	0.75	0.15	1.78	0.50	1.27	0.39	NS
Duration of hypothermia	hh:mm	05:01	00:10	02:53	00:13	01:39	00:09	<0.0001
Q_{10} values								
Pre-injection to last 30 min		3.63	0.17	3.53	0.30	4.40	1.00	NS
Post-injection to last 30 min		2.87	0.22	2.86	0.23	2.58	0.15	NS
T_b < 30°C to last 30 min		2.68	0.21	2.37	0.33			NS
Q_{10} to reach euthermic MR		2.19	0.05	2.00	0.11	1.95	0.06	NS
Extrapolated MR								
Based on individual Q_{10}	ml O_2/h	65.39	11.31	52.41	4.74	55.16	4.09	NS
Based on Q_{10} = 2.45	ml O_2/h	54.69	2.01	55.26	3.40	53.22	2.50	NS
Based on Q_{10} = 2.16	ml O_2/h	45.17	1.59	49.03	3.17	49.29	2.08	NS

Euthermic standard MR was calculated as the lowest 30 min average MR in a baseline light phase. Several MR averages (last 30 min pre-injection, first 30 min post-injection, last 30 min post-injection (before rewarming), MR between T_b of 29–30°C, MR between T_b of 26–27°C) were calculated for different purposes (see text).

p-values indicate significance level of repeated measures ANOVA or paired-*t* test, NS indicates non-significance

$T_a = 29°C$ data (see Table 31.1). These MR estimates were higher than the measured BMR data of 45.9 ml O_2/h (see Fig. 31.3, open large symbol), but this did not reach statistical significance (rmANOVA: $F_{3,26} = 1.66$, $p > 0.05$).

Using Q_{10} values of 2.45 (found for passive temperature effects of daily torpor; Heldmaier and Elvert 2004), or 2.16, derived from an equation for Q_{10} values of daily heterotherms ($^{10}\log(Q_{10}) = 0.453 - 0.078 \times {}^{10}\log^{(\text{body mass})}$; Geiser 2004), back-extrapolation did also not yield euthermic MR estimates significantly below the BMR data (see Table 31.1). To actually reach measured BMR, on average a Q_{10} of 2.06 (SEM 0.06) was calculated. Thus, no evidence was found for effects of metabolic suppression in 5′-AMP induced hypothermia, since all of the MR reduction could be explained by Q_{10} effects.

31.4 Conclusions

We now show that 5′-AMP induced hypothermia has an about 1.8 times faster cooling rate than torpor caused by simulated food shortage in house mice, quantitatively similar to the difference with fasting induced torpor (Swoap et al. 2007). Peripheral resistance appears to be relatively high in fasting induced torpor (Swoap and Gutilla 2009). The increased rate of heat loss may be mediated by enhanced vasodilatory activity of adenosine produced from 5′AMP by ecto-5′-nucleotidases. Potentially, also the lack of behavioral preparation, nest building and a sleep-like posture may add to the fast heat loss.

Furthermore, we investigated the relation between MR and T_b by measuring MR and T_b during 5′-AMP induced hypothermia at different T_a. Thus, we could compare hypothermic MR in steady state and transient hypothermic conditions. Since there was no difference between steady state and transient MR levels at the same T_b, T_b appears the primary factor determining MR during cooling, suggesting little room for metabolic suppression in 5′-AMP related MR reduction. Steady state hypothermic MR appeared similar or a little higher than that of fasting induced torpor MR, but about 50% higher than literature data on MR during spontaneous daily torpor (Geiser 2004). This may suggest that spontaneous torpor includes more metabolic suppression than torpor provoked by more artificial means.

To estimate passive temperature related effects of 5′-AMP induced hypothermia, we calculated Q_{10} values over a restricted range of the cooling curve ($T_b < 30°C$. This yielded biologically relevant Q_{10} levels around 2.57 (SEM 0.12). Since we also measured euthermic BMR, we could compare extrapolated hypothermic MR levels to BMR data: they were similar. The euthermic MR estimates thus revealed no additional contribution to the metabolic reduction besides passive Q_{10} effects. This contrasts to estimates of metabolic suppression in both daily torpor, estimated to be about 50% (Heldmaier and Elvert 2004; Song et al. 1997) and deep torpor in hibernation, estimated to be about 75% (Heldmaier and Elvert 2004).

We conclude that 5′-AMP induced hypothermia cannot be seen as a torpid state based on metabolic arguments: there is no apparent role for active metabolic suppression. This restricts the validity of the model system to research on effects of hypothermia and on temperature related metabolic effects.

References

Bouma HR, Verhaag EM, Otis JP, Heldmaier G, Swoap SJ, Strijkstra AM, Henning RH, Carey HV (2012) Induction of torpor: Mimicking natural metabolic suppression for biomedical applications. J Cell Physiol 227:1285–1290

Carey HV, Andrews MT, Martin SL (2003) Mammalian hibernation: cellular and molecular responses to depressed metabolism and low temperature. Physiol Rev 83:1153–1181

Dikic D, Heldmaier G, Meyer (2008) Induced torpor in different strains of laboratory mice. In: Lovegrove BG, McKechnie AE (eds) Hypometabolism in animals: hibernation, torpor and cryobiology. University of KwaZulu-Natal, Pietermartitzburg pp 223–230

Geiser F (2004) Metabolic rate and body temperature reduction during hibernation and daily torpor. Ann Rev Physiol 66:239–274

Geiser F, Ruf T (1995) Hibernation versus daily torpor in mammals and birds: physiological variables and classification of torpor patterns. Physiol Zool 68:935–966

Heldmaier G (2011) Life on low flame in hibernation. Science 331:866–867

Heldmaier G, Elvert R (2004) How to enter torpor: thermodynamic and physiological mechanisms of metabolic depression. In: Barnes BM, Carey HV (eds) Life in the cold: evolution, mechanisms, adaptation, and application. University of Alaska Fairbanks, Fairbanks, pp 185–198

Heldmaier G, Ortmann S, Elvert R (2004) Natural hypometabolism during hibernation and daily torpor in mammals. Respir Physiol Neurobiol 141:317–329

Heller HC (1979) Hibernation: neural aspects. Ann Rev Physiol 41:305–321

Hudson JW, Scott IM (1979) Daily torpor in the laboratory mouse, Mus musculus Var. albino. Physiol Zool 52:205–218

Hut RA, Pilorz V, Boerema AS, Strijkstra AM, Daan S (2011) Working for food shifts nocturnal mouse activity into the day. PLoS ONE 6:e17527

Lee CC (2008) Is human hibernation possible? Annu Rev Med 59:177–186

Oelkrug R, Heldmaier G, Meyer CW (2011) Torpor patterns, arousal rates, and temporal organization of torpor entry in wildtype and UCP1-ablated mice. J Comp Physiol B 181:137–145

Overton JM, Williams TD (2004) Behavioral and physiologic responses to caloric restriction in mice. Physiol Behav 81:749–754

Schubert KA, Boerema AS, Vaanholt LM, de Boer SF, Strijkstra AM, Daan S (2010) Daily torpor in mice: high foraging costs trigger energy-saving hypothermia. Biol Lett 6:132–135

Song X, Körtner G, Geiser F (1995) Reduction of metabolic rate and thermoregulation during daily torpor. J Comp Physiol B Biochem Syst Environ Physiol 165:291–297

Song X, Körtner G, Geiser F (1997) Thermal relations of metabolic rate reduction in a hibernating marsupial. Am J Physiol 273:R2097–R2104

Swoap SJ, Rathvon M, Gutilla M (2007) AMP does not induce torpor. Am J Physiol Regul Integr Comp Physiol 293:R468–R473

Swoap SJ, Gutilla M (2009) Cardiovascular changes during daily torpor in the laboratory mouse. Am J Physiol Regul Integr Comp Physiol 297:R769–R774

Tøjen Ø, Blake J, Edgar DM, Grahn DA, Helle HC, Barnes BM (2011) Hibernation in black bears: independence of metabolic suppression from body temperature. Science 331:906–909

Zhang J, Kaasik K, Blackburn MR, Lee CC (2006) Constant darkness is a circadian metabolic signal in mammals. Nature 439:340–343

Chapter 32
Potential Mechanisms of Metabolic Suppression Downstream of Central A_1AR Activation During Onset of Torpor

Tulasi R. Jinka, Zachary A. Barrickman, Lori K. Bogren, Trixie N. Lee, Jasmine M. Olson, Melanie M. Richter, Brady M. Salli, Timothy J. Stevenson, Øivind Tøien, C. Loren Buck and Kelly L. Drew

T. R. Jinka (✉) · L. K. Bogren · T. N. Lee · J. M. Olson · Ø. Tøien · K. L. Drew
Institute of Arctic Biology, University of Alaska Fairbanks, 902 N. Koyukuk Dr, P.O. Box 757000, Fairbanks, AK 99775, USA
e-mail: tjinka@alaska.edu

L. K. Bogren
e-mail: lkbogren@alaska.edu

T. N. Lee
e-mail: tnlee@alaska.edu

J. M. Olson
e-mail: jolson31@uw.edu

Ø. Tøien
e-mail: otoien@alaska.edu

K. L. Drew
e-mail: kdrew@alaska.edu

Z. A. Barrickman · B. M. Salli · T. J. Stevenson · C. Loren Buck
Department of Biological Sciences, University of Alaska Anchorage, 3101 Science Circle, CPSB 101, Anchorage, AK 99508, USA
e-mail: zabarrickman@alaska.edu

B. M. Salli
e-mail: bmsalli@alaska.edu

T. J. Stevenson
e-mail: tjstevenson@uaa.alaska.edu

C. Loren Buck
e-mail: afclb1@uaa.alaska.edu

M. M. Richter
Department of Biology and Wildlife, University of Alaska Fairbanks, 211 Irving 1, 902 N. Koyukuk Dr, Fairbanks, AK 99775, USA
e-mail: mmrichter@alaska.edu

T. Ruf et al. (eds.), *Living in a Seasonal World*,
DOI: 10.1007/978-3-642-28678-0_32, © Springer-Verlag Berlin Heidelberg 2012

Abstract Hibernating animals demonstrate a nadir in metabolic demand and body temperature (T_b) during torpor that is fundamental to adaptation to seasonal periods of reduced resource availability. A recent study shows how the brain regulates metabolic suppression during onset of torpor suggesting that central A_1 adenosine receptor signaling is both necessary and sufficient to trigger decreases in metabolic rate and T_b. This leads to an interesting question of how central signals are transduced to the periphery to elicit global suppression of metabolism and this chapter discusses relevant hypotheses.

List of Abbreviations

A_1AR	A_1 adenosine receptor
ANS	Autonomic nervous system
BMR	Basal metabolic rate
CHA	N^6-Cyclohexyladenosine
CNS	Central nervous system
HIT	Hibernation induction trigger
H_2S	Hydrogen sulfide
kD	Kilodalton
KO	Knockout
MR	Metabolic rate
PNS	Parasympathetic nervous system
ppm	Parts per million
Q_{10}	Temperature coefficient
RQ	Respiratory quotient
SNS	Sympathetic nervous system
T_a	Ambient temperature
T_b	Body temperature
TMR	Torpid metabolic rate
TNZ	Thermoneutral zone
T_1AM	3-Iodothyronamine

32.1 Introduction

One of the great "mysteries" in hibernation physiology is the identity of the signal that initiates induction of metabolic suppression to levels that are a fraction of basal metabolic rate (BMR) seen in entry into torpor. It has been recently shown in arctic ground squirrels (*Urocitellus parryii*) that central A_1 adenosine receptor (A_1AR) signaling is both necessary and sufficient to trigger decreases in metabolic rate and T_b consistent in magnitude to natural torpor (Jinka et al. 2011). Torpor induction by an A_1AR agonist depends on season—occurs only in winter—and spontaneous torpor is reversible with administration of an A_1AR antagonist.

The importance of central A_1AR signaling to initiate torpor is not only seen in obligate seasonal hibernators like the arctic ground squirrel but has also been identified in animals that express both facultative and fasting-induced torpor. Central A_1AR activation decreases T_b at torpor onset in Syrian hamster (*Mesocricetus auratus*), a facultative hibernator, (Shiomi and Tamura 2000; Shintani et al. 2005) and central administration of adenosine antagonist reverses fasting-induced torpor in mice (Iliff and Swoap 2010). This exciting discovery that stimulation of A_1AR in the brain regulates the onset of torpor leads to the question of the nature of the signal from the central nervous system (CNS) to the periphery that elicits global suppression of metabolism characteristic of the hibernation phenotype.

Whole animal metabolism is a composite of basal metabolic functions and the metabolic cost associated with thermoregulation, locomotion, post-prandial heat production etc. The BMR portion of whole animal metabolism represents the minimum rate of catabolism to sustain life while the animal is fasting (post-absorptive) at rest and thermal neutral; thus, by definition BMR excludes thermoregulatory heat production. BMR in arctic ground squirrels is estimated to be 0.40–0.61 ml O_2 g^{-1} h^{-1} (Scholander et al. 1950; Withers et al. 1979) and torpid metabolic rate (TMR) is 0.01–0.02 ml O_2 g^{-1} h^{-1} (Buck and Barnes 2000). The CNS governs thermoregulation (Nakamura 2011) and adenosine within the CNS tonically modulates basal T_b (Barros et al. 2006). Inhibiting thermogenesis through central inhibition of A_1AR could decrease metabolism to BMR, but how might A_1AR-dependent signaling within the CNS decrease MR to as low as 1–2% of BMR? To decrease metabolic rate to less than BMR, central A_1AR signaling may decrease organ-specific MR. In this contribution, we set out to critically examine hypotheses regarding the mechanism of transduction of a central signal to the periphery that ultimately leads to radical and controlled depression of metabolism.

32.2 A_1AR Agonist-Induced Torpor Versus Spontaneous Torpor

Torpor in hibernating animals and following fasting is a means of energy conservation that is distinguished from other conditions of T_b decline. First, torpor onset is characterized by a drop in metabolic rate which precedes a decrease in T_b. Second, torpor is distinguished from hypothermia by the endogenous ability to spontaneously rewarm without intervention (Lyman 1958; Carey et al. 2003; Drew et al. 2007). A_1AR agonist-induced torpor meets the two criteria outlined above for torpor in arctic ground squirrels (Jinka et al. 2011), but has not met or been tested for both of these criteria in other species. In Syrian hamsters A_1AR agonist-induced cooling mimicked spontaneous torpor in the magnitude of T_b decline; however, it is unclear if animals entered torpor or cooled to torpor-like T_b since MR was not measured (Shintani et al. 2005). Following A_1AR agonist-induced hypothermia, Syrian hamsters were unable to arouse from torpor and died (Shintani et al. 2005). However, in another study in which Syrian hamsters were acclimated to ambient temperature of 2°C prior to central administration of the A_1AR agonist animals aroused without incident (Miyazawa et al. 2008). Similarly, arctic ground squirrels were cold acclimated and rewarmed spontaneously after A_1AR agonist-induced torpor (Jinka et al. 2011). These observations suggest that cold adaptation may contribute to the capacity to rewarm following central A_1AR-induced torpor. Other differences between A_1AR agonist-induced and spontaneous torpor may point to important regulatory mechanisms.

Q_{10} of biological systems are predicted to be approximately two (Snyder and Nestler 1990). Literature values for Q_{10} of metabolism during steady-state torpor and entrance into torpor differ among species and across the temperature range tested. Q_{10} calculated during steady-state torpor ranges from 3.2 to 5.4 (Geiser 2004), but may vary widely between 1.0 and 14.1 depending on ambient temperature (Buck and Barnes 2000) and indicate substantial degrees of metabolic suppression that cannot be explained by pure temperature effects. During entrance into daily torpor in 25 g adult, stripe-faced dunnarts (*Sminthopsis macroura*) Q_{10} varied between 2.5 and 3.7 (Song et al. 1996). Values during A_1AR agonist-induced torpor in the arctic ground squirrel, calculated from data reported in Jinka et al. (2011), are also outside of the range expected if temperature alone was causing the decrease in MR. During A_1AR agonist-induced torpor, Q_{10} ranged between 2.4 and 7.1 with a mean $\pm$ SD of 4.1 $\pm$ 1.0, $n = 6$. By contrast to reports of spontaneous torpor, variability in Q_{10} values increased as T_b approached 5°C and MR neared TMR. Greater variation in Q_{10} in the later portion of the A_1AR agonist-induced cooling curve suggests greater variation in mechanisms responsible for metabolic suppression as TMR is approached that may vary between A_1AR agonist-induced and spontaneous torpor.

Subtle differences in the temporal profile of oxygen consumption and T_b during A_1AR agonist-induced and spontaneous torpor are also noted [Fig. 1 data from Karpovich et al. (2009); Jinka et al. (2011)]. Animals entering torpor spontaneously show a typical cooling curve from euthermic temperatures, with cooling rate decreasing exponentially. Occasional interruptions in the cooling curve correspond to spikes in oxygen consumption. Oxygen consumption in a spontaneous entrance decreases dramatically at onset of torpor and is at times followed by a few spikes in oxygen consumption that are sustained for up to a few hours. By contrast, mid-season A_1AR agonist-induced animals ($n = 6$ from Jinka et al. 2011) cool in a more linear fashion, with a significantly lower cooling rate between T_b's of 30–25°C compared with spontaneous entrance [$n = 6$, 1 from Jinka et al. (2011), 5 from Karpovich et al. (2009); means $\pm$ SD 2.71 $\pm$ 0.83°C h^{-1} (spontaneous) and 1.45 $\pm$ 0.20°C h^{-1} (CHA-induced), $n = 6$, $t_6 = 3.6$, $p < 0.05$]. T_b of animals during A_1AR agonist-induced torpor often level off at a minimal T_b rather suddenly, and oxygen consumption is more erratic. Oxygen consumption decreases at a similar rate during A_1AR agonist-induced torpor as in spontaneous torpor, but multiple spikes in oxygen consumption occur more consistently when torpor is induced (6 of 6 traces show erratic spikes in oxygen consumption) than when it

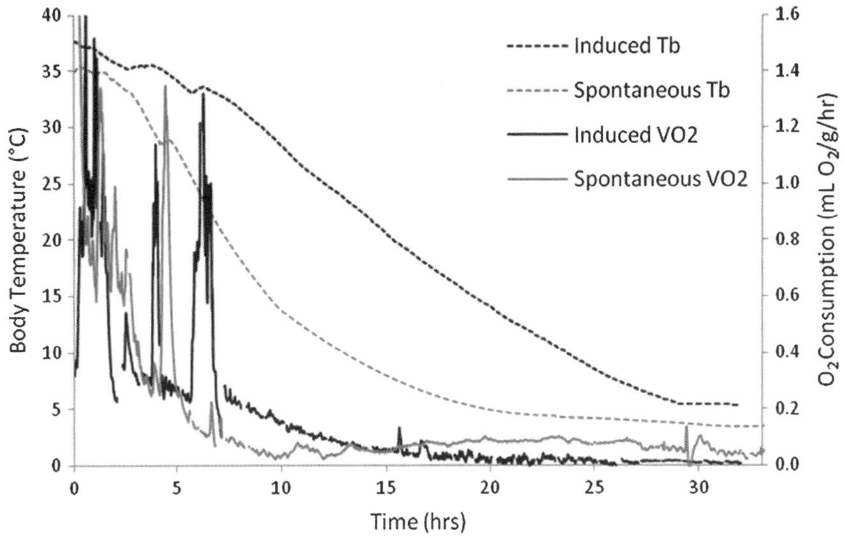

Fig. 32.1 Comparison of body temperature (T_b) and oxygen consumption in arctic ground squirrels during A_1AR agonist-induced and spontaneous entrance to torpor at an ambient temperature of 2°C [data from Jinka et al. (2011) and Karpovich et al. (2009), respectively]. *Dotted lines* represent T_b and *solid lines* represent oxygen consumption. *Black lines* represent the induced state while *gray lines* represent spontaneous torpor. This comparison is conservative by comparing the most similar torpor entrances between the two types

occurs spontaneously (3 of 6 traces show erratic spikes in oxygen consumption) (Fig. 32.1). These differences may reflect the action of factors that may operate in addition to or downstream of central A_1AR to regulate T_b and metabolic suppression during torpor entrance.

32.3 Modulation of the Autonomic Nervous System

One potential avenue for direct central-to-peripheral control of the metabolic suppression that is characteristic of torpor and could be influenced by central A_1AR activation is via the autonomic nervous system (ANS). Although there has been much controversy over the ANS's contribution to torpor in the past (Lyman and O' Brien 1960, 1963; Johansen et al. 1964; Twente and Twente 1978; Lyman 1982; Harris and Milsom 1995), the most recent studies indicate that it plays a significant role. The parasympathetic and sympathetic nervous systems (PNS and SNS, respectively) both control mechanisms that have the potential to assist in a regulated suppression of metabolism. These mechanisms include alterations in thermal conductance mediated by SNS modulation of vasomotor tone as well as organ-specific energy demand associated with excitability and ion homeostasis.

Evidence shows that the SNS is necessary for torpor. Pointedly, Braulke and Heldmaier (2010) found that in Djungarian hamsters, reversible ablation of SNS activity with the catecholaminergic neurotoxin 6-Hydroxydopamine (6-OHDA) resulted in complete disappearance of torpor for about 6 days. SNS activity may contribute to the increase in peripheral resistance suggested to occur during torpor onset although specific effects of the SNS on MR during torpor onset has not been studied. Blocking α-adrenergic signaling of the SNS early in arousal increases peripheral blood flow by 700% (Osborne et al. 2005). The SNS regulates essential brown adipose tissue thermogenesis during onset and emergence from fasting-induced torpor in mice and in this way increases MR (Swoap and Weinshenker 2008). Resistin, an adipokine, administered into the CNS enhances sympathetic nerve activity to the hindlimb but attenuates SNS activity to brown adipose tissue (Kosari et al. 2011).

The PNS is more likely to decrease MR by decreasing excitability and maintaining ion homeostasis than the SNS. PNS suppresses cardiac excitability. Inhibiting the PNS increases overall heart rate in daily torpor (Zosky 2002; Zosky and Larcombe 2003). However, PNS inhibition alone does not affect torpor occurrence or quality (Braulke and Heldmaier 2010). Generalizations regarding the influence of the ANS on torpor drawn from published findings are limited because studies were conducted on different species and on varying types of torpor (daily to fasting -induced). Whether or not the ANS mechanisms described are fundamental to the generalized, cross-species phenomenon of torpor or if they are species or torpor-type specific, are yet to be addressed. Although an exact mechanism by which the ANS contributes to metabolic suppression and the various stages of torpor are yet to be identified, Zosky and Larcombe (2003) suggest that an increasing inhibition of the ANS is characteristic of deep, steady-state torpor. The role of central A_1AR in regulating ANS function is not well characterized; however, A_1AR is distributed throughout the brainstem, plays a role in the regulation of breathing and vasomotor tone (Mosqueda-Garcia et al. 1989; Barraco and Phillis 1991). Stimulation of A_1AR located in the subpostremal nucleus of the solitary tract elicits vasoconstriction of the hindlimb vasculature (McClure et al. 2005).

32.4 Thermal Conductance

Autonomic control of vasomotor tone as well as centrally mediated effects on thermoregulatory behavior could be influenced by central A_1AR. Heat loss decreases as brain temperature cools during entrance into torpor in *Urocitellus lateralis* (Snapp and Heller 1981). Prior to and during early onset of torpor animals thermoregulate through postural adjustments and selection of ambient temperature. Most studies show that if given a choice, heterothermic mammals select a warm T_a for daily torpor (Song et al. 1998), and selection of warm T_a may supersede torpor (Wojciechowski et al. 2007). All of these factors may influence the rate of heat loss.

Within an animals' thermoneutral zone (TNZ) the basal rate of heat production is in equilibrium with the rate of heat loss to the external environment. The torpid state is distinguished from euthermy by a decrease in the threshold temperature that stimulates thermogenesis (Heller et al. 1977). Under these conditions a new equilibrium between heat loss and production defines the T_b and the torpid TNZ, but does a decrease in thermal conductance contribute to a decrease in MR? The influence of thermal conductance on TMR has been studied with somewhat equivocal results. Conductance calculated from torpid and euthermic metabolic rates from a wide range of species was found to be lower during torpor than during euthermy (Snyder and Nestler 1990). By contrast, thermal conductance measured by direct calorimetry was not found to differ between normothermic and torpid states in Djungarian hamsters (Heldmaier and Ruf 1992). Moreover, Ortmann and Heldmaier (2000) found no difference in thermal conductance between hibernating Alpine marmots and normothermic marmots exposed to very low T_a; however, this result appeared to depend on ambient temperature conditions when monochromic thermal conductance was measured. Finally, arguing against a role for thermal conductance in MR reduction are findings in torpid stripe-faced dunnarts (*Sminthopsis macroura*). Exposing these small (25 g) marsupials housed at 20°C to He–O_2 increased thermal conductance but decreased or had no effect on TMR (Geiser et al. 1996). A thorough review of findings on thermal conductance during hibernation is beyond the scope of this chapter; however, given discrepancy between studies results that do not measure thermal conductance directly must be interpreted with caution.

Another explanation for equivocal results may be due to an interaction between season and the T_a to which animals are acclimated. Jinka et al. (2011) show that A_1AR-induced cooling during onset of torpor is slower than A_1AR-induced cooling in summer when animals do not show spontaneous bouts of torpor. Summer and winter animals cool at different rates despite similar rates in initial decline of MR suggesting that slower rate of cooling in winter animals is due to a slower rate of heat loss (Jinka et al. 2011). We subsequently studied T_b, but not MR in arctic ground squirrels housed several months at 18°C and again found a seasonal difference in response to CHA. However, results in animals acclimated to this warmer T_a differed from those reported by Jinka et al. When CHA was administered to these animals acclimated to 18°C T_a the two winter animals (i.e., those who had begun to display spontaneous torpor bouts), cooled at a rate of 4°C h^{-1} and died with T_b's between 25 and 26°C (Fig. 32.2). For ethical reasons this dose of CHA was not repeated in animals acclimated to this warm T_a. Similarly, rapid cooling and subsequent death following CHA given icv was observed in hamsters, although animals died about 18 h after drug administration after reaching torpid-like T_b (Shintani et al. 2005). Rate of cooling may be decreased by mechanisms engaged through an interaction between season and ambient temperature such that the gain in the capacity to minimize heat loss is increased during the winter season. More work is required to systematically monitor A_1AR agonist induced decreases in MR, thermal conductance, and T_b across seasons in animals acclimated to different ambient

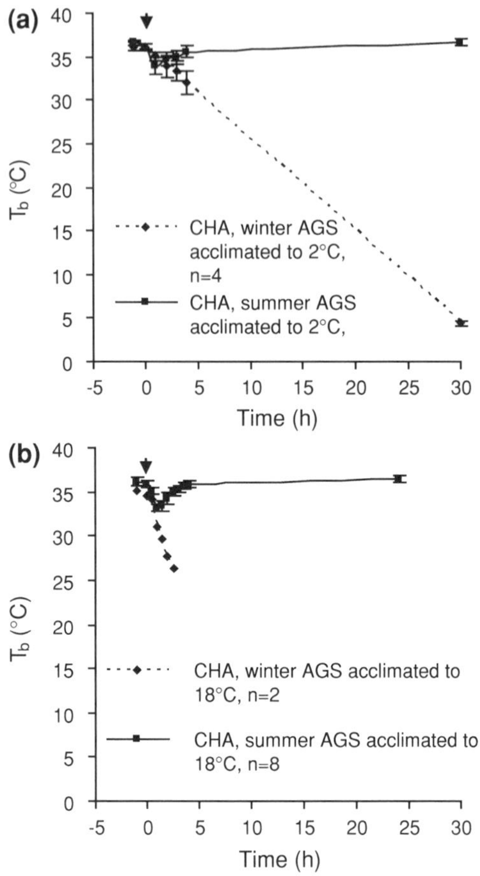

Fig. 32.2 Interaction between the effects of T_a and season on the rate of CHA (0.5 mg kg^{-1}, ip)-induced decline in T_b and CHA-induced torpor. CHA induced a torpor-like state from which animals rewarmed without extraneous heat or other intervention in winter animals, i.e., those showing spontaneous torpor that were housed continuously at T_a of 2°C. Summer animals, at T_a of 2°C, showed a typical hypothermic-like response (**a**). By contrast, CHA-induced a more rapid decline in T_b and cessation of vital signs in winter animals, i.e., those showing spontaneous torpor housed continuously at T_a of 18°C. Summer animals, at T_a of 18°C, showed a typical hypothermic-like response (**b**). Winter animals acclimated to 18°C died following CHA administration. Data shown in **a** are from Jinka et al. (2011). T_b (sc) was collected using IPTT-300 transponders (Bio Medic Data Systems). T_b at 30 h was monitored with a thermocouple (Model H H21 Microprocessor Thermometer, Type J-K-T Thermocouple, OMEGA Engineering) as the IPTT transponders are not reliable below 30°C. Studies were approved by the UAF Institutional Animal Care and Use Committee (IACUC)

temperatures to fully understand this relationship and its impact on TMR. Cooling at rates greater than 1°C h^{-1} appears to either compromise the capacity to survive CHA-induced torpor or high rates of cooling could be a symptom of these animals not being fully prepared for heterothermy.

32.5 Black Bears

Hibernation in black bears differs from that in smaller hibernators by a very moderate decrease in T_b to 30–35°C while reducing minimum metabolic rate to only 25% of BMR (Tøien et al. 2011). In bears hibernation is not interrupted by episodic arousals, presumably because T_b is maintained above 30°C, a threshold T_b found to be associated with interbout arousals (Dausmann et al. 2004). About 50% of the decrease in minimum metabolism in hibernating bears was found to be temperature independent, while the other half may be due to Q_{10} effects. Like ground squirrels, entry into a hypometabolic state is seasonal (Tøien et al. 2011). Little is known about A_1AR signaling in bears. Hibernating black bears sleep most of the time (Barnes et al. 1999); that may be initiated or maintained by high A_1AR stimulation. However, the reduction in minimum metabolism to only 25% in hibernating bears is not due to sleep alone. Tøien et al. (2011) showed that MR in sleeping bears after complete recovery from hibernation was significantly higher than minimum MR during the hibernating season and close that expected for a non-hibernating carnivore. Thus, differences in metabolic rates were not due to differences in vigilance state.

32.6 Other Triggers in Torpor Induction

Through influences within the CNS A_1AR activation could potentially influence the release of circulating factors capable of affecting metabolic rate of peripheral organs or production of other signaling molecules within peripheral tissues. Administration of the A_1AR agonist N^6-cyclohexyladenosine (CHA) to arctic ground squirrels in summer produces a decrease in T_b within 1 h of icv injection that subsides within 3 h. By contrast, in winter, CHA induces a torpor-like response, where a minimum T_b is not achieved until 30 h or more after the icv injection. T_b does not fall significantly in the first few hours after injection arguing against a significant role of temperature in prolonging the half-life or duration of action of the drug.

Investigations of blood-borne and tissue-borne factors date back to the 1930s. These studies reported decreases in T_b or MR when tissue extracts from hibernators were injected either i.p. or i.v. into both hibernating and non-hibernating species (Nietschke and Maier 1932; Wendt 1937; Hook 1940; Zirm 1956; Morrison and Allen 1962; Bigelow et al. 1964; Johannsen 1973). In many cases, however, these treatments failed to mimic the duration of natural torpor or were unable to be replicated. In the late 1960s, hibernation induction trigger (HIT) isolated from the blood of 13-lined ground squirrels (*Ictidomys tridecemlineatus*) was suggested to induce a torpor-like state through activation of opioid receptors (Dawe and Spurrier 1969; Margules et al. 1979; Bruce et al. 1987). Later, Wang et al. (1988) questioned the specificity and validity of the 13-lined ground squirrel bioassay since saline injections were equally effective as the "trigger" substance.

Many of these early studies lack convincing positive and negative control experiments. Jinka et al. (2011) illustrate several approaches to minimize confounding influences. These include tracking of the hibernation and "off-hibernation" or "summer" season by monitoring circannual rhythms in spontaneous torpor in individual animals. Positive control experiments can be more challenging to design than negative controls. When possible, however, administration of a substance known to produce a measurable response makes a negative response more convincing. The search for circulating factors will benefit from more carefully controlled experiments in the future.

Although different from spontaneous torpor, a state of metabolic suppression has been demonstrated convincingly in mice exposed to gaseous hydrogen sulfide (H_2S) at 80 ppm. This concentration of H_2S decreases metabolic rate prior to a drop in core T_b when ambient temperature is gradually decreased (Blackstone et al. 2005). H_2S is a neuromodulator (Kimura 2002) that inhibits cytochrome oxidase at exposure levels greater than 30 ppm (Dorman et al. 2002). Exogenous sulfide-induced cooling and metabolic suppression may find medicinal applications, although H_2S's role in hibernation induction remains to be demonstrated. An A_1AR-dependent mechanism cannot be ruled out in H_2S-induced metabolic suppression since H_2S at 80 ppm produces chemical hypoxia and hypoxia leads to an increase in adenosine release and suppresses MR and T_b (Dunwiddie and Masino 2001; Tattersall and Milsom 2003).

3-Iodothyronamine (T_1AM) is found in the brain and is considered to be an endogenous derivative of thyroid hormone. It is an agent that, when systemically administered, causes an immediate decrease in T_b and bradycardia in rats (Scanlan et al. 2004). When administered to mice or hamsters T_1AM produced a small but significant decrease in MR. Like torpor, T_1AM (50 mg kg^{-1}, ip) produced a decrease in respiratory quotient (RQ) from 0.9 to 0.7 within 3 h of injection (Braulke et al. 2008).

Ghrelin is an orexigenic hormone that is derived primarily in the stomach and is synthesized centrally within neurons in the hypothalamus (Gluck et al. 2006; Szentirmai et al. 2009; Healy et al. 2010). Ghrelin interacts with NPY/AgRP neurons in the arcuate nucleus (Gluck et al. 2006) and is known to stimulate appetite (Gluck et al. 2006; Szentirmai et al. 2009; Healy et al. 2010), and growth hormone release (Strassburg et al. 2008), and thus plays an important role in energy balance (Gluck et al. 2006; Healy et al. 2010). Ghrelin has been shown to induce torpor in mice through intraperitoneal injections (Gluck et al. 2006). Healy et al. (2010) conducted an extensive study of ghrelin levels in an obligate seasonal hibernator, the golden-mantled ground squirrel, and found that ghrelin levels during torpor were highest during entrance and exit from sleep and lowest during deep torpor. At this time, no study has been conducted to determine ghrelin concentrations in the blood upon entry into torpor.

The observations by Gluck et al. (2006) and Healy et al. (2010) suggest a potential role for ghrelin as a seasonal modulator of torpor. Buck et al. (unpublished) routinely observe that arctic ground squirrels fast for approximately 1 week prior to entering their first torpor bout in late summer and that arctic ground squirrels typically enter torpor during the sleep phase of their daily cycle. Thus, fasting and time of day would increase circulating ghrelin levels (Healy et al. 2010) coincident with the increase in endogenous adenosine release (Jinka et al. 2011) and could influence the propensity to enter torpor. However, ghrelin levels during deep torpor are lower than when animals are euthermic suggesting that if ghrelin does have a role in torpor of obligate hibernators its effects are likely limited to entry into torpor rather than maintenance of the torpid state. Stimulation of A_1AR increases permeability of the blood–brain barrier, allowing passage of molecules of up to 70kD (Carman et al. 2011); ghrelin is a 6kD protein and could easily pass through to elicit central effects.

32.7 Conclusion

In conclusion, the mechanism of transduction of a central A_1AR-mediated signal to the periphery that suppresses MR from BMR to TMR remains unknown and warrants further study. Hibernation is characterized by profound metabolic suppression and the search for the physiological mechanism(s) responsible for globally reduced metabolism has been a major research focus in hibernation biology for many decades. Central A_1AR activation is emerging as a common mechanism across species for induction of torpor; however, the mechanism by which central A_1AR activation suppresses metabolism remains unknown. Better knowledge of mechanisms of metabolic suppression would likely increase the utility of the hibernator as a model for translational research through application of metabolic suppression and low T_b to human disease states such as hemorrhagic shock, cerebral ischemia, and reperfusion injury following stroke, and multiorgan failure.

References

Barnes BM, Tøien Ø, Blake J, Grahn D, Heller CH, Edgar DM (1999) Hibernation in black bears: body temperature cycles and sleep. FASEB J 13:A740

Barraco RA, Phillis JW (1991) Subtypes of adenosine receptors in the brainstem mediate opposite blood pressure responses. Neuropharmacology 30:403–407

Barros RC, Branco LG, Carnio EC (2006) Respiratory and body temperature modulation by adenosine A1 receptors in the anteroventral preoptic region during normoxia and hypoxia. Respir Physiol Neurobiol 153:115–125

Bigelow W, Trimple AS, Schonbaum E, Kovats L (1964) A report on studies with Mannota monax. 1. Biochemical and pharmacological investigations of blood fat. Ann Acad Sci Fenn A 4:37–50

Blackstone E, Morrison M, Roth MB (2005) H_2S induces a suspended animation-like state in mice. Science 308:518

Braulke LJ, Heldmaier G (2010) Torpor and ultradian rhythms require an intact signalling of the sympathetic nervous system. Cryobiology 60:198–203

Braulke LJ, Klingenspor M, DeBarber A, Tobias SC, Grandy DK, Scanlan TS, Heldmaier G (2008) 3-Iodothyronamine: a novel hormone controlling the balance between glucose and lipid utilisation. J Comp Physiol B 178:167–177

Bruce DS, Cope GW, Elam TR, Ruit KA, Oeltgen PR, Su TP (1987) Opioids and hibernation. I. Effects of naloxone on bear HIT'S depression of guinea pig ileum contractility and on induction of summer hibernation in the ground squirrel. Life Sci 41:2107–2113

Buck CL, Barnes BM (2000) Effects of ambient temperature on metabolic rate, respiratory quotient, and torpor in an arctic hibernator. Am J Physiol Regul Integr Comp Physiol 279: R255–R262

Carey HV, Andrews MT, Martin SL (2003) Mammalian hibernation: cellular and molecular responses to depressed metabolism and low temperature. Physiol Rev 83:1153–1181

Carman AJ, Mills JH, Krenz A, Kim DG, Bynoe MS (2011) Adenosine receptor signaling modulates permeability of the blood-brain barrier. J Neurosci 31:13272–13280

Dausmann KH, Glos J, Ganzhorn JU, Heldmaier G (2004) Physiology: hibernation in a tropical primate. Nature 429:825–826

Dawe AR, Spurrier WA (1969) Hibernation induced in ground squirrels by blood transfusion. Science 163:298–299

Dorman DC, Moulin FJ, McManus BE, Mahle KC, James RA, Struve MF (2002) Cytochrome oxidase inhibition induced by acute hydrogen sulfide inhalation: correlation with tissue sulfide concentrations in the rat brain, liver, lung, and nasal epithelium. Toxicol Sci 65:18–25

Drew KL, Buck CL, Barnes BM, Christian SL, Rasley BT, Harris MB (2007) Central nervous system regulation of mammalian hibernation: implications for metabolic suppression and ischemia tolerance. J Neurochem 102:1713–1726

Dunwiddie TV, Masino SA (2001) The role and regulation of adenosine in the central nervous systemcentral nervous system. Annu Rev Neurosci 24:31–55

Geiser F (2004) Metabolic rate and body temperature reduction during hibernation and daily torpor. Annu Rev Physiol 66:239–274

Geiser F, Song X, Körtner G (1996) The effect of He-O_2 exposure on metabolic rate, thermoregulation and thermal conductance during normothermia and daily torpor. J Comp Physiol B 166:190–196

Gluck EF, Stephens N, Swoap SJ (2006) Peripheral ghrelin deepens torpor bouts in mice through the arcuate nucleus neuropeptide Y signaling pathway. Am J Physiol Regul Integr Comp Physiol 291:R1303–R1309

Harris MB, Milsom WK (1995) Parasympathetic influence on heart rate in euthermic and hibernating ground squirrels. J Exp Biol 198:931–937

Healy JE, Ostrom CE, Wilkerson GK, Florant GL (2010) Plasma ghrelin concentrations change with physiological state in a sciurid hibernator (*Spermophilus lateralis*). Gen Comp Endocrinol 166:372–378

Heldmaier G, Ruf T (1992) Body temperature and metabolic rate during natural hypothermia in endotherms. J Comp Physiol B 162:696–706

Heller HC, Colliver GW, Beard J (1977) Thermoregulation during entrance into hibernation. Pflügers Arch 369:55–59

Hook WE (1940) Effect of crude peanut oil extracts of brown fat on metabolism of white fat. Proc Sot Exp Biol Med 45:37–40

Iliff BW, Swoap SJ (2010) Central adenosine receptor signaling is necessary for torpor in mice. In: Biology. Williams College, Massachusetts. http://library.williams.edu/theses/pdf.php?id=409, http://www.fasebj.org/cgi/content/meeting_abstract/24/1_MeetingAbstracts/992.2

Jinka TR, Tøien Ø, Drew KL (2011) Season primes the brain in an arctic hibernator to facilitate entrance into torpor mediated by adenosine A1 receptors. J Neurosci 31:10752–10758

Johannsen BW (1973) Effects of drugs on hibernation. In: Schijnbaum E, Lomax P (eds) The pharmacology of thermoregulation. Karger, Basel, pp 364–381

Johansen K, Krog J, Reite O (1964) Autonomic nervous influence on the heart of the hypothermic hibernator. Ann Acad Sci Fenn 71:243–255

Karpovich SA, Tøien Ø, Buck CL, Barnes BM (2009) Energetics of arousal episodes in hibernating arctic ground squirrels. J Comp Physiol B 179(6):691–700. http://www.springerlink.com/content/878r6648j3mm2516/

Kimura H (2002) Hydrogen sulfide as a neuromodulator. Mol Neurobiol 26:13–19

Kosari S, Rathner JA, Chen F, Badoer E (2011) Centrally administered resistin enhances sympathetic nerve activity to the hindlimb but attenuates the activity to brown adipose tissue. Endocrinology 152:2626–2633

Lyman CP (1958) Oxygen consumption, body temperature and heart rate of woodchucks entering hibernation. Am J Physiol 194:83–91

Lyman CP (ed) (1982) The hibernating state, recent theories of hibernation. Academic Press, New York

Lyman CP, O' Brien RC (1960) Circulatory changes in the thirteen-lined ground squirrel during the hibernating cycle. Bull Mus Comp Zool 124:353–372

Lyman CP, O'Brien RC (1963) Autonomic control of circulation during the hibernating cycle in ground squirrels. J Physiol 168:477–499

Margules DL, Goldman B, Finck A (1979) Hibernation: an opioid-dependent state? Brain Res Bull 4:721–724

McClure JM, O'Leary DS, Scislo TJ (2005) Stimulation of NTS A1 adenosine receptors evokes counteracting effects on hindlimb vasculature. Am J Physiol Heart Circ Physiol 289: H2536–H2542

Miyazawa S, Shimizu Y, Shiina T, Hirayama H, Morita H, Takewaki T (2008) Central A1-receptor activation associated with onset of torpor protects the heart against low temperature in the Syrian hamster. Am J Physiol Regul Integr Comp Physiol 295:R991–R996

Morrison PR, Allen WT (1962) Temperature response of white mice to implants of brown fat. J Mammal 43:13–17

Mosqueda-Garcia R, Tseng CJ, Appalsamy M, Robertson D (1989) Modulatory effects of adenosine on baroreflex activation in the brainstem of normotensive rats. Eur J Pharmacol 174:119–122

Nakamura K (2011) Central circuitries for body temperature regulation and fever. Am J Physiol Regul Integr Comp Physiol 301:R1207–R1228

Nietschke A, Maier E (1932) Uber das Vergiftungbild nach Injektion von Extrakten aus lymphatischem Bewebe (P-Substanz) und seine Beziehung zum Winterschlaf. A Ges Exp Med 82:215–226

Ortmann S, Heldmaier G (2000) Regulation of body temperature and energy requirements of hibernating alpine marmots (Marmota marmota). Am J Physiol Regul Integr Comp Physiol 278:R698–704

Osborne PG, Sato J, Shuke N, Hashimoto M (2005) Sympathetic alpha-adrenergic regulation of blood flow and volume in hamsters arousing from hibernation. Am J Physiol Regul Integr Comp Physiol 289:R554–R562

Scanlan TS, Suchland KL, Hart ME, Chiellini G, Huang Y, Kruzich PJ, Frascarelli S, Crossley DA, Bunzow JR, Ronca-Testoni S, Lin ET, Hatton D, Zucchi R, Grandy DK (2004) 3-Iodothyronamine is an endogenous and rapid-acting derivative of thyroid hormone. Nat Med 10:638–642

Scholander PF, Hock R, Walters V, Irving L (1950) Adaptation to cold in arctic and tropical mammals and birds in relation to body temperature, insulation, and basal metabolic ratebasal metabolic rate. Biol Bull 99:259–271

Shintani M, Tamura Y, Monden M, Shiomi H (2005) Characterization of N(6)-cyclohexyladenosine-induced hypothermia in Syrian hamsters. J Pharmacol Sci 97:451–454

Shiomi H, Tamura Y (2000) Pharmacological aspects of mammalian hibernation: central thermoregulation factors in hibernation cycle. Folia Pharmacol Jpn 116:304–312

Snapp BD, Heller CH (1981) Suppression of Metabolism during Hibernation in Ground Squirrels (Citellus lateralis). Physiol Zool 54:297–307

Snyder GK, Nestler JR (1990) Relationships between body temperature, thermal conductance, Q_{10} and energy metabolism during daily torpor and hibernation in rodents. J Comp Physiol B 159:667–675

Song X, Kortner G, Geiser F (1996) Interrelations between metabolic rate and body temperature during entry into daily torpor in *Sminthopsis macroura*. In: Geiser F, Hulbert AJ, Nicol SC (eds) Adaptations to the cold: the tenth international hibernation symposium. University of New England Press, Armidale, pp 63–69

Song X, Kortner G, Geiser F (1998) Temperature selection and use of torpor by the marsupial Sminthopsis macroura. Physiol Behav 64:675–682

Strassburg S, Anker SD, Castaneda TR, Burget L, Perez-Tilve D, Pfluger PT, Nogueiras R, Halem H, Dong JZ, Culler MD, Datta R, Tschop MH (2008) Long-term effects of ghrelin and ghrelin receptor agonists on energy balance in rats. Am J Physiol Endocrinol Metab 295: E78–E84

Swoap SJ, Weinshenker D (2008) Norepinephrine controls both torpor initiation and emergence via distinct mechanisms in the mouse. PLoS ONE 3:e4038

Szentirmai E, Kapas L, Sun Y, Smith RG, Krueger JM (2009) The preproghrelin gene is required for the normal integration of thermoregulation and sleep in mice. Proc Natl Acad Sci U S A 106:14069–14074

Tattersall GJ, Milsom WK (2003) Transient peripheral warming accompanies the hypoxic metabolic response in the golden-mantled ground squirrel. J Exp Biol 206:33–42

Tøien Ø, Blake J, Edgar DM, Grahn DA, Heller HC, Barnes BM (2011) Hibernation in black bears: independence of metabolic suppression from body temperature. Science 331:906–909

Twente JW, Twente J (eds) (1978) Autonomic regulation of hibernation by Citellus and Eptesicus. Academic Press, New York

Wang LC, Belke D, Jourdan ML, Lee TF, Westly J, Nurnberger F (1988) The "hibernation induction trigger": specificity and validity of bioassay using the 13-lined ground squirrel. Cryobiology 25:355–362

Wendt CF (1937) Uber Wirkungen eines extraktes aus dem braunen Fettgewebe des Winterschlafenden Igels. Z Physiol Chem 249:182

Withers PC, Casey TM, Casey KK (1979) Allometry of respiratory and haematological parameters of arctic mammals. J Comp Physiol Behav 64:343–350

Wojciechowski MS, Jefimow M, Tegowska E (2007) Environmental conditions, rather than season, determine torpor use and temperature selection in large mouse-eared bats (*Myotis myotis*). Comp Biochem Physiol A Mol Integr Physiol 147:828–840

Zirm KL (1956) Ein Beitrag zur Kenntnis der naturlichen Winterschlafes und seines regulierenden Wirkstoffes II. Z Naturforsch 166:536–538

Zosky GR (2002) The parasympathetic nervous system: its role during torpor in the fat-tailed dunnart (*Sminthopsis crassicaudata*). J Comp Physiol B 172:677–684

Zosky GR, Larcombe AN (2003) The parasympathetic nervous system and its influence on heart rate in torpid western pygmy possums, Cercatetus concinnus (Marsupialia: Burramyidae). Zoology (Jena) 106:143–150

Chapter 33
Fast In, Slow Out: Patterns and Mechanisms of Mitochondrial Suppression in Hibernation

James F. Staples

Abstract Oxidative phosphorylation by *Ictidomys tridecemlineatus* skeletal muscle mitochondria is suppressed by 35% in the torpor phase of hibernation. This suppression is evident only when measured at 37°C and with succinate as respiratory substrate. During entrance into a torpor bout liver mitochondrial respiration is suppressed rapidly by 80%; however, this suppression is reversed only gradually during arousal. Oxaloacetate inhibition of succinate dehydrogenase can explain little of this suppression. This "fast in, slow out" pattern suggests acute, temperature-sensitive, enzyme-mediated downregulation of mitochondrial metabolism, perhaps by post-translational modification of key enzymes.

33.1 Introduction

The metabolic suppression characteristic of mammalian hibernation (Fig. 33.1) is not only a fascinating biological phenomenon but may also provide insights into important aspects of metabolic regulation. Experiments evaluating hibernation should consider the cyclical nature of torpor bouts, as well as seasonal acclimation and endogenous rhythms. The importance of this "two-switch" model has been highlighted recently by proteomic (Martin et al. 2008) and metabolomic (Serkova et al. 2007) data, but attempts to incorporate this approach into functional analyses of metabolic suppression per se have been relatively scarce.

If suppression of mitochondrial metabolism contributes significantly to energy conservation in hibernation one would predict that the pattern of suppression

J. F. Staples (✉)
Department of Biology, University of Western Ontario, London, ON, Canada
e-mail: jfstaple@uwo.ca

T. Ruf et al. (eds.), *Living in a Seasonal World*,
DOI: 10.1007/978-3-642-28678-0_33,

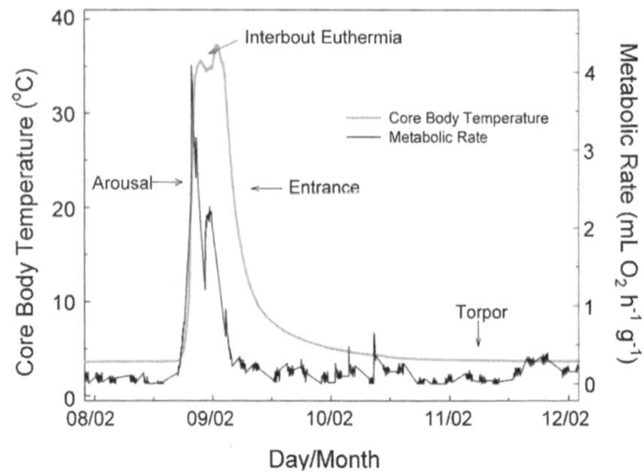

Fig. 33.1 Metabolic rate and body temperature of an individual *Ictidomys tridecemlineatus* in different stages of a torpor bout. Modified from Brown et al. (2012) with permission

would parallel that of whole-animal metabolic rate, i.e., a rapid suppression during entrance into a torpor bout and a rapid reversal during arousal. Data from earlier studies on liver mitochondria have supported this prediction, but to different degrees in different species (Roberts and Chaffee 1972; Martin et al. 1998; Muleme et al. 2006). This chapter describes experiments that further evaluate this prediction, using more time points during entrance and arousal to better discern the pattern of reversible metabolic suppression.

In contrast to liver, no suppression has been noted when skeletal muscle mitochondrial respiration is compared between torpid and summer-active ground squirrels (Barger et al. 2003; Muleme et al. 2006). Unfortunately, to my knowledge, no studies have compared skeletal muscle mitochondrial metabolism between the different phases of a torpor bout. This chapter describes experiments that compare the metabolism of skeletal muscle mitochondria isolated from ground squirrels in torpor and interbout euthermia.

Finally, mechanisms of mitochondrial metabolic suppression in hibernation remain elusive. In this chapter I present data from experiments that evaluate one potential mechanism, the inhibition of succinate dehydrogenase by oxaloacetate. For full descriptions of the experiments and data described below, see papers by Armstrong and Staples (2010), Chung et al. (2011), and Brown et al. (2012).

33.2 Materials and Methods

33.2.1 Animal Husbandry and Hibernation Monitoring

Thirteen-lined ground squirrels (*Ictidomys tridecemlineatus*) were either live-trapped (Carman MB, 49°29′N 98°0′W), or bred in captivity from wild-caught animals (Vaughan et al. 2006). Details of animal care can be found in recently published papers (Armstrong and Staples 2010; Chung et al. 2011). During the summer, while being held at natural photoperiods and 23°C, temperature-sensitive radiotelemeters (TA-F20, Data Sciences International, Arden Hills, MN) were implanted intraperitoneally under aseptic conditions to monitor core body temperature (T_b). Following at least 1 week of recovery some summer-active animals were sampled for liver mitochondrial studies (see below).

By early October most animals had ceased to gain weight and were transferred to environment chambers, where ambient temperature was decreased 3°C daily down to 5°C, and photoperiod was reduced to 2L:22D. Torpor bouts generally commenced within 1 week at 5°C, and animals were left undisturbed for 1 month before sampling began. During the hibernation season animals used for both skeletal muscle and liver studies were sampled while torpid (3–4 days into a torpor bout, T_b near 5°C and constant) and during interbout euthermia (T_b near 37°C and constant for 1–2 h). For liver studies animals were also sampled in the early (T_b 15°C) and late (T_b 30°C) stages of arousal, and early (T_b 30°C) and late (T_b 15°C) stages of entrance.

33.2.2 Isolation and Respiratory Characterization of Mitochondria

A crude liver mitochondrial fraction was isolated by gentle homogenization in isotonic buffer, followed by differential centrifugation (Muleme et al. 2006). This fraction was further purified by Percoll density centrifugation as described recently (Armstrong and Staples 2010). Crude muscle mitochondria were isolated following treatment with protease (Bhattacharya et al. 1991), and further purified using Percoll density centrifugation (Yoshida et al. 2007).

Liver mitochondrial respiration was measured at 37°C using a Clark-type oxygen electrode (Rank Brothers, Cambridge, UK). The incubation medium consisted of 225 mM sucrose, 20 mM HEPES, 10 mM KH_2PO_4, and 0.5 mM EGTA, pH 7.2. State 2 respiration was stimulated by adding succinate (6 mM) along with rotenone (2 μg ml^{-1} in ethanol), and 0.1 mM ADP was added to stimulate state 3 respiration. Respiration stabilized to state 4 when most of the ADP was phosphorylated. To evaluate potential inhibition of succinate dehydrogenase by the accumulation of oxaloacetate within the mitochondrial matrix, aliquots of liver mitochondria were preincubated with 2 mM isocitrate at 37°C for 10 min before state 3 respiration was determined.

Muscle mitochondrial respiration was determined at 10, 25, and 37°C using an Oroboros O2K high-resolution respirometer. The muscle assay buffer consisted of 110 mM sucrose, 0.5 mM EGTA, 3 mM $MgCl_2$, 60 mM potassium lactobionate, 20 mM taurine, 10 mM KH_2PO_4, 20 mM HEPES, 1% bovine serum albumin, and pH 7.1. State 3 respiration was measured with 10 mM glutamate (with 2 mM malate) and 6 mM succinate (with 2 μg ml^{-1} rotenone). After stable state 3 values were established, state 4 was estimated by adding oligomycin (2 μg ml^{-1} in ethanol). All respiration rates were standardized to total mitochondrial protein, as determined using Bio-Rad dye reagent, with bovine serum albumin as a standard.

33.3 Results

In skeletal muscle mitochondria using glutamate and malate as substrates, neither state 3 nor state 4 respiration differed between animals sampled in torpor versus interbout euthermia, regardless of in vitro assay temperature (Fig. 33.2a, b). Succinate oxidation produced higher state 3 respiration rates than glutamate with malate (compare Fig. 33.2a and c). Moreover, with succinate as a substrate, muscle state 3 was suppressed by 35% in torpor compared with interbout euthermia (Fig. 33.2c; $P < 0.05$, Tukey's following repeated measures mixed model). This difference, however, was only evident at an assay temperature of 37°C, and not at 25 or 10°C. While decreasing assay temperature both states 3 and 4 respiration rates decreased, this effect did not differ between torpor and interbout euthermia (Table 33.1).

During arousal liver, mitochondrial state 3 respiration increases, but only gradually. As T_b increases from approximately 5°C in torpor to 15°C in early arousal and 30°C in late arousal, state 3 respiration increases more than threefold (Fig. 33.3a). Despite this impressive increase, between late arousal and interbout euthermia state 3 increases almost twofold more, to values similar to the summer-active condition. By contrast, between interbout euthermia and early entrance ($T_b = 30°C$) state 3 respiration declines almost fivefold, and no further decline is seen through late entrance or into torpor. A similar pattern is noted for state 4 respiration (Fig. 33.3b).

Preincubating liver mitochondria with 2 mM isocitrate nearly doubled state 3 respiration in liver mitochondria from torpid ground squirrels (Fig. 33.4). There was little effect of isocitrate in other torpor bout stages.

33.4 Discussion

Previous studies found no difference in skeletal muscle state 3 respiration between torpid and summer-active ground squirrels (Muleme et al. 2006). By comparing mitochondria isolated from animals in torpor and interbout euthermia, the experiments described here controlled seasonal acclimation that could potentially mask effects attributable to hibernation per se. This innovative approach elucidated significant suppression of skeletal muscle succinate oxidation capacity in torpor (Brown et al. 2012). Another study compared torpid arctic ground squirrels (*Spermophilus parryi*) with "active" animals and found no difference in any of several bioenergetic properties of skeletal muscle mitochondria, including state 3 (Barger et al. 2003). The apparent disagreement between these studies may represent actual species-specific differences in mitochondrial properties, but more likely reflect differences in experimental design, highlighting the importance of rigorous monitoring of hibernation status.

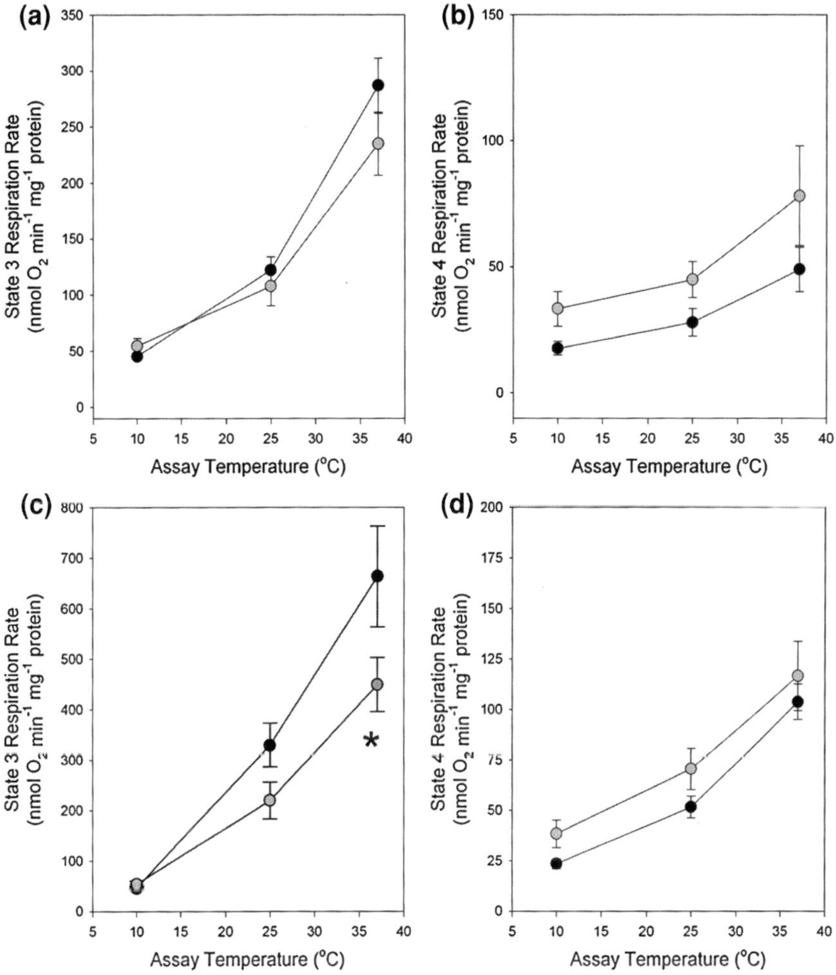

Fig. 33.2 Respiration of skeletal muscle mitochondria isolated from *Ictidomys tridecemlineatus* in different stages of a torpor bout. Respiration rates of interbout euthermic (*solid circles*) and torpid (*gray circles*) animals were measured using 10 mM glutamate with 2 mM malate (**a** and **b**) or 10 mM succinate with rotenone (**c** and **d**). Significant differences between interbout euthermic and torpid groups at a given temperature are indicated with *. Values are mean ± SEM. Sample sizes; torpid $n = 8$, interbout euthermia $n = 5$. From Brown et al. (2012) with permission

The suppression of skeletal muscle mitochondrial metabolism seen in *I. tridecemlineatus* is temperature-dependent, and significant differences between torpor and interbout euthermia are seen at 37°C, but not at 25 or 10°C, similar to the pattern observed for liver mitochondria (Muleme et al. 2006). This temperature dependence suggests that suppression of skeletal muscle mitochondria may

Table 33.1 Q_{10} values showing the effect of in vitro assay temperature on respiration rates of skeletal muscle mitochondria isolated from *Ictidomys tridecemlineatus* in different stages of a torpor bout

	State 3		State 4	
	Interbout euthermia	Torpor	Interbout euthermia	Torpor
Gultamate + malate	2.02 ± 0.08	2.72 ± 0.87	1.48 ± 0.09	1.48 ± 0.08
Succinate	2.01 ± 0.048	2.94 ± 0.99	1.77 ± 0.07	1.65 ± 0.09

Values are means ± SEM. Sample size for each group is 5. Adapted from Brown et al. (2012)

contribute most to reducing whole-animal metabolism during the early stages of entrance when T_b is relatively high. The relatively small 35% suppression (compared with up to 80% in liver) seen in skeletal muscle mitochondria is reversed during the few hours of arousal, presumably precluding major changes in mitochondrial biogenesis and expression of gene products. In fact T_b during most of arousal is low enough to inhibit protein synthesis (van Breukelen and Martin 2001). Such a rapid increase in mitochondrial metabolism argues for acute regulation by mechanisms including allosteric modulation or post-translational modification of enzymes. This pattern of reversible suppression is similar to that noted in liver (see below); however, the kinetics of induction during entrance and reversal during arousal are yet to be established for skeletal muscle.

Following the initial increase in metabolic rate at the beginning of arousal, T_b increases from approximately 5 to 15°C in 2–3 h, and to 30°C in 3–4 h. During these time spans liver mitochondrial respiration increases approximately threefold, but does not reach maximal levels until T_b is close to 37°C during interbout euthermia. By contrast, during entrance T_b falls relatively slowly, reaching 30°C only after approximately 4 h, and 15°C after 6 h. During this relatively slow decline in T_b, liver mitochondrial metabolism decreases fivefold and remains at low levels through late entrance. These results differ somewhat from an earlier study on golden-mantled ground squirrels, which showed significantly lower state 3 respiration in arousal (T_b 18–25°C), compared with interbout euthermia, but no difference during entrance (Martin et al. 1998). While these differences may represent species specificities, they more likely reflect the effect of in vitro assay temperatures. Data in Fig. 33.3 were collected at 37°C, whereas Martin et al. (1998) used 25°C, a temperature that greatly reduces the degree of metabolic suppression seen in *I. tridecemlineatus* liver mitochondria (Muleme et al. 2006).

The rapid decline of liver mitochondrial respiration during entrance supports a potential role for suppression of liver mitochondrial metabolism in reducing whole-animal metabolic rate during entrance. The relatively slow rate of T_b decline during entrance would delay temperature-dependent "Arrhenius" reductions in tissue metabolism, and mechanisms of active, regulated suppression of mitochondrial metabolism may have provided a selective advantage. Conversely the reversal of mitochondrial metabolic suppression during arousal happens only gradually, but whole-animal metabolic rate increases very rapidly. Therefore it

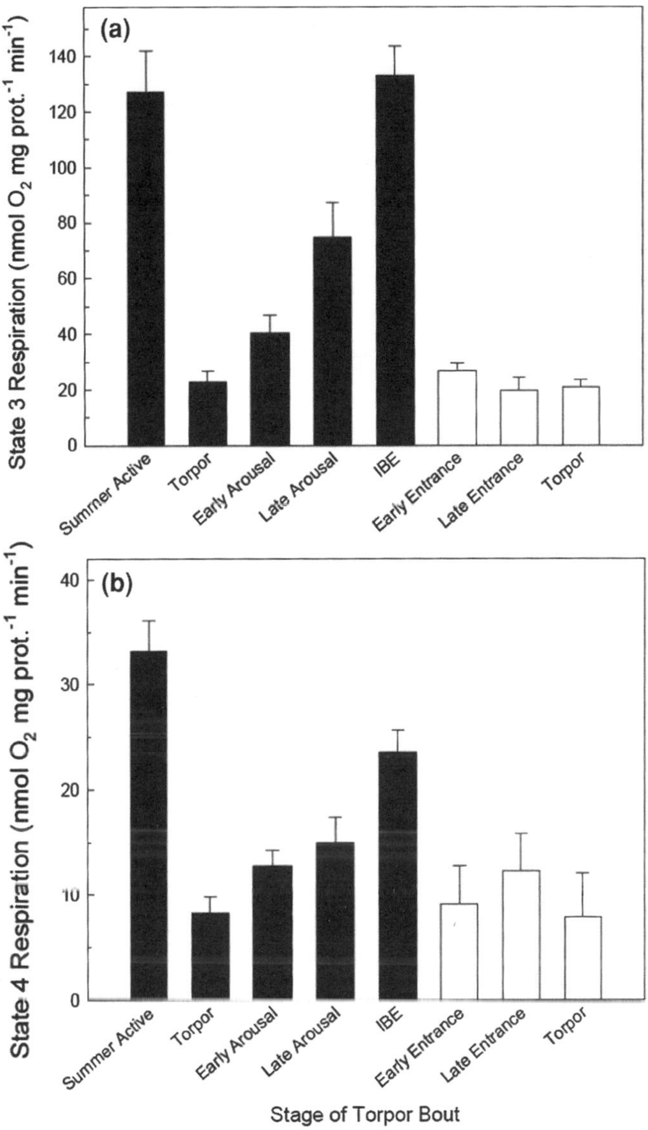

Fig. 33.3 State 3 (**a**) and state 4 (**b**) respiration rates (6 mM succinate) of liver mitochondria isolated from *Ictidomys tridecemlineatus* sampled in different stages of a torpor bout and in summer. Values are mean ± SEM. *Filled bars* adapted from Armstrong and Staples (2010; summer active $n = 6$, torpor $n = 6$, early arousal $n = 5$, late arousal $n = 5$, interbout euthermic $n = 4$), *open bars* adapted from Chung et al. (2011; early entrance $n = 4$, late entrance $n = 4$, torpor $n = 3$)

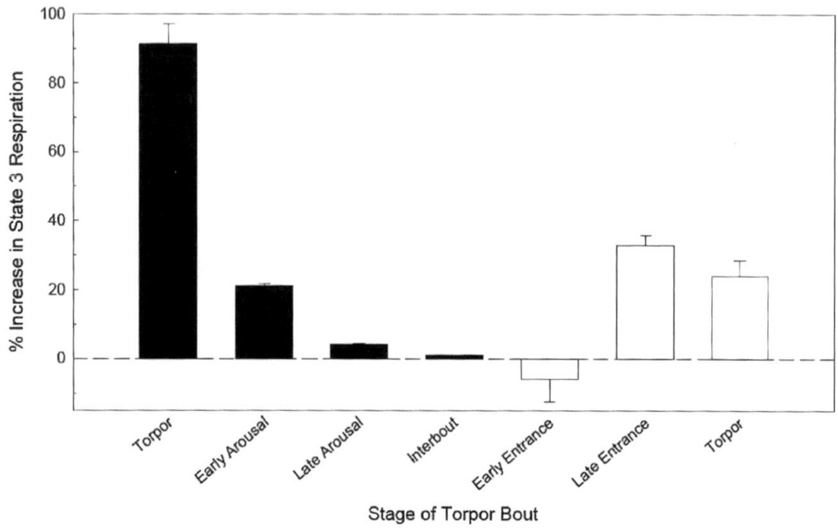

Fig. 33.4 The effect of preincubation with isocitrate (2 mM, 10 min at 37°C) on state 3 respiration of liver mitochondria isolated from *Ictidomys tridecemlineatus* in different stages of a torpor bout. *Filled bars* adapted from Armstrong and Staples (2010), *open bars* adapted from Chung et al. (2011). Values are mean ± SEM. For sample sizes see Fig. 33.3 caption

seems likely that tissues other than liver are capable of much more rapid metabolic upregulation during arousal. Indeed acute increases in GDP-binding in brown adipose tissue are observed during arousal in Richardson's ground squirrels, *Urocitellus richardsonii* (Milner et al. 1989), facilitating non-shivering thermogenesis, and increases in both whole-animal metabolic rate and T_b.

There are several potential mechanisms by which liver and muscle mitochondrial metabolism could be suppressed acutely during torpor. Succinate dehydrogenase can be inhibited by oxaloacetate (Tyler 1960; Ackrell et al. 1974), which may accumulate in liver mitochondria during torpor due to reduced activity of citrate synthase (Page et al. 2009). Preincubation with isocitrate, which alleviates oxaloacetate inhibition of succinate dehydrogenase (Fedotcheva et al. 1985), stimulated liver mitochondrial state 3 respiration rate (Fig. 33.4). Although this stimulation almost doubles respiration rates in early arousal, these rates are still 70% lower than interbout euthermia values (Armstrong and Staples 2010). Moreover this stimulation is much less pronounced in late arousal, interbout euthermia, and entrance. So it appears that any accumulation of oxaloacetate is quickly reversed early in entrance, and probably does not contribute significantly to acute regulation of mitochondrial metabolism in other stages of torpor bouts.

Mitochondrial metabolic suppression, in liver at least, is initiated early in entrance, when T_b is quite high, but is reversed only gradually during arousal, where T_b is quite low. This "fast in, slow out" pattern suggests that the mechanism of suppression is enzyme-mediated, and therefore sensitive to temperature.

The role of kinase-mediated protein phosphorylation in the regulation of mitochondrial metabolism is a rapidly emerging field. Recently protein kinase A and adenylate cyclase activity have been reported within mitochondria (Acin-Perez et al. 2009). Activation of this signaling pathway by CO_2–HCO_3^- leads to phosphorylation of mitochondrial proteins including subunits of cytochrome c oxidase, altering enzyme activity (Acin-Perez et al. 2009). Particularly relevant to our data, phosphorylation of a flavoprotein subunit of succinate dehydrogenase increases the activity of the holoenzyme (Tomitsuka et al. 2009). Inspired by these recent discoveries we have begun comparing the phosphoproteome of mitochondria purified from *I. tridecemlineatus* in summer, torpor, and interbout euthermia.

References

Acin-Perez R, Salazar E, Kamenetsky M, Buck J, Levin LR, Manfredi G (2009) Cyclic AMP produced inside mitochondria regulates oxidative phosphorylation. Cell Metab 9:265–276

Ackrell BAC, Kearney EB, Mayr M (1974) Role of oxalacetate in the regulation of mammalian succinate dehydrogenase. J Biol Chem 249:2021–2027

Armstrong C, Staples J (2010) The role of succinate dehydrogenase and oxaloacetate in metabolic suppression during hibernation and arousal. J Comp Physiol B 180:775–783

Barger J, Brand MD, Barnes BM, Boyer BB (2003) Tissue-specific depression of mitochondrial proton leak and substrate oxidation in hibernating arctic ground squirrels. Am J Physiol 284: R1306–R1313

Bhattacharya SK, Thakar JH, Johnson PL, Shanklin DR (1991) Isolation of skeletal muscle mitochondria from hamsters using an ionic medium containing ethylenediaminetraacetic acid and nagarse. Anal Biochem 192:344–349

Brown JCL, Chung DJ, Belgrave KR, Staples JF (2012) Mitochondrial metabolic suppression and reactive oxygen species production in liver and skeletal muscle of hibernating thirteen-lined ground squirrels. Am J Physiol 302:R15–R28

Chung D, Lloyd GP, Thomas RH, Guglielmo CG, Staples JF (2011) Mitochondrial respiration and succinate dehydrogenase are suppressed early during entrance into a hibernation bout, but membrane remodeling is only transient. J Comp Physiol B 181:699–711

Fedotcheva NJ, Sharyshev AA, Mironova GD, Kondrashova MN (1985) Inhibition of succinate oxidation and K^+ transport in mitochondria during hibernation. Comp Biochem Physiol 82B:191–195

Martin SL, Maniero GD, Carey C, Hand SC (1998) Reversible depression of oxygten consumption in isolated liver mitochondria during hibernation. Physiol Biochem Zool 72:255–264

Martin SL, Epperson LE, Rose JC, Kurtz CC, Ane C, Carey HV (2008) Proteomic analysis of the winter-protected phenotype of hibernating ground squirrel intestine. Am J Physiol 295:R316–R328

Milner R, Wang L, Trayhurn P (1989) Brown fat thermogenesis during hibernation and arousal in Richardson's ground squirrel. Am J Physiol 256:R42–R48

Muleme HM, Walpole AC, Staples JF (2006) Mitochondrial metabolism in hibernation: metabolic suppression, temperature effects, and substrate preferences. Physiol Biochem Zool 79:474–483

Page MM, Peters CW, Staples JF, Stuart JA (2009) Intracellular superoxide dismutases are not globally upregulated during hibernation in the 13-lined ground squirrel *Spermophilus tridecemlineatus*. Comp Biochem Physiol A 152:115–122

Roberts JC, Chaffee RR (1972) Suppression of mitochondrial respiration in hibernation and its reversal in arousal. In: Smith RE, Shields JC, Hannon PP, Horwitz BA (eds) Proceedings of the international symposium on environmental physiology: bioenergetics and temperature regulation, FASEB, Bethesda

Serkova NJ, Rose JC, Epperson LE, Carey HV, Martin SL (2007) Quantitative analysis of liver metabolites in three stages of the circannual hibernation cycle in 13-lined ground squirrels by NMR. Physiol Genom 31:15–24

Tomitsuka E, Kita K, Esumi H (2009) Regulation of succinate-ubiquinone reductase and fumarate reductase activities in human complex II by phosphorylation of its flavoprotein subunit. Proc Japan Acad B 85:258–265

Tyler DB (1960) Effect of citric acid-cycle intermediates on oxaloacetate utilization and succinate oxidation. Biochem J 76:293–297

van Breukelen F, Martin SL (2001) Translational initiation is uncoupled from elongation at 18°C during mammalian hibernation. Am J Physiol 281:R1374–R1379

Vaughan DK, Gruber AR, Michalski ML, Seidling J, Schlink S (2006) Capture, care, and captive breeding of 13-lined ground squirrels, *Spermophilus tridecemlineatus*. Lab Anim 35:33–40

Yoshida Y, Holloway GP, Ljubicic V, Hatta H, Spriet LL, Hood DA, Bonen A (2007) Negligible direct lactate oxidation in subsarcolemmal and intermyofibrillar mitochondria obtained from red and white rat skeletal muscle muscle. J Physiol 582:1317–1335

Chapter 34
Adjustments of Mitochondrial Energy Transduction in Response to Physiological and Environmental Challenge

Martin Jastroch

Abstract The energy metabolism of animals is shaped by the ecological niche and requires adaptation and acclimatisation to physiological and environmental challenge. These adjustments are complex at different systemic levels and involve regulation of ATP homeostasis at the cellular level. Mitochondria are central to the conversion of nutrient to cellular energy (ATP). Mitochondrial ATP production is not fully efficient, flexible and allows a certain degree of plasticity for physiological adjustments. As a result of an inefficient energy transduction, by-products such as mitochondrial reactive oxygen species and heat are formed. Thus, a quantifiable knowledge on mitochondrial efficiency is required to understand the significance of mitochondrial adjustments for the biology and fitness of the animal. This chapter serves as a general introduction on the principles of mitochondrial energy transduction and efficiency, how to measure mitochondrial energy transduction in isolated mitochondria, reviews past efforts to elucidate adjustments of mitochondrial mechanisms and suggests future perspectives of mitochondrial bioenergetics in integrative and comparative physiology. In particular novel technologies, such as non-invasive measurement of oxygen consumption and membrane potential with fluorescent probes, allow the assessment of energy transduction *in the living cell*, therefore becoming the next stage to study mitochondrial energy metabolism.

M. Jastroch (✉)
German Research Center for Environmental Health (GmbH),
Institute for Diabetes and Obesity, Helmholtz Zentrum Munich,
Ingolstädter Landstr. 1, Neuherberg, 85764 Munich, Germany
e-mail: martin.jastroch@helmholtz-muenchen.de

T. Ruf et al. (eds.), *Living in a Seasonal World*,
DOI: 10.1007/978-3-642-28678-0_34,

34.1 Introduction

34.1.1 Oxidative Phosphorylation

Homeostasis of energy metabolism, when energy supply is balanced with energy demand, is required for the survival of an animal in its environment. This balance is affected by multiple environmental and physiological factors that collectively cause 'stress' and require adjustments of energy metabolism. In cells, there are various pathways of nutrient metabolism providing substrates for cellular energy. We are still far from understanding the control and interaction of various metabolic pathways, as well as the complexity and impact on cellular energy metabolism. Considering the output into cellular energy, substrate-level phosphorylation is one way to produce ATP during glycolysis and the Krebs cycle, without the direct involvement of oxidative phosphorylation at the respiratory chain. Substrate-level phosphorylation has to be considered for energy production but will not be specifically discussed in this chapter.

The major part of ATP is synthesised during oxidative phosphorylation, a process that displays flexibility by altering mitochondrial efficiency. The energy of the substrates is harvested as electrons and is transported along the respiratory chain (Fig. 34.1). This energy is used to pump protons out of the mitochondrial matrix generating a proton-motive force that is coupled to the synthesis of ATP. In the final step of the respiratory chain, the electrons reduce oxygen to water—a reaction that can experimentally assessed by measuring mitochondrial oxygen consumption.

34.1.2 Electron and Proton Leaks

Two processes result in an inefficient energy transfer: single electrons abortively exit during transport (electron leak), resulting in superoxide formation, and protons also leak back to the matrix without the generation of ATP (Fig. 34.1; Jastroch et al. 2010).

The electron leak affects mitochondrial efficiency of the energy producer side, namely the respiratory chain, which produces proton-motive force as the central intermediate. Some exit sites of electrons that will produce superoxide have been defined in vitro (Brand 2010). The magnitude of electron leak depends on the stoichiometry of different "leaking" enzymes and complexes, and their redox state, as well as regulation by proton-motive force. The entry points of substrate energy are well-known: electrons are donated from $NADH_2$ at mitochondrial complex I, from flavin-linked electron donors [α-glycerophosphate dehydrogenase and the electron-transferring flavoprotein Q oxidoreductase (ETFQOR) of β-oxidation], and from oxidation of succinate at mitochondrial complex II. The capacity of these entry points can vary between cell type and physiological

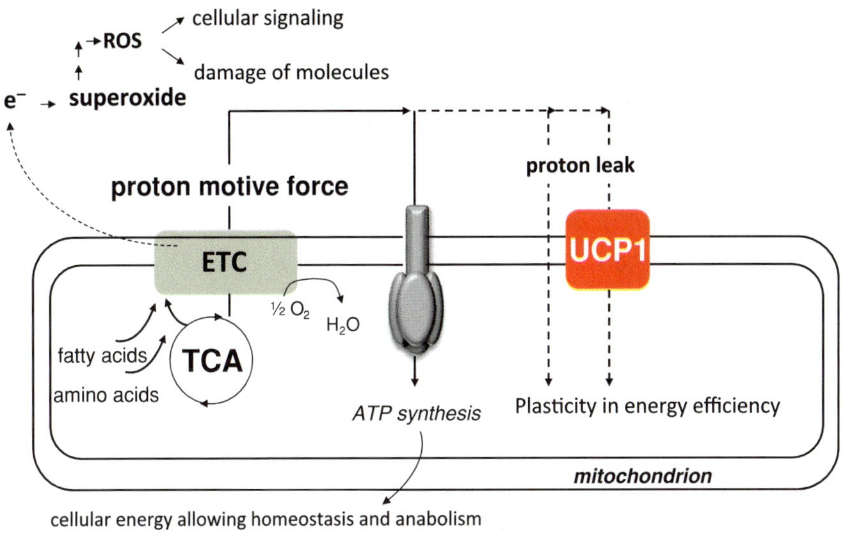

Fig. 34.1 Mitochondrial energy transduction: substrates are oxidised at the respiratory chain and the energy of electrons is harvested during electron transfer until finally oxygen is reduced to water. The energy is used to transport protons from the mitochondrial matrix to the intermembrane space generating a proton-motive force. The proton-motive force is used to drive the ATP synthase; and ATP production fuels cellular processes. This energy transduction is not fully efficient as electrons can leak from the respiratory chain forming superoxide that will subsequently form hydrogen peroxide and other reactive oxygen species that may serve as signalling molecules or damage cellular compounds. Second, protons also leak back to the matrix without the generation of ATP in an unregulated manner, a process that is collectively termed 'basal proton leak', or regulated by mitochondrial carrier proteins (e.g., UCP1)

condition resulting in variable substrate preferences. The enigmatic role of mitochondrial superoxide and resulting reactive oxygen species (ROS) attracts a growing number of studies investigating the cellular role of mitochondrial ROS production. It appears that mitochondrial ROS serves as a signalling molecule, affecting and controlling a variety of cellular processes. When produced over excessively, mitochondrial ROS can damage molecules such as lipids, proteins and DNA. Therefore, the cell neutralise ROS with anti-oxidant defence systems and repairs cellular damage.

The mitochondrial proton leak affects the efficiency of ATP production as the consumption of proton-motive force is partitioned between ATP synthesis and proton leak (Fig. 34.3). While a major interest in proton leak was promoted by brown adipose tissue, a specialised mammalian organ that utilises uncoupling protein 1 (UCP1) to dissipate proton-motive force as heat, a much wider physiological role of proton leak in other tissues, has just been occasionally studied. An unregulated proportion of proton leak is collectively termed 'basal proton leak'. Although the molecular nature is not conclusively revealed, some highly abundant mitochondrial membrane proteins such as the adenine nucleotide translocase

(ANT) and lipid composition may contribute to the magnitude of basal proton leak (Brand et al. 2005; Brookes et al. 1998). Potential roles of basal proton leak may be variation between the stoichiometry of oxidative phosphorylation to other metabolic pathways (e.g., Krebs cycle), or adjustments of ATP production (e.g., in mammalian pancreatic beta cells; Affourtit et al. 2011).

34.1.3 Interrelation Between Electron and Proton Leak

Since proton leak lowers the redox state of the respiratory chain by lowering proton-motive force, ROS production of proton-motive force-sensitive sites is also lowered. Generally, this is referred to as 'mild uncoupling'. However, when investigating dependence of ROS production to proton-motive force, one also has to consider that lowering substrate oxidation also lowers proton-motive force. Thus, control of mitochondrial ROS can work through two different molecular mechanisms.

34.2 Measurement of Energy Transduction

34.2.1 Mitochondrial Respiration States and Efficiency

How can we qualitatively and quantitatively measure mitochondrial energy transduction determining the magnitudes of mitochondrial adjustments? The most important measurement is oxygen consumption that determines the flux through the respiratory chain. Modified from the original definition of mitochondrial respiration states (Chance and Williams 1955), a more functional assay has been derived (Estabrook 1967). As a consequence, for example, 'state 2' which was originally measured in the presence of ADP but absence of exogenous substrate, resulting in very little residual respiration, is now widely used to describe a steady state where exogenous substrate is added before the addition of ADP (now state 3) (Nicholls and Ferguson 2002). In a typical mitochondrial respiration assay, state 2 respiration is induced by the addition of substrate but no exogenous ADP, state 3 respiration is induced by addition of ADP and when all ADP is converted to ATP, the residual respiration defends proton-motive force against depletion by proton leak (Fig. 34.2). Additionally, mitochondria can be completely uncoupled using protonophores such as FCCP (carbonyl cyanide-p-trifluoromethoxyphenylhydrazone). State 2 respiration is functionally the same as state 4. Practically, state 4 can also be induced by inhibition of ATP synthesis. This can be achieved by either blocking the exchange of ADP/ATP via the ANT (e.g., with carboxyatractylate) or by inhibition of the ATP synthase per se (e.g., with oligomycin). This induced state is usually referred to as state 4_{ol}. There are certain advantages to the use of oligomycin: (a) it prevents overestimation of membrane leakiness, in case of ATP

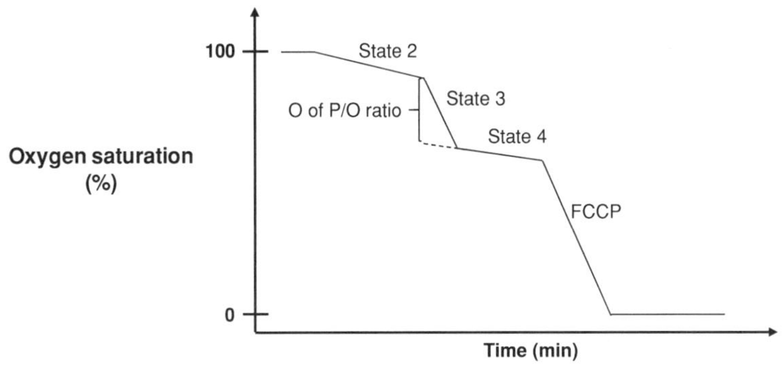

Fig. 34.2 Mitochondrial respiration states: mitochondrial respiration begins with substrate (commonly referred as state 2 respiration), subsequent addition of ADP induces state 3. When ADP is depleted or the ATP synthase is inhibited, respiration is shifted to state 4. FCCP fully uncouples respiration and serves as a measure for maximal substrate oxidation rates. P/O ratio can be determined by calculating the molecular ratio of phosphorylated ADP/atomic oxygen, either with or without correction for leak respiration

consumption within the mitochondrial matrix, and (b) oligomycin inhibits F_o proton channel activity of the F_1F_o-ATP-Synthase if the F_1 subunit has been disassembled during the mitochondrial isolation. In conclusion, the use of oligomycin assures a better estimate of the mitochondrial inner membrane 'leakiness'. State 3 respiration indicates the capacity of ATP synthesis. FCCP respiration measures the maximal substrate oxidation capacity. A conventional method to estimate the integrity of mitochondria is given by the respiratory control ratio (RCR) that is calculated by dividing state 3 by state 4 respiration. This also gives a rough estimate of mitochondrial efficiency. High RCRs are regarded as indicative of a 'good' mitochondrial isolation but should be taken carefully as physiological adjustments of ATP synthesis and substrate oxidation can lead to low RCRs (e.g., Martin et al. 1999). Also indicative of mitochondrial efficiency is the P/O ratio that describes the relation between ATP synthesis and oxygen consumption. Practically, a known amount of ADP is completely converted to ATP and the amount of consumed oxygen is measured. This method would assume a disappearance of the proton leak during state 3 but can be corrected for state 2 or 4 respiration (Fig. 34.2). Although this has been repeatedly reported to be a precise measure of efficiency, it remains estimation. The P/O ratio does not take into account the proton motive force-dependent behaviour of the mitochondrial proton leak (Fig. 34.3). With the addition of ADP, the proton-motive force decreases and the proton leak can decrease to unknown magnitude as the proton leak displays non-ohmic behaviour at high membrane potentials. Therefore, the correction using state 2 or 4 respiration may underestimate the oxygen that is consumed to drive the ATP synthase.

Taken together, the sole measurement of oxygen consumption can be very informative to describe mitochondrial oxidative phosphorylation and identify adjustments in mitochondrial efficiency. It is, however, not precise enough to identify

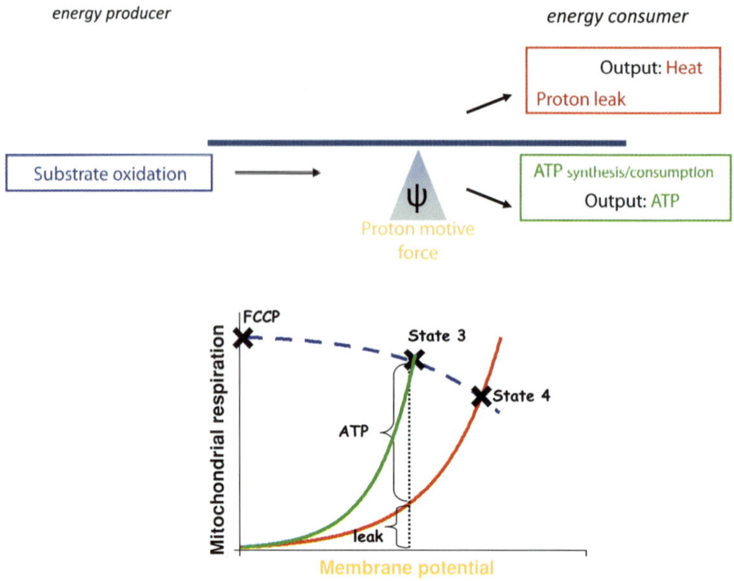

Fig. 34.3 Modular kinetic analysis of mitochondrial energy transduction: proton-motive force is the central intermediate of mitochondrial energy transduction that is produced by substrate oxidation and consumed by either ATP synthesis or proton leak. Measuring mitochondrial respiration and membrane potential simultaneously, all components of energy transduction can be assessed. The three kinetic modules substrate oxidation, ATP synthesis and proton leak are titrated using suitable inhibitors. Proton-motive force is measured as membrane potential when the pH gradient has been converted to membrane potential. For example, mitochondrial efficiency is determined as the proportion of oxygen consumption that is used to produce ATP at a certain membrane potential

the molecular mechanisms of adjustments. For example, one can manipulate mitochondria and decrease substrate oxidation. This can be directly measured as a decrease in FCCP-induced maximal substrate oxidation. Due to the modular kinetics of energy transduction, state 3 and state 4 respiration will also decrease although neither the ATP synthase nor the membrane leakiness has been affected. To address which of the three major modules of mitochondrial energy transduction have been adjusted, one has to measure the proton-motive force and the kinetics of substrate oxidation, ATP synthesis and proton leak.

34.2.2 Mitochondrial Membrane Potential

The proton-motive force consists of a pH gradient and the membrane potential. For measurements of membrane potential, the accumulation of lipophilic cations (such as TPMP+ or TPP+) in the negatively charged mitochondrial matrix according to the Nernst equation is utilised (Brand 1995). To address proton-motive force, the pH gradient can be converted to membrane potential by using high concentrations of phosphate ions or nigericin (a potassium–proton exchanger). As most reporters of mitochondrial membrane potential also bind to mitochondria in a membrane potential manner and decrease the amount of freely diffusing reporter ions, their concentration has to be corrected for unspecific binding. Another method to measure membrane potential is the use of fluorescent safranine O, a dye that forms quenching aggregates when accumulated in the mitochondrial matrix (Akerman and Wikström 1976). The measured potential can be calibrated versus other assays (TPMP+) or calibrated with potassium.

34.2.3 Modular Kinetics of Energy Transduction

By measuring mitochondrial oxygen consumption and membrane potential simultaneously, we are able to accurately describe the three major modules of energy transduction (substrate oxidation, ATP synthesis and proton leak kinetics) (Fig. 34.3). We can monitor which module of energy transduction has been adapted to environmental and physiological challenge, and which module has greatest control on the energetic output. Furthermore, we can precisely calculate how much substrate energy is used for ATP synthesis and how much is wasted as heat by the mitochondrial proton leak (Fig. 34.3).

34.3 Physiological Regulation of Mitochondrial Proton Leak

34.3.1 Adaptation and Plasticity of Mitochondrial Proton Leak

Using simultaneous measurements of proton-motive force and oxygen consumption, different strategies of energy adjustments during adaptation and acclimatisation were addressed. This chapter will be focussed on adjustments of mitochondrial proton leak and molecular mechanisms altering mitochondrial proton leak. We performed measurements of proton leak kinetics in a variety of species in response to temperature and nutritional status. To elucidate the molecular mechanisms of proton leak adjustments, we studied the evolution and regulation of uncoupling proteins (UCPs) in evolutionary distant species. Our experiments on mammalian UCP1 demonstrate an interrelationship of proton and electron leak in a physiologically relevant manner, possibly preventing oxidative stress (Oelkrug et al. 2010; Keipert et al. 2010).

34.3.2 Basal Proton Leak in Ecto- and Endothermic Vertebrates

Within vertebrates, the mitochondrial basal proton leak at the identical driving proton-motive force, mostly assessed in liver mitochondria, is lower in ectotherms than in endotherms. This difference in proton leak reflects the differences in standard (or for mammals and birds, in 'basal') metabolic rate. In endotherms, proton leak correlates with basal metabolic rate (Porter and Brand 1993), but this correlation is not the same between the major mammalian subgroups, eutherians and marsupials (Polymeropoulos et al. 2011a). Interestingly, the basal proton leak in marsupials is higher than expected considering a lower basal metabolic rate as compared to eutherians. The significance of these findings requires further experimentation and diversification of data collection from different species.

Environmental stresses such as temperature and nutritional status do not majorly impact the mitochondrial proton leak in most mammalian tissues. For example, isolated liver and skeletal muscle mitochondria of the small marsupial *Antechinus flavipes,* measured at 35°C (resembling body temperature), showed no changes in proton leak in response to cold acclimatisation (Fig. 34.4; Jastroch et al. 2009). The adjustments of energy transduction are more pronounced in ectothermic vertebrates. In the common carp, cold water temperature reduces the proton conductance in liver mitochondria (Jastroch et al. 2007). In cane toads, we showed that ambient temperature and long-term fasting have a major impact on proton leak kinetics of liver mitochondria possibly assisting in energy savings (Fig. 34.4; Trzcionka et al. 2008). In skeletal muscle of hibernating frogs kept in cold water, hypoxia induces a further decrease in ATP production by a decrease in substrate oxidation while the proton conductance remains unaltered (Boutilier and St-Pierre 2002).

34.3.3 Regulatory Proteins of Mitochondrial Proton Leak

In eutherian brown adipose tissue, proton leak can be regulated with UCP1 and serves to execute adaptive non-shivering thermogenesis (Nicholls and Locke 1984). Focussing on proton leak-catalysing UCPs, we have uncovered the evolutionary fate of UCP1 in vertebrates back to teleost fish. Inducible proton leak supporting a thermogenic function can be clearly addressed in all eutherians (Mzilikazi et al. 2007), where UCP1 is mainly expressed in brown adipose tissue. In marsupials, UCP1 is cold induced in the interscapular fat of the small dasyurid *Sminthopsis crassicaudata* (Jastroch et al. 2008). However, no evidence for classical adaptive non-shivering thermogenesis, assessed by noradrenaline tests, is found (Polymeropoulos et al. 2011b) and so far, we have no evidence that marsupial UCP1 catalyses proton leak. In the common carp, we demonstrated that carp UCP1 gene expression correlates with the activation of proton leak by known eutherian UCP1 activators (Jastroch et al. 2007). For a conclusive demonstration of a 'proton leak function' of marsupial and fish UCP1, further studies are required that measures solely UCP1 function without confounding differences of cellular background. Since we cannot get hold of knockout models for these species, other test systems such as proteoliposomes with reconstituted UCP1, or UCP1 overexpression in cell culture systems, are required. Considering incorporation of UCPs from

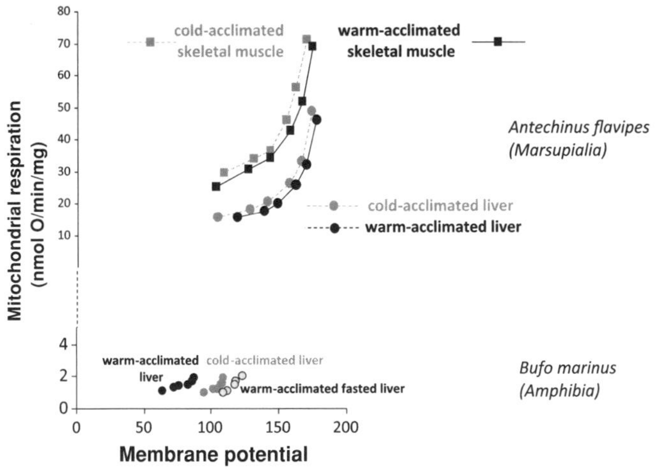

Fig. 34.4 Examples for adjustments of mitochondrial proton leak in an ectothermic (the cane toad) and an endothermic (the yellow-footed Antechinus) vertebrate: nutritional status (fasting) and ambient temperature (cold exposure) have minor effects on the proton leak of this marsupial species while in the amphibian, fasting and cold decrease the proton leak, thereby increasing efficiency of ATP production. Data were reconstructed from Jastroch et al. (2010) and Trzcionka et al. (2008)

different species in membranes, one has to consider that phospholipid composition will also affect protein activity (see also Wu et al. 2004).

34.3.4 *Novel Methods to Study Mitochondrial Function in the Living Cell*

While mitochondrial efficiency was majorly measured in vitro in isolated mitochondria, some adjustments may be lost during isolation. Therefore, one would prefer to measure mitochondrial function in its cellular environment or in the living cell.

Permeabilization of cells or skinned muscle fibres with saponin has been established as a method to measure mitochondrial bioenergetics in the cellular environment and reduce potential damage by the isolation process (summarised by Kuznetsov et al. 2008). When some caveats of the saponin treatment, such as over digestion of mitochondrial outer membrane and changes in calcium concentration by the permeabilization process, are considered, isolation artefacts and differential isolation of mitochondrial populations can be avoided. Another elegant method to assess mitochondria in skeletal muscle in vivo is the use of magnetic resonance and optical spectroscopy (Marcinek et al. 2004).

In the past few years, an extracellular flux analyser has been established (XF24/96; SEAHORSE BIOSCIENCE) that allows the measurement of cellular oxygen consumption in adherent cell culture. Mitochondrial respiration can be calculated by subtracting non-mitochondrial respiration. Non-mitochondrial respiration is measured by inhibiting electron flux through the mitochondrial complex I (e.g., with rotenone) and complex III (e.g., with antimycin A). The basal mitochondrial respiration (no inhibitors) is similar to state 3 of mitochondria, state 4 respiration can be assessed in the presence of oligomycin, and administration of FCCP allows an estimation of maximal substrate oxidation (Nicholls et al. 2010).

The assessment of mitochondrial membrane potential in the living cell is more difficult. Using radioactively labelled probes (e.g., TPMP+*), the mitochondrial membrane potential was measured in cell suspension. Advances in fluorescent microscopy allow the use of fluorescent probes in adherent cell culture. Although a variety of fluorescent probes promise an accurate measurement of mitochondrial membrane potential, changes in fluorescence can just give a vague to false estimation of mitochondrial membrane potential, as some peculiarities of the cell are not considered. First, some probes form fluorescent aggregates and diffuse slowly through the plasma membrane. The amplitude of aggregate formation depends on the diffusion distance and therefore, on the 'thickness of the cell'. Most illustrative, there appear mitochondria with high membrane potential in dendrites and cell protrusions—but these signals are areas of faster aggregating fluorescent probes and do not reflect higher mitochondrial membrane potentials (or not even presence of mitochondria). Other probes do distribute freely in a Nernstian manner. Often times, it is not considered that the plasma membrane also possesses a potential, and therefore, the fluorescence of the probe is convoluted by plasma and mitochondrial membrane potential. A major advance was the calibration of the mitochondrial probe TMRM+ and rhodamine123 in quench mode with a plasma membrane potential indicator (Nicholls 2006). This allowed the accurate measurement of mitochondrial membrane potential deviations, assuming a resting mitochondrial membrane potential. In an elegant follow-up development, TMRM+ is used in non-quench mode and the diffusion kinetics of the probe over the plasma membrane is used to fully calibrate the resting mitochondrial membrane potential to absolute values (Birket et al. 2011; Gerencser AA, Chinopoulos C, Birket MJ, Jastroch M, Vitelli C, Nicholls DG, Brand MD, unpublished work). In short, the mitochondrial volume fraction of the cytosolic volume is determined by stereology of confocal images, and the kinetics of TMRM+ release over the plasma membrane is recorded with time-lapse fluorescent microscopy. Depolarisation of mitochondria leads to the time-dependent release of TMRM+ over the plasma membrane; depolarising the plasma membrane is used to obtain the zero potential value of TMRM and the plasma membrane potential indicator.

The measurement of the absolute mitochondrial membrane potential is an essential requirement to apply modular kinetic analysis on mitochondrial energy transduction (Fig. 34.3). Similar to isolated mitochondria, parallel measurements of oxygen consumption in the extracellular flux analyser and mitochondrial membrane potential with fluorescence allow titration of all three modules of energy transduction in the living cell. Practically, the titration using microscopic methods is time intensive and limited by technical constraints of the fluorescent microscope. We currently develop a fast multi-well microplate reader-based assay to replace microscopic measurements (Jastroch M, Gerencser AA, Brand MD, unpublished work) which allows the parallel measurements of membrane potentials for mitochondrial substrate oxidation, ATP synthesis and proton leak.

34.4 Conclusion and Perspective

Mitochondrial energy transduction can be adjusted to physiological and environmental challenge to meet the energy demands of the cell and of the organism. In the past, studies show that adaptation and acclimatisation of energy metabolism is reflected in isolated mitochondria. The origin of the adjustment can be determined by modular kinetics of energy transduction that requires the measurement of not only mitochondrial respiration but also of mitochondrial membrane potential (or proton-motive force). Under these in vitro conditions, some adjustments in the energy transduction may be lost. Novel cell-based technologies have been developed to assess mitochondrial function in a native scenario with the surrounding cytosol of an intact cell. The integration of these novel technologies into physiological studies will have major impact to pinpoint molecular strategies of energy adjustments on the mitochondrial and cellular level.

References

Affourtit C, Jastroch M, Brand MD (2011) Uncoupling protein-2 attenuates glucose-stimulated insulin secretion in INS-1E insulinoma cells by lowering mitochondrial reactive oxygen species. Free Rad Biol Med 50:609–616

Akerman KE, Wikström MK (1976) Safranine as a probe of the mitochondrial membrane potential. FEBS Lett 68(2):191–197

Birket MJ, Orr AL, Gerencser AA, Madden DT, Vitelli C, Swistowski A, Brand MD, Zeng X (2011) A reduction in ATP demand and mitochondrial activity with neural differentiation of human embryonic stem cells. J Cell Sci 124(Pt 3):348–358

Boutilier RG, St-Pierre J (2002) Adaptive plasticity of skeletal muscle energetics in hibernating frogs: mitochondrial proton leak during metabolic depression. J Exp Biol 205(Pt 15): 2287–2296

Brand MD (1995) Measurement of mitochondrial protonmotive force. In: Brown GC, Cooper CE (eds) Bioenergetics: a practical approach. Oxford University Press, Oxford, pp 39–62

Brand MD (2010) The sites and topology of mitochondrial superoxide production. Exp Gerontol 45(7–8):466–472

Brand MD, Pakay JL, Ocloo A, Kokoszka J, Wallace DC, Brookes PS, Cornwall EJ (2005) The basal proton conductance of mitochondria depends on adenine nucleotide translocase content. Biochem J 392(Pt 2):353–362

Brookes PS, Buckingham JA, Tenreiro AM, Hulbert AJ, Brand MD (1998) The proton permeability of the inner membrane of liver mitochondria from ectothermic and endothermic vertebrates and from obese rats: correlations with standard metabolic rate and phospholipid fatty acid composition. Comp Biochem Physiol B Biochem Mol Biol 119(2):325–334

Chance B, Williams GR (1955) Respiratory enzymes in oxidative phosphorylation. III. The steady state. J Biol Chem 217:409–427

Estabrook R (1967) Mitochondrial respiratory control and the polarographic measurement of ADP:O ratios. Methods Enzymol 10:41–47

Jastroch M, Buckingham JA, Helwig M, Klingenspor M, Brand MD (2007) Functional characterisation of UCP1 in the common carp: uncoupling activity in liver mitochondria and cold-induced expression in the brain. J Comp Physiol B 177(7):743–752

Jastroch M, Withers KW, Taudien S, Frappell PB, Helwig M, Fromme T, Hirschberg V, Heldmaier G, McAllan BM, Firth BT, Burmester T, Platzer M, Klingenspor M (2008) Marsupial uncoupling protein 1 sheds light on the evolution of mammalian nonshivering thermogenesis. Physiol Genomics 32(2):161–169

Jastroch M, Withers KW, Stoehr S, Klingenspor M (2009) Mitochondrial proton conductance in skeletal muscle of a cold-exposed marsupial, *Antechinus flavipes*, is unlikely to be involved in adaptive nonshivering thermogenesis but displays increased sensitivity toward carbon-centered radicals. Physiol Biochem Zool 82(5):447–454

Jastroch M, Divakaruni AS, Mookerjee S, Treberg JR, Brand MD (2010) Mitochondrial proton and electron leaks. Essays Biochem 47:53–56

Keipert S, Klaus S, Heldmaier G, Jastroch M (2010) UCP1 ectopically expressed in murine muscle displays native function and mitigates mitochondrial superoxide production. Biochim Biophys Acta 1797:324–330

Kuznetsov AV, Veksler V, Gellerich FN, Saks V, Margreiter R, Kunz WS (2008) Analysis of mitochondrial function in situ in permeabilized muscle fibers, tissues and cells. Nat Protoc 3:965–976

Marcinek DJ, Schenkman KA, Ciesielski WA, Conley KE (2004) Mitochondrial coupling in vivo in mouse skeletal muscle. Am J Physiol Cell Physiol 286:C457–C463

Martin SL, Maniero GD, Carey C, Hand SC (1999) Reversible depression of oxygen consumption in isolated liver mitochondria during hibernation. Physiol Biochem Zool 72:255–264

Mzilikazi N, Jastroch M, Meyer CW, Klingenspor M (2007) The molecular and biochemical basis of nonshivering thermogenesis in an African endemic mammal, *Elephantulus myurus*. Am J Physiol Regul Integr Comp Physiol 293(5):R2120–R2127

Nicholls DG (2006) Simultaneous monitoring of ionophore- and inhibitor-mediated plasma and mitochondrial membrane potential changes in cultured neurons. J Biol Chem 281(21): 14864–14874

Nicholls DG, Ferguson SJ (2002) Bioenergetics 3. Academic Press, London, p 287

Nicholls DG, Locke RM (1984) Thermogenic mechanisms in brown fat. Physiol Rev 64(1):1–64

Nicholls DG, Darley-Usmar VM, Wu M, Jensen PB, Rogers GW, Ferrick DA (2010) Bioenergetic profile experiment using C2C12 myoblast cells. J Vis Exp (46), pii: 2511. doi:10.3791/2511

Oelkrug R, Kutschke M, Meyer CW, Heldmaier G, Jastroch M (2010) Uncoupling protein 1 decreases superoxide production in brown adipose tissue mitochondria. J Biol Chem 285(29):21961–21968

Polymeropoulos ET, Heldmaier G, Frappell PB, McAllan BM, Withers KW, Klingenspor M, White CR, Jastroch M (2011a) Phylogenetic differences of mammalian basal metabolic rate are not explained by mitochondrial basal proton leak. Proc Biol Sci. doi:10.1098/rspb.2011.0881

Polymeropoulos ET, Jastroch M, Frappell PB (2011b) Absence of adaptive nonshivering thermogenesis in a marsupial, the fat-tailed dunnart (*Sminthopsis crassicaudata*). J Comp Physiol B. doi:10.1007/s00360-011-0623-x

Porter RK, Brand MD (1993) Body mass dependence of H+ leak in mitochondria and its relevance to metabolic rate. Nature 362(6421):628–630

Trzcionka M, Withers KW, Klingenspor M, Jastroch M (2008) The effects of fasting and cold exposure on metabolic rate and mitochondrial proton leak in liver and skeletal muscle of an amphibian, the cane toad *Bufo marinus*. J Exp Biol 211(Pt 12):1911–1918

Wu BJ, Hulbert AJ, Storlien LH, Else PL (2004) Membrane lipids and sodium pumps of cattle and crocodiles: an experimental test of the membrane pacemaker theory of metabolism. Am J Physiol Regul Integr Comp Physiol 287:R633–R641

Chapter 35
Redox Metabolism During Tropical Diapause in a Lepidoptera Larva

Daniel Carneiro Moreira, Débora Pires Paula and Marcelo Hermes-Lima

Abstract Many studies on metabolic rate depression and redox metabolism exist in the literature; however, virtually none focuses on tropical insect diapause. Thus, our aim was to evaluate peculiarities of the metabolism of reactive oxygen species (ROS) between diapausing and non-diapausing insects in a tropical region. The lepidopteran *Chlosyne lacinia* undergoes diapause as larva at the third instar prior to the dry season in middle-west Brazil. We measured the activity of metabolic and anti-oxidant enzymes at day 20 of diapause. The activity of citrate synthase decreased by 81% in whole-body extracts as compared with larvae sampled before diapause entry. Moreover, total-glutathione content and lipid peroxidation dropped significantly (by 82 and 24%, respectively) in diapausing insects. On the other hand, the activities of catalase and glucose 6-phosphate dehydrogenase (G6PDH) were unchanged. These results indicate a diminished oxidative metabolism and suggest important roles for catalase and G6PDH in ROS control in diapause and, possibly, during arousal. The diminished glutathione levels could be related to its depletion by glutathione-dependent systems or by its diminished biosynthesis.

D. C. Moreira · M. Hermes-Lima (✉)
Laboratório de Radicais Livres, Departamento de Biologia Celular,
Universidade de Brasília, Brasilia, DF 70910-900, Brazil
e-mail: m_hermes_lima@yahoo.com.br

D. P. Paula
Laboratório de Ecologia Molecular, Embrapa Recursos Genéticos e Biotecnologia, Brasilia,
DF 70770-917, Brazil

T. Ruf et al. (eds.), *Living in a Seasonal World*,
DOI: 10.1007/978-3-642-28678-0_35,

35.1 Introduction

Our research group has been engaged over the years in the investigation of free radical metabolism in animals that are naturally exposed to environmental stresses. Surviving natural insults often involves a reduction of metabolic rate leading to high stress tolerance and reduced metabolic demand. Several works observed common biochemical mechanisms (Guppy and Withers 1999; Storey 2002) and marked modulation of free radical metabolism during conditions such as freezing, dehydration, hypoxia, anoxia, aestivation, hibernation, and diapause (Hermes-Lima and Zenteno-Savin 2002; Bickler and Buck 2007; Jovanovic-Galovic et al. 2007; Gorr et al. 2010).

Insects and other arthropods use diapause as an anticipatory response to the advent of unfavorable conditions including low temperature and dry periods (Denlinger 2002). Diapause is an endogenously programed arrest in development characterized by suppressed metabolic rate and high stress tolerance. It is triggered by environmental cues, such as changes in photoperiod and temperature (Denlinger 2002). The developmental stage in which diapause occurs is species specific and it can occur in eggs, different larval instars, pupae, or adults. More than a simple halt in development, diapause is an alternative physiological state consisting of several successive stages (Kostal 2006). During diapause, differential gene expression and metabolic regulation occur (Denlinger 2002; MacRae 2010). As a result of the selective reduction of cellular processes, overall metabolic demand is decreased. Thus, animals are able to survive long periods without feeding during diapause. In temperate regions, diapausing insects take advantage of low temperatures during winter to decrease their energetic requirements, since metabolic rate is closely related to environmental temperature in ectotherms (Hahn and Denlinger 2011). On the other hand, tropical diapausing insects depress their metabolic rate without the assistance of low temperatures (Denlinger 1986).

Many studies in comparative biology have been dedicated to the investigation of the metabolism of reactive oxygen species (ROS) and other radical/reactive species. Although hazardous at high concentrations, the highly reactive nature of these compounds is employed by aerobic organisms in many cellular processes (Droge 2001; Hermes-Lima 2004). ROS are formed continuously as by-products of mitochondrial respiration and other biochemical "routes" in the animal kingdom. In the case of phytophagous insects, the metabolism of allelochemicals is an important source of ROS (Felton and Summers 1995; Barbehenn 2002). These highly reactive compounds participate in several vital cellular pathways. However, when ROS production rises to levels that jeopardize redox control and signaling, oxidative stress occurs (Jones 2006). Excess ROS may induce oxidation of virtually all groups of biomolecules, including proteins, lipids, and nucleic acids (Sies 1993; Halliwell and Gutteridge 2007). Hence, aerobic organisms rely on a suite of anti-oxidants that include enzymatic and non-enzymatic components able to manage both pro-oxidant activity and signaling networks mediated by ROS (Hermes-Lima 2004; Pamplona and Costantini 2011).

In insects, as well as in other taxa, several anti-oxidant enzymes have been characterized, for example, superoxide dismutase (SOD), catalase, glutathione S-tranferases (GSTs), and peroxiredoxins (Felton and Summers 1995; Hermes-Lima 2004; Halliwell and Gutteridge 2007). Auxiliary enzymes, such the NADPH generating glucose 6-phosphate dehydrogenase (G6PDH), provide reducing power for the regeneration of reduced glutathione (GSH) from glutathione disulfide (GSSG) by glutathione reductase (GR). GSH a tripeptide known as "the most abundant non-protein thiol in eukaryotic cells" plays important roles in ROS metabolism that include redox buffering, cell signaling, as substrate for several enzymes, and direct scavenging of reactive species (Dalle-Donne et al. 2007; Halliwell and Gutteridge 2007). Most studies concerning anti-oxidant defense responses in insects are limited to exposure to plant pro-oxidants, insecticides or pollutants (Ku et al. 1994; Zaman et al. 1995; Krishnan and Kodrik 2006; Barbehenn et al. 2008; Augustyniak et al. 2009, 2011). Anti-oxidant defenses may also be important during shifts from anaerobic toward aerobic metabolism, when oxygen uptake rapidly increases, resulting in a potential ROS overproduction ultimately reflected as oxidative damage (Hermes-Lima and Zenteno-Savin 2002).

The sunflower caterpillar, *Chlosyne lacinia* (Geyer) (Lepidoptera: Nymphalidae), is widely distributed, ranging from northern Argentina to the southern USA (Drummond et al. 1970). It is the main defoliator of sunflower (*Helianthus annuus*) crops in Brazil, representing an important economical issue (Boiça and Vendramin 1993). *C. lacinia* features five larval instars and the generation time ranges between 35 and 40 days under field conditions in Texas, USA (Drummond et al. 1970) or 30–40 days in laboratory cages in middle-west Brazil (Paula DP, personal communication). Both winter and summer diapauses have been observed specifically at the third larval instar in *C. lacinia* at Texas, USA (Drummond et al. 1970; Scott 1986). Unfed animals can be maintained in laboratory in a quiescent state for over a year (Drummond et al. 1970). Diapausing *C. lacinia* larvae aggregate, do not feed, remain inactive, and form a thin silk layer surrounding animals. Changes in color patterns may occur. Environmental regulators of *C. lacinia* diapause are still unknown; however, potential regulators of diapause in tropical insects include photoperiod, temperature, rainfall, and nutrition (Denlinger 1986).

Several observations suggest a relationship between the modulation of free radical metabolism and biochemical adaptations of cold-hardy and diapausing insects of temperate regions (Grubor-Lajsic et al. 1997; Joanisse and Storey 1998; Stanic et al. 2004; Kojic et al. 2009). Few studies have compared diapausing animals to their active counterparts (Jovanovic-Galovic et al. 2004, 2007; Sim and Denlinger 2011) and data on free radical metabolism and redox balance during diapause in insects inhabiting tropical regions is scarce. Thus our aim was to identify particularities of the anti-oxidant apparatus between non-diapausing and diapausing *C. lacinia* larvae prior to the dry season in middle-west Brazil. We determined the activities of citrate synthase (CS), G6PDH, and catalase and total glutathione concentration in whole-body extracts of active and 20-day diapausing third instar *C. lacinia* larvae.

35.2 Methods

Eggs of *C. lacinia* were collected near the University of Brasilia (in preserved areas of cerrado vegetation, 15°43′S, 47°53′W, approximately at 1,000 m altitude) in January-2010, prior to the dry season, from the leaves of Mexican sunflower (*Tithonia diversifolia*, Asteraceae). The eggs were maintained at 25 ± 2°C, 55 ± 10% R.H., and 16:8 h photophase in 3.5 L plastic cages. After hatching, the caterpillars were kept at the same conditions and fed daily with fresh leaves of *T. diversifolia*. Once transformed to the third instar, a group of larvae were frozen in liquid nitrogen to be used as the control group and kept at −80°C until analysis. The remaining animals, kept without food, entered diapause (at third instar) and stayed in this state for 20 days, when they were frozen in liquid nitrogen and stored at −80°C until analysis. Whole-body extracts were prepared from a pool of three larvae and represented one experimental unit per assay.

Frozen samples were homogenized using an OMNI Tissue Master homogenizer (Omni International, Marietta, GA) in ice cold 50 mM potassium phosphate, pH 7.2, containing 0.5 mM EDTA and 10 μM phenylmethylsulfonyl fluoride (PMSF). Extracts were centrifuged at 10,000×*g* for 15 min at 5°C and supernatants were collected for enzyme assays. Citrate synthase (CS) activity was measured as described by Srere (1969). Catalase and G6PDH activities were measured as described by Ramos-Vasconcelos and Hermes-Lima (2003). One unit of CS, catalase, or G6PDH activity is defined as the amount that converts 1 μmol substrate into product per min. Protein concentration was measured with Coomassie Brilliant Blue G-250 (Bradford 1976) using bovine serum albumin as a standard. Enzyme activities were expressed as units per milligram of soluble protein.

To measure total glutathione, frozen samples were homogenized in ice cold 10% (w/v) trichloroacetic acid, centrifuged at 10,000×*g* for 6 min at 5°C, and supernatants were collected for analysis. Total glutathione, reduced + oxidized forms, was determined as total-glutathione equivalents (GSH-eq) based on Griffith (1980). GSH-eq was quantified by following the reduction rate of DTNB by GSH catalyzed by GR and comparing this rate to a standard curve (see details in Ramos-Vasconcelos and Hermes-Lima 2003). Thiobarbituric acid reactive substances (TBARS) were measured as an index of lipid peroxidation as described by Buege and Aust (1978). Statistical analyses used the software GraphPad Prism 5 (San Diego, USA). The comparisons between groups were performed by a two-tailed Student's t-test and a significance level at $P < 0.05$ was considered. Results are presented as means ± standard error (SEM).

35.3 Results and Discussion

The activity of CS in 20-day diapausing larvae decreased by 81% (Fig. 35.1a), indicating that the TCA cycle works at a lower rate, causing diminished oxidative capacity of diapausing animals. Citrate and fumarate—aerobic intermediates—are

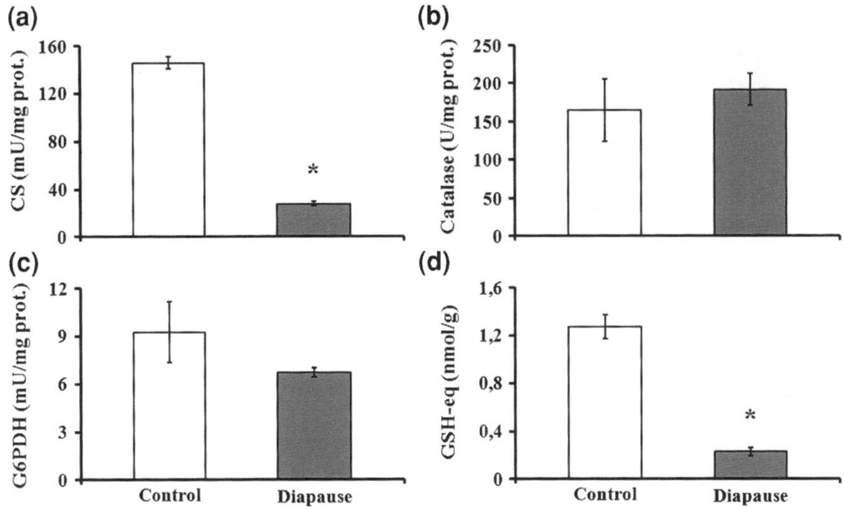

Fig. 35.1 Whole-body CS, catalase and G6PDH activities per milligram of protein and GSH-eq levels per gram wet weight of active, and 20-day diapausing third instar *C. lacinia* larvae. (*) Significantly different from corresponding control values. Data are means ± SEM, $n = 3$ for enzyme activities, and $n = 7$ for total-glutathione (GSH-eq) levels

markedly decreased in whole-body extracts of 30-day diapausing flesh fly (*Sarcophaga crassipalpis*) pupae at 20°C also indicating a reduction in TCA cycle activity (Michaud and Denlinger 2007). In addition, whole-body transcript profiling of *S. crassipalpis* diapausing pupae revealed that even though the transcript abundance of important TCA cycle enzymes increased, the initial reactions of the cycle are suppressed (Ragland et al. 2010). Moreover, increased gene expression of phosphoenolpyruvate carboxykinase (PEPCK) and sorbitol dehydrogenase was observed in diapausing larvae of the mosquito, *Wyeomyia smithii* (Emerson et al. 2010). It has been proposed that an anticipatory physiological response for low oxygen consumption is associated with diapause (Emerson et al. 2010). During diapause there is a trend toward anaerobic metabolism, even when animals are not under hypoxia or anoxia, suggesting that the shift away from aerobic metabolism is a preprogramed component of the diapause phenomenon (Hahn and Denlinger 2011). The shift from aerobic toward anaerobic metabolism is among the shared physiological features of animals in hypometabolic states (Guppy and Withers 1999; Storey 2002; Storey and Storey 2007).

Catalase activity remained unchanged in 20-day diapausing larvae (Fig. 35.1b). This result differs from previous observations by Jovanovic–Galovic et al. (2004) in which catalase activity was lower in whole-body homogenates of diapausing compared to non-diapausing larvae of the European corn borer, *Ostrinia nubilalis*. Catalase activity was also reduced in isolated mitochondria of diapausing *O. nubilalis* larvae (Jovanovic-Galovic et al. 2007). The maintenance of catalase activity under low metabolic rate indicates that the control of hydrogen peroxide

(H_2O_2) is somehow important during diapause. In diapausing mosquitoes (*Culex pipiens*) catalase plays important roles in both lifespan extension and stress tolerance, as evidenced by an upregulation of catalase expression in young diapausing females (Sim and Denlinger 2011). Moreover, high levels of apoptosis in ovaries and increased mortality were observed in diapausing animals in which catalase expression was suppressed by RNA interference (Sim and Denlinger 2011). It has been shown that in *Caenorhabditis elegans* larvae under metabolic depression there is an upregulation of catalase and SOD expression (Houthoofd et al. 2002). Sima et al. (2011) suggest that changes in H_2O_2 levels are involved in initiation and termination of diapause in eggs of the silkworm *Bombyx mori*. H_2O_2 may also be associated with the release of a neuropeptide hormone related to diapause in *B. mori* as evidenced by peculiarities of H_2O_2 metabolism between univoltine and polyvoltine strains (Zhao and Shi 2009). Thus, the maintenance of catalase activity in *C. lacinia* could be related to the control of H_2O_2 levels possibly preventing a premature break of diapause.

G6PDH activity remained unchanged after 20 days of diapause (Fig. 35.1c). This result indicates the ability to maintain NADPH production via the pentose phosphate pathway. This pathway would be essential to provide reducing power for biosynthetic pathways and for the enzyme-catalyzed reduction of GSSG to GSH at the cost of NAPDH by the glutathione cycle (Hermes-Lima 2004). A supply of NADPH and reduced glutathione is also important for the functioning of glutathione-dependent enzymes, such as glutathione S-transferases (GSTs). Other NADPH consuming antioxidants, such as peroxiredoxins (see Conclusion section), could also take advantage of this supply. This, together with unaltered catalase activity, may be especially important during the recovery from hypometabolism, because awakening is considered a condition of potential ROS overproduction (Ferreira-Cravo et al. 2010). In the garden snail, *Helix aspersa*, the awakening from winter aestivation is accompanied by increased levels of lipid peroxidation and GSSG/GSH-eq ratio (a key index of cellular redox status and oxidative stress), suggesting augmented ROS production (Ramos-Vasconcelos and Hermes-Lima 2003). Increased SOD activity and lipid peroxidation in arousing snails *Otala lactea* also indicate oxidative stress in the recovery from metabolic depression (Hermes-Lima and Storey 1995).

In many hypometabolic states metabolism is not constant. For example, aestivating snails, *Otala lactea*, show intermittent increases in metabolic rate associated with opening the pneuomostome for gas exchange (Barnhart and McMahon 1987). Moreover, periodic arousal events are characteristic of all mammalian hibernators (Carey et al. 2003). During insect diapause there are also cycles of enhanced metabolic rate, as observed by monitoring oxygen uptake in pupal diapause of the tropical flesh flies, *Sarcophaga inzi* and *Poecilometopa spilogaster* (Denlinger 1979). Periodic cycles of oxygen uptake were also observed in diapausing pupae of the lepidopterans *Pieris brassicae* and *Papilio machaon* (Crozier 1979). It is unclear if diapausing *C. lacinia* larvae exhibit these cycles; however, the maintenance of anti-oxidant activities (catalase and G6PDH) may be very important for the maintenance of redox homeostasis preventing oxidative stress from any bursts in oxygen consumption.

Another critical role of the pentose phosphate pathway is the supply of reducing equivalents for the synthesis of polyols (Storey and Storey 1991). The accumulation of low molecular weight sugars and polyols such as glycerol and sorbitol are known to confer not only cold hardiness but also enhanced dehydration tolerance in dormant insects (Kostal et al. 1998; Benoit 2010). A comparison between diapausing and non-diapausing *Pyrrohocoris apterus* revealed increased G6PDH activity related to polyol biosynthesis in diapausing animals (Kostal et al. 2004). Although increased levels of polyols were reported in diapausing tropical insects (Pullin and Wolda 1993) the levels of low molecular weight sugars and polyols, as well as activities of enzymes involved in polyol production, are unknown in *C. lacinia*. Nevertheless, it would be critical to maintain the water balance during diapause through the dry season in the Brazilian savanna-like cerrado.

The GSH-eq levels in diapausing animals fell to 18% of control levels (Fig. 35.1d). This reduction is consistent with the lowered metabolic rate and agrees with another study (Meng et al. 2011) in which diapausing silkworm eggs showed decreased total-glutathione concentration as compared to non-diapausing eggs. The lowered GSH-eq levels in *C. lacinia* could be a result of decreased GSH biosynthesis, increased protein glutathionylation, or increased GSH-eq depletion caused, for example, by GST-catalyzed conjugation. Molecular analysis identified a GST from *Choristoneura fumiferana* as a diapause-associated protein, with levels markedly increased in diapausing second instar larvae as compared to control second instar larvae (Feng et al. 1999). It is possible that GST activity is increased in diapausing *C. lacinia* and that this could account for the observed depletion in total glutathione. Moreover, TBARS levels decreased by 24% in diapausing animals (control animals: 19.0 ± 0.94 nmol/g; $n = 7$). The diminished lipid peroxidation is in accordance with the suppressed metabolic rate and the maintenance of the enzymatic anti-oxidant potential.

35.4 Conclusion and Perspectives

To our knowledge, this is the first study that investigates redox metabolism during diapause in an insect inhabiting a tropical region. Our results showed an expected diminished oxidative metabolism, the maintenance of catalase and G6PDH activities, the depletion of total glutathione, and reduced lipid peroxidation in diapausing animals. The potential to maintain both the NADPH supply and the control of H_2O_2 metabolism may be crucial at the moment of arousal from diapause toward larval development. Moreover, gene expression analyses of peroxiredoxins (Prxs), a family of enzymes that could take advantage of the potential NADPH supply and plays important roles in ROS signaling (Cox et al. 2010), should be a further step in our investigation. The Prxs family is being studied in several insect systems and their roles in oxidative stress tolerance and life span extension have been reported (Kim et al. 2005; Lee et al. 2009; Hu et al. 2010). Furthermore, the investigation of polyol biosynthesis may compose a clearer picture of the relationship between carbohydrate and free radical metabolism in diapausing *C. lacinia*. The reasons for the depletion of glutathione may be further elucidated by the determination of glutathionylated proteins in addition to the assessment of the activities and expression of GST isozymes; those experiments are currently in progress. The assessment of enzymes involved in GSH biosynthesis could also aid the understanding the depletion of GSH-eq. The study of diapause in the tropical zone, where seasonal variations of temperature are not as wide as in temperate regions, is important not only to unveil the characteristics of tropical insect diapause, but also to assist the investigation of the biochemical mechanisms related to metabolic depression in all of its forms.

Acknowledgments This work was supported by grants from FINATEC (Brasília, Brazil), Projeto Universal (CNPq, Brazil), and INCT-Processos Redox em Biomedicina (Redoxoma, CNPq). Daniel C. Moreira is a recipient of an undergraduate fellowship from CNPq. We thank graduate student Renata Timbó (UnB) for taking good care of our "sleeping" bugs and Prof. Élida G. Campos (UnB) for revising this manuscript. We also thank an anonymous reviewer for insightful comments. This study is in honor of Cláudio Mário Guimarães da Silva (Rio de Janeiro, Brazil), retired biology teacher and an inspiring mind.

References

Augustyniak M, Babczynska A, Augustyniak M (2009) Does the grasshopper *Chorthippus brunneus* adapt to metal polluted habitats? A study of glutathione-dependent enzymes in grasshopper nymphs. Insect Sci 16:33–42

Augustyniak M, Babczynska A, Augustyniak M (2011) Oxidative stress in newly-hatched *Chorthippus brunneus*—the effects of zinc treatment during diapause, depending on the female's age and its origins. Comp Biochem Physiol C 154:172–179

Barbehenn RV (2002) Gut-based antioxidant enzymes in a polyphagous and a graminivorous grasshopper. J Chem Ecol 28:1329–1347

Barbehenn RV, Maben RE, Knoester JJ (2008) Linking phenolic oxidation in the midgut lumen with oxidative stress in the midgut tissues of a tree-feeding caterpillar *Malacosoma disstria* (Lepidoptera: Lasiocampidae). Environ Entomol 37:1113–1118

Barnhart MC, McMahon BR (1987) Discontinuous carbon dioxide release and metabolic depression in dormant land snails. J Exp Biol 128:123–138

Benoit JB (2010) Water management by dormant insects: comparisons between dehydration resistance during summer aestivation and winter diapause. In: Navas CA, Carvalho JE (eds) Aestivation: molecular and physiological aspects. Progress in molecular and subcellular biology, vol 49. Springer, Heidelberg, pp 209–229

Bickler PE, Buck LT (2007) Hypoxia tolerance in reptiles, amphibians, and fishes: life with variable oxygen availability. Annu Rev Physiol 69:145–170

Boiça AL Jr, Vendramin JD (1993) Infestação de girassol pela lagarta *Chlosyne lacinia saundersii* em duas épocas de cultivo. Sci Agric 50:244–253

Bradford MM (1976) A rapid and sensitive method for the quantitation of microgram quantities of protein utilizing the principle of protein-dye binding. Anal Biochem 72:248–254

Buege JA, Aust SD (1978) Microsomal lipid peroxidation. Methods Enzymol 52:302–310

Carey HV, Andrews MT, Martin SL (2003) Mammalian hibernation: cellular and molecular responses to depressed metabolism and low temperature. Physiol Rev 83:1153–1181

Cox AG, Winterbourn CC, Hampton MB (2010) Mitochondrial peroxiredoxin involvement in antioxidant defence and redox signalling. Biochem J 425:313–325

Crozier AJG (1979) Diel oxygen uptake rhythms in diapausing pupae of *Pieris brassicae* and *Papilio machaon*. J Insect Physiol 25:647–652

Dalle-Donne I, Rossi R, Giustarini D, Colombo R, Milzani A (2007) S-glutathionylation in protein redox regulation. Free Radic Biol Med 43:883–898

Denlinger DL (1979) Pupal diapause in tropical flesh flies: environmental and endocrine regulation, metabolic rate and genetic selection. Biol Bull 156:31–46

Denlinger DL (1986) Dormancy in tropical insects. Annu Rev Entomol 31:239–264

Denlinger DL (2002) Regulation of diapause. Annu Rev Entomol 47:93–122

Droge W (2001) Free radicals in the physiological control of cell function. Physiol Rev 82:47–95

Drummond III BA, Bush GL, Emmel TC (1970) The biology and laboratory culture of *Chlosyne lacinia* Geyer (Nymphalidae). J Lep Soc 24:135–142

Emerson KJ, Bradshaw WE, Holzapfel CM (2010) Microarrays reveal early transcriptional events during the termination of larval diapause in natural populations of the mosquito, *Wyeomyia smithii*. PLoS One 5:e9574

Felton GW, Summers CB (1995) Antioxidant systems in insects. Arch Insect Biochem Physiol 29:187–197

Feng QL, Davey KG, Pang ASD, Primavera M, Ladd TR, Zheng SC, Sohi SS, Retnakaran A, Palli SR (1999) Glutathione S-transferase from the spruce budworm, *Choristoneura fumiferana*: identification, characterization, localization, cDNA cloning, and expression. Insect Biochem Mol Biol 29:779–793

Ferreira-Cravo M, Welker AF, Hermes-Lima M (2010) The connection between oxidative stress and estivation in gastropods and anurans. In: Navas CA, Carvalho JE (eds) Aestivation: molecular and physiological aspects. Progress in molecular and subcellular biology, vol 49. Springer, Heidelberg, pp 47–61

Gorr TA, Wichmann D, Hu J, Hermes-Lima M, Welker AF, Terwilliger N, Wren JF, Viney M, Morris S, Nilsson GE, Deten A, Soliz J, Gassmann M (2010) Hypoxia tolerance in animals: biology and application. Physiol Biochem Zool 83:733–752

Griffith OW (1980) Determination of glutathione and glutathione disulfide using glutathione reductase and 2-vinylpyridine. Anal Biochem 106:207–212

Grubor-Lajsic G, Block W, Telesmanic M, Jovanovic A, Stevanovic D, Baca F (1997) Effect of cold acclimation on the antioxidant defense system of two larval Lepidoptera (Noctuidae). Arch Insect Biochem Physiol 36:1–10

Guppy M, Withers P (1999) Metabolic depression in animals: physiological perspectives and biochemical generalizations. Biol Rev Camb Philos Soc 74:1–40

Hahn DA, Denlinger DL (2011) Energetics of insect diapause. Annu Rev Entomol 56:103–121

Halliwell B, Gutteridge JMC (2007) Free radicals in biology and medicine. University Press, Oxford

Hermes-Lima M (2004) Oxygen in biology and biochemistry: role of free radicals. In: Storey KB (ed) Functional metabolism: regulation and adaptation. Wiley, New York, pp 319–368

Hermes-Lima M, Storey KB (1995) Antioxidant defenses and metabolic depression in a pulmonate land snail. Am J Physiol 268:R1386–R1393

Hermes-Lima M, Zenteno-Savin T (2002) Animal response to drastic changes in oxygen availability and physiological oxidative stress. Comp Biochem Physiol C 133:537–556

Houthoofd K, Braeckman BP, Lenaerts I, Brys K, Vreese A, Eygen SV, Vanfleteren JR (2002) Ageing is reversed, and metabolism is reset to young levels in recovering dauer larvae of *C. elegans*. Exp Gerontol 37:1015–1021

Hu Z, Lee KS, Choo YM, Yoon HJ, Lee SM, Lee JH, Kim DH, Sohn HD, Jin BR (2010) Molecular cloning and characterization of 1-Cys and 2-Cys peroxiredoxins from the bumblebee *Bombus ignites*. Comp Biochem Physiol B 155:272–280

Joanisse DR, Storey KB (1998) Oxidative stress and antioxidants in stress and recovery of cold-hardy insects. Insect Biochem Mol Biol 28:23–30

Jones DP (2006) Redefining oxidative stress. Antioxid Redox Signal 8:1865–1879

Jovanovic-Galovic A, Blagojevic DP, Grubor-Lajsic G, Worland R, Spasic MB (2004) Role of antioxidant defense during different stages of preadult life cycle in European corn borer (*Ostrinia nubilalis*, Hubn.): diapause and metamorphosis. Arch Insect Biochem Physiol 55:79–89

Jovanovic-Galovic A, Blagojevic DP, Grubor-Lajsic G, Worland MR, Spasic MB (2007) Antioxidant defense in mitochondria during diapause and postdiapause development of European corn borer (*Ostrinia nubilalis*, Hubn.). Arch Insect Biochem Physiol 64:111–119

Kim I, Lee KS, Hwang JS, Ahn MY, Li J, Sohn HD, Jin BR (2005) Molecular cloning and characterization of a peroxiredoxin gene from the mole cricket, *Gryllotalpa orientalis*. Comp Biochem Physiol B 140:579–587

Kojic D, Spasojevic I, Mojovic M, Blagojevic D, Worland MR, Grubor-Lajsic G, Spasic MB (2009) Potential role of hydrogen peroxide and melanin in the cold hardiness of *Ostrinia nubilalis* (Lepidoptera: Pyralidae). Eur J Entomol 106:451–454

Kostal V (2006) Eco-physiological phases of insect diapause. J Insect Physiol 52:113–127

Kostal V, Sula J, Simek P (1998) Physiology of drought tolerance and cold hardiness of the Mediterranean tiger moth *Cymbalophora pudica* during summer diapause. J Insect Physiol 44:165–173

Kostal V, Tollarova M, Sula J (2004) Adjustments of the enzymatic complement for polyol biosynthesis and accumulation in diapausing cold-acclimated adults of Pyrrhocoris apterus. J Insect Physiol 50:303–313

Krishnan N, Kodrik D (2006) Antioxidant enzymes in *Spodoptera littoralis* (Boisduval): are they enhanced to protect gut tissues during oxidative stress. J Insect Physiol 52:11–20

Ku C, Chiang F, Hsin C, Yao Y, Sun C (1994) Glutathione transferase isozymes involved in insecticide resistance of diamondback moth larvae. Pestic Biochem Physiol 50:191–197

Lee K, Iijima-Ando K, Iijima K, Lee W, Lee JH, Yu K, Lee D (2009) JNK/FOXO-mediated neuronal expression of fly homologue of peroxiredoxin II reduces oxidative stress and extends life span. J Biol Chem 284:29454–29461

MacRae TH (2010) Gene expression, metabolic regulation and stress tolerance during diapause. Cell Mol Life Sci 67:2405–2424

Meng G, Yao JM, Wang L, Zhao L (2011) Variation in glutathione status associated with induction and initiation of diapause in eggs of the bivoltine strain of the silkworm *Bombyx mori*. Physiol Entomol 36:173–179

Michaud MR, Denlinger DL (2007) Shifts in the carbohydrate, polyol, and amino acid pools during rapid cold-hardening and diapause-associated cold-hardening in flesh flies (*Sarcophaga crassipalpis*): a metabolomic comparison. J Comp Physiol B 177:753–763

Pamplona R, Costantini D (2011) Molecular and structural antioxidant defenses against oxidative stress in animals. Am J Physiol Regul Integr Comp Physiol 301:R843–R863

Pullin AS, Wolda H (1993) Glycerol and glucose accumulation during diapause in a tropical beetle. Physiol Entomol 18:75–78

Ragland GJ, Denlinger DL, Hahn DA (2010) Mechanisms of suspended animation are revealed by transcript profiling of diapause in the flesh fly. Proc Natl Acad Sci U S A 107:14909–14914

Ramos-Vasconcelos GR, Hermes-Lima M (2003) Hypometabolism, antioxidant defenses and free radical metabolism in the pulmonate land snail Helix aspersa. J Exp Biol 206:675–685

Scott JA (1986) The butterflies of North America: a natural history and field guide. University Press, Stanford

Sies H (1993) Strategies of antioxidant defense. Eur J Biochem 215:213–219

Sim C, Denlinger DL (2011) Catalase and superoxide dismutase-2 enhance survival and protect ovaries during overwintering diapause in the mosquito *Culex pipiens*. J Insect Physiol 57:628–634

Sima Y, Yao J, Hou Y, Wang L, Zhao L (2011) Variations of hydrogen peroxide and catalase expression in *Bombyx* eggs during diapause initiation and termination. Arch Insect Biochem Physiol 77:72–80

Srere PA (1969) Citrate synthase. Methods Enzymol 13:3–11

Stanic B, Jovanovic-Galovic A, Blagojevic DP, Grubor-Lajsic G, Worland R, Spasic MB (2004) Cold hardiness in *Ostrinia nubilalis* (Lepidoptera: Pyralidae): glycerol content, hexose monophosphate shunt activity, and antioxidative defense system. Eur J Entomol 101:459–466

Storey KB (2002) Life in the slow lane: molecular mechanisms of estivation. Comp Biochem Physiol A 133:733–754

Storey KB, Storey JM (1991) Glucose-6-phosphate dehydrogenase in cold hardy insects: kinetic properties, freezing stabilization, and control of hexose monophosphate shunt activity. Insect Biochem 21:157–164

Storey KB, Storey JM (2007) Tribute to P. L. Lutz: putting life on 'pause'—molecular regulation of hypometabolism. J Exp Biol 210:1700–1714

Zaman K, MacGill RS, Johnson JE, Ahmad S, Pardini RS (1995) An insect model for assessing oxidative stress related to arsenic toxicity. Arch Insect Biochem Physiol 29:199–209

Zhao L, Shi L (2009) Metabolism of hydrogen peroxide in univoltine and polyvoltine strains of silkworm (*Bombyx mori*). Comp Biochem Physiol B 152:339–345

Chapter 36
Biochemical Regulation of Carbohydrate Metabolism in Hibernating Bats

Kenneth B. Storey

Abstract Glycolysis is the core pathway of carbohydrate metabolism in cells; it is strongly regulated to mediate the use of sugar fuels for energy production (especially when oxygen is limiting) and biosynthesis as well as to allow opposite carbon flow during gluconeogenesis. Control of glycolysis should be a central part of metabolic suppression during torpor. Regulatory enzymes of carbohydrate catabolism (glycogen phosphorylase, 6-phosphofructo-1-kinase [PFK-1], pyruvate kinase, pyruvate dehydrogenase) were evaluated, along with levels of fructose-2,6-P_2, a potent PFK-1 activator, in tissues of little brown bats (*Myotis lucifugus*) comparing aroused and torpor states of winter-collected animals. The data show substantial changes in enzyme activities and properties indicating differential regulation via reversible protein phosphorylation between aroused and torpid states. Torpor also triggered strong increases at the mRNA and protein level of the hypoxia-inducible transcription factor (HIF-1) (that regulates several glycolytic enzymes) in bat skeletal muscle and liver and the study documented for the first time the involvement of microRNA (miR-106b) and antisense RNA in the regulation of a transcription factor in a hibernating species.

36.1 Introduction

The little brown bat (*Myotis lucifugus* Le Conte, 1831) is abundant over the forested areas of Canada and the U.S., ranging north to Alaska and Labrador. This species is a well-known hibernator and multiple aspects of its hibernation ecology,

K. B. Storey (✉)
Institute of Biochemistry, Carleton University, Ottawa K1S 5B6, Canada
e-mail: kenneth_storey@carleton.ca

T. Ruf et al. (eds.), *Living in a Seasonal World*,
DOI: 10.1007/978-3-642-28678-0_36,

physiology, and biochemistry have been studied by many authors (e.g. Thomas 1993; Thomas and Geiser 1997; Speakman and Thomas 2003; Boyles et al. 2007; Townsend et al. 2008; McGuire et al. 2009; Matheson et al. 2010). *M. lucifugus* has also received significant attention recently as a species of concern with respect to two issues: climate change and the white-nosed fungus epidemic (Humphries et al. 2002; Willis et al. 2011; Meteyer et al. 2011).

The profound suppression of metabolic rate during torpor episodes is a prominent feature of hibernation. In *M. lucifugus*, metabolic rate of during torpor may be just 3–4% of basal metabolic rate while euthermic (Geiser 2004). My laboratory has spent many years analyzing the regulation of metabolic rate depression in hibernation and other forms of hypometabolism (for review: Storey and Storey, 2004, 2007, 2010). We have documented conserved regulatory mechanisms including the use of reversible protein phosphorylation to coordinate global metabolic suppression and reprioritize activities of key enzymes and functional proteins so that homeostasis is re-established in the hypometabolic state. Several of our studies have used *M. lucifugus* as a model species. These include differential regulation of mitochondrial genes and antioxidant enzymes during torpor (Eddy et al. 2005, 2006), up-regulation of fatty acid binding proteins (Eddy and Storey 2004), and torpor-responsive changes in signaling by protein kinases including PKA, PKC, Akt, and p38 MAPK (Mehrani and Storey 1997; Holden and Storey 1998; Eddy and Storey 2003, 2007).

Suppression of carbohydrate catabolism via reversible phosphorylation of regulatory enzymes is a prominent feature of animal hypometabolism (Storey and Storey 2004, 2007). This contributes both to global suppression of biosynthesis during torpor and to carbohydrate sparing by some organs that saves glucose for tissues (e.g., brain, erythrocytes) that need this sugar fuel. Studies of the effects of torpor in the jumping mouse, *Zapus hudsonius*, showed coordinated regulation of glycolytic enzymes including glycogen phosphorylase (GP), 6-phosphofructo-1-kinase (PFK-1) and pyruvate kinase (PK) to suppress carbohydrate catabolism in liver (Storey 1987a) and selective inhibition of GP and PFK-1 in other tissues (Storey 1987b). Inhibition of pyruvate dehydrogenase (PDH) also occurred to shut down carbohydrate entry into mitochondrial catabolism (Storey 1989). In all cases the mechanism was covalent modification via phosphorylation or dephosphorylation. Strong PDH suppression also occurred in thirteen-lined ground squirrels, *Ictidomys* (*Spermophilus*) *tridecemlineatus*, during torpor but there was less evidence for phosphorylation of glycolytic enzymes (Brooks and Storey 1992). It has been suggested that small body mass hibernators need stronger controls to suppress metabolic rate during entry into torpor; indeed, small mass hibernators often show Q_{10} values of 3–4 in torpor compared with euthermia whereas larger species show values closer to 2. This might account for the more prominent controls on glycolytic enzymes in liver of *Z. hudsonius* (body mass 12–25 g, maximally 35 g at the start of hibernation) versus *I. tridecemlineatus* (body mass 130–180 g, maximally 220–240 pre-hibernation). The present chapter examines the responses and regulation of glycolytic enzymes and PDH in selected tissues of little brown bats, one of the smallest hibernating species with a body mass of just 6–8 g.

For all of our studies of bat hibernation, my laboratory is deeply indebted to the generosity of Dr. Donald W. Thomas (Université de Sherbrooke), colleague and friend, who allowed me and different students over the years to invade his laboratory in midwinter to collect *M. lucifugus* samples. Don's death in 2009 at too young an age has left a hole in Canadian hibernation research for the impressive ecophysiological studies by his group greatly improved our understanding of the winter energetics and behavior of small hibernators, both bats (e.g. Thomas 1993; Thomas and Geiser 1997; Speakman and Thomas 2003; Humphries et al. 2002) and chipmunks (e.g. Humphries et al. 2003; Landry-Cuerrier et al. 2008). Our bat research is totally Don-derived with animals from his study sites as our guide. His generous access to his laboratory for conducting torpor/arousal studies was always appreciated along with the midnight food runs. I dedicate this paper to Don's memory.

36.2 Materials and Methods

Bats (6–8 g) were collected in mid-January from abandoned mines near Sherbrooke, PQ where they had been hibernating for 3–4 months. Bats were aroused by collection and remained aroused during transport to Université de Sherbrooke. Upon arrival, half of the bats were kept at 23–24°C and remained aroused for 48 h post-collection before being euthanized by cervical dislocation. The others were placed in an environmental chamber at 5°C, allowed to reenter torpor, and sampled after 36–38 h while torpid. Rectal temperatures at sampling were a mean of 5.6°C for torpid bats and 36–37°C for aroused bats. Tissues were rapidly excised, frozen in liquid nitrogen, transported to Carleton University and stored at −80°C.

Preparation of tissue extracts and standard assays for GP, PFK-1, PK, and PDH, as well as the assay method for fructose-2,6-bisphosphate ($F2,6P_2$) were as described in Brooks and Storey (1992). Analysis of HIF-1α protein by immunoblotting, and levels *hif-1*α mRNA, antisense *hif-1*α transcripts, and microRNA 106b by PCR were as in Maistrovski et al. (2012). Data are means ± SEM, $n = 3$–4 samples from different animals. Statistical testing used the Student's *t* test.

36.3 Results and Discussion

The PDH enzyme complex gates the entry of carbohydrate fuels into mitochondrial oxidative reactions. The complex is tightly controlled not only to regulate the catabolism of carbohydrate versus lipid fuels but also to mediate carbohydrate use as a biosynthetic precursor in tissues that do significant synthesis of fatty acids and amino acids. PDH is strongly suppressed not just during torpor in hibernators (Storey 1989; Brooks and Storey 1992) but also during daily torpor

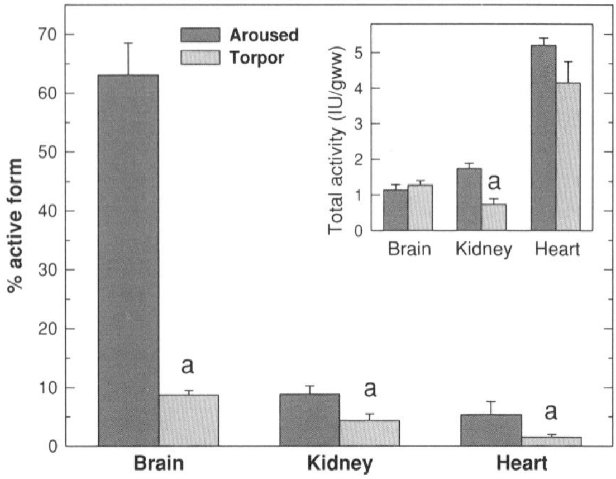

Fig. 36.1 Percentage of pyruvate dehydrogenase (PDH) in the dephosphorylated active a form in tissues from aroused and torpid little brown bats. *Insert* shows total PDH activity determined after in vitro dephosphorylation of the enzyme. *a* significantly different from the corresponding aroused value, $P < 0.01$

(Heldmaier et al. 1999). The major control mechanism of PDH activity is reversible phosphorylation; the enzyme is active in its dephosphorylated form and shut off by phosphorylation via pyruvate dehydrogenase kinase (PDK). Multiple PDK isoforms exist that are responsive to different signals in different tissues; PDK4 is strongly up-regulated during hibernation in ground squirrel heart (Buck et al. 2002). Figure 36.1 shows the strong suppression of PDH in selected bat tissues (those with high aerobic capacity) during torpor. Total activity of PDH (measured after in vitro dephosphorylation of the enzyme) did not change in brain and heart between euthermia and torpor but dropped by 58% in kidney of torpid bats (Fig. 36.1 inset). A similar decrease in total PDH activity of 45% was seen in ground squirrel heart during torpor (Brooks and Storey 1992). However, the major effect of torpor on PDH was a strong decrease in the percentage of total enzyme activity in the active dephosphorylated a form. In bat brain, the percent PDHa dropped from 63% in euthermia to 9% during torpor, kidney PDHa decreased from 9 to 4% and heart PDHa dropped from 5.3 to 1.5%. Similar data were obtained for *Z. hudsonius* (%PDHa decreased from 15% in kidney and 29% in heart to just 1% during torpor in both tissues) (Storey 1989) and *I. tridecemlineatus* (PDH-a activity in kidney and heart of torpid animals was only 3–4% of the euthermic value) (Brooks and Storey 1992).

Another measure of the state of carbohydrate metabolism in tissues is the activity of GP that gives an estimate of the relative dependence on stored glycogen as a fuel. Table 36.1 shows the total activities of GP in six tissues of aroused and torpid bats along with the percentage in the active a form. Total GP activity changed significantly during torpor in only two tissues, increasing by 2.4-fold in heart and decreasing to just 59% of the aroused value in liver. However, the percentage of phosphorylated active GPa changed in all tissues, decreasing by about one-half in liver, brain, kidney, and brown fat whereas % a doubled in skeletal muscle and increased by 1.6-fold in heart. These results for four tissues suggest reduced reliance on carbohydrate fuels during torpor. By contrast, data for

Table 36.1 Total activity of glycogen phosphorylase (measured in the presence of 1.6 mM AMP) and percent in the active a form (measured without AMP) in tissues of aroused and torpid bats

	% a		Total (units/gww)	
	Aroused	Torpor	Aroused	Torpor
Skeletal muscle	20.2 ± 0.9	42.0 ± 4.6[a]	6.05 ± 0.29	5.68 ± 0.68
Heart	20.0 ± 1.4	32.4 ± 6.2[b]	3.63 ± 0.90	8.70 ± 0.37[b]
Liver	84.7 ± 3.2	34.8 ± 4.2[a]	9.60 ± 0.42	5.68 ± 0.75[a]
Brain	65.9 ± 8.3	35.9 ± 3.8[a]	7.00 ± 0.54	6.9 ± 0.21
Kidney	62.5 ± 7.3	38.5 ± 4.0[a]	1.48 ± 0.31	0.95 ± 0.10
Brown fat	55.3 ± 3.9	35.9 ± 1.8[b]	1.73 ± 0.22	2.35 ± 0.44

Significantly different from the corresponding aroused value
[a] $P < 0.01$
[b] $P < 0.05$

heart indicate a prominent increase in dependence on glycogen fuel in heart of hibernating bats; when changes in total GP and %GPa are combined, a nearly fourfold increase in the amount of GPa activity was seen (a rise from 0.73 to 2.82 U/gww). Other species show similar regulation of GP in torpor. Both total GP and %GPa also decreased strongly in *Z. hudsonius* liver giving a net 11-fold reduction in activity during torpor (Storey 1987a) and the amount of active GP also decreased by 60% in torpid brain (Storey 1987b). Liver GPa content decreased by one-half in torpid *I. tridecemlineatus* (Brooks and Storey 1992). Liver is the primary source of carbohydrate fuels for export to organs that depend on glucose; other organs have endogenous reserves of glycogen for their own use but do not export glucose. Hence, it makes sense that glycogenolysis is suppressed in liver during torpor to place strong controls on the consumption of this major glycogen depot. Furthermore, liver is the site of most gluconeogenesis and during torpor would receive significant input of glycerol (derived from triglyceride hydrolysis) for conversion to glucose or glycogen. Inhibition of GP in liver is key to promoting the opposing glycogen synthase reaction.

PFK-1 is the major regulatory enzyme of glycolysis and the ATP-dependent committed step in converting hexose phosphates into triose phosphates for use as aerobic fuels or in biosynthetic reactions. Brigham et al. (1990) found a seasonal decrease in PFK-1 activity (by 60%) in gastrocnemius muscle of hibernating (November) *M. lucifugus* compared with active bats in September correlated with a shift to a higher percentage of fast twitch oxidative fibers in the hibernating season. Seasonal decreases in PFK-1 activity also occurred in pectoralis muscle of *M. lucifugus* and *Eptesicus fuscus* (Yacoe 1983; Brigham et al. 1990). Table 36.2 shows that selected kinetic parameters of liver and skeletal muscle PFK-1 also change significantly over the torpor-arousal cycle (but no differences were found for heart PFK-1). Liver PFK-1 showed significantly reduced affinity for both of its substrates, ATP and fructose-6-phosphate (F6P) during torpor; substrate affinity ($S_{0.5}$) values increased by 67 and 50%, respectively, compared with aroused values. Sensitivity to activation by inorganic phosphate (Pi) was also reduced.

Table 36.2 Kinetic properties of PFK-1 from tissues of aroused and torpid bats

	Liver		Skeletal muscle		Heart
	Aroused	Torpid	Aroused	Torpid	
V_{max} (U/gww)	2.29 ± 0.18	2.11 ± 0.38	2.99 ± 0.38	6.00 ± 0.46[a]	4.90 ± 0.37
$S_{0.5}$ Mg.ATP (μM)	31.0 ± 2.9	51.9 ± 2.1[a]	31.6 ± 0.1	30.7 ± 03.7	22.8 ± 2.4
$S_{0.5}$ F6P (mM)	7.07 ± 0.83	10.6 ± 0.79[b]	2.37 ± 0.18	1.71 ± 0.04[b]	1.74 ± 0.20
I_{50} Mg.ATP (mM)	≫25	≫25	0.78 ± 0.03	0.88 ± 0.09	≫15
I_{50} Mg.citrate (μM)	4,940 ± 580	4,240 ± 490	29.0 ± 1.6	31.8 ± 2.7	134 ± 12
K_a F2,6P_2 (nM)	103 ± 20	84 ± 8	33 ± 4	19 ± 2[b]	162 ± 23
K_a AMP (μM)	86.7 ± 1.2	88.3 ± 2.2	13.3 ± 0.84	5.72 ± 0.34[a]	12.5 ± 0.62
K_a Pi (mM)	1.57 ± 0.37	2.88 ± 0.19[a]	0.95 ± 0.06	0.49 ± 0.05[a]	0.68 ± 0.08
K_a NH_4^+ (mM)	3.59 ± 0.25	4.25 ± 0.18	2.16 ± 0.08	2.31 ± 0.40	2.52 ± 0.45

Data for heart are means of aroused and torpid values since no changes were found between the two states for any parameter. $S_{0.5}$ values, [substrate] producing half-maximal velocity, for ATP and F6P were determined at co-substrate levels of 5 mM F6P (10 mM for liver) and 0.5 mM Mg.ATP, respectively. I_{50} is the [inhibitor] that reduced activity by 50% and K_a is the [activator] giving half-maximal activation. For I_{50} ATP determinations, F6P was 3 mM for muscles and 10 mM for liver; I_{50} Mg. was measured at 3 mM F6P (10 mM for liver) and 0.3 mM Mg.ATP. K_a values were determined at 1 mM F6P (2 mM for liver) and 0.3 mM Mg.ATP. Significantly different from the corresponding aroused value

[a] $P < 0.01$

[b] $P < 0.05$

All of these kinetic changes would contribute to inhibiting liver PFK-1 during torpor. The enzyme from *Z. hudsonius* liver also showed altered kinetic properties during torpor that were inhibitory; PFK-1 was less activated by fructose-2,6-P_2 (K_a 2.5-fold higher than controls) and more strongly inhibited by ATP and citrate (I_{50} values were ~fourfold lower than controls) (Storey 1987a). By contrast, skeletal muscle PFK-1 showed changes in kinetic parameters that would generally enhance enzyme function during torpor. Maximal activity increased by twofold, F6P substrate affinity increased ($S_{0.5}$ decreased by 30%) and sensitivity to activation by F2,6P_2, AMP and Pi all increased (K_a values reduced by 40–60%). Combined with the data for GP this could suggest a relative shift in favor of carbohydrate catabolism in bat skeletal muscle during torpor. Stable changes in properties of PFK-1 are often due to changes in the phosphorylation state of the enzyme and in vitro incubation studies that stimulated the actions of protein kinases A and C or treated the enzyme with alkaline phosphatase indicated that this was also true of bat skeletal muscle PFK-1. When the K_a F2,6P_2 was used as a sensitive measure of PFK-1 modification, incubations of PFK-1 from muscle of aroused bats with cAMP + ATP + PKA or with Ca^{2+} + ATP + phorbol 12-myristate 13-acetate (to stimulate PKC) reduced K_a F2,6P_2 by 50–55%, similar to the effect of torpor seen in Table 36.2. By contrast, phosphatase treatment of PFK-1 from torpid bats had the opposite effect, increasing K_a F2,6P_2 by 1.9–2.4-fold.

F2,6P_2 is the most potent allosteric activator of PFK-1 and is generally considered to regulate PFK-1 with respect to anabolic signals; F2,6P_2 typically rises under conditions where carbohydrates are being used for biosynthesis and decreases under conditions (e.g., hypoxia, starvation) when glycolysis is reserved for energy production (Okar and Lange 1999). $F2,6P_2$ levels rose significantly ($P < 0.05$) during torpor in liver by fivefold (from 0.71 ± 0.16 nmol/gww in aroused to 3.86 ± 0.19 nmol/gww in torpid bats) and in skeletal muscle by 2.6-fold (from 0.50 ± 0.05 to 1.31 ± 0.21 nmol/gww). Levels did not change in other tissues; overall means were 0.58 ± 0.20, 2.85 ± 0.79, 1.36 ± 0.20, and 45.6 ± 1.90 nmol/gww in heart, brain, kidney, and brown fat, respectively. $F2,6P_2$ did not change in either liver or muscle during torpor in *Z. hudsonius* or *I. tridecemlineatus* (Storey 1987a, b). Combined with the opposite decrease in K_a $F2,6P_2$ of skeletal muscle PFK-1 that makes the enzyme more sensitive to K_a $F2,6P_2$ during torpor, this adds further evidence for a relative enhancement of glycolysis in muscle.

The enzyme 6-phosphofructo-2-kinase (PFK-2) synthesizes $F2,6P_2$. PFK-2 was assessed in bat skeletal muscle to determine if altered properties between aroused and torpid states or temperature change might be linked with the altered $F2,6P_2$ levels between the two states. K_m values for F6P substrate did not differ between aroused and torpid states but were ~twofold higher when assayed at 37°C (128 ± 29 vs. 119 ± 33 μM for PFK-2 from aroused vs. torpid bats) than at lower temperatures; values at 25°C were 70 ± 27 and 52 ± 11 μM and at 5°C were 75 ± 25 and 67 ± 13 μM, respectively. Hence, temperature effects in lowering the K_m F6P could help to promote the rise in skeletal muscle [$F2,6P_2$] during torpor.

A previous study of PK from *M. lucifugus* found seasonal differences (early April hibernating vs. August active) in isozyme pattern, activity, and temperature effects on PK from liver and pectoralis muscle that optimized PK for function at lower and variable body temperatures (Borgmann and Moon 1976). Table 36.3 shows that *M. lucifugus* also adjusts PK properties over the torpor-arousal cycle. The bat liver enzyme was differentially regulated between aroused and torpid states and so was heart PK but skeletal muscle PK was not. This latter result agrees with known properties of mammalian muscle PK and a lack of PK modification in both *Z. hudsonius* and *I. tridecemlineatus* muscle during torpor (Storey 1987b; Brooks and Storey 1992). Heart PK showed a 1.7-fold increase in maximal activity during torpor along with a small change in affinity for phosphoenolpyruvate (PEP) substrate and greater inhibition by ATP that could make PK primarily responsive to cellular energy levels during torpor. Bat liver PK showed several significant modifications during torpor. Maximal activity of the enzyme doubled, sensitivity to inhibition by ATP and L-alanine decreased (I_{50} values approximately doubled) and sensitivity to feed-forward activation by $F1,6P_2$ (the product of the PFK-1 reaction) increased (K_a dropped fourfold). All of these changes argue for a more active liver PK in the torpid state although this is contrary to the evidence of less active GP and PFK-1. However, it might be proposed that this data would fit with a model where glycerol derived from triglyceride hydrolysis in other organs was being utilized as a fuel by liver; hence, the triose phosphate portion of glycolysis could be favored while the hexose phosphate portion is relatively suppressed. In vitro incubations with liver PK suggested that reversible phosphorylation was

Table 36.3 Kinetic properties of pyruvate kinase from tissues of aroused and torpid bats

	Liver		Heart		Skeletal muscle
	Aroused	Torpid	Aroused	Torpid	
Vmax (U/gww)	7.53 ± 1.04	16.0 ± 1.37[a]	54.4 ± 8.3	93.2 ± 6.4[a]	53.8 ± 10.0
$S_{0.5}$ Mg.ADP (μM)	410 ± 30	460 ± 30	630 ± 20	660 ± 30	820 ± 40
$S_{0.5}$ PEP (μM)	430 ± 70	480 ± 50	66 ± 11	85 ± 2[a]	56 ± 5
I_{50} Mg.ATP (mM)	19.6 ± 2.2	41.0 ± 6.1[a]	29.1 ± 4.2	19.3 ± 0.2[a]	20.2 ± 3.0
I_{50} L-alanine (mM)	0.37 ± 0.02	0.61 ± 0.05[a]	≫25	≫25	≫50
K_a F1,6P_2 (μM)	4.84 ± 1.50	1.18 ± 0.32[a]	n.a.d.	n.a.d.	n.a.d.

Data for skeletal muscle are means of aroused and torpid values since no changes were detected between the two conditions. $S_{0.5}$ values for ADP and PEP were determined at co-substrate levels of 5 mM PEP and 3 mM ADP, respectively. For I_{50} determinations, PEP was 0.08 mM for muscle and 0.5 mM for liver and heart; for K_a analysis, PEP was 0.02 mM for muscles and 0.20 mM for liver. n.a.d. = no activation detected. Significantly different from the corresponding aroused value

[a] $P < 0.01$

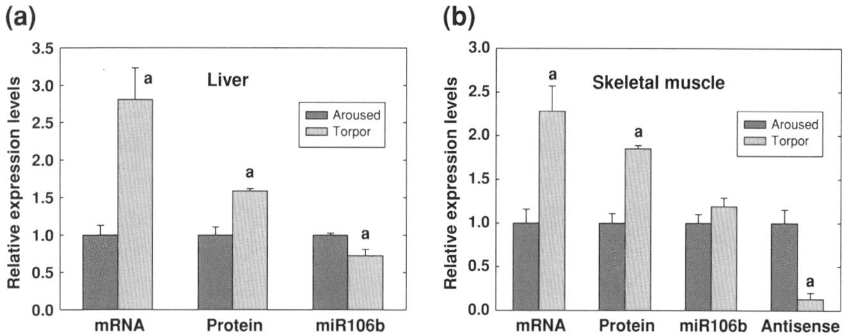

Fig. 36.2 Responses of the hypoxia-inducible transcription factor (HIF-1) system to torpor in *M. lucifugus* liver and skeletal muscle. Relative levels are shown for *hif-1α* mRNA, HIF-1α protein, microRNA 106b, and *hif-1α* antisense RNA. *a* significantly different from the corresponding aroused value, $P < 0.05$. Data compiled from Maistrovski et al. (2012)

responsible for the changes in PK properties between aroused and torpid states. Using I_{50} for L-alanine as an indicator of enzyme modification, the results showed that incubation of PK from aroused bats under conditions that stimulated PKA or PKC action resulted in a 2.1–2.7-fold increase in the I_{50} value, similar to the effect of torpor on the enzyme whereas alkaline phosphatase treatment had no effect (studies with liver of torpid bats were not done). Hence, as for muscle PFK-1, the data for liver PK are consistent with the enzyme from aroused bats being a low phosphate form that is modified by protein kinase action in torpor to create a high phosphate form with altered properties.

Multiple glycolytic enzymes are under transcriptional control by the hypoxia-inducible transcription factor (HIF-1), a dimer made of a constitutively expressed ß subunit and an inducible α subunit that is stable under low oxygen conditions but targeted for rapid degradation when oxygen is high (Semenza 2007). Previous studies showed HIF-1 involvement in ground squirrel hibernation; HIF-1α protein increased 60–70% in thermogenic organs (brown adipose, skeletal muscle) during torpor and HIF-1 binding to DNA in nuclear extracts increased sixfold indicating greater HIF-1-mediated gene expression during torpor (Morin and Storey 2005). We recently evaluated the HIF-1 system in bat liver and skeletal muscle, finding strong evidence of HIF-1 involvement in torpor (Maistrovski et al. 2012). Figure 36.2 shows that transcript levels of *hif-1*α rose in both tissues (by 2.3–2.8-fold) during torpor compared with aroused bats as did HIF-1α protein (1.6–1.9-fold increase). In a unique discovery about hibernation, HIF-1α expression during torpor was linked with the actions of two kinds of non-coding RNA. In liver, levels of microRNA 106b that negatively regulates *hif-1*α translation were reduced, thereby contributing to increased HIF-1α synthesis during torpor. We previously linked differential expression of other microRNA species with hibernation in *I. tridecemlineatus* (Morin et al. 2008). In muscle, the first evidence of translation control by antisense RNA in hibernation was found. Levels of antisense *hif-1*α RNA decreased sharply during torpor to just 13% of the euthermic value. Antisense transcripts suppress translation by binding to the 3′UTR of a mRNA, both interfering with translation by overlapping ribosome binding sites and recruiting proteins such as the RNA-induced silencing complex (RISC) that targets the mRNA for degradation (Good 2003). Reduced antisense RNA would therefore potentiate *hif-1*α translation to enhance HIF-1α protein levels and lead to selective expression of HIF-1 regulated genes during torpor.

In conclusion, these studies with bats show organ-specific differences in the control of carbohydrate catabolism in torpor versus arousal and key roles for protein phosphorylation in enzyme control and non-coding RNA in HIF-1 regulation.

Acknowledgments I am forever grateful to Dr. D.W. Thomas, Université de Sherbrooke for bat collection and use of his laboratory for experimental hibernation studies. Thanks to students from my laboratory (R. Whitwam, J. Duncan, M. de la Roche, Y. Maistrovski, K. Biggar) for their contributions to the bat research reported here and to J.M. Storey for editorial review of the manuscript. Supported by the N.S.E.R.C. Canada and the Canada Research Chairs program.

References

Borgmann AI, Moon TW (1976) Enzymes of the normothermic and hibernating bat, *Myotis lucifugus*: temperature as a modulator of pyruvate kinase. J Comp Physiol 107:185–199

Boyles JG, Dunbar MB, Storm JJ, Brack V (2007) Energy availability influences microclimate selection of hibernating bats. J Exp Biol 210(24):4345–4350

Brigham RM, Ianuzzo CD, Hamilton N, Fenton MB (1990) Histochemical and biochemical plasticity of muscle fibers in little brown bat (*Myotis lucifugus*). J Comp Physiol B 160:183–186

Brooks SPJ, Storey KB (1992) Mechanisms of glycolytic control during hibernation in the ground squirrel *Spermophilus lateralis*. J Comp Physiol B 162:23–28

Buck MJ, Squire TL, Andrews MT (2002) Coordinate expression of the PDK4 gene: a means of regulating fuel selection in a hibernating mammal. Physiol Genom 8(1):5–13

Eddy SF, Storey KB (2003) Differential expression of Akt, PPAR-γ and PGC-1 during hibernation in bats. Biochem Cell Biol 81:269–274

Eddy SF, Storey KB (2004) Up-regulation of fatty acid-binding proteins during hibernation in the little brown bat, *Myotis lucifugus*. Biochim Biophys Acta 1676:63–70

Eddy SF, Storey KB (2007) p38 MAPK regulation of transcription factor targets in muscle and heart of bats, *Myotis lucifugus*. Cell Biochem Funct 25:759–765

Eddy SF, McNally JD, Storey KB (2005) Up-regulation of a thioredoxin peroxidase-like protein, proliferation associated gene, in hibernating bats. Arch Biochem Biophys 435:101–111

Eddy SF, Morin P, Storey KB (2006) Differential expression of selected mitochondrial genes in hibernating little brown bats, *Myotis lucifugus*. J Exp Zool A 305:620–630

Geiser F (2004) Metabolic rate and body temperature reduction during hibernation and daily torpor. Annu Rev Physiol 66:239–274

Good L (2003) Translation repression by antisense sequences. Cell Mol Life Sci 60:854–861

Heldmaier G, Klingenspor M, Werneyer M, Lampi BJ, Brooks SP, Storey KB (1999) Metabolic adjustments during daily torpor in the Djungarian hamster. Am J Physiol 276:E896–E906

Holden CP, Storey KB (1998) Protein kinase A catalytic subunit from bat skeletal muscle: a kinetic study of the enzyme from a hibernating mammal. Arch Biochem Biophys 358: 243–250

Humphries MM, Thomas DW, Speakman JR (2002) Climate-mediated energetic constraints on the distribution of hibernating mammals. Nature 418(6895):313–316

Humphries MM, Kramer DL, Thomas DW (2003) The role of energy availability in mammalian hibernation: an experimental test in free-ranging eastern chipmunks. Physiol Biochem Zool 76(2):180–186

Landry-Cuerrier M, Munro D, Thomas DW, Humphries MM (2008) Climate and resource determinants of fundamental and realized metabolic niches of hibernating chipmunks. Ecology 89(12):3306–3316

Maistrovski Y, Biggar KK, Storey KB (2012) HIF-1α regulation in mammalian hibernators: role of non-coding RNA in HIF-1α control during torpor in ground squirrels and bats. J Comp Physiol (in press)

Matheson AL, Campbell KL, Willis CKR (2010) Feasting, fasting and freezing: energetic effects of meal size and temperature on torpor expression by little brown bats *Myotis lucifugus*. J Exp Biol 213(12):2165–2173

McGuire LP, Fenton MB, Guglielmo CG (2009) Effect of age on energy storage during prehibernation swarming in little brown bats (*Myotis lucifugus*). Can J Zool 87(6):515–519

Mehrani H, Storey KB (1997) Protein kinase C from bat brain: the enzyme from a hibernating mammal. Neurochem Intl 31:139–150

Meteyer CU, Valent M, Kashmer J et al (2011) Recovery of little brown bats (*Myotis lucifugus*) from natural infection with *Geomyces destructans*, white-nose syndrome. J Wildlife Dis 47(3):618–626

Morin P, Storey K (2005) Cloning and expression of hypoxia-inducible factor 1α from the hibernating ground squirrel, *Spermophilus tridecemlineatus*. Biochim Biophys Acta 1729: 32–40

Morin P, Dubuc A, Storey KB (2008) Differential expression of microRNA species in organs of hibernating ground squirrels: a role in translational suppression during torpor. Biochim Biophys Acta 1779:628–633

Okar DA, Lange AJ (1999) Fructose-2,6-bisphosphate and control of carbohydrate metabolism in eukaryones. Biofactors 10:1–14

Semenza GL (2007) Hypoxia-inducible factor-1 (HIF-1) pathway. Sci STKE 407:cm8

Speakman JR, Thomas DW (2003) Physiological ecology and energetics of bats. In: Kunz TH, Fenton MB (eds) Bat ecology. University of Chicago Press, Chicago, pp 430–490

Storey KB (1987a) Regulation of liver metabolism by enzyme phosphorylation during mammalian hibernation. J Biol Chem 262:1670–1673

Storey KB (1987b) Investigations of the mechanisms of glycolytic control during hibernation. Can J Zool 65:3079–3083

Storey KB (1989) Integrated control of metabolic rate depression via reversible phosphorylation of enzymes in hibernating mammals. In: Malan A, Canguilhem B (eds) Living in the cold. John Libbey Eurotext, Montrouge, pp 309–319

Storey KB, Storey JM (2004) Metabolic rate depression in animals: transcriptional and translational controls. Biol Rev Camb Philos Soc 79:207–233

Storey KB, Storey JM (2007) Putting life on 'pause'—molecular regulation of hypometabolism. J Exp Biol 210:1700–1714

Storey KB, Storey JM (2010) Metabolic rate depression: the biochemistry of mammalian hibernation. Adv Clin Chem 52:77–108

Thomas DW (1993) Lack of evidence for a biological alarm clock in bats (*Myotis* spp.) hibernating under natural conditions. Can J Zool 71(1):1–3

Thomas DW, Geiser F (1997) Periodic arousals in hibernating mammals: is evaporative water loss involved? Funct Ecol 11(5):585–591

Townsend KL, Kunz TH, Widmaier EP (2008) Changes in body mass, serum leptin, and mRNA levels of leptin receptor isoforms during the premigratory period in *Myotis lucifugus*. J Comp Physiol B 178(2):217–223

Willis CKR, Menzies AK, Boyles JG, Wojciechowski MS (2011) Evaporative water loss is a plausible explanation for mortality of bats from white-nose syndrome. Integ Comp Biol 51(3):364–373

Yacoe ME (1983) Adjustments of metabolic pathways in the pectoralis muscles of the bat *Eptesicus fuscus* related to carbohydrate sparing during hibernation. Physiol Zool 56:648–658

Chapter 37
Theme and Variation: Proteomic Changes Across Three Organs in Hibernation Cycles of the 13-Lined Ground Squirrel

Katharine R. Grabek and Sandra L. Martin

Abstract Circannual hibernation is characterized by two cycles—the annual cycle between summer homeothermy and winter heterothermy as well as the individual torpor-arousal cycles within winter heterothermy. We hypothesized that proteins and pathways that promote or enhance survival throughout hibernation cycles are concordantly regulated in organs throughout the body. In this study, we applied a meta-analysis to previously described proteomic changes in six identical seasonal and physiological states of heart, skeletal muscle, and kidney in order to identify common protective mechanisms underlying resistance to challenges of hibernation. Unexpectedly, we detected little overlap among organs. In heart, few protein spots changed in abundance among the six states, while in kidney more than half of the spots detected differed among these same states. Only two significantly changing proteins were shared among all three datasets, and of proteins common between any two organs, many were discordant for abundance pattern changes among states. In sum, most proteomic changes accompanying hibernation appear to be organ-specific.

Abbreviations

DAVID	Database for annotation, visualization and integrated discovery
E-Ar	Early arousing
Ent	Entrance
FAC	Functional annotation cluster
HDL	High-density lipoprotein
IBA	Interbout aroused
I/R	Ischemia–reperfusion
LT	Late torpor

K. R. Grabek · S. L. Martin (✉)
Human Medical Genetics and Genomics Program,
Department of Cell and Developmental Biology,
University of Colorado School of Medicine, Aurora, CO, USA
e-mail: Sandy.Martin@ucdenver.edu

T. Ruf et al. (eds.), *Living in a Seasonal World*,
DOI: 10.1007/978-3-642-28678-0_37, © Springer-Verlag Berlin Heidelberg 2012

SpD	Spring dark
SA	Summer active
2D DiGE	Two-dimensional difference gel electrophoresis

37.1 Introduction

Circannual hibernators such as the 13-lined ground squirrel (*Ictidomys tridecemlineatus*) conserve energy during cold winter seasons by entering into a state of reversible metabolic depression referred to as torpor. While in torpor, body temperature (T_b) is lowered to near freezing, and heart, respiratory and metabolic rates are dramatically reduced to 1–4% of euthermic values (Carey et al. 2003). Torpor is not continuous throughout the period of hibernation, however, and animals intermittently arouse back to euthermia. For circannual hibernators, the ability to become torpid only exists in the winter (Kortner and Geiser 2000), and a recent model proposes that two "switches" underlie the phenomena (Serkova et al. 2007). The first switch defines the annual cycle, in which a seasonal reprogramming allows animals to shift from summer homeothermy to winter heterothermy, while the second switch underlies the individual torpor-arousal cycles that give rise to winter heterothermy.

In order to survive a season of hibernation, there must be a coordination of the hibernator's organs that permit successful entry into, continuance of, and arousal from multiple bouts of torpor. Because the demand for function while cold and metabolically depressed depends on the organ, the energetic and metabolic demands of each organ must be addressed and regulated in order to minimize damage yet maintain support of essential functions. Additionally, while arousing from torpor for each brief euthermic period, some organs are highly metabolically active while others are relatively quiescent (e.g., heart vs. kidney). Rates of reperfusion during arousal also vary among organs (Lyman 1982). Hence, the sets of biochemical adjustments that enhance function and protect from damage likely differ among organs. However, we hypothesize that there are global adjustments in organs that promote or enhance survival throughout hibernation cycles.

In order to better identify the components of the two cycles, we previously screened for protein abundance changes among distinct seasonal and torpor-arousal states in multiple organs, including heart (Grabek et al. 2011), skeletal muscle (Hindle et al. 2011) and kidney (manuscript submitted), of the 13-lined ground squirrel. By virtue of our research design, these individual organ-specific studies shared six identical, precisely defined groups, making it now possible to elucidate global protein changes that underlie the cycles of hibernation. Here, we applied a meta-analysis to the previously described individual proteomic datasets of these three organs to identify common protective mechanisms underlying their shared resistance to challenges of mammalian homeostasis in hibernation.

37.2 Methods

Proteomic datasets previously generated in our lab were compiled in order to compare seasonal protein abundance changes across multiple organs. These datasets included the multi-state analyses of heart (Grabek et al. 2011), skeletal muscle (Hindle et al. 2011), and kidney (manuscript submitted). The methods for organ collection, protein abundance quantification, and protein identification were described previously, but briefly, follow this general scheme: organs were collected from 13-lined ground squirrels representing different seasonal and physiological groups ($n = 5$–6 per group) as defined by body temperature telemetry. The groups used for this analysis included: Early Arousing (E-Ar; $T_b = 6$–12.8°C); Interbout Aroused (IBA; $T_b \sim 37$°C, 3 h after inflection point demarking the end of rapid rewarming), Entrance (Ent; $T_b = 27$–23°C); Late Torpor (LT; $T_b = 5$–6°C, time torpid = 80–95% of previous bout); Spring Dark (SpD; $T_b = 37$°C for >10 days after spontaneous termination of hibernation); and Summer Active (SA; animals in July–early August). Extracts were prepared from each organ sample and proteins separated using two-dimensional difference gel electrophoresis (2D DiGE) as described previously. The resulting 2D gels were evaluated with DeCyder 2D 7.0 software (GE Healthcare), where protein spots were matched across gels for each dataset and spot intensities were normalized to their respective internal standard. After statistical analyses, protein spots of interest were robotically picked from gels, digested into peptides and identified using LC MS/MS as previously described (Epperson and Martin 2011).

In the present analysis, a one way ANOVA followed by a Benjamini Hochberg false discovery rate correction (Benjamini and Hochberg 1995) was applied to protein spot intensities for spots present in $n-1$ samples per group across all groups, where n represents the minimum number of samples that constitute a defined group. This reanalysis of the original data was done in order to identify spots that changed significantly in abundance among the same six groups in the kidney, heart, and skeletal muscle datasets by the same criteria. Thus, spots present in five samples/group in heart and skeletal muscle, and four samples/group in kidney were used. Post hoc Tukey pairwise comparisons were performed using R (R Development Core Team 2011) for all spots that were significant by one-way ANOVA ($\alpha = 0.05$).

Significantly changing spots that were previously identified by mass spectrometry were compared for shared changes among the datasets. Proteins that were shared in a minimum of two organs are reported by their official gene symbol from the Entrez Gene database at NCBI. This database was also used to assign functions to individual proteins. A Spearman's rank test was used to identify the types of protein abundance pattern relationships between each pair of organs. The mean protein abundance value for each state was first normalized to the highest mean abundance value of every individual protein tested in order to account for differences in intensities across datasets. Protein abundance patterns between any two organs were considered either positively correlated for a resulting Spearman's rank correlation coefficient rho (ρ) > 0.5, negatively correlated for $\rho < -0.5$, and non-correlated for $-0.5 < \rho < 0.5$.

To identify common biological themes underlying seasonal and physiological changes of the hibernating phenotype, the functional annotation clustering (FACs) tool in Database for Annotation, Visualization, and Integrated Discovery (DAVID) (Huang da et al. 2009) was used. Official gene symbols of identified proteins that were shared between at least two organs were submitted to the DAVID database; FACs were generated using a *Homo sapiens* background and clustering stringency set to "high", with default settings for all other parameters. The terms that provided the most biological insight for FACs with a score >1.3 are reported in the text. The Kegg Pathway component of DAVID was also used to reveal additional functional pathways not identified by the FACs. The same input used for generating the FACs was used to create Kegg functional annotation charts with parameters set to default. Additional Kegg annotations not previously identified by the FACs and significant for enrichment after a Benjamini–Hochberg correction ($q < 0.05$) are reported in the text.

37.3 Results

After testing for spot abundance changes among the six states (SpD, SA, IBA, Ent, LT, E-Ar), 420/820 (51%) spots in kidney, 244/927 (26%) spots in skeletal muscle, and 35/432 (8%) spots in heart were significant ($q < 0.05$). In skeletal muscle and heart, most spots appeared to change seasonally (summer homeothermic versus winter heterothermic states), while in kidney, changes within the winter cycle were also evident (Fig. 37.1a). These patterns were confirmed statistically via post hoc Tukey tests. Counts of significant Tukey values for each pairwise comparison revealed that most abundance changes occurred between the summer (SA and SpD) and winter groups (IBA, ENT, LT, and E-Ar) for both heart and skeletal muscle, while in kidney, there was an equally large number of changes between E-Ar and the other winter groups (Fig. 37.1b). The numbers of significant spots with proteins identified by mass spectrometry were as follows: 30 (25 unique proteins) in heart (Grabek et al. 2011), 142 (69 unique proteins) in skeletal muscle (Hindle et al. 2011), and 124 (87 unique proteins) in kidney (manuscript submitted).

When compared for overlap, only two identified proteins (Fig. 37.2) were shared among all three datasets: apolipoprotein A1 (APOA1) and fatty acid binding protein 3 (FABP3). A plot of the relative abundances among groups for these two proteins (Fig. 37.3) revealed a similar trend of changes among all three organs for APOA1—its pattern most closely matched between heart and kidney with highest abundance present in LT and E-Ar. In contrast, FABP3 appeared to be somewhat similar in heart and skeletal muscle with the greatest abundance occurring during winter, however, an inverse pattern was apparent in kidney.

Additional overlap was found between pairs of organs. Four proteins were shared between either skeletal muscle and heart, or kidney and heart, yet 17 proteins were shared between skeletal muscle and kidney (Fig. 37.2). All four

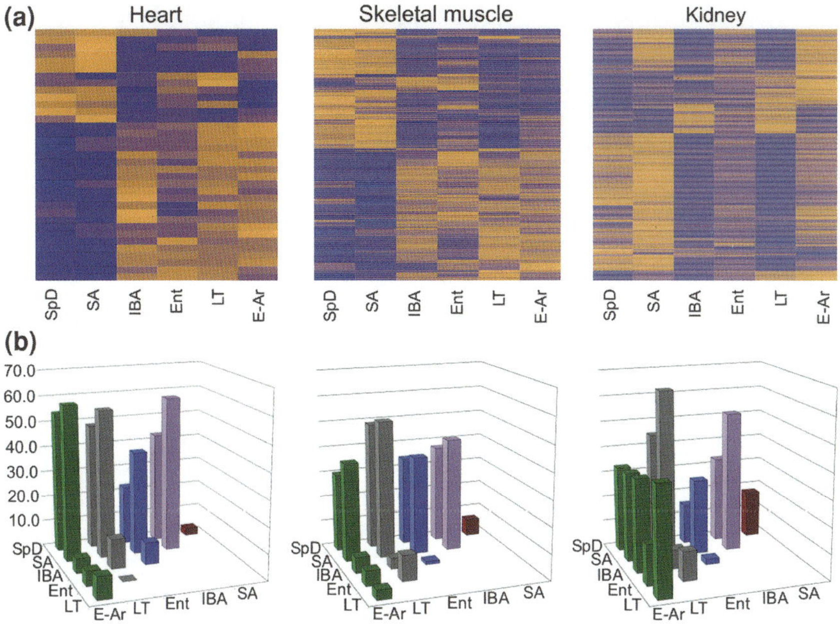

Fig. 37.1 Protein abundance changes in heart and skeletal muscle are mainly seasonal while in kidney, proteins also change with the torpor-arousal cycle. **a** Heat maps showing mean intensity values of significantly changing protein spots ($q < 0.05$) in heart (*left*), skeletal muscle (*middle*) and kidney (*right*). *Blue* and *orange* indicate decreased or increased intensity, respectively. Hierarchically clustered individual spots are displayed in rows; sample groups are listed along the *bottom*. **b** Percentages of differing protein spots in pairwise comparisons of physiological groups are represented by 3D bars in heart (*left*), skeletal muscle (*middle*) and kidney (*right*)

Fig. 37.2 Most significantly changing proteins are organ-specific. Venn diagram shows the number of unique and shared identified significant ($q < 0.05$) proteins between and among the three organs (heart, skeletal muscle and kidney)

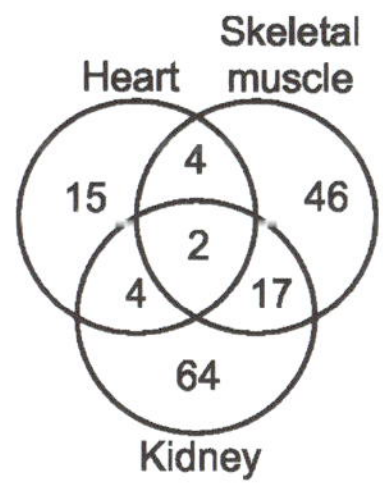

proteins common to the heart and kidney datasets were positively correlated and decreased in winter compared to summer (Table 37.1). In contrast, only 4/17 proteins were positively correlated between skeletal muscle and kidney, whereas 7/17 were negatively correlated (Table 37.1). Examination of these negatively correlated proteins revealed a general trend in abundance patterns: most of these proteins increased in winter in skeletal muscle but decreased in winter in kidney.

FAC of shared proteins between the two-organ comparisons yielded several enriched FACs. Between heart and kidney, there was an enrichment for the FACs

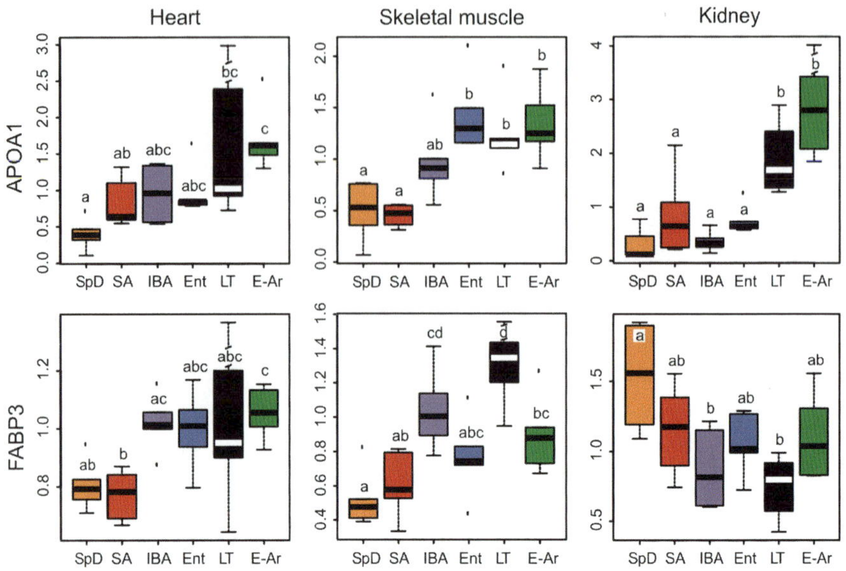

Fig. 37.3 *Box plots* show the relative abundances of the shared proteins APOA1 (*top*) and FABP3 (*bottom*) in heart (*left*), skeletal muscle (*middle*) and kidney (*right*) among the six groups. *Colored boxes* represent the region between the twenty-fifth and seventy-fifth percentiles, *bold horizontal lines* the median, *outside horizontal lines* the hundreth percentile and *circles* the outliers. Significantly different abundance values ($q < 0.05$) are indicated by letters above boxes. Patterns were positively correlated for APOA1 between heart-skeletal muscle and heart-kidney. For FABP3, heart-skeletal muscle were positively correlated while heart-kidney and skeletal muscle-kidney were negatively correlated; only the latter correlation was significant ($p < 0.05$)

"Valine, isoleucine and leucine degradation" and "Response to hormone stimulus." In the comparison between heart and skeletal muscle, the FAC "Generation of precursor metabolites and energy" was enriched. When shared proteins between skeletal muscle and kidney were analyzed, the FACs "Mitochondrion", "Oxidoreductase activity", "Glucose metabolism", "Purine nucleotide binding", and "Vesicle" were enriched. Using the Kegg pathways component in DAVID, two additional metabolic pathways were significantly enriched ($q < 0.05$) for proteins shared between skeletal muscle and kidney. These pathways were "Valine, isoleucine and leucine degradation", and "Lysine degradation". No additional enriched pathways were revealed by Kegg in the other two-organ analyses.

37.4 Discussion

Comparisons of the heart, kidney, and skeletal muscle proteomics multi-state datasets revealed several organ-specific patterns. In the heart, few protein spots changed in abundance among the six seasonal and physiological states, while in

Table 37.1 Significantly changing proteins shared between two organs

H	SM	K	Protein	Function	Concordance
X	–	X	A2M	Protease inhibitor	C[a]
X	–	X	PCCA	Amino acid degradation	C[a]
X	–	X	DBT	Branched-chain amino acid catabolism	C[a]
X	–	X	BCKDHB	Branched-chain amino acid catabolism	C[a]
X	X	–	ACO2	TCA cycle	R
X	X	–	CRAT	Fatty acid β-oxidation	C
X	X	–	PDHA1	Glycolysis, TCA cycle	C
X	X	–	CKM	Phosphocreatine metabolism	R
–	X	X	ALB	Transport	C
–	X	X	ALDH2	Glycolysis and gluconeogenesis	R
–	X	X	ALDH9A1	Glycolysis and gluconeogenesis	D
–	X	X	ETFA	Fatty acid β-oxidation	D[a]
–	X	X	ETFDH	Electron transport chain	D[a]
–	X	X	GPD1	Glycerophospholipid metabolism	C
–	X	X	HIBADH	Branched-chain amino acid catabolism	D
–	X	X	HSPA8	Protein folding, clathrin-coated vesicle disassembly	R
–	X	X	HSPD1	Mitochondrial protein folding	D
–	X	X	IMMT	Mitochondrial structure and function	R
–	X	X	MCCC1	Branched-chain amino acid catabolism	R
–	X	X	OGDH	TCA cycle	R
–	X	X	PRDX3	Antioxidant, peroxide reductase	D
–	X	X	TF	Iron transport	C
–	X	X	TUBB2C	Microtubule-based movement, polymerization	R
–	X	X	UBA1	Ubiquitin conjugation, protein degradation	C
–	X	X	VCL	F-actin membrane anchoring	D

Proteins significant ($q < 0.05$) for abundance changes among six states and shared between two organs are listed. The "X" in rows denotes protein changes with hibernation physiology in heart (H), skeletal muscle (SM), or kidney (K). The official gene symbol identifies the protein (Protein); its possible function is based on NCBI "gene" information (Function). The correlation of protein abundance patterns between organs (Concordance) is classified as either concordant (C; $\rho > 0.5$), discordant (D; $\rho < -0.5$) or random (R; $-0.5 < \rho < 0.5$)

[a] indicates a significant correlation ($\rho < 0.05$)

the kidney, more than half of the spots detected differed among these same states. The difference in protein reprogramming for hibernation in these two organs likely reflects their unique functions. The heart must beat continuously throughout the torpor-arousal cycles of hibernation, and thus a dramatic reprogramming of its proteome may not be temporally feasible or efficient. In contrast, renal function is effectively eliminated during torpor but restored during arousal (Tempel et al. 1977; Zatzman and South 1981). Thus, changes in protein spot abundance, including post-translational modification, may help facilitate the inhibition and restoration of function. In the heart and skeletal muscle, most protein abundance changes occur between the summer and winter seasons, while the kidney exhibits a distinct pattern, with many abundance differences detected both seasonally and within the torpor-arousal cycle, particularly in E-Ar. Further, more proteins increased in the winter in skeletal muscle and heart, while more kidney proteins decreased. Similar to heart, skeletal muscle is metabolically active at lower temperatures during arousal in order to facilitate whole-body rewarming via shivering thermogenesis (Hindle et al. 2011). The seasonal reprogramming of proteins are likely needed to enhance the continual function of the heart during the winter, and poise skeletal muscle for a burst of metabolic activity during late arousal.

In contrast to our expectations, only two proteins were identified as significantly changing among six hibernation states in all three organs; both of these are transporters. APOA1 is a major component of plasma high-density lipoprotein (HDL). HDL promotes cholesterol efflux from peripheral cells (Kwiterovich 2000). Further, HDL is associated with enhanced cardio- and renal protection via its role in clearing lipid plaques from arterial walls. HDL also protects against ischemia–reperfusion (I/R) injury by rapidly removing I/R-induced pro-inflammatory factors such as TNF-α (Shi and Wu 2008; Calabresi et al. 2003). The arousing period was previously found to be transiently hypoxic in hibernator brains (Ma et al. 2005); hence, increased APOA1 in heart and kidney during E-Ar may serve to play a protective role against I/R injury during arousal. FABP3 transports lipids and other hydrophobic moieties within cells (Storch and Thumser 2000). Its increase during the winter states in skeletal muscle and heart is likely related to the need for enhanced intercellular lipid transport because lipids are the primary metabolic fuel throughout winter heterothermy. On the other hand, FABP3 in the kidney is highest during spring and is relatively decreased in the winter; its suppression may be related to a general depression of renal function during hibernation.

Of the limited overlapping proteins identified in the two-organ comparisons, metabolic pathways appear to be a central theme. Specifically, the shared proteins between heart or skeletal muscle and kidney are both enriched for the degradation of branched-chain amino acids. A winter decrease of proteins involved in this pathway was previously reported in the individual analyses of the heart (Grabek et al. 2011) and liver (Epperson et al. 2010). Here, three of the four proteins shared between heart and kidney are involved in branched-chain amino acid catabolism and are concordant for a pattern of decreased abundance during the winter. Thus, the enrichment in this analysis lends further support that preservation of essential amino acids during the fasting period of winter heterothermy is a critical component for success in surviving a season of hibernation. However, in contrast, when skeletal muscle and kidney are compared, although there is also an enrichment for pathways involved in branched-chain and lysine amino acid catabolism, the associated proteins show a discordant (e.g. ALDH9A1 and HIBADH) or random (e.g. MCCC1, ALDH2, and OGDH) pattern of abundance among the six states. Given that ALDH9A1, ALDH2, and OGDH catabolize intermediate metabolites originating from multiple substrates, the discordance of protein abundance seen here may imply that differing pathways are modified for enhanced function of each organ, but the small overlap of proteins artificially enriches for the catabolism of amino acids. However, MCCC1 and HIBADH specifically function in the pathways of leucine and valine catabolism, respectively, therefore an alternate interpretation is that during the winter, branched-chain amino acid catabolism is enhanced in skeletal muscle while suppressed in heart and kidney. It is not clear why this pathway would be selectively enhanced, especially because skeletal muscle appears to be conserved, and hence protein degradation should be suppressed during the winter (Cotton and Harlow 2010), but perhaps it is needed for a limited role in gluconeogenesis (Galster and Morrison 1975).

The paucity of common protein changes may be due to technical limitations of protein identification by the methods used, however, considering the similarity of heart and muscle cell types compared to kidney, one would expect, at the least, to find greater protein overlap between these two organs. Instead, equal changes in the number of proteins between heart versus kidney and heart versus skeletal muscle were detected. Additionally, we observed concordance for abundance patterns in only 2/4 proteins shared between heart and skeletal muscle, whereas all of the shared proteins between heart and kidney were significantly concordant. Thus, the segregation of the subset of protein changes detected here, which lead to enhanced function of each organ individually during hibernation, is likely representative of the adjustments to the proteome in organs throughout the body.

In summary, common themes of proteomic changes among seasonal and physiological states in heart, skeletal muscle, and kidney of 13-lined ground squirrels were not found. Instead, our proteomics analyses of six precisely defined states using DiGE primarily revealed organ-specific adjustments that may contribute to the maintenance of homeostasis and promote survival despite the prolonged winter fast and challenges of heterothermy.

Acknowledgments This work was supported by National Institutes of Health Grants HL-089049 to S. L. Martin.

References

Benjamini Y, Hochberg Y (1995) Controlling the false discovery rate—a practical and powerful approach to multiple testing. J Roy Stat Soc B Met 57(1):289–300

Calabresi L, Rossoni G, Gomaraschi M, Sisto F, Berti F, Franceschini G (2003) High-density lipoproteins protect isolated rat hearts from ischemia-reperfusion injury by reducing cardiac tumor necrosis factor-alpha content and enhancing prostaglandin release. Circ Res 92(3): 330–337

Carey HV, Andrews MT, Martin SL (2003) Mammalian hibernation: cellular and molecular responses to depressed metabolism and low temperature. Physiol Rev 83 (4):1153–1181. doi:10.1152/physrev.00008.2003, 83/4/1153 [pii]

Cotton CJ, Harlow HJ (2010) Avoidance of skeletal muscle atrophy in spontaneous and facultative hibernators. Physiol Biochem Zool 83(3):551–560. doi:10.1086/650471

Epperson LE, Martin SL (2011) Proteomic strategies to investigate adaptive processes. In: Eckersall PD, Whitfield PD (eds) Methods in animal proteomics. Wiley-Blackwell, Oxford, pp 191–209. doi:10.1002/9780470960660.ch8

Epperson LE, Rose JC, Carey HV, Martin SL (2010) Seasonal proteomic changes reveal molecular adaptations to preserve and replenish liver proteins during ground squirrel hibernation. Am J Physiol Regul Integr Comp Physiol 298(2):R329–340. doi:10.1152/ajpregu.00416.2009, 00416.2009 [pii]

Galster W, Morrison PR (1975) Gluconeogenesis in arctic ground squirrels between periods of hibernation. Am J Physiol 228(1):325–330

Grabek KR, Karimpour-Fard A, Epperson LE, Hindle AG, Hunter LE, Martin SL (2011) A multistate proteomics analysis reveals novel strategies used by a hibernator to precondition the heart and conserve ATP for winter heterothermy. Physiol Genomics. doi:10.1152/physiolgenomics.00125.2011

Hindle AG, Karimpour-Fard A, Epperson LE, Hunter LE, Martin SL (2011) Skeletal muscle proteomics: carbohydrate metabolism oscillates with seasonal and torpor-arousal physiology of hibernation. Am J Physiol Regul Integr Comp Physiol 301(5):R1440–R1452. doi:10.1152/ajpregu.00298.2011, ajpregu.00298.2011 [pii]

Huang da W, Sherman BT, Lempicki RA (2009) Systematic and integrative analysis of large gene lists using DAVID bioinformatics resources. Nat Protoc 4(1):44–57. doi:10.1038/nprot.2008.211, nprot.2008.211 [pii]

Kortner G, Geiser F (2000) The temporal organization of daily torpor and hibernation: circadian and circannual rhythms. Chronobiol Int 17(2):103–128

Kwiterovich PO Jr (2000) The metabolic pathways of high-density lipoprotein, low-density lipoprotein, and triglycerides: a current review. Am J Cardiol 86(12A):5L–10L. doi:10.1016/S0002914900014612, S0002914900014612 [pii]

Lyman CP, Willis JS, Malan A, Wang LCH (1982) Hibernation and torpor in mammals and birds. Physiological ecology. Academic Press, New York

Ma YL, Zhu X, Rivera PM, Toien O, Barnes BM, LaManna JC, Smith MA, Drew KL (2005) Absence of cellular stress in brain after hypoxia induced by arousal from hibernation in Arctic ground squirrels. Am J Physiol Regul Integr Comp Physiol 289(5):R1297–1306. doi:10.1152/ajpregu.00260.2005, 00260.2005 [pii]

Serkova NJ, Rose JC, Epperson LE, Carey HV, Martin SL (2007) Quantitative analysis of liver metabolites in three stages of the circannual hibernation cycle in 13-lined ground squirrels by NMR. Physiol Genomics 31(1):15–24. doi:10.1152/physiolgenomics.00028.2007, 00028.2007 [pii]

Shi N, Wu MP (2008) Apolipoprotein A-I attenuates renal ischemia/reperfusion injury in rats. J Biomed Sci 15(5):577–583. doi:10.1007/s11373-008-9258-7

Storch J, Thumser AE (2000) The fatty acid transport function of fatty acid-binding proteins. Biochim Biophys Acta 1486(1):28–44. doi:10.1016/S1388-1981(00)00046-9, S1388-1981(00)00046-9 [pii]

R Development Core Team (2011) R: a language and environment for statistical computing

Tempel GE, Musacchia XJ, Jones SB (1977) Mechanisms responsible for decreased glomerular filtration in hibernation and hypothermia. J Appl Physiol 42(3):420–425

Zatzman ML, South FE (1981) Circannual renal function and plasma electrolytes of the marmot. Am J Physiol 241(1):R87–91

Chapter 38
Putting the Brakes on Protein Synthesis in Mammalian Hibernation

Frank van Breukelen, Jenifer C. Utz, Michael Treat and Peipei Pan

Abstract The roles for both passive depression and active suppression of energy-consuming processes in effecting the metabolic depression of a hibernator are poorly understood. The metabolic depression is likely the result of both active and passive processes. An understanding of what processes precipitate the decline in energy utilization may prove critical in our understanding of how hibernators go torpid. In this review, we will analyze existing data on protein synthesis to understand how its regulation may inform on the processes of torpor utilization.

38.1 Introduction

Mammalian hibernation is characterized by oscillations of torpor and euthermy. During torpor, animals may experience very low body temperatures (T_b) to as low as −2.9°C (Barnes 1989) and markedly reduced oxygen consumption to as low as 1/100th of active rates (Wang and Lee 2000). This period of torpor may last for as long as 3 weeks depending on the time of year, ambient temperature, and species. For the typical hibernator, torpor bouts are shorter in the beginning and end of a hibernation season and longest in the middle of that season. Ambient temperature has a profound effect on torpor bout duration with colder temperatures resulting in longer torpor bouts except when the animal must actively thermoregulate e.g. animals will allow their T_b to approach that of ambient until ambient temperature gets too cold (typically near 0°C in ground squirrels but this may vary by species).

F. van Breukelen (✉) · J. C. Utz · M. Treat · P. Pan
School of Life Sciences, University of Nevada Las Vegas, 4505 Maryland Parkway, Las Vegas, NV 89154, USA
e-mail: frank.vanbreukelen@unlv.edu

T. Ruf et al. (eds.), *Living in a Seasonal World*,
DOI: 10.1007/978-3-642-28678-0_38,

As evidenced by oxygen consumption, these animals will actively defend T_b at a temperature greater than their passive minimal temperature (Buck and Barnes 2000; Barnes 1989). Species vary tremendously in their use of torpor with some species being facultative hibernators e.g. black versus white-tailed prairie dogs and others utilizing shallower depressions of both temperature and oxygen consumption. Many mammals use torpor on a daily basis. Typically, these daily torpor bouts result in much more moderate depressions of T_b and oxygen consumption. However, several of these species are known to have multi-day bouts of torpor (Génin et al. 2003; Schmid and Ganzhorn 2009). Further, some species may have fairly profound depressions of T_b and oxygen consumption suggesting that perhaps, the distinction between daily torpor and hibernation may not be robust (Geiser 2004; Schmid 2000; Schmid et al. 2000). The notion that hibernation or torpor use is restricted to just a few mammals is false. Rather, torpor and/or hibernation use is observed in most orders of mammals including all three basal groups of mammals, the placental mammals, monotremes, and marsupials (Geiser 2004; McKechnie and Mzilikazi 2011). Such a distribution would imply that either torpor use is readily exploited i.e. evolved independently multiple times or that torpor use is an inherent part of the typical mammalian bauplan.

An interesting discussion then is what allows a hibernator to hibernate? In other words, if metabolic activities produce heat and the result of torpor is a decrease in heat production, then what metabolic processes are depressed or suppressed? An understanding of depressed (may be passive inhibition; see below) versus suppressed (active inhibition) events may yield significant insight into the mechanisms of torpor. A hierarchy of ATP consuming events reveals that the bulk ($\sim$90% in many cells) of standard metabolic rate is the result of ion pumping and protein synthesis (Rolfe and Brown 1997). Not surprisingly then, both of these processes are vital to homeostatic regulation. Understanding what vital homeostatic processes are actually suppressed may be the key to understanding what effects a metabolic off-switch like hibernation.

38.1.1 Hibernation: Applying the Brakes?

To understand the mechanisms of effecting hibernation, one must understand the relative roles of metabolic suppression versus depression in effecting reductions of metabolism. Effecting hibernation may be analogous to driving a car. When driving a car, one requires the ability to *reversibly* stop the car much like a hibernator must be able to *reversibly* reduce its metabolism. There is more than one way to stop a car effectively. Both active and passive processes may be made to work in synergy to stop the car. Similarly, hibernation may be the summation of mechanisms to actively reduce metabolism and/or a consequence of reduced metabolism.

To stop a car, the driver may apply the brakes. This is what we typically consider when thinking about *active metabolic suppression*. Processes like phosphorylation signaling cascades may occur which result in a reduced ability to produce and consume ATP. However, a car may also slow when the driver simply releases the accelerator. Disruption of the flow of oxygen to the tissues of a hibernator would necessarily result in reduced rates of oxygen utilization. Furthermore, if a hibernator allows itself to cool, metabolic processes may be diminished through simple temperature considerations of enzymatic rates. When driving a car, one normally takes one's foot off of the accelerator to apply the brake. Hibernation is likely a similar process i.e. it is likely that animals exploit the conditions of torpor to depress metabolic processes that would otherwise generate heat and require energy. For instance, the bulk of initiation of protein synthesis appears to be blocked as ground squirrels reach 18°C during entrance into torpor (van Breukelen and Martin 2001). Since protein synthesis accounts for approximately 12–25% of a typical cellular energy budget (Rolfe and Brown 1997), a large component of the heat supply to the squirrel is removed. However, it is also likely that there must be precipitating events that allow for the reduction of body temperature or the flow of oxygen to provide conditions that may be exploited by the hibernator. The question then arises as to if hibernators employ mechanisms to actively down-regulate heat-generating activities (or even the flow of oxygen) and thereby decrease body temperature or did a passive event allow the hibernator to cool down and thereby reduce metabolism. In other words, a *passive metabolic depression* is a consequence of cooling or disruption of fuel as opposed to a causative event leading to metabolic suppression. Irrespective of if mechanisms are active or passive, mechanisms to enact a metabolic off-switch pose significant issues. Issues with slowing a car include (1) constant driver input is required to ensure the car is not slowed too rapidly or slowly, (2) the wheels must be slowed as appropriate for each individual wheel, and (3) hills represent variable conditions for which the driver must adjust. Each of these analogous problems exists for mammalian hibernation; (1) Signals must be used to effect the metabolic depression; there must be signals to differentiate a torpor cycle from sleep cycle from an active cycle. At what level(s) are the signals generated and processed? (2) Organ diversity and associated variable metabolic needs dictates some tissue-specific regulation; more active tissues that must continue to work during torpor like heart and lung may experience different regulation than skeletal muscle during deep torpor. How do hibernators supply metabolically active tissues like the heart in the face of an effective overall metabolic depression? and (3) Hibernators live in variable environments; burrow temperatures may suddenly decrease with weather conditions. Low ambient temperatures may require active thermoregulation even during torpor- how might this to be effected mechanistically?

Not all of the questions posed above may even be addressed with existing data. However, we have accumulated significant data on the role of passive and active mechanisms in the context of protein synthesis. Here, we will discuss these data and how they help to answer the questions of what makes a hibernator hibernate. Protein synthesis is the translation of mRNA to proteins. The process of translation consists of initiation wherein an mRNA is recruited to ribosomes, elongation of the nascent polypeptide, and termination wherein the ribosome is released from the translated mRNA. During steady state conditions, translation is thought to be regulated primarily at the level of initiation (Gray and Wickens 1998). It is important to note, however, that torpor is not a steady state condition; the requirements for homeostasis are not met (van Breukelen et al. 2008).

38.1.2 Radiolabel Incorporation Studies

Radiolabel incorporation studies were key to a basic understanding of the role of protein synthesis in hibernators. The approach is simple; animals are administered radiolabeled amino acids at various stages of a torpor bout usually through an intraperitoneal injection. Following a sufficient labeling period ($\sim$2 h for most studies), the animals are killed and tissues are collected. The amount of radiolabel incorporation indicates the amount of protein synthesis. There are several technical issues that must be factored in interpreting these data. Most labeling periods are relatively short to try and limit the effect of protein degradation which will result in the liberation of recently incorporated radiolabeled amino acids from the protein pool. Protein degradation rates change during torpor (Velickovska et al. 2005) thus making interpretation difficult. Furthermore, there are significant issues with availability of the radiolabeled amino acids. First, there may be much reduced heart rate to 1/100th of active rates in torpid hibernators (Dawe and Morrison 1955). This reduction means that mobilization of radiolabeled amino acids following injection is relatively slow as compared to euthermy. Second, there are altered blood flow patterns during hibernation e.g. there is a prioritization to the central circulation during torpor that does not exist during euthermy (Bullard and Funkhouser 1962). As such, the distribution patterns of the radiolabel may not be the same for both active and torpid states. Third, it is presumed that effective transporters are present in torpid hibernators. This may be an erred assumption in that many amino acid transporters belong to the SLC6 or SLC38 family of transporters. These transporters are coupled to Na^+ movements and may or may not also be associated with an exchange for H^+ (Boron and Boulpaep 2005). It is uncertain how cold body temperatures would affect these transporters or even the ion gradients present. Importantly, torpid hibernators experience an apparent acidosis as while a tremendous alkalosis should be expected based on simple temperature effects, there is no increase in pH_i indicating a role for acidotic mechanisms (Clausen and Ersland 1968; Bock et al. 2002). Such acid/base issues as well as uncertainty on ion movements preclude a simple assumption that amino acid transporters would be fully functional during torpor. Finally, there are documented changes in endogenous free amino acids with as much as a twofold difference between torpid and aroused hibernators (Klain and Whitten 1968). Such changes may have dramatic effects on calculated protein synthesis rates. For instance, if 1/100 methionine molecules are labeled in state A and 1/200 methionine molecules are labeled in state B then equivalent incorporation rates actually denote a twofold difference in protein synthesis rates. All of these factors may be accounted for by calculating the specific radioactivity given as the amount of a radiolabeled amino acid per total cellular concentration of that amino acid for a given tissue. Ideally, such data are collected using high pressure liquid chromatography (HPLC) to isolate and quantify individual amino acids in tissues and then determining the radioactivity of collected fractions for the labeled amino acid. Unfortunately, most radiolabel incorporation assays in hibernators have ignored this important but tedious component (Gulevsky et al. 1992; Zhegunov et al. 1988; Frerichs et al. 1998). Even when attempts were made to control for specific radioactivity, assumptions that one tissue would be the same as another tissue were made. As such, the actual extent of translational depression is imprecisely known in hibernation. Nevertheless, reported values of radiolabel incorporation during torpor range from 0.13 to 0.5% of normothermic values depending on the study and tissue (Gulevsky et al. 1992 and Zhegunov et al. 1988). Even a twofold underestimate suggests almost negligible protein synthesis in the hibernating rodent. Equally important, data suggest a 150–200% hyperactivation of protein synthesis during the interbout arousal (Zhegunov et al. 1988). It is tempting to speculate that such a hyperactivation may be compensatory given the lack of protein synthesis during a torpor bout. However, no mechanism for this compensatory increase has been demonstrated.

38.1.3 Polysome Analyses

With the knowledge that concomitant with energetic demands of torpor, there is little protein synthesis, we can then ask how such a depression is accomplished. Significant insight for mechanism is provided by polysome analyses. In polysome analyses, the number of ribosomes that associate with a given mRNA is related to the amount of translation. In actively translating systems, as many as 20–30 ribosomes may be associated with an mRNA (Petersen and McLaughlin 1974). By fractioning polysomes across a sucrose density gradient, several investigators demonstrated disruption of translational initiation during torpor (Whitten and Klain 1968, Frerichs et al. 1998, van Breukelen and Martin 2001). Indeed, by exploiting polysome analyses, we recognized an important role for passive depression by temperature. As ground squirrels enter into torpor and precisely at a core body temperature of 18°C, the bulk of initiation was uncoupled from elongation (van Breukelen and Martin 2001). Similarly, it was not until body temperature reached 18°C during the arousal process that the bulk of initiation and elongation were recoupled. The disruption and recoupling coincided at a body temperature of 18°C, strongly suggests passive mechanisms account for this block of initiation. However, this same study revealed more than simple temperature-dependent regulation; polysome profiles from animals killed during the early arousal phase (T_b below 18°C) were more elongated than those from late torpor suggesting that some new initiation had occurred (Fig. 38.1). In other words, while passive events may explain most of the observed inhibition, there is still a role for active processes as we will see later.

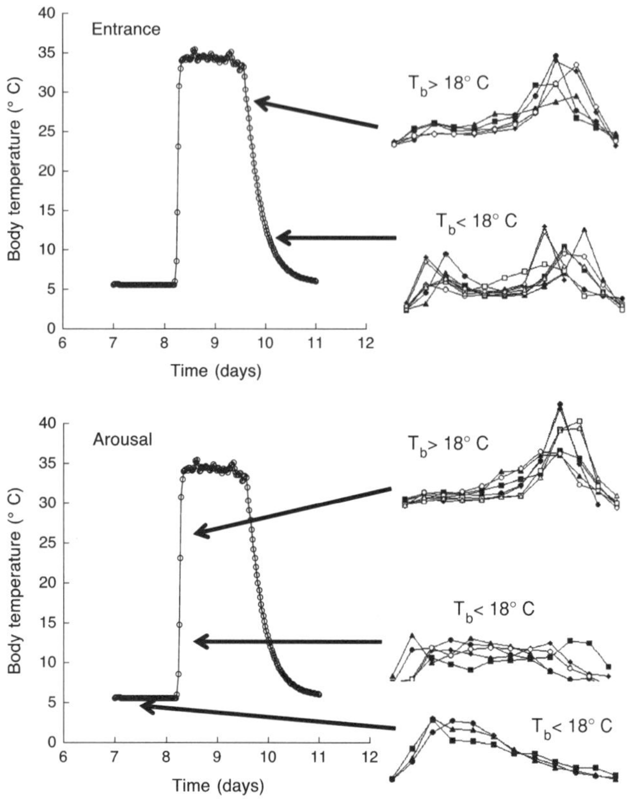

Fig. 38.1 Polysome profiles for hibernating ground squirrels. The data are modified from van Breukelen and Martin (2001). In the *top*, entrance into torpor is indicated. At body temperatures (T_b) above 18°C, polysome profiles are shifted to the right (each *line* represents an individual animal). These data indicate active translation similar to what is observed during summer active and interbout aroused states. At precisely $T_b = 18$°C, there is a block of translational initiation and an uncoupling from elongation processes. In the *bottom*, the processes is somewhat reversed wherein at T_b above 18°C, polysome profiles are restored to the active condition. However, there is a clear distinction between late torpor (*bottom* most polysome trace) and the early arousing animals (*middle* polysome trace of *bottom*). Whereas, the late torpor animals demonstrate a leftward shift typical of little polysome material or protein synthesis, the early arousing animals have a more distributed pattern in their polysome profiles. These data indicate a role for more than simple passive temperature effects in depressing protein synthesis in hibernators

38.1.4 Initiation Factor regulation

The next insights into translational control during hibernation come from studies of the availability and phosphorylation status of the eukaryotic initiation factors (eIFs) through western blotting (for review, see van Breukelen et al. 2004). If inhibiting translation initiation were key to effecting a reduction in metabolism, then it seems likely that active inhibition of the eIFs would play a significant role. However, the competition of both phosphorylation and dephosphorylation processes may make data interpretation very difficult. eIF4E phosphorylation is usually assumed to be associated with enhanced translation rates. However, eIF4E may be phosphorylated during times of dramatic decreases in translation (Gingras et al. 1999) e.g. when translational initiation is depressed via distinct mechanisms such as eIF2α phosphorylation. The interpretation then is that eIF4E phosphorylation may be a compensatory mechanism induced by stress as an *attempt* to stimulate translation. Whether that attempt is realized should be considered in interpreting the data. Importantly, there may be viable alternative interpretations of the data. One could readily interpret those data to imply that perhaps the lack of engagement of eIF4E in the initiation process led to its ultimate phosphorylation (see Morley and McKendrick 1997; Johnson and Warner 1987). In effect, one may alternatively interpret the same data set as (1) indicating that there was an increase in translation initiation (eIF4E was phosphorylated), (2) that there was a compensatory mechanism applied to increase translation that may or may not have been effective (eIF4E was phosphorylated but initiation was being repressed by another mechanism), or (3) that eIF4E phosphorylation has nothing to do with the regulation of translation initiation because eIF4E was simply phosphorylated as a result of not being engaged in the regulatory process under those conditions. Thus, understanding the precise timing of translational activity in the context of phosphorylation changes is required. Unfortunately, no study to date in hibernation has effectively provided such data (including work by the lead author of this paper). Nevertheless, important insight is provided by these studies. Significant eIF2α phosphorylation from 2 to 13% of the total eIF2α present was noted in ground squirrel brains as they entered torpor (Frerichs et al. 1998). Phosphorylation of eIF2α is known to result in inhibition of initiation in other systems. These authors also reported reduced incorporation of a radiolabeled amino acid in a single animal as it entered torpor, at temperatures above the extremely cold body temperatures (T_b) of torpor. A very reasonable conclusion then was that an active inhibition of initiation, presumably the observed phosphorylation of eIF2α, was responsible for the down-regulation of protein synthesis. Importantly, the single entrance animal used by Frerichs et al. had a core T_b of 19°C at the onset of the experiment. Further, its T_b decreased to 7.5°C by the conclusion of the labeling period. Thus, the majority of the labeling period for the one entrance animal was spent with a T_b below 18°C. Later, we demonstrated that, at least in liver, there is a profound block of the bulk of translational initiation as animals entered torpor and T_b reached 18°C (van Breukelen and Martin 2001). Tissue-specific variability in eIF2α phosphorylation is also apparent; eIF2α is phosphorylated in kidney and brain but not in liver or brown adipose tissue (Frerichs et al. 1998; Hittel and Storey 2002, van Breukelen et al. 2004). Such variability between tissue types and the limited degree of phosphorylation (e.g. 87% of eIF2α is in the active form in the brain during torpor) begs the question if active inhibition via eIF2α is actually an integral component of depressing metabolism. Or is the phosphorylation of eIF2α simply a consequence of a depressed initiation status? At the very least, additional work is required before an ascribed role for inhibition is made. Since liver metabolism accounts for ~20% of standard metabolic rate (Rolfe and Brown 1997) and no changes in eIF2α

phosphorylation status were detected (van Breukelen et al. 2004), the variability between tissue types would at the very least suggests a limited role of active suppression of translation initiation in downregulating metabolism on the global scale.

38.1.5 Qualitative Versus Quantitative Properties of Protein Synthesis During Hibernation

An interesting question is if alteration of eIF activity could result in more than changes in the rate of protein synthesis. We examined five eIFs and three other regulatory proteins in liver (van Breukelen et al. 2004). Only one locus of regulation was found: eIF4E activity. eIF4E binds to the 7-methylguanosine cap of an mRNA on its 5′ end and helps recruit the ribosomal components to that mRNA. Enhanced phosphorylation of eIF4E normally increases translation initiation and translation rates. Surprisingly, winter animals experienced greater phosphorylation of eIF4E as compared to squirrels sampled in summer. These data obviously need to be interpreted with caution in light of the previously mentioned data wherein increased phosphorylation of eIF4E occurs without associated increases in rate of translation (Gingras et al. 1999). Although this may be the case in hibernators, an interesting idea is that perhaps this eIF4E phosphorylation is what allows for the observed hyperactivation of protein synthesis during the interbout arousal. However, such a mechanism must be confirmed experimentally. Both interbout aroused and torpid squirrels experienced enhanced phosphorylation of eIF4E. However and very importantly, eIF4E activity is also regulated by association with 4E-binding protein 1, 2, or 3 (4E-BP1, 2, 3). 4E-BP1 is absent in squirrels sampled in summer but present in winter squirrels. 4E-BP1 phosphorylation status changes as a function of torpor. All of these data indicate a regulation consistent with depression of cap-dependent translation during torpor. One question that arises from the eIF4E and 4E-BP1 data then is that if squirrels regulate cap-dependent initiation of translation, what happens to cap-independent initiation of translation? In recent years, the canonical view of eIF4E binding to the 5′ cap of mRNA as a key step toward initiation of translation has been augmented by recognition of Internal Ribosome Entry Site (IRES) mediated initiation of translation [IRESmt; (see Pan and van Breukelen 2011 and references therein)]. Poorly predicted but structurally conserved secondary structures in the 5′ untranslated regions of the mRNA allow for recruitment of key eIFs and initiation of translation. The use of IRES had been noted in stress such as mild hypothermia, oxidative stress, and glucose deprivation. IRES containing transcripts preferentially encode for stress proteins. Several proteins that are known to increase in abundance during hibernation are encoded by known IRES bearing transcripts. Importantly, there had to be a mechanism that provided for their expression at a time of limited transcription. Very recently we showed that as a squirrel progresses through a torpor bout, ribosomes are preferentially loaded with IRES-containing transcripts (Pan and van Breukelen 2011). The use of these IRESes allows a squirrel an opportunity to express proteins that may be critical to the survival of the hardships associated with torpor utilization.

38.1.6 Application of this Model to Other Models

An interesting question to ask is if the lessons that we learned from hibernating ground squirrels would be applicable to other model systems such as a mouse going through daily torpor. We contend that the answer is not clear. Certainly, a mouse that experiences a more modest reduction in body temperature or metabolic rate is not likely to experience the same extent and duration of translational depression as a torpid ground squirrel. For that reason, we probably should not expect to find identical mechanisms. However, effecting an adequate metabolic depression requires depression of key energy-consuming processes. The widespread phylogenetic distribution suggests that torpor use was either readily acquired or basal to mammals. Either way, experimental evidence indicates few unique molecular adaptations (van Breukelen and Martin 2002). Identification of conservation of mechanisms per se will likely prove critical in elucidating the evolutionary aspects of torpor use. However, robust identification of molecular control mechanisms *specific and unique* to offswitches of metabolism is required before such questions may be adequately answered.

38.1.7 Driving the Car

An overview of regulation of protein synthesis during hibernation was presented to gain perspective of the roles for both passive depression and active suppression in effecting the reduced metabolism of torpor. Concordant with energetic demands, there is a near arrest of protein synthesis during deep torpor. In the original analogy of hibernation as being similar to driving a car, this part of the car came to a practical standstill. Polysome profiles reveal a major role for passive depression due to declining temperature in depressing initiation of translation. However, there are modifications to the process that imply some active regulation. Cautious data interpretation is required but the extant data would suggest that active suppression of translation is not a major cause of the depressed metabolic rate associated with hibernation. Despite its highly energetically consumptive role, protein synthesis does not represent the hibernator's brakes. Rather, other energy consumptive events must be depressed to reduce heat production. Identification of these processes may be critical to our understanding of what makes hibernators hibernate.

Acknowledgments This work was supported by NSF grant IOB 0448396. We thank the other members of the laboratory for their diligent efforts.

References

Barnes BM (1989) Freeze avoidance in a mammal: body temperatures below 0°C in an arctic hibernator. Science 244:1593–1595

Bock C, Strijkstra AM, Pörtner HO (2002) MRI and ^{31}P NMR studies of brain metabolism in European ground squirrels during hibernation and arousal. Abstract for international society for magnetic resonance in medicine, tenth scientific meeting and exhibition, Honolulu, Hawaii

Boron WF, Boulpaep EL (2005) Medical physiology: a cellular and molecular approach. Elsevier Saunders, Philadelphia

Buck CL, Barnes BM (2000) Effects of ambient temperature on metabolic rate, respiratory quotient, and torpor in an Arctic hibernator. Am J Physiol Regul Integr Comp Physiol 279:R255–R262

Bullard RW, Funkhouser GE (1962) Estimated regional blood flow by rubidium 86 distribution during arousal from hibernation. Am J Physiol 203:266–270

Clausen G, Ersland A (1968) The respiratory properties of the blood of the hibernating hedgehog *Erinaceus europaeus*. Respir Physiol 5:221–233

Dawe AR, Morrison PR (1955) Characteristics of the hibernating heart. Am Heart J 49:367–384

Frerichs KU, Smith CB, Brenner M, DeGracia DJ, Krause GS, Marrone L, Dever TE, Hallenbeck JM (1998) Suppression of protein synthesis in brain during hibernation involves inhibition of protein initiation and elongation. Proc Natl Acad Sci 95:14511–14516

Geiser F (2004) Metabolic rate and body temperature reduction during hibernation and daily torpor. Annu Rev Physiol 66:239–274

Génin F, Nibbelink M, Galand M, Perret M, Ambid L (2003) Brown fat and nonshivering thermogenesis in the gray mouse lemur (*Microcebus murinus*). Am J Physiol Regul Integr Comp Physiol 284:R811–R818

Gingras A-C, Raught B, Sonenberg N (1999) eIF4 initiation factors: effectors of mRNA recruitment to ribosomes and regulators of translation. Ann Rev Biochem 68:913–963

Gray NK, Wickens M (1998) Control of translation initiation in animals. Annu Rev Cell Dev Biol 14:399–458

Gulevsky AK, Zagnoiko VI, Mishneva LG (1992) The intensity of protein synthesis in organs of heterothermal animals on hibernation. Cryo-Letters 13:99–108

Hittel D, Storey KB (2002) The translation state of differentially expressed mRNAs in the hibernating 13-lined ground squirrel (*Spermophilus tridecemlineatus*). Arch Biochem Biophys 401:244–254

Johnson SP, Warner JR (1987) Phosphorylation of the *Saccharomyces cerevisiae* equivalent of ribosomal protein S6 has no detectable effect on growth. Mol Cell Biol 7:1338–1345

Klain GJ, Whitten BK (1968) Plasma free amino acids in hibernation and arousal. Comp Biochem Physiol 27:617–619

Mckechnie AE, Mzilikazi N (2011) Heterothermy in Afrotropical mammals and birds: a review. Integr Comp Biol 51:349–363

Morley SJ, McKendrick L (1997) Involvement of stress-activated protein kinase and p38/RK mitogen-activated protein kinase signaling pathways in the enhanced phosphorylation of initiation factor 4E in NIH 3T3 cells. J Biol Chem 272:17887–17893

Pan P, van Breukelen F (2011) Preference of IRES-mediated initiation of translation during hibernation in golden-mantled ground squirrels, *Spermophilus lateralis*. Am J Physiol Regul Integr Comp Physiol 301:R370–R377

Petersen NS, McLaughlin CS (1974) Polysome metabolism in protein synthesis mutants of yeast. Mol Gen Genet 129:189–200

Rolfe DFS, Brown GC (1997) Cellular energy utilization and molecular origin of standard metabolic rate in mammals. Physiol Rev 77:731–758

Schmid J (2000) Daily torpor in the gray mouse lemur (*Microcebus murinus*) in Madagascar: energetic consequences and biological significance. Oecologia 123:175–183

Schmid J, Ganzhorn JU (2009) Optional strategies for reduced metabolism in gray mouse lemurs. Naturwissenschaften 96:737–741

Schmid J, Ruf T, Heldmaier G (2000) Metabolism and temperature regulation during daily torpor in the smallest primate, the pygmy mouse lemur (*Microcebus myoxinus*) in Madagascar. J Comp Physiol B 170:59–68

van Breukelen F, Martin SL (2001) Translational initiation is uncoupled from elongation at 18°C during mammalian hibernation. Am J Physiol Regul Integr Comp Physiol 281:R1374–R1379

van Breukelen F, Martin SL (2002) Molecular biology of thermoregulation invited review: molecular adaptations in mammalian hibernators: unique adaptations or generalized responses? J App Physiol 92:2640–2647

van Breukelen F, Sonenberg N, Martin S (2004) Seasonal and state dependent changes of eIF4E and 4E-BP1 during mammalian hibernation: implications for the control of translation during torpor. Am J Physiol Regul Integr Comp Physiol 287:R349–R353

van Breukelen F, Pan P, Rausch CM, Utz JC, Velickovska V (2008) Homeostasis on hold: implications of imprecise coordination of protein metabolism during mammalian hibernation. In: Lovegrove BG, McKechnie AE (eds) Hypometabolism in animals. Interpak Books, Pietermaritzburg, pp 163–170

Velickovska V, Lloyd BP, Qureshi S, van Breukelen F (2005) Proteolysis is depressed during hibernation at the level of the 20S core protease. J Comp Physiol B 175:329–335

Wang LCH, Lee TF (2000) Perspectives on metabolic suppression during mammalian hibernation and daily torpor. In: Heldmaier G, Klingenspor M (eds) Life in the cold. Springer, Berlin, pp 152–158

Whitten BK, Klain GJ (1968) Protein metabolism in hepatic tissue of hibernating and arousing ground squirrels. Am J Physiol 214:1360–1362

Zhegunov GF, Mikulinsky YE, Kudokotseva EV (1988) Hyperactivation of protein synthesis in tissues of hibernating animals on arousal. Cryo-Letters 9:236–245

Part IV
Energy Balance and Thermoregulation

Chapter 39
Brown Adipose Tissue: A Seasonal Tissue in Mammals, Including Humans?

Jan Nedergaard and Barbara Cannon

Abstract Brown adipose tissue is no longer considered merely as another kind of adipose tissue. Rather, deriving as it does from precursors that may also develop into muscle cells, it is an organ in its own right. It produces heat due to the action of the—only—thermogenic uncoupling protein, uncoupling protein-1 (UCP1), and this heat is the sole basis for both thermoregulatory nonshivering thermogenesis and metaboloregulatory thermogenesis (diet-induced thermogenesis), i.e., all adaptive thermogenesis; its function is probably also mandatory for the thermogenesis needed for arousal from hibernation. The increase in brown-fat capacity associated with (seasonal) adaptation ("recruitment") occurs through two processes: one resulting from chronic sympathetic nervous stimulation, obligatorily coupled to heat production, and another, principally unknown, non-thermogenic process, probably experimentally mimicked by thiazolidinediones. This latter process is probably responsible for recruitment in hibernators, for recruitment induced by short-day, and for precocial prenatal recruitment. Brown adipose tissue is active in a large fraction of adult humans, and, also in humans, it may be seasonally regulated.

39.1 The Origin of the Brown Adipocyte

Traditionally, brown adipose tissue has been understood as being a special type of adipose tissue. Today we realize that this is not the case: brown adipocytes rather derive from the same cell lineage as do skeletal muscle cells (Atit et al. 2006; Timmons et al. 2007; Seale et al. 2008). Similar to muscle cells, brown adipocytes

J. Nedergaard (✉) · B. Cannon
The Wenner-Gren Institute, Stockholm University, Stockholm, Sweden
e-mail: jan@metabol.su.se

T. Ruf et al. (eds.), *Living in a Seasonal World*,
DOI: 10.1007/978-3-642-28678-0_39,

have a dense complement of mitochondria. However, thanks to the presence of Uncoupling Protein-1, UCP1, the energy-transforming function of the brown-fat mitochondria can be used to directly generate heat without going via the synthesis and breakdown of ATP, as is the case in muscle. This understanding of the muscle origin of brown adipocytes can explain a series of features of brown adipose tissue that have earlier been considered to be remarkably different from what is seen in white adipose tissue. Enzyme isoforms that are found in brown adipose tissue are often the same as those known to occur in muscle, whereas white adipocytes more often possess those isoforms found in liver. This includes the presence of the heart isoform—to a higher extent than the adipocyte isoform—of the fatty acid-binding protein (H-FABP/FABP3, not A-FABP/aP2/FABP4) (Daikoku et al. 1997), as well as the presence of the muscle isoform of carnitine palmitoyl transferase rather than the liver isoform found in white fat (Daikoku et al. 2000). The regulation of lipoprotein lipase in brown fat (Carneheim et al. 1984) is similar to the regulation of this enzyme in muscle and heart but different from its regulation in white adipose tissue.

39.2 The Origin and Action of the Uncoupling Protein-1

A large family of transporter proteins, the mitochondrial carrier protein family, is required to serve the needs of the mitochondria for the molecular exchange of substrates and building materials with the cytosolic environment. One member of this family has evolved into Uncoupling Protein-1 (UCP1), the protein necessary to allow dissipation in brown fat mitochondria of the inner membrane proton motive force and thus allow for heat generation (Nedergaard et al. 2001). This protein is in a sequence-defined subfamily of the carriers, which also includes UCP2 and UCP3, whose functions are still being clarified—but there is little reason to think that these functions include thermogenesis, rather that they do not (e.g. Nabben et al. 2011).

An evolutionarily early ortholog of UCP1, a protoUCP1, is found in fish (Jastroch et al. 2005) and frogs etc., although its function in these organisms remains unclear, but it is again unlikely that the function would include thermogenesis, in these ectothermic animals. Its evolution into the heat-producing form we know today in mammals seems to have occurred with the evolution of placental mammals (Saito et al. 2008).

Although it is clear that the heat-producing property of UCP1 requires translocation of protons or their equivalents across the inner mitochondrial membrane, the molecular species that is actually transported and the site of the transport within the protein are still unclear. The translocation process can be inhibited on the outer face of UCP1 by purine nucleotides, probably ATP in the cell, although GDP is often used experimentally (Cannon and Nedergaard 2004).

A significant number, possibly all, of the mitochondrial carrier proteins can be forced, in the absence of their normal substrates, to transport fatty acid anions across the membrane. If the fatty acids are able to return across the membrane in an undissociated form, i.e., are "flip-flopable", then this process is an uncoupling process, it increases respiration and dissipates energy as heat. However, this is not a specific process. In general it appears to have rather low capacity and can theoretically occur in any kind of mitochondria, including brown fat mitochondria (Shabalina et al. 2006). Whether UCP1 functions through an enhanced version of this mechanism or in a principally different way is still unknown.

39.3 The Special Case of the Brite Adipocytes

As indicated above, the brown adipocytes found in the classical depots of brown adipose tissue derive from precursor cells that also give rise to muscle cells. However, low amounts of UCP1, particularly readily detectable as UCP1 mRNA, can also be found in various subcutaneous depots of what is generally classified as white adipose tissue (Young et al. 1984; Esterbauer et al. 1998; Guerra et al. 1998). The adipocytes that express the UCP1 in these depots do not arise from muscle-related precursors (Xue et al. 2007; Seale et al. 2008; Petrovic et al. 2010). Since these are brown-like cells in white depots, they may be termed "brite" adipocytes (Petrovic et al. 2010). The levels of expression of UCP1 in these cells can change dramatically as a result of changes in physiological parameters such as cold (Walden et al. 2012) and diet (Madsen et al. 2010); however, since the unstimulated levels are almost unmeasurable, then these dramatic increases are mainly a result of the extremely low initial values (Walden et al. 2012).

Consequently, while manipulation of these depots to become UCP1-expressing appears to be a dramatic and exciting means of potentially altering the thermogenic potential of an organism, it is most likely that these depots physiologically are of rather limited systemic importance because, at least in rodents, the total amount of UCP1 in the brite depots is rather small compared to that in the classical depots (our unpublished observations); certain studies do, however, imply a significant role of these depots, particularly when they are arteficially induced (e.g. Madsen et al. 2010; Vegiopoulos et al. 2010).

39.4 The Physiological Significance of Brown Adipose Tissue

Brown adipose tissue thermogenesis was proposed from the outset (Smith 1961; Donhoffer et al. 1964) to contribute significantly to nonshivering thermogenesis, i.e., the mechanism found to develop in mammals after exposure for periods of weeks to a cold environment and to replace shivering as a source of heat to defend body temperature. However, its role as the sole source of cold-induced adaptive nonshivering thermogenesis could only be unequivocally demonstrated after the introduction of mice in which the UCP1 gene was ablated (Enerbäck et al. 1997). Such UCP1-knock-out mice continue to shiver even after prolonged exposure to cold, demonstrating the absence of nonshivering thermogenesis (Golozoubova et al. 2001), and they can therefore not reach the maximal heat production rate of wild-type mice (Meyer et al. 2010) (i.e. the sum of shivering and nonshivering thermogenesis). However, both mice with and without UCP1 display an adaptation phenomenon (Golozoubova et al. 2001; Meyer et al. 2010) regarding the highest rate of heat production (Meyer et al. 2010); the nature of this adaptation is not as yet clarified.

A process termed diet-induced thermogenesis, a form of extra energy dissipation that follows adaptation to a high-fat or cafeteria diet, was also proposed to occur in brown adipose tissue (Rothwell and Stock 1979) and to have as its function to diminish the body weight gain that results from such diets. That the site of this type of adaptive thermogenesis is indeed only brown adipose tissue was subsequently confirmed by the use of UCP1-knock-out mice (Feldmann et al. 2009). However, the existence of the phenomenon of diet-induced thermogenesis (and thus evidently its localization to brown adipose tissue) has been repeatedly disputed (Maxwell et al. 1984; Kozak 2010).

Thus, brown adipose tissue uniquely underlies the major adaptive processes of classical nonshivering thermogenesis and of diet-induced thermogenesis, i.e., in reality all adaptive adrenergic thermogenesis defined in mammals.

A question of whether the mere presence of brown adipose tissue and UCP1 has any influence on basal metabolic rate has gained recent interest in view of possibilities to develop brown adipose tissue as an anti-obesity treatment. It must therefore be emphasized that the basal metabolic rate is unchanged in an animal whether or not UCP1 is present. The reason for this is that in its non-activated state, UCP1 does not slip protons, i.e., it is not leaky, and consequently it does not contribute to the basal metabolism (Shabalina et al. 2010).

Arousal from hibernation requires elevation of body temperature by almost 40°C against a thermal gradient. It was in this situation that Smith originally recognized that brown adipose tissue was particularly active (Smith and Hock 1963). At temperatures close to zero, shivering is not functional, and brown fat can be seen to raise body temperature in the thorax (Hayward and Lyman 1967). Indeed, in curarized bats, arousal is not delayed compared to control animals, indicating that shivering in this species is not a requirement to re-attain body temperature (Hayward and Lyman 1967). In other species, such as golden hamster, arousal is delayed but is nonetheless achieved, indicating that shivering can hasten the arousal process but is not an absolute requirement (Hayward and Lyman 1967). To date, no genetically modified hibernators are available to confirm the role of UCP1 in these animals but the indirect evidence for an obligatory role seems robust.

39.5 Seasonality and the Recruitment of Brown Adipose Tissue

Brown adipose tissue is a highly dynamic tissue, the amount of which can increase several-fold in response to physiological stimuli such as cold, diet, or day-length. The increase in the amount of tissue is the result of cell proliferation (and probably a decrease in the normal rate of apoptosis) and enhanced cellular differentiation, a combined process referred to as recruitment. (Lipid stores also influence the weight of the tissue, which can confuse understanding, as lipid content may change in the opposite direction from the functional part of the tissue.)

39.5.1 Classical Adrenergic Recruitment

Two cellular pathways for the recruitment phenomenon may today be discussed (Fig. 39.1). In the first, hypothalamic nuclei respond to afferent signals mediating cold and diet information to activating sympathetic outflows to brown adipose tissue. Under these circumstances, precursor cells in the tissue are stimulated to divide and differentiate (Cameron and Smith 1964; Bukowiecki et al. 1982; Rehnmark and Nedergaard 1989). This sympathetically mediated recruitment process based on norepinephrine release thus underlies the ability of an animal to acclimate to cold or to adapt to certain diets. This can principally explain seasonal adaptation that it is secondary to decreased temperature. It can also explain the recruitment of brown adipose tissue that occurs in the first postnatal week in altricial newborns such as mice and rats (Nedergaard et al. 1986), again being secondary to decreased environmental temperature (Obregon et al. 1989).

39.5.2 Another Recruitment Pathway

Experimentally, treatment of animals or cells with the insulin sensitizer rosiglitazone or similar thiazolidinediones results in the recruitment of brown adipocytes both in classical brown adipose tissue depots and in brite depots (Thurlby et al. 1987; Petrovic et al. 2008, 2010). This is thus a cell-autonomous response to the drug and is not the result of indirect stimulation of the sympathetic nervous system. Rosiglitazone does not stimulate thermogenesis. Whether there is a physiological equivalent to rosiglitazone is not known, although a need for such is evident.

This is because, physiologically, a pathway is needed to understand the recruitment process under two particular circumstances. One is that occurring prenatally in animals born with a well-developed brown adipose tissue so-called precocial neonates, such as the guinea pig, lambs, calves, etc. (Nedergaard et al. 1986). If brown fat recruitment in these cases was due to constant sympathetic

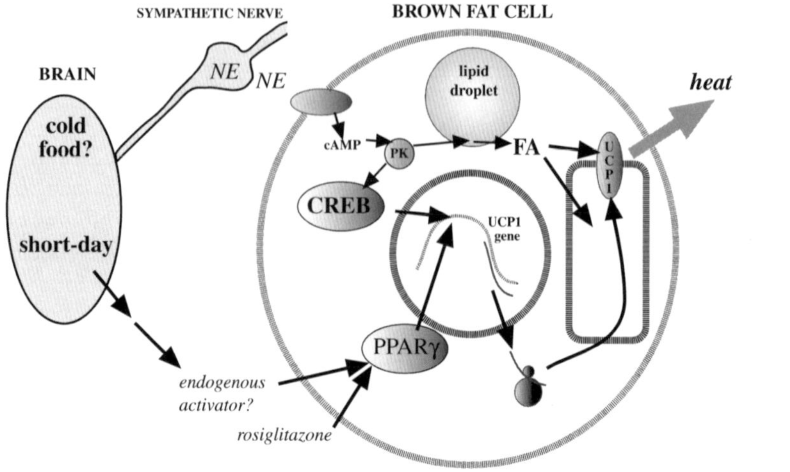

Fig. 39.1 The two pathways for recruitment of brown adipose tissue. NE: norepinephrine; FA: fatty acid; UCP1: Uncoupling protein-1; PK: protein kinase A; CREB: cyclic AMP response element binding protein

stimulation of the tissue within the womb, thermogenesis would necessarily also be stimulated and this would increase heat stress in the dam.

The other is the recruitment resulting from a seasonal adaptation that is not understandable as being cold-induced (and where an adrenergic pathway that is necessarily associated with ongoing thermogenesis would be counter-productive for energy accumulation for the winter season). It is likely that this process is caused by short day-length. Seasonal darkness is known to stimulate recruitment even in the absence of cold. This is believed to be mediated via melatonin (Heldmaier and Hoffman 1974; Heldmaier et al. 1981) but the further mechanism is unclear. It has been discussed that the recruitment may occur indirectly via sympathetic activation (Tan et al. 2011). However, this would entail chronic stimulation of the tissue and a chronically increased "basal" metabolism. There is no convincing evidence in the literature that short-day/melatonin increases basal metabolic rate, and the possibility that the recruitment seen could be due to chronic sympathetic stimulation is therefore rather small, but the issue is not settled.

There are thus indications of the existence of a non-sympathetic recruitment process that leads to tissue growth but without concomitant activation of thermogenesis, and the possibility exists that it is this pathway that is experimentally stimulated by rosiglitazone (Fig. 39.1).

39.5.3 Tissue Involution

Since the recruitment of the tissue can occur so dynamically, it is also evident that the opposite process, that of involution or atrophy, must be equally dynamic. During spring, as environmental temperatures rise, sympathetic stimulation to the tissue ceases, acute thermogenesis ceases, and the tissue involutes. The half-life of mitochondria is much decreased in the warm compared with that in the cold, reflecting the loss of ongoing sympathetic stimulation (Bukowiecki and Himms-Hagen 1971). Furthermore, removal of sympathetic stimulation results in a higher rate of cellular apoptosis (Lindquist and Rehnmark 1998). Both these processes will contribute to involution of the tissue.

An involution of brown adipose tissue also occurs with age and this cannot simply be understood as being the result of diminished stimulation via the adrenergic pathway, as this would be expected to remain constant. At present, the mechanism behind this age-related involution has not been clarified. It has been shown that glucocorticoids have an inhibitory effect on the tissue, decreasing, for example, UCP1 expression. However, glucocorticoid levels are not believed to change during ageing, so that this pathway could only contribute to the involution if the relative effectiveness of the glucocorticoids were to increase. A possible pathway to enhance glucocorticoid effectiveness could perhaps be a decreasing stimulatory component from sex hormones. If the counteracting effects of glucocorticoids and sex hormones are in balance in young animals, this balance could be lost with increasing age (Nedergaard and Cannon 2010).

39.6 Brown Adipose Tissue in Adult Humans

Brown adipose tissue was classically demonstrated to be present in small rodents, hibernators, and mammalian newborns. This included the human infant (Aherne and Hull 1966), where numerous depots of various sizes were localized. It became an accepted contention that active brown adipose tissue was not present in the adults of larger species, including man. However, doubts about this could have been voiced following the demonstration of brown adipose tissue in phaeochromocytoma patients, which clearly implied that stem cells for brown adipocytes were still present in adults and responsive to adrenergic stimulation (Rona 1964; English et al. 1973; Ricquier et al. 1982). In the 1990s, reports started appearing in the radiology literature of cancer patients who had undergone positron-emission tomography-scanning (PET-scanning) with the tracer 18F-fluorodeoxyglucose and who demonstrated symmetrical uptake areas in the neck, shoulder, and upper body regions, which were not tumor metastases and which were initially proposed to be areas of tense muscle (Barrington and Maisey 1996). However, computer-assisted tomography (CT) indicated that the regions had the composition of a fatty tissue (Hany et al. 2002), and it became evident that a fatty tissue that had the capacity to take up significant amounts of a glucose tracer must be brown and not white adipose tissue (Hany et al. 2002). The radiology literature was reviewed in this respect in 2007 (Nedergaard et al. 2007), and it was clear that indeed adult humans did possess active brown adipose tissue, most notably in the neck and supraclavicular regions, although other depots of lower activity were also evident.

Since the glucose uptake only occurs if the brown adipose tissue is acutely active and thermogenic, it is difficult from retrospective patient scans in hospital archives to determine the prevalence of brown fat in adult man. It is impossible to know whether the examination rooms were experienced as being cool or not and therefore whether the patient would have activated any existing brown fat or not. The earliest reports that summarized retrospective studies indicated a prevalence of some 5%, a value so low as to be essentially uninteresting from a human metabolic point of view (Nedergaard et al. 2007, 2010). Gradually a number of dedicated studies of volunteer subjects started to appear and showed significantly higher prevalence, varying between 33 and 100%, depending upon the cohort (Saito et al. 2009; Virtanen et al. 2009; van Marken Lichtenbelt et al. 2009) ("dedicated" PET studies simply indicate that the subjects were actively exposed to the cold). Also, analysis by immunohistochemistry of the presence of the UCP1 protein in biopsy samples of patients undergoing thyroid surgery showed a frequency of about 33% (Zingaretti et al. 2009). Thus, a large fraction of adult humans possess brown adipose tissue.

39.7 Seasonality in Human Brown Adipose Tissue?

Regarding seasonality of brown adipose tissue in humans, nearly all reports from retrospective studies (i.e. studying the hospital archives from routine examinations of cancer patients) (Cohade et al. 2003; Kim et al. 2008; Cheng et al. 2009; Cypess et al. 2009; Au-Yong et al. 2009; Ouellet et al. 2011) document a higher prevalence of detecting brown adipose tissue when the examination is performed in the winter rather than in the summer months. Thus, at first sight, this would indicate that human brown adipose tissue shows seasonality just as brown adipose tissue does in animals exposed to the temperature and light fluctuations of nature.

However, the temperatures of the examination rooms used for the routine study of cancer patients are not necessarily well controlled, and there is thus good reason to think that the seasonality observed could be fully due to a few degrees' difference in examination room temperatures between summer and winter. All these studies could therefore be dismissed on technical grounds. However, there is one dedicated study (Saito et al. 2009) that indeed demonstrates seasonality in humans. The cause of this is not readily understandable, as it would normally be considered that the amount of cold that normal people are exposed to does not vary much between winter and summer (as we try to keep ourselves practically thermoneutral, with clothes and buildings and heating). Similarly, light intensities would not vary much over the year, as most people are exposed to indoor light during most of the day in any season. Nonetheless, these (or other) factors would seem to be sufficient to cause seasonality of brown adipose tissue.

39.8 Is Increased Human Obesity Caused by Diminished Recruitment of Brown Adipose Tissue, due to Diminished Experienced Seasonality?

The correlation between the total amount/capacity of brown adipose tissue in the body and the diet-induced thermogenesis caused e.g. by a meal is not necessarily simple. Rather, it would seem that the brain decides how much of the capacity of brown adipose tissue should be used in a given situation for e.g. thermoregulatory or for metaboloregulatory thermogenesis. Thus, people with (more) brown adipose tissue should not necessarily be protected against obesity—although there seems to be a clear correlation between less brown adipose tissue and more obesity in reports from dedicated studies of humans (Saito et al. 2009; Virtanen et al. 2009; van Marken Lichtenbelt et al. 2009; Zingaretti et al. 2009). It would also seem that the absence of brown adipose tissue is associated with a decreased thermogenic response to a meal (Saito et al. 2011).

If this is the case, a decreased recruitment of brown adipose tissue, resulting from our lower exposure to cold and darkness, could be one of the causes of the present obesity epidemic, as has indeed been suggested (Johnson et al. 2011; Tan et al. 2011).

References

Aherne W, Hull D (1966) Brown adipose tissue and heat production in the newborn infant. J Pathol Bacteriol 91:223–234

Atit R, Sgaier SK, Mohamed OA, Taketo MM, Dufort D, Joyner AL, Niswander L, Conlon RA (2006) Beta catenin activation is necessary and sufficient to specify the dorsal dermal fate in the mouse. Dev Biol 296:164–176

Au-Yong IT, Thorn N, Ganatra R, Perkins AC, Symonds ME (2009) Brown adipose tissue and seasonal variation in humans. Diabetes 58:2583–2587

Barrington SF, Maisey MN (1996) Skeletal muscle uptake of fluorine-18-FDG: effect of oral diazepam. J Nucl Med 37:1127–1129

Bukowiecki L, Himms-Hagen J (1971) Decreased half-life of some mitochondrial proteins in skeletal muscle and brown adipose tissue of cold-acclimated rats. Can J Physiol Pharmacol 49:1015–1018

Bukowiecki LJ, Collet AJ, Follea N, Guay G, Jahjah L (1982) Brown adipose tissue hyperplasia: a fundamental mechanism of adaptation to cold and hyperphagia. Am J Physiol 242: E353–E359

Cameron IL, Smith RE (1964) Cytological responses of brown fat tissue in cold-exposed rats. J Cell Biol 23:89–100

Cannon B, Nedergaard J (2004) Brown adipose tissue: function and physiological significance. Physiol Rev 84:277–359

Carneheim C, Nedergaard J, Cannon B (1984) Beta-adrenergic stimulation of lipoprotein lipase in rat brown adipose tissue during acclimation to cold. Am J Physiol 246:E327–E333

Cheng WY, Zhu ZH, Ouyang M (2009) Patterns and characteristics of brown adipose tissue uptake of 18F-FDG position emission tomograph/computed tomography imaging. Zhongguo Yi Xue Ke Xue Yuan Xue Bao 31:370–373

Cohade C, Mourtzikos KA, Wahl RL (2003) "USA-Fat": prevalence is related to ambient outdoor temperature-evaluation with 18F-FDG PET/CT. J Nucl Med 44:1267–1270

Cypess AM, Lehman S, Williams G, Tal I, Rodman D, Goldfine AB, Kuo FC, Palmer EL, Tseng YH, Doria A, Kolodny GM, Kahn CR (2009) Identification and importance of brown adipose tissue in adult humans. N Engl J Med 360:1509–1517

Daikoku T, Shinohara Y, Shima A, Yamazaki N, Terada H (1997) Dramatic enhancement of the specific expression of the heart-type fatty acid binding protein in rat brown adipose tissue by cold exposure. FEBS Lett 410:383–386

Daikoku T, Shinohara Y, Shima A, Yamazaki N, Terada H (2000) Specific elevation of transcript levels of particular protein subtypes induced in brown adipose tissue by cold exposure. Biochim Biophys Acta 1457:263–272

Donhoffer S, Sardy F, Szegvari G (1964) Brown adipose tissue and thermoregulatory heat production in the rat. Nature 203:765–766

Enerbäck S, Jacobsson A, Simpson EM, Guerra C, Yamashita H, Harper M-E, Kozak LP (1997) Mice lacking mitochondrial uncoupling protein are cold-sensitive but not obese. Nature 387:90–94

English JT, Patel SK, Flanagan MJ (1973) Association of phaeochromocytomas with brown fat tumours. Radiology 107:279–283

Esterbauer H, Oberkofler H, Liu Y, Breban D, Hell E, Krempler F, Patsch W (1998) Uncoupling protein-1 mRNA expression in obese human subjects: the role of sequence variations at the uncoupling protein-1 gene locus. J Lipid Res 39:834–844

Feldmann HM, Golozoubova V, Cannon B, Nedergaard J (2009) UCP1 ablation induces obesity and abolishes diet-induced thermogenesis in mice exempt from thermal stress by living at thermoneutrality. Cell Metab 9:203–209

Golozoubova V, Hohtola E, Matthias A, Jacobsson A, Cannon B, Nedergaard J (2001) Only UCP1 can mediate adaptive nonshivering thermogenesis in the cold. FASEB J 15:2048–2050

Guerra C, Koza RA, Yamashita H, Walsh K, Kozak LP (1998) Emergence of brown adipocytes in white fat in mice is under genetic control. Effects on body weight and adiposity. J Clin Invest 102:412–420

Hany TF, Gharehpapagh E, Kamel EM, Buck A, Himms-Hagen J, von Schulthess GK (2002) Brown adipose tissue: a factor to consider in symmetrical tracer uptake in the neck and upper chest region. Eur J Nucl Med Mol Imaging 29:1393–1398

Hayward JS, Lyman CP (1967) Nonshivering heat production during arousal from hibernation and evidence for the contribution of brown fat. In: Fisher KC (ed) Mammalian hibernation III. American Elsevier, New York

Heldmaier G, Hoffman K (1974) Melatonin stimulates growth of brown adipose tissue. Nature 247:224–225

Heldmaier G, Steinlechner S, Rafael J, Vsiansky P (1981) Photoperiodic control and effects of melatonin on nonshivering thermogenesis and brown adipose tissue. Science 212:917–919

Jastroch M, Wuertz S, Kloas W, Klingenspor M (2005) Uncoupling protein 1 in fish uncovers an ancient evolutionary history of mammalian nonshivering thermogenesis. Physiol Genomics 22:150–156

Johnson F, Mavrogianni A, Ucci M, Vidal-Puig A, Wardle J (2011) Could increased time spent in a thermal comfort zone contribute to population increases in obesity? Obes Rev 12:543–551

Kim S, Krynyckyi BR, Machac J, Kim CK (2008) Temporal relation between temperature change and FDG uptake in brown adipose tissue. Eur J Nucl Med Mol Imaging 35:984–989

Kozak LP (2010) Brown fat and the myth of diet-induced thermogenesis. Cell Metab 11:263–267

Lindquist JM, Rehnmark S (1998) Ambient temperature regulation of apoptosis in brown adipose tissue: Erk 1/2 promotes norepinephrine-dependent cell survival. J Biol Chem 273:30147–30156

Madsen L, Pedersen LM, Lillefosse HH, Fjaere E, Bronstad I, Hao Q, Petersen RK, Hallenborg P, Ma T, De Matteis R, Araujo P, Mercader J, Bonet ML, Hansen JB, Cannon B, Nedergaard J, Wang J, Cinti S, Voshol P, Doskeland SO, Kristiansen K (2010) UCP1 induction during recruitment of brown adipocytes in white adipose tissue is dependent on cyclooxygenase activity. PLoS One 5:e11391

Maxwell GM, Nobbs S, Bates DJ (1984) Diet-induced thermogenesis in cafeteria-fed rats: a myth? Am J Physiol 253:E264–E270

Meyer CW, Willershauser M, Jastroch M, Rourke BC, Fromme T, Oelkrug R, Heldmaier G, Klingenspor M (2010) Adaptive thermogenesis and thermal conductance in wild-type and UCP1-KO mice. Am J Physiol Regul Integr Comp Physiol 299:R1396–R1406

Nabben M, Shabalina IG, Moonen-Kornips E, Van Beurden D, Cannon B, Schrauwen P, Nedergaard J, Hoeks J (2011) Uncoupled respiration, ROS production, acute lipotoxicity and oxidative damage in isolated skeletal muscle mitochondria from UCP3-ablated mice. Biochim Biophys Acta 1807:1095–1105

Nedergaard J, Connolly E, Cannon B (1986) Brown adipose tissue in the mammalian neonate. In: Trayhurn P, Nicholls DG (eds) Brown adipose tissue. Edward Arnold, London, pp 152–213

Nedergaard J, Golozoubova V, Matthias A, Asadi A, Jacobsson A, Cannon B (2001) UCP1: the only protein able to mediate adaptive non-shivering thermogenesis and metabolic inefficiency. Biochim Biophys Acta 1504:82–106

Nedergaard J, Bengtsson T, Cannon B (2007) Unexpected evidence for active brown adipose tissue in adult humans. Am J Physiol 293:E444–E452

Nedergaard J, Bengtsson T, Cannon B (2010) Three years with adult human brown adipose tissue. Ann N Y Acad Sci 1212:E20–E36

Nedergaard J, Cannon B (2010) The changed metabolic world with human brown adipose tissue: therapeutic visions. Cell Metab 11:268–272

Obregon MJ, Jacobsson A, Kirchgessner T, Schotz MC, Cannon B, Nedergaard J (1989) Postnatal recruitment of brown adipose tissue is induced by the cold stress experienced by the pups. An analysis of mRNA levels for thermogenin and lipoprotein lipase. Biochem J 259:341–346

Ouellet V, Routhier-Labadie A, Bellemare W, Lakhal-Chaieb L, Turcotte E, Carpentier AC, Richard D (2011) Outdoor temperature, age, sex, body mass index, and diabetic status determine the prevalence, mass, and glucose-uptake activity of 18F-FDG-detected BAT in humans. J Clin Endocrinol Metab 96:192–199

Petrovic N, Shabalina IG, Timmons JA, Cannon B, Nedergaard J (2008) Thermogenically competent nonadrenergic recruitment in brown preadipocytes by a PPARgamma agonist. Am J Physiol Endocrinol Metab 295:E287–E296

Petrovic N, Walden TB, Shabalina IG, Timmons JA, Cannon B, Nedergaard J (2010) Chronic peroxisome proliferator-activated receptor gamma (PPAR{gamma}) activation of epididymally derived white adipocyte cultures reveals a population of thermogenically competent, UCP1-containing adipocytes molecularly distinct from classic brown adipocytes. J Biol Chem 285:7153–7164

Rehnmark S, Nedergaard J (1989) DNA synthesis in mouse brown adipose tissue is under beta-adrenergic control. Exp Cell Res 180:574–579

Ricquier D, Néchad M, Mory G (1982) Ultrastructural and biochemical characterization of human brown adipose tissue in pheochromocytoma. J Clin Endocrinol Metab 54:803–807

Rona G (1964) Changes in adipose tissue accompanying pheochromocytoma. Can Med Ass J 91:303–305

Rothwell NJ, Stock MJ (1979) A role for brown adipose tissue in diet-induced thermogenesis. Nature 281:31–35

Saito M, Okamatsu-Ogura Y, Matsushita M, Watanabe K, Yoneshiro T, Nio-Kobayashi J, Iwanaga T, Miyagawa M, Kameya T, Nakada K, Kawai Y, Tsujisaki M (2009) High incidence of metabolically active brown adipose tissue in healthy adult humans: effects of cold exposure and adiposity. Diabetes 58:1526–1531

Saito M, Yoneshiro T, Aita S (2011) Postprandial thermogenesis and brown adipose tissue in humans. Obesity (Silver Spring) 19(Suppl 1):S80

Saito S, Saito CT, Shingai R (2008) Adaptive evolution of the uncoupling protein 1 gene contributed to the acquisition of novel nonshivering thermogenesis in ancestral eutherian mammals. Gene 408:37–44

Seale P, Bjork B, Yang W, Kajimura S, Chin S, Kuang S, Scime A, Devarakonda S, Conroe HM, Erdjument-Bromage H, Tempst P, Rudnicki MA, Beier DR, Spiegelman BM (2008) PRDM16 controls a brown fat/skeletal muscle switch. Nature 454:961–967

Shabalina IG, Kramarova TV, Nedergaard J, Cannon B (2006) Carboxyatractyloside effects on brown-fat mitochondria imply that the adenine nucleotide translocator isoforms ANT1 and ANT2 may be responsible for basal and fatty-acid-induced uncoupling respectively. Biochem J 399:405–414

Shabalina IG, Ost M, Petrovic N, Vrbacky M, Nedergaard J, Cannon B (2010) Uncoupling protein-1 is not leaky. Biochim Biophys Acta 1797:773–784

Smith RE (1961) Thermogenic activity of the hibernating gland in the cold-acclimated rat. Physiologist 4:113

Smith RE, Hock RJ (1963) Brown fat: thermogenic effector of arousal in hibernators. Science 140:199–200

Tan DX, Manchester LC, Fuentes-Broto L, Paredes SD, Reiter RJ (2011) Significance and application of melatonin in the regulation of brown adipose tissue metabolism: relation to human obesity. Obes Rev 12:167–188

Thurlby PL, Wilson S, Arch JR (1987) Ciglitazone is not itself thermogenic but increases the potential for thermogenesis in lean mice. Biosci Rep 7:573–577

Timmons JA, Wennmalm K, Larsson O, Walden TB, Lassmann T, Petrovic N, Hamilton DL, Gimeno RE, Wahlestedt C, Baar K, Nedergaard J, Cannon B (2007) Myogenic gene expression signature establishes that brown and white adipocytes originate from distinct cell lineages. Proc Natl Acad Sci U S A 104:4401–4406

van Marken Lichtenbelt WD, Vanhommerig JW, Smulders NM, Drossaerts JM, Kemerink GJ, Bouvy ND, Schrauwen P, Teule GJ (2009) Cold-activated brown adipose tissue in healthy men. N Engl J Med 360:1500–1508

Vegiopoulos A, Muller-Decker K, Strzoda D, Schmitt I, Chichelnitskiy E, Ostertag A, Berriel Diaz M, Rozman J, Hrabe de Angelis M, Nusing RM, Meyer CW, Wahli W, Klingenspor M, Herzig S (2010) Cyclooxygenase-2 controls energy homeostasis in mice by de novo recruitment of brown adipocytes. Science 328:1158–1161

Virtanen KA, Lidell ME, Orava J, Heglind M, Westergren R, Niemi T, Taittonen M, Laine J, Savisto NJ, Enerbäck S, Nuutila P (2009) Functional brown adipose tissue in healthy adults. N Engl J Med 360:1518–1525

Walden TB, Hansen IR, Timmons JA, Cannon B, Nedergaard J (2012) Recruited vs. nonrecruited molecular signatures of brown, "brite", and white adipose tissues. Am J Physiol Endocrinol Metab 302:E19–E31

Xue B, Rim JS, Hogan JC, Coulter AA, Koza RA, Kozak LP (2007) Genetic variability affects the development of brown adipocytes in white fat but not in interscapular brown fat. J Lipid Res 48:41–51

Young P, Arch J, Ashwell M (1984) Brown adipose tissue in the parametrial fat pad of the mouse. FEBS Lett 167:10–14

Zingaretti MC, Crosta F, Vitali A, Guerrieri M, Frontini A, Cannon B, Nedergaard J, Cinti S (2009) The presence of UCP1 demonstrates that metabolically active adipose tissue in the neck of adult humans truly represents brown adipose tissue. FASEB J 23:3113–3120

Chapter 40
Systematic Screening for Mutant Mouse Lines with Defects in Body Temperature Regulation

M. Willershäuser, N. Ehrhardt, R. Elvert, E. K. Wirth, U. Schweizer, V. Gailus-Durner, H. Fuchs, M. Hrabě de Angelis, J. Rozman and M. Klingenspor

Abstract Spontaneous daily torpor as well as cold- and fasting-induced torpor is observed in many mammals and birds. The frequency and intensity of torpor are largely influenced by well characterized environmental conditions. The inherited

---

M. Willershäuser and N. Ehrhardt—equally contributed to this paper.

M. Willershäuser · J. Rozman (✉) · M. Klingenspor (✉)
Molecular Nutritional Medicine, Else Kröner-Fresenius Center and ZIEL Research Center for Nutrition and Food Sciences, Technische Universität München, Gregor-Mendel-Str. 2, 85350, Freising-Weihenstephan, Germany
e-mail: jan.rozman@helmholtz-muenchen.de

M. Klingenspor
e-mail: mk@tum.de

M. Willershäuser · V. Gailus-Durner · H. Fuchs · M. Hrabě de Angelis · J. Rozman
German Mouse Clinic, Helmholtz Zentrum München, German Research Center for Environmental Health, Institute of Experimental Genetics, Neuherberg, Germany

N. Ehrhardt · R. Elvert
Department of Animal Physiology, Faculty of Biology, Philipps-Universität Marburg, Marburg, Germany

E. K. Wirth · U. Schweizer
Institut für Experimentelle Endokrinologie, Charité-Universitätsmedizin Berlin, Berlin, Germany

M. Hrabě de Angelis
Lehrstuhl für Experimentelle Genetik, Technische Universität München, Freising-Weihenstephan, Germany

Present Address
N. Ehrhardt
Department of Medical Genetics, Cedars-Sinai Medical Center, Los Angeles, CA, USA

R. Elvert
Sanofi-Aventis GmbH, Frankfurt am Main, Germany

T. Ruf et al. (eds.), *Living in a Seasonal World*,
DOI: 10.1007/978-3-642-28678-0_40,

component of phenotypic variation in body temperature regulation is, however, only poorly understood. The identification of mutations affecting physiological body temperature regulation may elucidate novel molecular mechanisms contributing to metabo- and thermoregulatory behavior. Therefore, we monitored rectal body temperature (T_b) in six inbred mouse strains (129/SvJ, AKR/J, C57BL/6J, C57BL/6N, Balb/cJ, and SWR/J) and in 146 mutant mouse lines and age-matched controls. Measurements were taken under ad libitum conditions and in response to either food restriction or food deprivation. Inbred mice showed considerable strain-specific differences in T_b under ad libitum feeding as well as after food restriction and deprivation. T_b was on average $\approx$0.4°C higher in females as compared to males in all inbred lines. The thermoregulatory response to food restriction reflected a trade-off between changes in T_b and body mass with the least body mass loss in mice exhibiting the largest drop in T_b. In mutant mouse lines under ad libitum feeding T_b of 36 out of 146 was significantly higher (15 lines) or lower (21 lines) as compared with controls. In response to food restriction and deprivation, 13 mutant mouse lines with augmented T_b reduction were found two of which showed rectal T_b lower than 31°C. In conclusion, T_b under ad libitum as well as in response to food deprivation is an important indicator for genotype-related differences on systemic functions. Depending on the test conditions 25–32% of all mutant lines showed significant alterations in T_b.

40.1 Introduction

The maintenance of energy homeostasis in endotherm organisms requires fine tuning of complex metabolic pathways and physiological functions. Only subtle deviations in the balance between dietary energy intake and partitioning of food energy to different components of the daily energy expenditure budget may result in altered body temperature, body mass, and body composition. Especially in small sized mammals such as the laboratory mouse, the generation of heat to maintain a stable body temperature (T_b) at room temperature contributes up to 30% of daily energy expenditure (Cannon and Nedergaard 2011). Therefore, it could be expected that mutations causing variation in T_b may be associated with alterations in body mass and body composition. Indeed, reduced T_b was been observed in obese (e.g. ob/ob mice, Trayhurn et al. 1977; Sma1 growth hormone deficiency, Meyer et al. 2004) and lean mouse models (ghrelin/ghrelin receptor knock out mouse, Pfluger et al. 2008). Knowledge about the role of thermoregulation in the maintenance of energy balance will impact on clinically oriented research aiming to understand the causes of human metabolic diseases as well as basic research addressing physiological adaptations of energy balance regulation.

In response to cold and food shortage, small mammals and birds use daily torpor to reduce energy expenditure. Daily torpor relies on coordinated metabolic suppression during entry and maintenance of the torpid state and reactivation of suppressed metabolic pathways to initiate rapid arousal (Geiser 2004; Heldmaier et al. 2004). In the laboratory mouse short phases of daily torpor can be reliably induced by exposure to food restriction or deprivation and low ambient temperature (Hudson and Scott 1979; Webb et al. 1982; Dikic et al. 2008). The induction of torpor is associated with changes in the levels of circulating hormones involved in energy balance regulation (Swoap 2008). The serum levels of the adipostatic hormone leptin are decreased whereas the orexigenic hormone ghrelin is increased. Both hormones modulate the expression and secretion of orexigenic and anorexigenic peptides in neuronal populations in the hypothalamus. Under conditions of food shortage and increased demand for thermoregulatory heat production, these central neuroendocrine signaling pathways are gating metabolic suppression followed by a decrease in body temperature (Swoap 2008).

In the search of genes involved in metabolic suppression, we screened the metabolic phenotype of mutant mice. In this screen, T_b was taken as an indicator for primary defects in body temperature regulation and dysregulated energy partitioning. Phenotyping data of six inbred mouse strains and of 146 mutant mouse lines and corresponding controls were analyzed regarding variation in rectal body temperature under ad libitum feeding conditions and in response to two different food shortage challenges.

40.2 Materials and Methods

Phenotyping procedures followed standardized and validated protocols ensuring the comparability of data sets (Gailus Durner et al. 2005, 2009). We first analyzed data obtained from six inbred mouse lines (129/SvJ, AKR/J, C57BL/6J, Balb/cJ, C3HeB/FeJ, SWR/J; $n = 370$ mice) to evaluate the extent of phenotypic variance in body temperature regulation. In a second step, first-line phenotyping data of male and female mutant and age-matched control mice ($n = 3{,}945$ mice) were investigated. Control mice comprised wildtype littermates, in a few cases het erozygous littermates, as well as age-matched breeding stock mice with the same genetic background as the corresponding mutant cohort.

All mice, inbred and mutant lines, were single caged at room temperature (23±1°C) in a 12:12 h light:dark cycle in individually ventilated type II macro-lon cages supplemented with a mouse igloo and two to three pieces of soft paper tissue as enrichment. Body mass and food intake were monitored regularly. Rectal body temperature was measured between 8:00 and 10:00 a.m. by careful insertion of a thermocouple (ALMEMO 2390-1, probe N856-1, Ahlborn, Germany) into the rectum for up to about 2 cm.

The present study compiles data gathered by standardized phenotyping of mutant mouse lines delivered to the German Mouse Clinic in the past 10 years. During this period several adjustments of the workflow and phenotyping protocols occurred. The selection of mutant mouse lines subjected to this workflow was random. In a first trial, 83 (16–18 weeks of age) of 146 mutant mouse lines were monitored during 1 week ad libitum feeding (standard chow Altromin 1314, Lage, Germany). Body temperature and body mass were measured every day. In the following week, 35 of these mutant lines were challenged by food restriction (60% of their daily ad libitum food intake) for 7 days and 31 mutant lines were challenged by food deprivation for 2 days. During food restriction and food deprivation, T_b and body mass were measured daily. Values of the last day of the challenge tests were used for analysis. In a second trial with 63 mutant mouse lines, T_b and body mass were taken under ad libitum feeding at 10–14 weeks of age under ad libitum feeding only.

Mutant lines were generated by different technologies (chemical mutagenesis, knock in, knock out, gene trapping, Gailus-Durner et al. 2005). At the present stage gene identity cannot be disclosed due to confidentiality agreements with collaboration partners. Cohorts of mutant and control mice consisted of at least five mice per sex and genotype. Within cohorts, effects on T_b were analyzed by 2-way-ANOVA (inbred strains: main effects of strain and sex, and sex:strain interaction; mutant lines: main effects of genotype and sex, and sex:genotype interaction; post hoc testing: Bonferroni correction for multiple testing). A significance level of $p < 0.05$ was accepted as statistically significant.

40.3 Results

40.3.1 Phenotypic Variation of Body Temperature in Inbred Mouse Lines

Under ad libitum conditions, strain and sex variation in T_b was found in the six inbred mouse lines (Table 40.1 and Fig. 40.1). Both in males and females, T_b was the lowest in C3HeB/FeJ (vs. 129/SvJ $p < 0.001$, vs. AKR/J, C57BL/6J and SWR/J $p < 0.05$, Balb/cJ n.s.) and Balb/cJ (SWR/J $p < 0.05$, others n.s.), and the highest in SWR/J. Balb/cJ showed the highest individual variability in T_b. In all strains, female T_b was on average $\approx$0.4°C higher compared with males. This sex difference was most pronounced in C3HeB/FeJ (+0.6°C, $p < 0.001$) and least in SWR/J (+0.3°C, $p < 0.05$). In Balb/cJ, this effect of sex did not attain statistical significance. We found no correlation between T_b and body mass.

In response to food restriction, males and females of five out of six inbred lines showed a significant reduction in T_b (Table 40.2 and Fig. 40.1). While the degree of T_b reduction varied between −1.4 and −0.6°C in these five inbred strains, in Balb/cJ females showed only a mild, statistically non-significant reduction by −0.2°C whereas males increased T_b by 0.5°C. The heterothermic response to food restriction was associated with a corresponding reduction in body mass ranging from −4.2 to −8.7 g between strains (Table 40.2). Body mass loss was highest in

Table 40.1 Rectal body temperature and body mass of six inbred mouse lines under ad libitum feeding (mean ± SEM, coefficient of variance)

	129/SvJ		AKR/J		C57BL/6J		Balb/cJ		C3HeB/FeJ		SWR/J	
ad libitum	♂	♀	♂	♀	♂	♀	♂	♀	♂	♀	♂	♀
Sample size	12	10	22	13	71	43	18	17	67	32	34	31
Rectal T_b (°C)	36.8	37.0	36.7	37.1	36.8	37.1	36.5	37.0	36.3	36.9	36.9	37.2
SEM	±0.1	±0.1	±0.1	±0.1	±0.0	±0.0	±0.3	±0.2	±0.1	±0.1	±0.1	±0.1
CV%	0.8	0.7	0.8	0.7	1.1	0.8	3.0	2.1	1.8	1.7	1.6	1.2
Body mass (g)	31.1	25.9	38.5	36.9	29.5	23.0	34.2	26.6	39.3	36.2	28.7	23.1
SEM	±0.3	±0.3	±0.5	±0.7	±0.3	±0.2	±0.8	±0.7	±0.3	±0.6	±0.4	±0.5
CV%	4.0	4.3	8.1	7.2	8.3	7.0	10.1	10.6	7.1	9.4	8.0	11.3

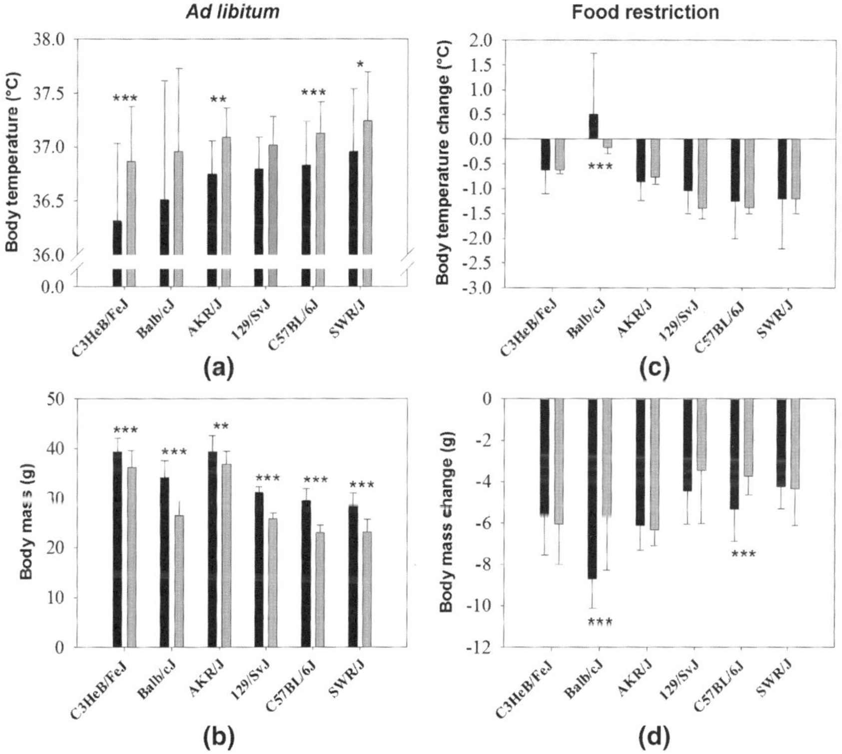

Fig. 40.1 Rectal body temperature and body mass as well as changes in rectal body temperature and body mass in males (*black bars*) and females (*gray bars*) of six inbred mouse lines either fed ad libitum or food restricted. **a** Body temperature, ad libitum, **b** body mass, ad libitum, **c** Δ body temperature, food restricted, **d** Δ body mass, food restricted

Balb/cJ with −8.7 g in males and −5.7 g in females. Overall an inverse relation was found between T_b reduction and body mass loss ($r^2 = 0.80$, $p < 0.001$, $n = 12$, Fig. 40.2).

Table 40.2 Rectal body temperature, body temperature reduction, body mass and body mass reduction of six inbred mouse lines in response to food restriction (mean ± SEM)

	129/SvJ		AKR/J		C57BL/6J		Balb/cJ		C3HeB/FeJ		SWR/J	
	♂	♀	♂	♀	♂	♀	♂	♀	♂	♀	♂	♀
Sample size	12	10	15	6	71	43	13	11	67	32	21	19
Rectal T_b (°C)	35.8 ± 0.1	35.6 ± 0.2	36.0 ± 0.1	36.5 ± 0.1	35.6 ± 0.1	35.7 ± 0.1	36.6 ± 0.2	36.8 ± 0.3	35.7 ± 0.1	36.2 ± 0.1	35.9 ± 0.3	36.1 ± 0.3
Δ rectal T_b (°C)	−1.0 ± 0.1	−1.4 ± 0.2	−0.9 ± 0.1	−0.8 ± 0.1	−1.2 ± 0.1	−1.4 ± 0.1	+0.5 ± 0.3	−0.2 ± 0.1	−0.6 ± 0.1	−0.6 ± 0.1	−1.2 ± 0.2	−1.2 ± 0.3
Body mass (g)	26.6 ± 0.5	22.4 ± 0.8	33.4 ± 0.7	30.4 ± 1.1	24.2 ± 0.3	19.2 ± 0.3	25.6 ± 1.04	20.3 ± 1.09	33.6 ± 0.5	30.2 ± 0.7	24.9 ± 0.5	19.5 ± 0.5
Δ body mass (g)	−4.5 ± 0.5	−3.5 ± 0.8	−6.1 ± 0.3	−6.3 ± 0.8	−5.3 ± 0.2	−3.7 ± 0.1	−8.7 ± 0.4	−5.7 ± 0.8	−5.7 ± 0.2	−6.1 ± 0.3	−4.2 ± 0.2	−4.3 ± 0.4

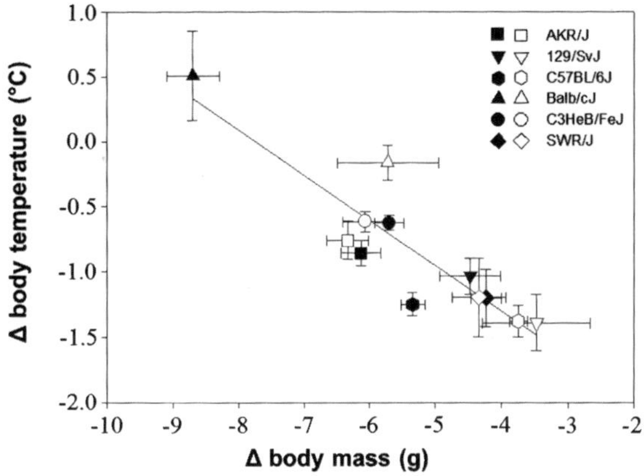

Fig. 40.2 Correlation of Δ rectal body temperature and Δ body mass of six inbred mouse lines after food restriction for 7 days. $T_{b,\ red} = -2.692 - 0.348 * BM_{red}$, $r^2 = 0.80$, $p < 0.001$, $n = 12$. Symbol definitions for inbred strains are shown in figure, *open* symbols = females, *solid* = males

40.3.2 T_b *Phenotypes in Mutant Lines*

In accordance with the inbred strains, T_b of mutant lines under ad libitum feeding was ≈0.4°C higher in females compared with males. In 36 of 146 mutant lines, T_b of either males or females was significantly different between mutants and controls (21 lower and 15 higher T_b). The total number of 146 mutant lines comprised 123 mutant lines with both sexes, 12 mutant lines with males only, and 11 with females only. When separated for sex, 31 mutant mouse lines showed significant deviations in T_b for males (19 lower vs. 12 higher, Table 40.3) and 30 mutant mouse lines for females (20 lower vs. 10 higher, Table 40.3). Thus, the overall frequency of mutant lines exhibiting T_b phenotypes was independent of sex. However, within five mutant lines a significant difference in T_b between mutant and control was observed only in one sex (three males, two females). In six mutant lines only one sex was analyzed (three males, three females) so that sex specificity could not be addressed.

In nearly all cohorts, food restricted and food deprived mice showed a pronounced reduction of T_b compared to ad libitum fed mice. In response to food restriction males and females of three mutant lines showed augmented decrease in T_b ranging between −3.4 and −0.6°C when compared to controls (Table 40.3). In one mutant line both sexes showed a lesser decrease in T_b in response to food restriction with +2.8°C (females) and +1.0°C (males). In one other line, only females exhibited an augmented reduction in T_b. In two mutant lines with a larger T_b depression only one sex was analyzed (one male, one female) so that sex specificity could not be addressed.

In response to food deprivation males and females of four mutant lines showed an augmented decrease in T_b ranging between −3.5 and 0.4°C when compared with controls. In six mutant lines, only one sex exhibited a significant genotype difference with decreased (two females, one male) as well as increased (two females, one male) T_b in three mutant lines, respectively (Table 40.3).

Table 40.3 Number of mutant lines with statistically significant difference ($p \leq 0.05$) in T_b between mutant and control mice

Treatment		Males				Females			
		Lower	Higher	Regular	Total	Lower	Higher	Regular	Total
ad libitum	n	19	12	104	135	20	10	104	134
	%	14.1	8.9	77.0	100	14.9	7.5	77.6	100
Food restriction	n	4	1	26	31	5	1	24	30
	%	12.9	3.2	83.9	100	16.7	3.3	80.0	100
Food deprivation	n	5	1	25	31	6	2	22	30
	%	16.1	3.2	80.7	100	20.0	6.7	73.3	100

ad libitum, food restriction, and food deprivation, 146 mutant lines in total, in 135 line both males and females were tested, in 12 only males, in 11 only females

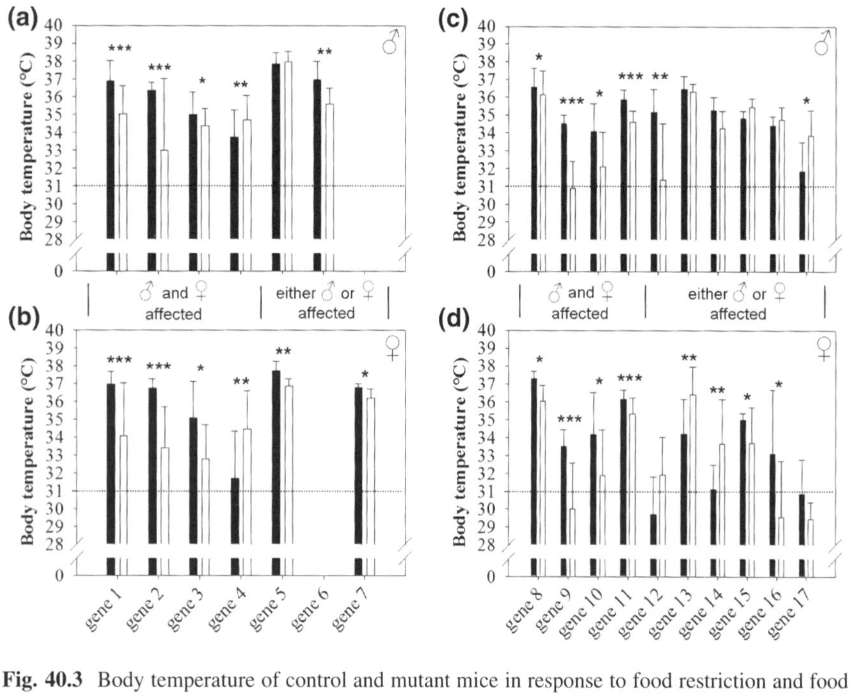

Fig. 40.3 Body temperature of control and mutant mice in response to food restriction and food deprivation. **a** Males, food restriction, **b** females, food restriction, **c** males, food deprivation, **d** females, food deprivation

To summarize these findings, targeted mutation of 13 genes caused a significant augmentation of T_b depression in response to food shortage whereas four gene mutations caused diminished T_b depression (Fig. 40.3). When applying an absolute T_b threshold of 31°C first suggested by Hudson and Scott (1979) to define torpor, only two mutant lines can be categorized as torpid (Fig. 40.3, genes 9 and 16). In one of these lines both sexes showed torpor whereas in the other only females were torpid. In one additional line only controls showed torpor. No torpor was found in cohorts exposed to the food restriction challenge.

40.4 Discussion

The measurement of rectal body temperature (T_b) was used as one phenotyping tool in the systematic first-line characterization of metabolic functions in inbred mouse strains and mouse mutants. Even though specific differences in T_b between inbred strains were detected, the range between highest and lowest mean T_b was rather narrow with 0.6°C in males and 0.3°C in females. Females consistently showed a sex-specific increase of T_b by about 0.4°C. In contrast to this, in about 25% of all mutant lines significant alterations in body temperature regulation were detected either under ad libitum conditions or in response to food shortage. Notably, three mutant mouse lines showed daily torpor in response to food deprivation.

So far, only few reports are available regarding systematic comparison of T_b in inbred mouse lines (e.g. the *Crabbe*2 study, Jackson Mouse Phenome Database, (Crabbe et al. 1998)). Similar to our study, Balb/cJ mice showed relatively low T_b during ad libitum feeding but were sensitive to hypothermia during food restriction and preferably depleted fat reserves instead of decreasing T_b. In the *Eumorphia*5 study (http://phenome.jax.org), mice of the sub-strain Balb/cAnNCrl maintained higher T_b in response to 6 h exposure to 4°C compared to mice of the strains 129S2/SvPas, C3HeB/FeJ, and C57BL/6J. Our findings and similar published reports highlight the importance of controlling for the effects of genetic background in mouse studies (e.g. Champy et al. 2008). The consistent sex effect on T_b which is also supported by published data, e.g. in C57BL/6 mice (Sanchez-Alavez et al. 2011), is another important confounder in studies on energy partitioning in mice. In addition to these more general aspects, findings of this study may help to evaluate the link between variation in T_b and obesity. Mice with lower T_b were hypothesized to be prone to obesity whereas mice with higher T_b, e.g. due to increased non-shivering thermogenesis, are protected (Jürgens et al. 2006). In line with this hypothesis, highest T_b within inbred strains was found in SWR/J which are resistant to diet induced obesity (DIO, Hesse et al. 2010). In contrast, DIO susceptible AKR/J did not reveal any marked difference in T_b (Hesse et al. 2010). Overall, no correlation between body temperature and body mass could be shown when the statistical model included sex as main factor. This was true for mice of the inbred lines as well as for individuals from mutant and control cohorts.

Under ad libitum conditions, a surprisingly high proportion of about 25% of all mutant lines showed a significant deviation in T_b compared with mice from the control cohort. The number of mutant lines showing increased or decreased T_b compared with controls was equally balanced. Similarly as in other studies (e.g. Reed et al. 2008), in our screen about 30% of the mutant mouse lines were underweight, only 3–4% overweight and the majority showed no genotype effect on body mass. This imbalance strikingly mismatched the proportion that was found in T_b variation being increased or decreased in equal parts. Therefore, on the level of mutant lines, a direct link between the variation in T_b and effects on body mass could not be supported. Interestingly, even though T_b variation in humans was linked to the risk to develop obesity (Landsberg et al. 2009), in recently published studies (e.g. Heikens et al. 2011) such association could not be confirmed.

Food restriction or deprivation proved to be efficient challenge to detect new T_b phenotypes. Ten of the 17 mutant lines that showed a T_b deviation in response to food shortage were inconspicuous under ad libitum feeding. As an additional filter mechanism to detect gene effects on energy partitioning and body temperature regulation, we searched for daily torpor in response to food shortage as an extreme phenotype. In contrast to rodent species that deliberately enter states of torpor, e.g. as part of seasonal acclimatization, mice need to be challenged by low ambient temperature in combination with food deprivation to reliably induce torpor (Dikic et al. 2008). In our study, food deprivation for 2 days under room temperature induced torpor in two mutant lines and also in one control cohort. No daily torpor could be detected in response to the food restriction challenge. The two hits provide novel gene candidates that may play a role in the regulation of metabolic suppression during torpor and may be linked to hypothalamic and peripheral functions of energy balance regulation. Interestingly, females were slightly more likely to enter torpor in response to food shortage as previously reported by Swoap and Gutilla (2009). Together with our finding of higher T_b in females, it is important to emphasize that substantial additional understanding of gene functions can be expected when sex $\sim$ gene interactions are considered.

Acknowledgments The excellent technical assistance of Ann-Elisabeth Schwarz, and Martin Taube, as well as the animal caretaker team at the GMC, is gratefully acknowledged. The study was supported by grants of the German Federal Ministry of Education and Research to the GMC (NGFN-Plus, grant Nos.: 01GS0850, 01GS0851, 01GS0869), to the German Center for Diabetes Research (DZD e.V.), and by the European Commission (EUMODIC LSHG-2006-037188).

References

Cannon B, Nedergaard J (2011) Nonshivering thermogenesis and its adequate measurement in metabolic studies. J Exp Biol 214:242–253

Champy MF, Selloum M, Zeitler V, Caradec C, Jung B, Rousseau S, Pouilly L, Sorg T, Auwerx J (2008) Genetic background determines metabolic phenotypes in the mouse. Mamm Genome 19(5):318–331

Dikic D, Heldmaier G, Meyer CW (2008) Induced torpor in different strains of laboratory mice. In: Lovegrove BG, McKechnie AE (eds) Hypometabolism in animals—hibernation, torpor and cryobiology. University of KwaZulu-Natal, Pietermaritzburg

Gailus-Durner V, Fuchs H, Becker L, Bolle I, Brielmeier M, Calzada-Wack J, Elvert R, Ehrhardt N, Dalke C, Franz TJ, Grundner-Culemann E, Hammelbacher S, Hölter SM, Hölzlwimmer G, Horsch M, Javaheri A, Kalaydjiev SV, Klempt M, Kling E, Kunder S, Lengger C, Lisse T, Mijalski T, Naton B, Pedersen V, Prehn C, Przemeck G, Racz I, Reinhard C, Reitmeir P, Schneider I, Schrewe A, Steinkamp R, Zybill C, Adamski J, Beckers J, Behrendt H, Favor J, Graw J, Heldmaier G, Höfler H, Ivandic B, Katus H, Kirchhof P, Klingenspor M, Klopstock T, Lengeling A, Müller W, Ohl F, Ollert M, Quintanilla-Martinez L, Schmidt J, Schulz H, Wolf E, Wurst W, Zimmer A, Busch DH, Hrabě de Angelis M (2005) Introducing the German Mouse Clinic: open access platform for standardized phenotyping. Nat Methods 2(6):403–404

Gailus-Durner V, Fuchs H, Adler T, Aguilar Pimentel A, Becker L, Bolle I, Calzada-Wack J, Dalke C, Ehrhardt N, Ferwagner B, Hans W, Hölter SM, Hölzlwimmer G, Horsch M, Javaheri A, Kallnik M, Kling E, Lengger C, Mörth C, Mossbrugger I, Naton B, Prehn C, Puk O,

Rathkolb B, Rozman J, Schrewe A, Thiele F, Adamski J, Aigner B, Behrendt H, Busch DH, Favor J, Graw J, Heldmaier G, Ivandic B, Katus H, Klingenspor M, Klopstock T, Kremmer E, Ollert M, Quintanilla-Martinez L, Schulz H, Wolf E, Wurst W, Hrabě de Angelis M (2009) Systemic first-line phenotyping. Methods Mol Biol 530:463–509

Geiser F (2004) Metabolic rate and body temperature reduction during hibernation and daily torpor. Annu Rev Physiol 66:239–274

Heikens MJ, Gorbach AM, Eden HS, Savastano DM, Chen KY, Skarulis MC, Yanovski JA (2011) Core body temperature in obesity. Am J Clin Nutr 93(5):963–967

Heldmaier G, Ortmann S, Elvert R (2004) Natural hypometabolism during hibernation and daily torpor in mammals. Respir Physiol Neurobiol 141(3):317–329

Hesse D, Dunn M, Heldmaier G, Klingenspor M, Rozman J (2010) Behavioural mechanisms affecting energy regulation in mice prone or resistant to diet-induced obesity. Physiol Behav 99(3):370–380

Hudson JW, Scott IM (1979) Daily torpor in the laboratory mouse *Mus musculus* var albino. Physiol Zool 52(2):205–218

Jürgens HS, Schürmann A, Kluge R, Ortmann S, Klaus S, Joost HG, Tschöp MH (2006) Hyperphagia, lower body temperature, and reduced running wheel activity precede development of morbid obesity in New Zealand obese mice. Physiol Genomics 25:234–241

Landsberg L, Young JB, Leonard WR, Linsenmeier RA, Turek FW (2009) Do the obese have lower body temperatures? A new look at a forgotten variable in energy balance. Trans Am Clin Climatol Assoc 120:287–295

Meyer CW, Klingenspor M, Rozman J, Heldmaier G (2004) Gene or size: metabolic rate and body temperature in obese growth hormone-deficient dwarf mice. Obes Res 12(9):1509–1518

Pfluger PT, Kirchner H, Günnel S, Schrott B, Perez-Tilve D, Fu S, Benoit SC, Horvath T, Joost HG, Wortley KE, Sleeman MW, Tschöp MH (2008) Simultaneous deletion of ghrelin and its receptor increases motor activity and energy expenditure. Am J Physiol Gastrointest Liver Physiol 294(3):G610–G618

Reed DR, Lawler MP, Tordoff MG (2008) Reduced body weight is a common effect of gene knockout in mice. BMC Genet 9:4

Sanchez-Alavez M, Alboni S, Conti B (2011) Sex- and age-specific differences in core body temperature of C57Bl/6 mice. Age (Dordr) 33(1):89–99

Swoap SJ (2008) The pharmacology and molecular mechanisms underlying temperature regulation and torpor. Biochem Pharmacol 76(7):817–824

Swoap SJ, Gutilla MJ (2009) Cardiovascular changes during daily torpor in the laboratory mouse. Am J Physiol Regul Integr Comp Physiol 297:R769–R774

Trayhurn P, Thurlby PL, James WP (1977) Thermogenic defect in pre-obese ob/ob mice. Nature 266(5597):60–62

Webb GP, Jagot SA, Jakobson ME (1982) Fasting-induced torpor in *Mus musculus* and its implications in the use of murine models for human obesity studies. Comp Biochem Physiol A Comp Physiol 72(1):211–219

Chapter 41
Diurnal Changes in Metabolic Rate in Pygmy Marmosets: Implications for Sleep, Torpor, and Basal Metabolism in Primates

Glenn J. Tattersall

Abstract The pygmy marmoset (*Cebuella pygmaea*) is the smallest New World Monkey and presents an interesting primate comparison to evolutionary experiments in body size evolution. Previous research has been suggested that pygmy marmosets exhibit a lower basal metabolism relative to mammals of similar size, with the adaptive reasons involving dietary constraints. Indeed, this hypometabolism hypothesis has been evoked as a correlate to the pygmy marmoset's proposed phyletic dwarfism. However, these earlier assessments of basal metabolic rates were conducted on sleeping animals. Pygmy marmosets are capable of quite pronounced diurnal changes in body temperature (4–5°C), with consequent effects on estimates of basal metabolism. Since comparison of a particular species' metabolism to standard mammalian curves for basal metabolism requires similar assessment parameters, I chose to re-evaluate these measurements in the context of the changes in metabolism that accompany the onset and maintenance of the diurnal inactive period when marmosets typically sleep. $\dot{V}_{O_2}$ estimates at thermoneutrality in captive raised pygmy marmosets ranged from 34 to 11 ml O_2 kg^{-1} min^{-1} during active periods to sleep periods. During this same time period, body temperature ranged from 39.2 to 35.5°C. Q_{10} for metabolism, assessed at similar thermal conductances, during these natural transitions was ~4.6, which suggests that the process of sleep and subsequent metabolic adjustments are associated with a suppression of metabolism below the normothermic basal value. These results have implications for the understanding of primate metabolism, and reinforce the notion that large diurnal changes in body temperature and metabolism are common strategies for energy conservation in small mammals.

G. J. Tattersall (✉)
Department of Biological Sciences, Brock University, St. Catharines, ON L2S 3A1, Canada
e-mail: gtatters@brocku.ca

T. Ruf et al. (eds.), *Living in a Seasonal World*,
DOI: 10.1007/978-3-642-28678-0_41, © Springer-Verlag Berlin Heidelberg 2012

41.1 Introduction

The Pygmy marmoset (*Cebuella pygmaea*), a New World Monkey, belongs to the family *Callitrichidae* and is the smallest simian primate (average mass 119 g; Soini 1993). Although more closely related to Hominids than similarly small prosimians, the genus represents a derived primate lineage (Perelman et al. 2011). One consequence is that the ancestral group is predicted to have been much larger; as such, pygmy marmosets have undergone phyletic dwarfism, a phenomenon sometimes observed in situations when energy conservation is deemed to have selective advantages (Stein et al. 2010), such as in various island or isolated fauna (Weston and Lister 2009). Indeed, Genoud et al. (1997) concluded, based on studies during sleep, that pygmy marmosets and other Callitrichids are 'hypometabolic' with respect to other mammals (BMR is 63–72% of value predicted from their body mass). These conclusions, however, rely upon measurements of basal metabolic rate (BMR) among mammals being truly comparable.

Relative to the ease of measurement, however, the concept of basal metabolic expenditure in mammals remains contentious (Cooper and Withers 2009, 2010). One of the major reasons concerns the standardisation required for appropriate determination of an animal in a commonly assessed, but resting metabolic state. Nevertheless, for interspecific comparisons of energy expenditure, attempts to standardise energetic assessment is essential (Blaxter 1989). BMR itself has been argued as a 'marker' of intrinsic metabolic turnover of endotherms below which normothermic life cannot be maintained (torpor metabolic rate is exclusive of this definition). BMR, however, needs to be measured under appropriate, comparable conditions: post-absorptive, non-pregnant, thermally neutral, resting and free of emotional stress. Indeed, the handbook of thermophysiology defines BMR as the 'heat production or oxygen consumption in an organism in a rested, awake, fasting sufficiently long to be in post-absorptive state, and thermoneutral zone' (Anonymous 2001). Some implications result from this. A resting animal is one exhibiting minimal muscular activity; however, rest does not equate to sleep. Nevertheless, some studies have assessed BMR during the animal's natural rest (i.e. sleep) phase (Genoud et al. 1997), which may complicate comparisons to animals in awake states.

Achieving a post-absorptive state may be viewed as quite challenging in small mammals, given their overall higher tissue metabolism, often elevated excitability, and lower tolerance to starvation; if the respiratory exchange ratio (RER) is assessed simultaneously, however, the transition to fat metabolism can be used as an indicator of post-absorptive mechanisms. Since a thermoneutral environment is one in which regulatory changes adjustments in heat production should not be evoked, and BMR is a standard measure within many awake animals, it is implicit that BMR should be compared among animals that are maintaining a normothermic body temperature, rather than a torpid or hypothermic body temperature (McNab 1988).

Despite these challenges, the objectives of this study were simple: to test for the magnitude of the diurnal changes in metabolism, body temperature and activity in pygmy marmosets in order to ascertain the most appropriate level for comparison to other normothermic primates. I compare the continuously measured MR and T_b obtained from captive pygmy marmosets between day and night in order to ascertain whether nighttime values represent metabolic suppression common in small, heterothermic mammals, and/or a simple reduction in thermoregulatory control.

41.2 Materials and Methods

41.2.1 Animals

Pygmy marmosets ($N = 10$; 5 male, 5 female, Body mass = 138 g) belonging to a captive-bred colony residing at Kent State University were examined. They were fed a mixture of marmoset chow (Zu-Preem®) and various fruits ad libidum. Under normal housing conditions, animals are usually kept in same-sex cages, being removed from their cages for the short periods of experimentation. Within their home cages, small nest boxes (approximately 2–4 l plastic containers) were placed, into which more than one marmoset would enter to sleep at night. Animals were maintained on a 12 h light–dark schedule from 0800 to 2000 hours, with the ambient temperature kept within thermoneutrality at 28°C (Genoud et al. 1997; personal observations) and the relative humidity maintained between 40 and 60%. Since marmosets are diurnal and engage in multiple stages of sleep during their inactive period of time (Issa and Wang 2008), metabolism during the periods of darkness was assumed to be related to overall sleep state. All experimental procedures were approved by the Kent State University's Institutional Animal Care and Use Committee.

41.3 Metabolic Rate Determination

Marmosets were transferred to a nest-box sized respirometer constructed from non-compressible, 3/8-inch plexiglass (total volume 4.5 l) for the purposes of measuring metabolic rate. This size allowed for small changes in oxygen and carbon dioxide (0.1–0.5%) at a gas flow rate of 1,500 ml min^{-1}, while also not confining the marmosets unnecessarily. The respirometer was held inside a constant-temperature environmental chamber set to regulate temperature at 28 ± 0.05°C and equipped with a television camera. Gases were mixed using a gas mixing flow meter (Cameron Instruments® Model GF-3/MP) to produce an accurate 21% oxygen mixture. Water vapour from the excurrent air was completely eliminated using a drierite column placed before the sub-sampled gas entered the CO_2 and O_2 analysers (Li-Cor® and Sable Systems®). $\dot{V}_{O_2}$ and $\dot{V}_{CO_2}$ were determined as described by Withers (1977), taking into account the change in gas flow rate introduced from CO_2 production $(\dot{V}_{CO_2})$. Both $\dot{V}_{CO_2}$ and $\dot{V}_{O_2}$ were expressed as ml kg^{-1} min^{-1}, corrected to STPD. The RER was calculated as the ratio of $\dot{V}_{CO_2} : \dot{V}_{O_2}$. In order to account for the delay in the turnover of respiratory gases within the respirometer, the raw CO_2 and O_2 signals were time-corrected as described by Boutilier and Shelton (1986), compensating for the wash-out effect of the respirometer using a time constant (approximately 3 min) determined empirically and the rate of change of the CO_2 and O_2 signals. Trial experiments of switching from 21 to 10% O_2 without an animal in the respirometer resulted in a near-perfect transformation of the O_2 trace into a square-wave signal with this correction. Before collecting data, marmosets were placed in the respirometer on numerous occasions in order to habituate them to the experimental conditions. The analyser was calibrated every 15 min by diverting the analyser gas flow to a subsample of the dried, inlet air supply. $\dot{V}_{O_2}$, $\dot{V}_{CO_2}$, respirometer chamber temperature (T_a), flow rate and chamber relative humidity were averaged and collected via the data acquisition system every 0.5 min, and later averaged over 5 min intervals.

41.4 Body Temperature Measurements

One month prior to taking any measurements, battery-less telemeters (PDT-4000 E-Mitters®, Mini-Mitter®) were implanted under halothane anaesthesia (2%). Animals were first injected with 0.2 mg Butorphanol to serve as a muscle relaxant and analgesic. A mid-line incision was made along the abdomen large enough to allow the insertion of the 1 g telemeters, which had been sterilized with ethylene oxide. All surgeries were performed under sterile conditions, and care was taken to ensure that anaesthesia was maintained. No marmosets developed fevers following the surgeries, nor were any complications encountered in the months following implantations. Body temperature and activity (arbitrary units provided from the telemeters based on changes in signal strength) was measured every 0.5 min using Vitalview® software from Minimitter®.

41.5 Experimental Protocol

Animals were minimally handled throughout the procedure. At 4 p.m., one animal was caught from its home cage using a trap-door cage baited with a mealworm or marshmallow. Once coaxed into the trap-door cage, the marmoset was transferred to the nest-box-sized respirometer. Light levels and timing followed the same protocol as occurred in the holding environment: lights on between 0800 and 2000 hours. Measurements commenced at 6 p.m. and lasted until the next morning at 9:30 a.m. Average values were obtained every 5 min for the following parameters: $\dot{V}_{O_2}, \dot{V}_{CO_2}$, T_b, T_a. Unless otherwise stated, all data are reported as mean ± SEM ($N = 10$).

41.6 Results

Metabolic rates followed a profile coincident with the photoperiod, activity and sleep period. After a 2 h period of habituation, and the commencement of recording, metabolic rates slowly rose to a maximum around 7:30 p.m., coincident with an elevated activity prior to the lights going out at 8 p.m. (Fig. 41.1a). Even prior to the beginning of the inactive period, metabolic rate began to fall until reaching a plateau approximately 30 min later. RER initially started at values of ~0.8–0.85, declining to a value of ~0.65 during the inactive sleep period (Fig. 41.1b). Interestingly, during periodic, brief time points when marmosets appeared to wake up and begin to warm back up, RER rose transiently towards 0.7. Nevertheless, values of 0.65 were sustained throughout most of the sleep period, rising to 0.7 and remaining there immediately after the lights came on at 8 a.m., coincident with the normal wakeful period.

T_b followed suit with changes in $\dot{V}_{O_2}$, in that it was elevated during the active period of time between 6 and 8 p.m., although a noticeable, slow decline in T_b was initiated prior to the normal beginning of the scotophase at 8 pm (Fig. 41.1c). After this time, T_b declined exponentially from a value >39°C to a steady state level of ~35.5°C during the inactive, sleep period, requiring 2–3 h to reach this constant level. At approximately 6 a.m., T_b began to rise slowly, until the lights came on at 8 a.m., after which T_b rapidly returned close to the previous daytime values of 39°C.

Assessment of the apparent thermosensitivity for metabolism was conducted, using only $\dot{V}_{O_2}$ data where conductance was within 10% of one another; a Q_{10} value of 4.57 ± 0.88 resulted (Fig. 41.2). Finally, to establish the relationship between activity and metabolism, attempts were made to assess correlations between $\dot{V}_{O_2}$ and activity (Fig. 41.3). During periods with the lights on, and presumably during wakefulness, activity strongly predicted $\dot{V}_{O_2}$. During sleep and inactivity, however, the relative changes in $\dot{V}_{O_2}$ were much greater than could be explained by similar changes in activity. I estimate that BMR calculated from daytime resting animals ($\dot{V}_{O_2}$ extrapolated to zero activity) would be 19.6 ± 0.7 ml kg^{-1} min^{-1}, while BMR calculated from minimum nighttime values can fall as low as 11.4 ± 0.5 ml kg^{-1} min^{-1}, which are 142 and 82% of the predicted value for eutherian mammals, respectively.

41.7 Discussion

Pygmy marmosets show a profound diurnal change in $\dot{V}_{O_2}$ and T_b that appears to accompany their natural sleep pattern. Although this is not deep enough to be considered torpor, the metabolic savings associated can be substantial when compared to daytime values. Indeed, if the MR is assessed only during stable

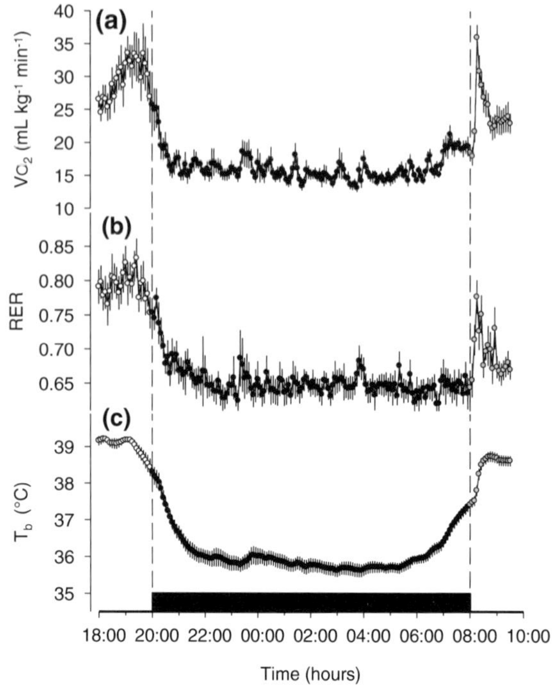

Fig. 41.1 Diurnal changes (mean ± SEM) in metabolic rate (**a**: $\dot{V}_{O_2}$), respiratory exchange ratio (**b**) and T_b(**c**) in pygmy marmosets. The *black bar* indicates when the lights were turned off, corresponding to the animals' normal period of sleep. *Open symbols* represent points in the evening before lights were turned off, *black symbols* correspond to data obtained during the sleep phase, and *grey symbols* represent the data following lights coming back on in the morning

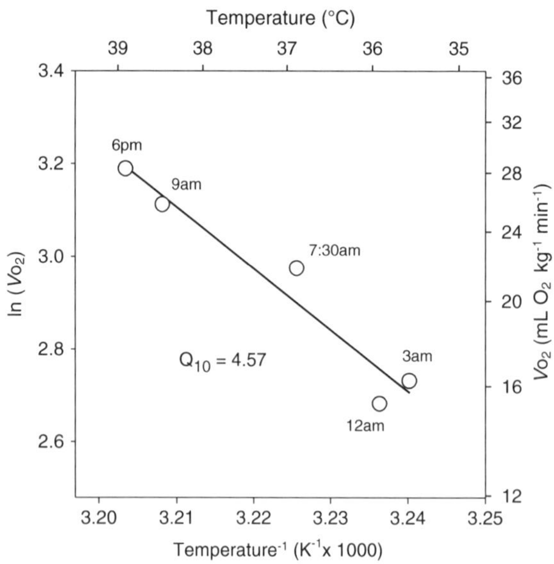

Fig. 41.2 Correlation between $\dot{V}_{O_2}$ and T_b determined from mean values at time points where whole animal conductance [$\dot{V}_{O_2}/(T_b - T_a)$] values were within 10% of each other (6 p.m., 12 a.m., 3 a.m., 7:30 a.m., 9 a.m.). Error bars have been removed for clarity. The mean Q_{10} value (calculated from *plots* from each individual marmoset) was 4.57 ± 0.89

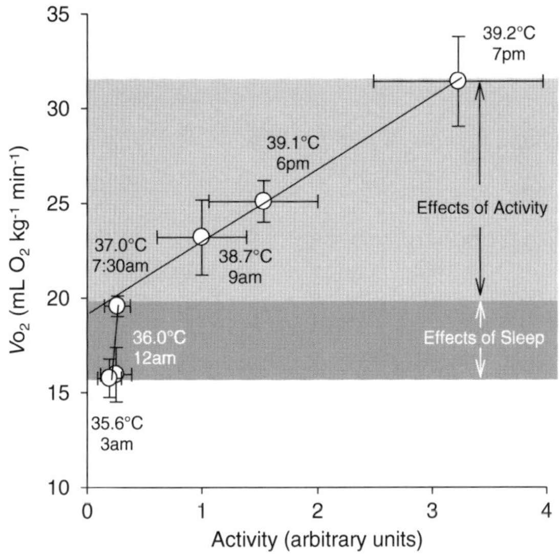

Fig. 41.3 Summary $\dot{V}_{O_2}$ and activity data (mean ± SEM) obtained from pygmy marmosets at critical time points to assess the appropriate 'resting' value for the estimate of basal metabolism during normothermia. Times of day and corresponding T_b data accompany the symbols. Data in the *light grey* region reflect strongly activity dependent $\dot{V}_{O_2}$ during the active/awake phase, whereas data in the *dark grey* region reflect the relatively small changes in activity observed during sleep. Active phase $\dot{V}_{O_2}$ during normothermia is extrapolated to the *x*-axis to estimate BMR at zero activity

periods where conductance is equivalent, a Q_{10} of 4.6 is observed. Arguably, using Q_{10} as a sole indicator of metabolic suppression in mammals has its challenges (Nestler 1990), however, since the marmosets show a clear tendency to regulate T_b during sleep (Genoud et al. 1997), then these declines in metabolic rate may be due to metabolic suppression. From where, however, does this nightly metabolic suppression originate?

Studies from hibernating mammals, as well as numerous non-hibernating rodents, indicate a strong reliance on central hypothalamic thermosensitivity for thermoregulatory processes in small mammals (Heller 1978). If the relatively larger brain size observed in primates (primarily neocortex) is associated with changes in sleep metabolism, and the brain represents a large portion of metabolic rate, then entry into different states (slow wave or REM sleep) could lead to large declines in MR.

This leads to the question of how MR differs among wake, rest and sleep? It has been estimated in humans that the metabolic savings associated with sleep itself are ~10% (Mason and Benedict 1934). Similar specific effects of sleep state are known to occur in other mammals; the brain's central intrinsic thermosensitivity is reduced in sleep, progressively so from wake to slow wave to paradoxical (REM) sleep (Heller and Glotzbach 1977). Given the steep central thermosensitivity known to occur in small mammals (Heller 1978), and that this sensitivity is influenced by higher brain centres and peripheral feedback (Glotzbach and Heller 1984), any change in this activity would alter the activation of thermoeffectors, and thus lead to reduced heat production and a decline in T_b. For example, if the normal hypothalamic temperature for initiating heat defenses is ~37°C, then the diurnal changes in T_b can be explained by a change in the 'set-point' brain temperature threshold for activation of cold responses, and/or a reduction in the sensitivity of thermosensitive neurons.

Nevertheless, based on data obtained during sleep, Genoud et al. (1997) estimated that pygmy marmosets exhibit a BMR that is 72% of that predicted for their body mass, and suggested that this hypometabolism may be an evolutionary strategy to dietary constraints. The arguments regarding Callitrichid hypometabolism have implications for the evolution of energy saving strategies in small endothermic gum feeders as well as for understanding evolutionary dwarfism. If phyletic dwarfism reflects an energetic strategy related to their specialisation in plant exudates, then assessing the relevant measure for metabolic rate is imperative.

The results of this study raise the question of whether nighttime MR (inactive phase) is an appropriate measure of BMR? One corollary to this are the RER results demonstrating that throughout most of the night, RER approaches values that are slightly less than expected for a fasted animal engaged in fat oxidation. It is possible that sleep in the pygmy marmoset is associated with CO_2 retention, similar to other mammals engaged in torpor (Malan et al. 1985; Nestler 1990; Snapp and Heller 1981), although such retention is unlikely to explain the sustained reduction in RER. Other explanations for low RER values range from 'altered metabolic processes' (Grigg 1978), to ketone body formation (Chaui-Berlinck and Bicudo 1995), or to gluconeogenesis (Hawley et al. 1933). In most cases, these low RERs are either associated with torpor or animals undergoing fasting. The very rapid return of RER to normal values that occurred during transient re-warming periods during the night, or during the morning re-warming period strongly implicates physiological state changes. Is the low RER at night associated with a dietary energetic strategy? Pygmy marmosets are known to be plant exudate specialists, and exhibit long gut transit times, suggesting they possess mechanisms to elevate digestive efficiencies (Power and Oftedal 1996), and thus, it is possible that nighttime metabolism reflects aspects of their feeding regime, although how this could explain why the RER immediately returns to 0.7 upon awakening is unknown.

In summary, the diurnal fluctuations in metabolic rate described here support the notion that pygmy marmosets undergo profound changes in metabolism despite relatively small changes in T_b. Although much of this is attributable to changes in activity, the portion of metabolism not based on activity demonstrates profound temperature sensitivity, suggesting a metabolic suppression occurring at night. Since fasted marmosets exhibit elevated metabolic rates and T_b, and hibernation and torpor have since been described in numerous African prosimian primates (Blanco and Rahalinarivo 2010; Dausmann et al. 2005), it is parsimonious to suggest that a form of nightly torpor is occurring in these small New World Monkeys.

Acknowledgments I would like to acknowledge Steve Wood for providing access to equipment, Jim Blank for kindly providing an environment within which to conduct the research and Suzette Tardif for allowing me to work with her colony of pygmy marmosets. This research was funded through a SUMMA Health System research fellowship to GJT.

References

Anonymous (2001) Glossary of terms for thermal physiology—third edition. Jpn J Physiol 51:245–280

Blanco MB, Rahalinarivo V (2010) First direct evidence of hibernation in an eastern dwarf lemur species (*Cheirogaleus crossleyi*) from the high-altitude forest of Tsinjoarivo, central-eastern Madagascar. Naturwissenschaften 97:945–950

Blaxter K (1989) Energy metabolism in animals and man. Cambridge University Press, Cambridge

Boutilier RG, Shelton G (1986) Gas exchange, storage and transport in voluntarily diving *Xenopus laevis*. J Exp Biol 126:133–155

Chaui-Berlinck JG, Bicudo JEPW (1995) Unusual metabolic shifts in fasting hummingbirds. Auk 112:774–778

Cooper CE, Withers PC (2009) Effects of measurement duration on the determination of basal metabolic rate and evaporative water loss of small marsupials: how long is long enough? Physiol Biochem Zool 82:438–446

Cooper CE, Withers PC (2010) Effect of sampling regime on estimation of basal metabolic rate and standard evaporative water loss using flow-through respirometry. Physiol Biochem Zool 83:385–393

Dausmann KH, Glos J, Ganzhorn JU, Heldmaier G (2005) Hibernation in the tropics: lessons from a primate. J Comp Physiol B 175:147–155

Genoud M, Martin RD, Glaser D (1997) Rate of metabolism in the smallest simian primate, the pygmy marmoset (*Cebuella pygmaea*). Am J Primatol 41:229–245

Glotzbach SF, Heller HC (1984) Changes in the thermal characteristics of hypothalamic neurons during sleep and wakefulness. Brain Res 309:17–26

Grigg GC (1978) Metabolic rate, Q_{10} and repiratory quotient (RQ) in *Crocodylus porosus*, and some generalizations about low RQ in reptiles. Physiol Zool 51:354–360

Hawley EE, Johnson CW, Murlin JR (1933) The possibility of gluconeogenesis from fat II. The effect of high fat diets on the respiratory metabolism and ketosis of man. J Nutr 6:523–557

Heller HC (1978) Hypothalamic thermosensitivity in mammals. Experientia Supp 32:267–276

Heller HC, Glotzbach SF (1977) Thermoregulation during sleep and hibernation. In: Robertshaw D (ed) Environmental physiology II, vol 15. University Park Press, Baltimore

Issa EB, Wang XQ (2008) Sensory responses during sleep in primate primary and secondary auditory cortex. J Neurosci 28:14467–14480

Malan A, Rodeau JL, Daull F (1985) Intracellular pH in hibernation and respiratory acidosis in the European hamster. J Comp Physiol B 156:251–258

Mason ED, Benedict FG (1934) The effect of sleep on human basal metabolism, with particular reference to South Indian women. Am J Physiol 108:377–383

McNab BK (1988) Complications inherent in scaling the basal rate of metabolism in mammals. Q Rev Biol 63:25–54

Nestler JR (1990) Relationships between respiratory quotient and metabolic rate during entry to and arousal from daily torpor in deer mice (*Peromyscus maniculatus*). Physiol Zool 63:504–515

Perelman P, Johnson WE, Roos C, Seuanez HN, Horvath JE, Moreira MA, Kessing B, Pontius J, Roelke M, Rumpler Y et al (2011) A molecular phylogeny of living primates. PLoS Genet 7:e1001342

Power ML, Oftedal OT (1996) Differences among captive callitrichids in the digestive responses to dietary gum. Am J Primatol 40:131–144

Snapp BD, Heller HC (1981) Suppression of metabolism during hibernation in ground squirrels (*Citellus lateralis*). Physiol Zool 54:297–307

Soini P (1993) The ecology of the pygmy marmoset, *Cebuella pygmaea:* some comparisons with two sympatric tamarins. In: Rylands AB (ed) Marmosets and tamarins. Systematics, behavior, and ecology. Oxford University Press, Oxford, pp 257–261

Stein K, Csiki Z, Rogers KC, Weishampel DB, Redelstorff R, Carballido JL, Sander PM (2010) Small body size and extreme cortical bone remodeling indicate phyletic dwarfism in Magyarosaurus dacus (Sauropoda: Titanosauria). Proc Natl Acad Sci U S A 107:9258–9263

Weston EM, Lister AM (2009) Insular dwarfism in hippos and a model for brain size reduction in Homo floresiensis. Nature 459:U85–U88

Withers PC (1977) Measurement of $\dot{V}_{O_2}$, $\dot{V}_{CO_2}$, and evaporative water loss with a flow-through mask. J Appl Physiol 42:120–123

Chapter 42
Torpor Use and Body Mass Gain During Pre-Hibernation in Late-Born Juvenile Garden Dormice Exposed to Food Shortage

Sylvain Giroud, Christopher Turbill and Thomas Ruf

Abstract Prior to hibernation, juvenile hibernators have to sustain both somatic growth and fattening to reach a sufficient body mass to survive the following winter season. This high demand for energy is especially challenging for juveniles born late in the season, since they might already experience reduced food availability and decreasing temperatures. In this study, we asked whether late-born juvenile garden dormice can use torpor to counteract intermittent food shortages and, if so, whether the use of torpor enables them to compensate the energy deficit and to maintain rates of body mass gain similar to well-fed juveniles. We measured daily torpor use, food intake, and body mass in weaned late-born juvenile dormice exposed to intermittent fasting ($n = 5$) or fed ad libitum ($n = 4$) under natural photoperiod and ambient temperature during 6 weeks prior to hibernation (13 September–1 November). We found that fasted juveniles frequently used torpor and despite eating less food grew at a significantly greater rate compared to juveniles fed ad libitum. Torpor frequency was positively related to body mass gain, but this effect was statistically significant only in the fasted group. Consequently, fasted juvenile dormice reached a critical body mass and entered into hibernation 1 week earlier than ad libitum fed juveniles. If fasted juveniles reached a similar pre-hibernation fat content compared to animals fed ad libitum, and how any difference in fat versus fat-free mass might affect hibernation propensity and winter survival, remains to be determined.

S. Giroud (✉) · C. Turbill · T. Ruf
Department of Integrative Biology and Evolution, University of Veterinary Medicine, Savoyentrasse 1, 1160 Vienna, Austria
e-mail: sylvain.giroud@fiwi.at

C. Turbill
University of Western Sydney, Locked Bag 1797, Penrith, NSW 2751, Australia

T. Ruf et al. (eds.), *Living in a Seasonal World*,
DOI: 10.1007/978-3-642-28678-0_42,

42.1 Introduction

Many small mammals survive the winter season by entering hibernation, which involves a series of prolonged torpor bouts characterized by a substantial reduction in body temperature and metabolic rate (Geiser and Ruf 1995). Most hibernators do not eat during the winter but rely entirely on stores of body fat to support their metabolism (Dark 2005). Hence, pre-hibernation body mass has been reported to be a good predictor of winter survival. Rapidly gaining sufficient fat before hibernation is critical for hibernators in order to survive the winter. This is particularly true for juveniles since they must allocate energy to both growth and fat storage (Lenihan and Van Vuren 1996). The situation becomes even more challenging for them when born late in the season, since they then might have to cope with uncertain and reduced food availability. To counteract these constraints, juveniles have to develop energetic strategies to reach a sufficient body mass, especially in terms of fat reserves, in order to survive the following winter season and to have an opportunity to reproduce (Armitage 1976).

One possible strategy to reduce maintenance energy costs is the use of short bouts of torpor during the pre-hibernation fattening period. Some hibernating species use short bouts of torpor outside of the hibernation season, especially to cope with short-term food shortages. Presently, we know very little about the use of short torpor bout by juveniles of hibernating species, and to what extent this might affect their energetics, growth and development (Geiser 2008).

Therefore, in the present study we aimed to determine (1) whether late-born juveniles of garden dormice (*Eliomys quercinus*) use torpor when food-deprived, and (2) if torpor use enables juveniles to compensate for an energy deficit and to sustain rates of pre-hibernation fattening similar to those juveniles fed ad libitum. The garden dormouse is a small rodent that exhibits strong seasonal changes in body mass and enters deep hibernation bouts during 7 months in winter (Lachiver and Boulouard 1965; Vogel 1997). In captivity, two litters per year are possible, and late-born juveniles (born in August) face a challenge to grow rapidly and store sufficient fat to survive hibernation in the subsequent winter.

42.2 Methods

42.2.1 Animals

The garden dormouse is a nocturnal rodent widely distributed in Europe, which colonizes various habitats, ranging from coniferous and mixed forest to hilly and cultivated land (Storch 1978; Amori et al. 1994). It shelters in hollow trees, in the branches of trees or shrubs, and in rocky crevices (Vaterlaus 1998; Bertolino et al. 2003). In winter, it hibernates often in large nests of moss and fine minced vegetable material placed in natural or artificial cavities. The nine juvenile garden dormice

(*Eliomys quercinus*; three males, six females) investigated in this study were l ate born (August) from a captive colony kept at the Research Institute of Wildlife Ecology (University of Veterinary Medicine, Vienna, Austria; latitude 48° 15′N, longitude 16° 22′E). Dormice were weaned at an age of 5 weeks and housed individually under natural photoperiod and ambient temperature, with an ad libitum access to food and water. Food consisted of cat pellets, which contained 30% protein, 10% fat, and 60% carbohydrate, and sufficient vitamins and minerals. Each cage (60 × 40 × 40 cm) included numerous branches and one custom made, insulated nest box filled with straw. The nest box was attached to the outside of the cage's wire door in a way such that the entrance of the nest was accessible for the animal from the inside of the cage.

After 5 days of habituation to cages and nests, all dormice entered the experiments. Four animals had access to ad libitum food (control group) and five were intermittently fasted (fasted group). Continuous ad libitum access to water was provided to all animals.

42.2.2 Protocol Overview

Each animal was studied during the pre-hibernating period (13 September–01 November) until its first prolonged (>24 h) torpor bout was recorded. In the fasted group, food was removed for 24 h twice a week starting on Monday and Wednesday evenings, resulting in two non-consecutive nights of fasting per week. This intermittent fasting was maintained until most of the fasted animals entered prolonged torpor, i.e. hibernation on week 6, after which all animals were fed under ad libitum. Body mass and food intake were measured once a week, and torpor use was monitored daily. Ambient temperature was recorded during the entire experiment using a thermometer placed next to the animal cages.

42.2.3 Body Mass and Food Intake Measurements

At the beginning of each week, each animal was removed from its nest box and weighed to the nearest 0.1 g (Mettler Toledo, PM34, Delta range). For each individual, body mass gain was calculated as percentage of mass change compared to the first week of the experiment.

Weekly food intake was estimated by the difference between the initial and the remaining food masses. Grams of food intake were converted into kilojoules using equivalents of 16.7, 37.2, and 16.7 kJ/g of protein, fat, and carbohydrate, respectively. The nutrient composition of the food (cat pellet) was provided by the manufacturing company (Topix, Saturn Petfood, GmbH, Bremen, Germany) and a coefficient of hydration of 20% was assumed. For each group, mass-specific food intake was also calculated as kilojoules of energy per gram of animal body mass.

42.2.4 Measurement of Body Temperature and Torpor Use

Body temperature (T_b) was measured using miniature PIT tags (Destroy Fearing, Life Chip Bio-thermo) injected subcutaneously in each animal at the back. At three times per weekday (between 8.30 and 9.30 a.m., 11.30 and 12.30 a.m., 2.30 and 3.30 p.m.), each nest was remotely scanned using a PIT tag reader (10 cm range) to record the animal's subcutaneous T_b. Torpor was conservatively defined by the animals reaching a T_b below 30°C. The PIT tags used do not record temperature below ~25°C, hence we could not record minimum T_b in torpor. We were principally interested in torpor frequency, which we calculated for each animal as the number of torpor bouts displayed per week.

42.2.5 Statistics

For all response variables, data analyses were conducted separately for the first 6 weeks of the experiment, when dormice were growing and subjected to either fasted or under ad libitum food, and the last 2 weeks of the experiment when dormice had begun to hibernate and all were fed ad libitum.

To adjust for repeated measurements, data were analyzed using linear mixed-effect models with time (weeks) and treatment (fasting vs. ad libitum) as main explanatory factors, and different intercepts per animal entered as a random factor. When the timecourse was nonlinear, a second degree polynomial was computed (for the analysis of food intake normalized by body weight). A t test was used to compare between animal groups (data and model residuals were normally distributed). Linear mixed-effects models were also used to test for significant relations between body mass gain and torpor frequency as well as food intake in fasted and non-fasted groups.

All reported values are means ± SEM, and $p < 0.05$ was considered significant. All statistical computations were performed using the program R (v.2.10.1) (R Development Core Team 2009).

42.3 Results

42.3.1 Torpor Frequency

Linear model analyses revealed a significant effect of fasting on the timecourse of torpor frequency over the 6 weeks of treatment ($t = 3.76$, $p < 0.001$; Fig. 42.1). Fasted dormice immediately displayed torpor bouts in the first week (1.0 ± 0.3 bouts/week) and torpor use increased in frequency over the 6 weeks of fasting, to reach a value of 2.4 ± 0.2 bouts/week on week 6. Three of the five

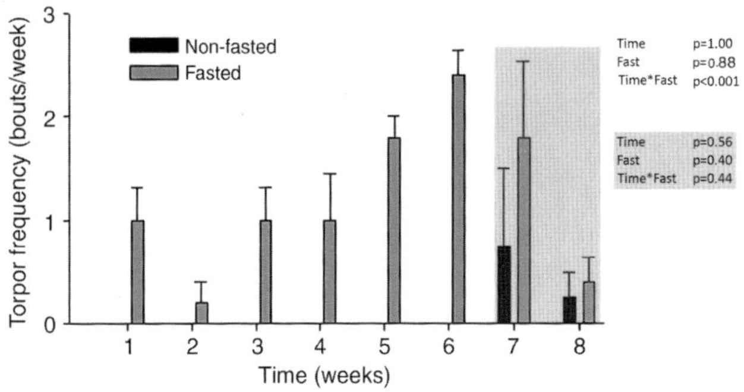

Fig. 42.1 Torpor frequency: Torpor frequency of individual juvenile garden dormice (means ± SEM) in response to intermittent fasting ($n = 5$) or ad libitum ($n = 4$) food supply. All dormice received ad libitum food in the last two-week period (*shaded area*). Statistics from a fitted linear model show a significant interaction effect of time and food supply during the growing phase (i.e. first 5 weeks) of the experiment

fasted garden dormice entered prolonged torpor (>24 h) on week 6 (data not shown). Conversely, dormice fed ad libitum did not show any torpor during the first 6 weeks of the experiment. During the six first weeks of the experiments, the ambient temperature decreased continuously from 19 down to 6°C.

During the second phase (weeks 7 and 8), no significant differences in torpor frequency were found between animal groups and across weeks. Only two of the fasted dormice entered torpor during week 7, but these were prolonged torpor bouts (>24 h). Garden dormice fed ad libitum displayed bouts of torpor for the first time in week 7, and one animal entered prolonged (3 days) torpor in week 7 (data not shown). During this 2-week period, the ambient temperature gradually increased from 6 to 13°C.

42.3.2 Food Intake

The time course of food intake differed significantly between fasted and non-fasted animals (time × treatment, $t = 2.89$, $p < 0.01$). This was mainly due to differences during weeks five and six when food intake was higher in non-fasted dormice (Fig. 42.2a). When normalized by body mass, the differential effect of the treatment across time on food intake remained significant ($t = 3.05$, $p < 0.01$; Fig. 42.2b). The fasted dormice showed a 70% reduction in their mass-specific food intake on week 6 (compared to the first week of experiment), whereas animals fed ad libitum displayed only a 30% decrease of their food intake over the same period of time.

During the second phase (week 7 and 8), we found no significant effects of time and treatment on food intake or mass-specific food intake (Fig. 42.2).

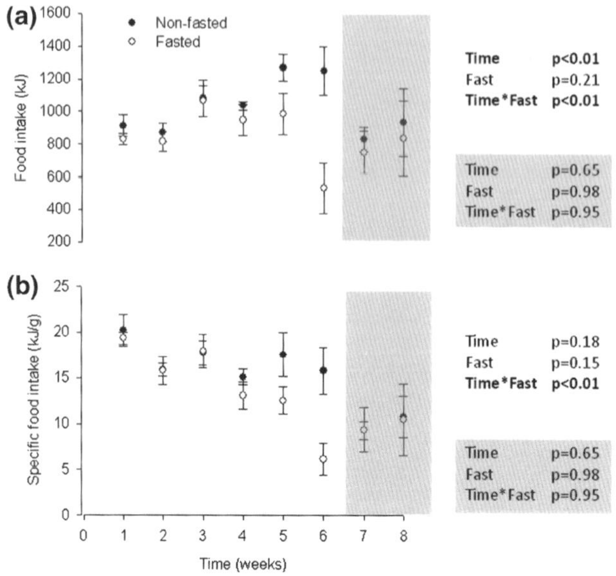

Fig. 42.2 Food intake (**a**) and mass-specific food intake (**b**): Energy intake estimated from the mass of food eaten each week (see the Methods section for conversion factors; means ± SEM). Statistics from fitted linear models show a significant interaction effect of time and food supply during the growing phase (i.e. first 5 weeks) of the experiment

42.3.3 Body Mass

At the onset of the experiment, body masses did not differ between fasted and ad libitum groups (43 ± 2 vs. 46 ± 2 g, respectively; $t = 0.30$, $p = 0.77$). The timecourse of body mass gain significantly differed between treatment groups ($t = 2.59$, $p < 0.05$; Fig. 42.3), indicating that the fasted dormice gained weight at a faster rate than the non-fasted animals. Indeed, during the 5 first weeks of the experiment, fasted and non-fasted juveniles grew at respective rates of +20 ± 2 and +12 ± 2% mass-change/week, values that significantly differed from each other ($t = -3.05$, $p < 0.05$).

42.3.4 Determinants of Body Mass Gain

We found a significant positive relationship between food intake and body mass gain only in non-fasted dormice ($t = 3.10$, $p < 0.01$, $r^2 = 0.31$; Fig. 42.4a). This relationship was not significant in fasted animals ($t = 0.60$, $p = 0.55$; Fig. 42.4b).

Instead, body mass gain in fasted dormice was significantly associated with torpor frequency ($t = 3.98$, $p < 0.001$, $r^2 = 0.36$; Fig. 42.5). Given the absence of torpor in non-fasted animals, no such relationship could be detected in this treatment group.

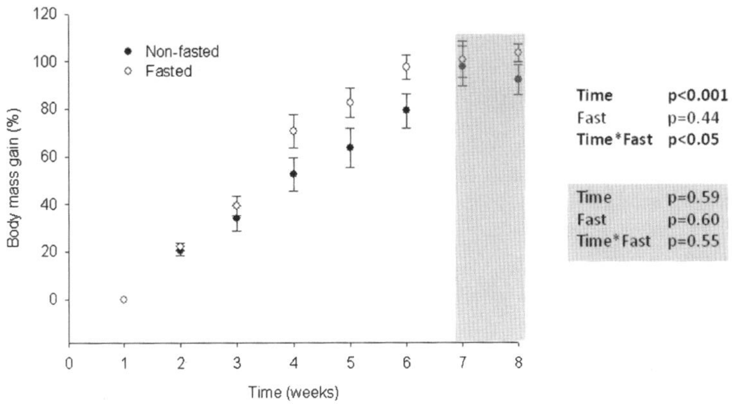

Fig. 42.3 Percent of body mass gain relative to initial body mass: Statistics from a fitted linear model show a significant interaction between time and food supply on the percent gain in body mass

42.4 Discussion

42.4.1 Use of Torpor by Juvenile Dormice

When provided with food ad libitum, juvenile dormice did not enter torpor during body mass gain in the autumn pre-hibernation period despite a continuous decrease in the ambient temperature to 6°C. These juvenile dormice started to use torpor and subsequently entered hibernation only after reaching maximum body mass (at an age of about 12 weeks). Similarly, juvenile Djungarian hamsters fed ad libitum and maintained under a short-day photoperiod express spontaneous daily torpor for the first time only after completing growth at ~13 weeks of age (Bae et al. 2003). We also observed an expected positive relation between body mass gain and food intake in non-fasted juvenile dormice. These results suggest that, when food is plentiful, the use of torpor might not play a major role in the early development and pre-hibernation fattening of late-born garden dormice. They are also in accordance with the notion that torpor seems to have both benefits and costs (Humphries et al. 2003), and is often avoided if food resources are unlimited.

Conversely, fasted juvenile dormice displayed torpor immediately, starting in the first week of the experiment, and torpor use gradually increased throughout the following 5 weeks until animals entered hibernation. Torpor might have evolved in young animals to facilitate energy conservation through a substantial reduction of energy expenditure during periods of food shortage (Geiser et al. 2006; Geiser 2008). It seems that in our experiment, when faced with a shortage of food and decreasing ambient temperature, the use of torpor indeed allowed juvenile dormice to reduce their energy expenditure. Similarly, juvenile Siberian hamsters born late in the breeding season can conserve energy by undergoing torpor when they experience

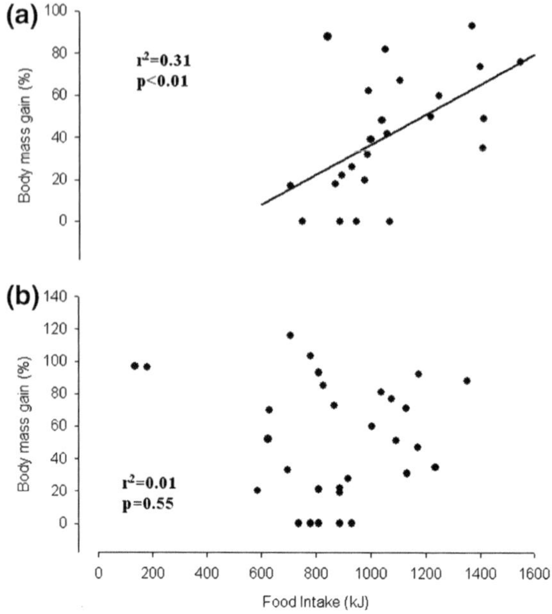

Fig. 42.4 Body mass gain as a function of food intake in non-fasted (**a**) and fasted (**b**) garden dormice: Body mass gain during the first 6 weeks of the experiment was significantly related to food intake in non-fasted (**a**) juvenile garden dormice (linear mixed effects model with individual as random factor), but not for fasted individuals (**b**)

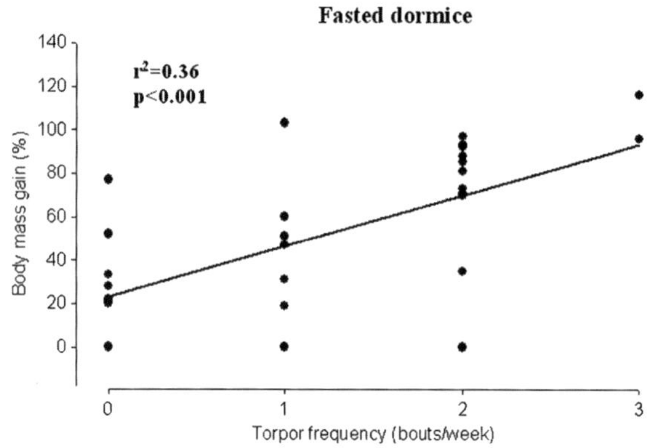

Fig. 42.5 Body mass gain as a function of torpor frequency in fasted juvenile garden dormice: Body mass gain was significantly related to torpor frequency during the first 6 weeks of the experiment (linear mixed-effects model with individual as random factor)

food shortages (Bae et al. 2003). Torpor use by juvenile dormice might have been enhanced by the constant decrease of ambient temperature, which acts as a facilitator for torpor occurrence and torpor depth in numerous species (Wolff and Bateman 1978; Vogt and Lynch 1981; Aujard et al. 1998). Exposure to decreasing ambient temperatures allows mouse lemurs to reduce their body temperature during torpor (Aujard et al. 1998) and triggers an increase in torpor frequency and duration in small rodents (Wolff and Bateman 1978). Whether or not changes in torpor depth and duration occur in fasted juvenile dormice during body mass gain still remains to be assessed.

42.4.2 Torpor Allows Over-Compensation of Energy Intake Restrictions

When intermittently fasted, juvenile garden dormice increased body mass at a higher rate compared to animals fed ad libitum. Hence, fasted juvenile dormice were not only able to avoid a negative energy balance, but also over-compensated so that mass gain was accelerated. It is possible that the unreliable food source also acted as a trigger to hasten fattening for hibernation. Captive rufous hummingbirds facing periods of food restriction prior to migration show increases in both body mass and use of torpor; this occurrence of torpor is dependant on food availability (Hiebert 1991). Food restriction would therefore act as a predictor of prolonged reduction in food availability, triggering accelerated fattening prior to the subsequent winter. Similar effects of intermittent food restriction were also found in birds that do not use torpor (Witter 1995).

We found a positive relation between body mass gain and torpor frequency. To our knowledge, this is the first study to show a link between torpor use and body mass gain in a juvenile hibernator. However, the use of torpor by juveniles to counteract energetic challenges during growth has been predicted before (Geiser 2008). Our results suggest that torpor-derived energy savings can largely explain the higher rate of body mass gain in fasted juveniles, although we do not know the extent of possible additional energy savings achieved by behavioral means such as a reduction in activity.

Juveniles of hibernating species must allocate energy both into growth and development of their body tissues (i.e. fat-free mass) and into accumulation of energy reserves for hibernation (fat mass). Growth and development as well as fattening of juveniles must occur in a short time (Neal 1965; Clark 1970; Morton and Tung 1971). For juveniles, investing energy into growth is important, since suboptimal nutrition at the early stage of life can significantly affect health over the lifespan (Metcalfe and Monaghan 2001). Furthermore, it is also crucial for juvenile hibernators to allocate energy into fattening, because survival of the first winter is a key bottleneck in the fitness of hibernating mammals, especially if they are born late in the season. Therefore, whether fasted juvenile dormice preferentially allocate torpor-derived energy savings to body growth, sacrificing optimal fattening, or vice versa still remains to be determined. However, it is likely that juvenile dormice preferentially allocate energy into fattening during the pre-hibernation period because sufficient body energy reserves seem crucial to ensure winter survival (Phillips 1981; Lenihan and Van Vuren 1996).

42.4.3 Conclusion and Perspectives

Our study provides new information about the role of torpor in regulating pre-hibernation body mass gain in juvenile hibernators. We show for the first time a significant relation between torpor use and body mass gain. Fasted juvenile dormice overcompensated for the energy intake deficit, presumably by their use of torpor, and showed accelerated body mass gain in anticipation of the unfavorable winter season. Further studies are needed to assess the partitioning of mass accretion between fat mass and lean mass, both in fasted and in non-fasted juvenile dormice.

Acknowledgment The authors thank Peter Steiger for help with animal handling and maintenance.

References

Amori G, Cantini M et al (1994) Distribution and conservation of Italian dormice. Hystrix It J Mamm (NS) 6:331–336

Armitage KB (1976) Seasonal changes in weights of marmots. Am Midl Nat 96:36–51

Aujard F, Perret M et al (1998) Thermoregulatory responses to variations of photoperiod and ambient temperature in the male lesser mouse lemur: a primitive or an advanced adaptive character? J Comp Physiol B 168(7):540–548

Bae HH, Larkin JE et al (2003) Juvenile Siberian hamsters display torpor and modified locomotor activity and body temperature rhythms in response to reduced food availability. Physiol Biochem Zool 76(6):858–867

Bertolino S, Cordero N et al (2003) Home ranges and habitat use of the garden dormouse (*Eliomys quercinus*) in a mountain habitat in summer. Acta Zool Acad Sci Hung 49:31–39

Clark TW (1970) Early growth, development, and behavior of the Richardson Ground Squirrel (*Spermophilus richardsoni elegans*). Am Midl Nat 83(1):197–205

Dark J (2005) Annual lipid cycles in hibernators: integration of physiology and behavior. Annu Rev Nutr 25:469–497

Geiser F (2008) Ontogeny and phylogeny of endothermy and torpor in mammals and birds. Comp Biochem Physiol A Mol Integr Physiol 150(2):176–180

Geiser F, Ruf T (1995) Hibernation versus daily torpor in mammals and birds: physiological variables and classification of torpor patterns. Physiol Zool 68(6):935–966

Geiser F, Westman W et al (2006) Development of thermoregulation and torpor in a marsupial: energetic and evolutionary implications. J Comp Physiol B 176(2):107–116

Hiebert SM (1991) Seasonal differences in the response of rufous hummingbirds to food restriction: body mass and the use of torpor. Condor 93:526–537

Humphries MM, Thomas DW et al (2003) The role of energy availability in mammalian hibernation: a cost-benefit approach. Physiol Biochem Zool 76(2):165–179

Lachiver F, Boulouard R (1965) Evolution du poids corporel chez un hibernant le lérot (*Eliomys quercinus L.*) durant la période hivernale et au cours de l'exposition au froid en été. Mammalia 29:343–366

Lenihan C, Van Vuren D (1996) Growth and survival of juvenile yellow-bellied marmots (*Marmota flaviventris*). Can J Zool 74:297–302

Metcalfe NB, Monaghan P (2001) Compensation for a bad start: grow now, pay later? Trends Ecol Evol 16(5):254–260

Morton LM, Tung HL (1971) Growth and development in the Belding ground squirrel (*Spermophilus beldingi*). J Mammal 52(3):611–616

Neal BJ (1965) Growth and development of the round-tailed and Harris Antelope ground squirrels. Am Midl Nat 73(2):479–489

Phillips JA (1981) Growth and its relationship to the initial annual cycle of the golden-mantled ground squirrel, *Spermophilus lateralis*. Can J Zool 59:865–871

R Development Core Team (2009) R: a language and environment for statistical computing. R Foundation for Statistical Computing, Vienna

Storch G (1978) *Eliomys quercinus* (Linnaeus, 1766)—Gartenschläfer. In: Niethammer J, Krapp F (eds) Handbuch der Saügetiere Europas, vol 1. Akademische Verlagsgesellschaft, Wiesbaden, pp 208–225

Vaterlaus C (1998) Der Gartenschläfer (*Eliomys quercinus*). Ökologie, Populationstruktur und die Verbreitung in der Schweiz. Ph.D. thesis, University of Basel

Vogel P (1997) Hibernation of recently captured Muscardinus, Eliomys and Myoxus: a comparative study. Nat Croat 6(2):217–231

Vogt FD, Lynch GR (1981) Influence of ambient temperature, nest availability, huddling, and daily torpor on energy expenditure in the white-footed mouse, *Peromyscus leucopus*. Physiol Zool 55(1):56–63

Witter MS (1995) Periodic food availability and strategic regulation of body mass in the European Starling, *Sturnus vulgaris*. Func Ecol 9:568–574

Wolff JO, Bateman GC (1978) Effects of food availability and ambient temperature on torpor cycles of *Perognathus flavus* (*Heteromyidae*). J Mammal 59(4):707–716

Chapter 43
Seasonal Variations in Energy Turnover and Body Temperature in Free-Living Edible Dormice, *Glis glis*

Joanna Fietz, Jutta Schmid and John R. Speakman

Abstract Edible dormice (*Glis glis*) face low food availability after emergence from hibernation, and during years when there is no seed production from their main feeding trees. In this field study we aimed to investigate seasonal changes of energy turnover and body temperature (T_b) in edible dormice during periods of high and low food availability, to understand energy saving strategies, besides the use of hibernation and torpor. We therefore measured daily energy expenditure of nine adults using the doubly labelled water and resting metabolic rate of 20 euthermic individuals with portable gas analysers. Additionally, we analysed T_b patterns of 118 individuals during years of high and low food availability. All measurements were carried out in the field during the active months of edible dormice between 1996 and 2006 in south western Germany. Results of this study demonstrate that energy turnover and T_b in edible dormice vary markedly over their active season and among years. Variations of food intake and the heat increment of feeding are likely to cause a substantial part of these variations together with the reactivation of organs that are atrophied during hibernation. T_b patterns strongly suggest that edible dormice reduce their T_b to save energy during periods of limited food supply.

J. Fietz (✉)
Animal Husbandry and Animal Breeding, University of Hohenheim,
Garbenstr. 17, 70599 Stuttgart, Germany
e-mail: Joanna.Fietz@uni-hohenheim.de

J. Fietz · J. Schmid
Department of Experimental Ecology, University of Ulm,
Albert-Einstein-Allee 11, 89069 Ulm, Germany

J. R. Speakman
Institute of Biological and Environmental Sciences,
University of Aberdeen, AB24 2TZ Aberdeen, UK

T. Ruf et al. (eds.), *Living in a Seasonal World*,
DOI: 10.1007/978-3-642-28678-0_43,

Abbreviations

APE	Atom percent excess
DEE	Daily energy expenditure
DLW	Doubly labelled water
HE	Hohenentringen
HL	Hagelloch
RMR	Resting metabolic rate
T_a	Ambient temperature
T_b	Body temperature
T_{nest}	Nest box temperature
TNZ	Thermoneutral zone

43.1 Introduction

Many mammals show marked seasonal fluctuations in their metabolic rate (MR) and body temperature (T_b) as a response to climatic changes and food availability (e.g., Nagy and Gruchacz 1994; Humphries et al. 2005; Arnold et al. 2006; Turbill et al. 2011). A substantial part of these annual changes in energy consumption and T_b can be attributed to different levels of locomotor activity, variations in food intake and the heat increment of feeding (e.g., Arnold et al. 2006), as MR is known to be affected substantially by food ingestion, particularly under protein-rich diet (Heldmaier 2004).

However, MR and T_b not only vary in response to an increase of food intake and activity, they may also be actively down-regulated as an endogenously controlled physiological adjustment to save energy during periods of food scarcity. Heterothermy represents the most extreme physiological energy-saving mechanism and MR is reduced to a small fraction of euthermic MR (Heldmaier and Ruf 1992; Geiser and Ruf 1995). Besides these extreme reductions of energy turnover rates, non-hibernating small mammals are known to compensate restricted energy intake by reducing their RMR and T_b (Hambly and Speakman 2005; Speakman and Mitchell 2011) and as well large endotherms like the red deer (*Cervus elaphus*), the Przewalski horse (*Equus ferus przewalskii*), sheep (*Ovis aries*) and king penguins (*Aptenodytes patagonicus*) use hypometabolism associated with peripheral cooling as an overwintering strategy (Blaxter 1962; Arnold et al. 2004, 2006; Eichhorn et al. 2011; Turbill et al. 2011). Accordingly, red deer show a 30% reduction of the MR in winter that is not explained by seasonal differences in food intake but rather represents an endogenously controlled physiological adjustment to seasonal food scarcity (Turbill et al. 2011).

Another well-known energy saving strategy in mammals and birds represents the reduction of organ size during phases of low or absent use (Hume and Biebach 1996; Piersma and Gill 1998; Hume et al. 2002; Piersma and Drent 2003). One of the energetically most expensive organs to maintain is the gastrointestinal tract (Stevens and Hume 1995) and its contribution to total oxygen consumption is estimated to range from 12% in rats up to 25% in pigs (McBride and Kelly 1990). Alpine marmots (*Marmota marmota*) are obligate hibernators and their gastrointestinal tract was shown to be reduced in tissue mass and length before the onset of hibernation and reactivated as a response to ingested food afterwards (Hume et al. 2002). Thus, immediately after emergence from hibernation in spring, activity, tissue mass and consequently energetic requirements of the gastrointestinal tract are minimal but increase substantially in response to ingested food until midsummer (Hume et al. 2002).

The edible dormouse (*G. glis*) is an obligate fat-storing hibernator and in central Europe it occurs preferentially in deciduous mixed forests dominated by European beech (*Fagus sylvatica*; Schlund et al. 1997; Schlund 2005). They are arboreal and strongly specialised on beechnuts, which they predominantly feed on during offspring raising and pre-hibernal fattening. Before beechnuts become ripe in August, they also consume leaves, buds and fruits (Storch 1978; Fietz et al. 2005). In Germany, adult edible dormice hibernate for 7–8 months from September until June and have only one litter per year. Mating takes place at the end of June and juveniles are born after 30 days of gestation. However, in years without beechnut production dormice skip gonadal growth and reproduction (Bieber and Ruf 2004; Fietz et al. 2009). After emergence from hibernation, especially reproductively active male dormice show a considerable and significant decline in their body mass, which has been attributed to high energetic costs of reproduction when food supply is still limited (Bieber 1998; Fietz et al. 2004, 2009; Sailer and Fietz 2009). Previous studies have revealed that male edible dormice incur high thermoregulatory costs because high levels of testosterone block the ability to enter torpor (Jallageas and Assenmacher 1983; Fietz et al. 2004), which might explain part of the lost body mass. However, as soon as acorns and beechnuts become ripe in August body mass increases enormously and may double by the onset of hibernation in September (Bieber 1998; Fietz et al. 2009; Sailer and Fietz 2009).

In this field study we aimed to investigate seasonal changes of energy turnover and T_b in edible dormice during periods of high and low food availability, to understand energy-saving strategies besides the use of hibernation and torpor.

43.2 Materials and Methods

43.2.1 Study Areas

Field data were collected between 1996 and 2006 in two study areas located on the southern rim of the "Schönbuch" Nature Reserve in southwest Germany (Baden-Württemberg: 48°33′N, 9°00′E). Both are mixed deciduous forests that are located 1 km apart. Hohenentringen (HE) comprised an area of 12 ha with 128 nest boxes, while Hagelloch (HL) had an area of 8 ha with 80 nest boxes (for details see Schlund et al. 1993). Nest boxes were installed at the intersections of a grid system.

43.2.2 Individual Marking and Morphological Measurements

We performed regular checks of nest boxes from June to October during daytime in both study areas. Upon first capture, we marked individuals by subdermal injection of transponders (Trovan, EURO I.D. Usling, Weilerswist, Germany) that carried unique identification numbers. We recorded sex and assessed body mass for each capture to the nearest gram using a 300 g spring balance (Pesola, Baar, Switzerland; division: 2 g, accuracy: 99.7%). At the end of the procedure, animals were returned to their nest boxes. By definition, individuals that had overwintered at least 2 years were classified as adults and as yearlings if they had overwintered only once. Adults and yearlings were discriminated by their body size and the colour of their fur (Schlund 1997). Males were regarded as reproductive if they had visible testes, and females if they were known to have been gestating or lactating during the respective year.

43.2.3 Oxygen Consumption

Oxygen consumption of 20 males was measured with two portable gas analysers (Oxbox; designed and constructed by Thomas Ruf, Research Institute of Wildlife Ecology, Vienna; for further information see Schmid et al. 2000) during 2006, a year of high seed production (Kager and Fietz 2009). Measurements were conducted during the daytime for at least 6 h between June and September 2006, using the nest box (3SV; Schwegler; Schorndorf; volume 1.5 l) as a metabolic chamber. During the experiments air flow through the chamber was controlled at about 100 l h^{-1}.

We defined the lowest measured 10 min moving averages within the species-specific thermoneutral zone (TNZ: 22–29°C Heldmaier and Elvert 2004) of *G. glis* as resting metabolic rate (RMR in ml O_2 $*h^{-1}$). Dormice in our study were removed from the nest box, weighed with a spring balance and returned into the nest box immediately (5–10 min) before the measurement started.

Oxygen consumption was measured only in euthermic individuals because we wanted to investigate seasonal effects on energy consumption besides the phenomenon of hibernation and torpor. Individuals were regarded as euthermic, if they did not feel cold to the touch and showed normal locomotor activity and behaviour, when taken out of the nest box.

43.2.4 Doubly Labelled Water Measurements

We measured the DEE (kJ day^{-1}) using the doubly labelled water (DLW) technique (Lifson and McClintock 1966; Butler et al. 2004) during 2001–2003 in HE. 2001 and 2003 were years with good seed production of the beech, while seed production in 2002 was comparatively low (Kager and Fietz 2009). Dormice were sexed, weighed and injected intraperitoneally with 0.3–0.4 ml water containing $H_2{}^{18}O$ (9.2 Atom percent excess: APE) and D_2O (4.6 APE). After injection dormice were held for 1 h in small cages to allow complete equilibration of the isotopes (Speakman and Krol 2005). Animals were then anaesthetized with 0.2–0.3 Ketavet (100 mg ml^{-1}) and an initial blood sample of approximately 50 μl was taken by puncture of the lower lip. Dormice were then released into their nest box of capture. Upon recapture dormice were weighed and a second blood sample was taken. After recovering from anaesthesia, animals were released at their capture site. Blood samples were flame-sealed into calibrated 50 μl capillaries (Vitrex, Camlab, Cambridge, UK). To determine background isotope levels, we took blood samples from five dormice that were not labelled (method C: Speakman and Racey 1988).

Capillaries that contained the blood samples were vacuum-distilled using the pipette method of Nagy (1983), and water from the resulting distillate was used to produce CO_2 and H_2 [methods in Speakman et al. (1990) for CO_2 and Speakman and Krol (2005) for H_2]. The analyses of the isotope ratios $^{18}O/^{16}O$ and $^2H/^1H$ were performed using a gas source isotope ratio mass spectrometry (Optima, Micromass IRMS for H_2 and Isochrom μG, Manchester, UK for CO_2). Isotope enrichments were converted to values of CO_2 production using the single pool model equation (Eq. 7.17 in Speakman 1997) as recommended by Visser and Schekkerman (1999). Conversion to energy expenditure was made by means of an assumed RQ of 0.8. All calculations were made using the Natureware DLW software (Speakman and Lemen 1999).

A total of eight dormice during early summer and one animal during late summer, respectively, were injected with $D_2{}^{18}O$ and successfully recaptured for blood sampling. For each individual body mass obtained at initial and final capture was averaged.

43.2.5 T_b Measurement

In both study areas T_b was measured during a high food year in 1999 and a year of low food availability in 1996 (Kager and Fietz 2009). T_b was measured orally with a digital thermocouple to the nearest 0.1°C directly after taking the individual out of the nest box. Thermocouples were inserted 1–1.5 cm into the mouth of the dormouse until constant values were reached. Corresponding ambient temperature (T_a) within the nest box (T_{nest}) was measured by inserting the thermocouple into the nest box, before opening it, until constant values were reached. As we wanted to investigate T_b values of euthermic individuals, we restricted our analyses to T_b values above 30°C collected between 06:00 and 14:00 hours because edible dormice generally start their torpor bouts in the early morning (02:00 hours) and terminate them in the afternoon (15:00 hours). Thus it is unlikely that if we encountered an individual with a T_b above 30°C within that time frame that it had previously been torpid. Additionally we analysed only T_b data of yearling and adult males to avoid effects of gestation and lactation on T_b.

43.2.6 Statistics

The effect of season on RMR was investigated with a linear mixed effect model (type III; SPSS 2004) with RMR as the dependent variable, "season" as factor and "body mass" as a covariate. As RMR of 5 out of 20 individuals was measured in both seasons, we entered "individual" as random effect into the model.

The effect of "season" and "year" on T_b was also investigated with linear mixed effect models (type III) with T_b as the dependent variable "study area" and either "year" or "month" as factors. We further entered time of day (in hours) as a factor into the models, to correct for circadian variations in T_b known from former studies (Fietz et al. 2004). To test for effects of "T_{nest}" on "T_b" we additionally entered "T_{nest}" as a covariate into the model. However, in cases were T_{nest} and time were correlated either of them was entered alternatively into the model. As T_b of several individuals was measured several times, we included "individual" as random effect into the model.

43.3 Results

43.3.1 RMR in Free-Living Dormice

During a high food year, RMR of edible dormice showed extreme seasonal variations (Fig. 43.1; Table 43.1) with comparatively low values measured in early summer (early summer; 92.4 ± 22.2 ml O_2 $*h^{-1}$; $n = 12$) and more than 68 percent higher rates in late summer (late summer; 155.8 ± 45.6 ml O_2 $*h^{-1}$; $n = 13$). A significant relationship was observed between body mass and RMR during early summer [linear regression: RMR = 43.15 + 0.51*body mass (g); $r^2 = 0.35$, $n = 12$; $p = 0.045$]. However, during late summer body mass did not explain a significant amount of variance in RMR ($r^2 = 0.26$, $n = 13$; $p = 0.07$). Furthermore, RMR measured during early summer significantly increased with date (Pearson's correlation: $r_s = 0.66$; $p = 0.019$; $n = 12$; Fig. 43.1).

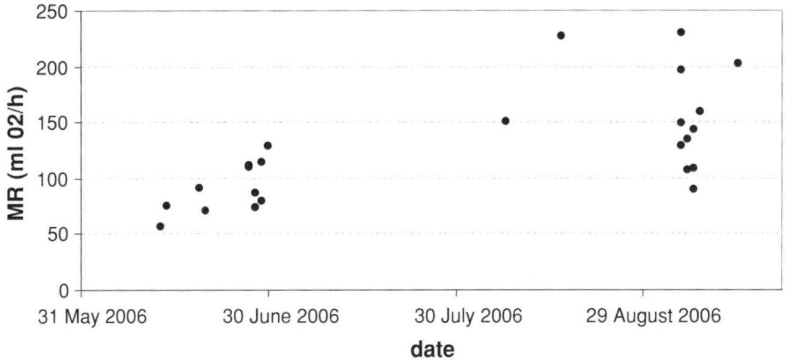

Fig. 43.1 RMR (ml O_2 $*h^{-1}$) of male edible dormice measured during their activity season in 2006. Note that oxygen consumption was measured in 20 individuals, with five individuals measured in both seasons

Table 43.1 Tests and estimates of fixed effects of the linear mixed effect model

Parameter	Num df	Den df	*F*	Significance	Estimate	Standard error	df	*T*	*p*
Intercept	1	14	7.14	0.018	86.5	25.90	17	3.33	–
Season	1	18	16.00	0.001	–	–	–	–	–
Early summer	–	–	–	–	−51.7	12.83	18	−4.03	0.001
Late summer	–	–	–	–	0[a]		–	–	–
Body mass	1	12	8.11	0.014	0.6	0.21	13	2.85	–

Formula: RMR ~ body mass+season+(1 | individual); Number of observations: 25, individuals: 20; AIC = 230

[a] This parameter is set to zero because it is redundant

43.3.2 Daily Energy Expenditure and Water Flux Rate in Free-Living Dormice

The time between the initial and final blood sample averaged 2.6 days (range: 1.7–5.9 , $n = 9$). DEEs were highly variable between dormice measured at different time periods of the year (Table 43.2). The median DEE of dormice measured in early summer was 83.3 kJ day^{-1} (range: 55.8–202.3; $n = 8$) about 372% lower than the value of 393.41 kJ day^{-1} measured in the single dormouse during late summer. The median DEE in early summer was 1.9 times the RMR (range: 1.4–4.2; $n = 8$) measured on dormice in the present study (Table 43.2).

Water flux rates averaged 38.05 ml day^{-1} (±11.14 ml day^{-1}, $n = 8$) in dormice obtained during early summer, which was very similar to the single value measured in the dormice during late summer (38.09 ml day^{-1}, Table 43.2).

Table 43.2 Summary of DEE and water turnover in *G. glis* in the deciduous mixed forest HE in southwest Germany during 2001–2003

Animal ID	Mean mass (g)	Reproductive status	Year	Daily energy expenditure		Water flux rate (ml day^{-1})
				(KJ day^{-1})	Relative to RMR	
Data of dormice obtained in early summer						
m-B71A	79.0	0	2001	55.77	1.4	38.53
f-2505	111.0	1	2003	83.81	1.8	39.38
f-DB32	101.0	1	2003	99.72	2.2	12.40
f-B053	80.0	1	2003	73.22	1.8	38.89
f-B3E0	81.0	1	2003	82.80	2.0	36.50
m-1364	115.5	1	2003	93.29	1.9	47.51
m-AD74	110.5	1	2003	202.33	4.2	46.87
m-BC4F	101.5	1	2003	73.82	1.6	44.31
Data of dormice obtained in late summer						
m-9229	90.0	0	2002	393.41	–	38.09

m, male; f, female; 0, reproductively inactive; 1, reproductively active, for further details see Sect. 43.2

RMR during early summer was calculated using following regression for dormice (measured in this study): RMR = 43.154 + (0.511 * body mass (g)). These resting values were converted into units of energy equivalence (kJ day^{-1}) assuming the factor 20.083 kJ l^{-1} O_2 (Schmidt-Nielsen 1997)

43.3.3 T_b Patterns of Free-Living Dormice

After emergence form hibernation in June, T_b values did not differ between the two study years (Fig. 43.2; ANOVA; $p > 0.1$; T_b mean = 32.7 ± 1.2°C; $n = 22$). T_b values were also not correlated with "time", "T_{nest}" and "study area" ($p > 0.1$).

During a high food year, T_b values significantly increased between June and July (Table 43.3) and were significantly higher during July and August (T_b mean = 35.1 ± 1.4°C; $n = 116$) than within the same period of time during a low food year (T_b mean = 32.4 ± 1.4°C; $n = 49$; Table 43.4). "Time" of the day, "T_{nest}" and "study area" did not show significant correlations with T_b.

During the low food year, T_b values were low throughout the whole year with only small seasonal but nevertheless significant changes (Table 43.5). Here, lowest values were measured before the onset of hibernation. "Study area", "T_{nest}" or "time" did not show significant correlations with T_b.

43.4 Discussion

Results of our study revealed that MR and T_b of edible dormice measured in the field showed distinct seasonal variations. After emergence from hibernation edible dormice had their lowest RMRs that subsequently increased by nearly 70% until the end of June. This increase could possibly be a response to food intake after hibernation was terminated and further be explained by the reactivation of

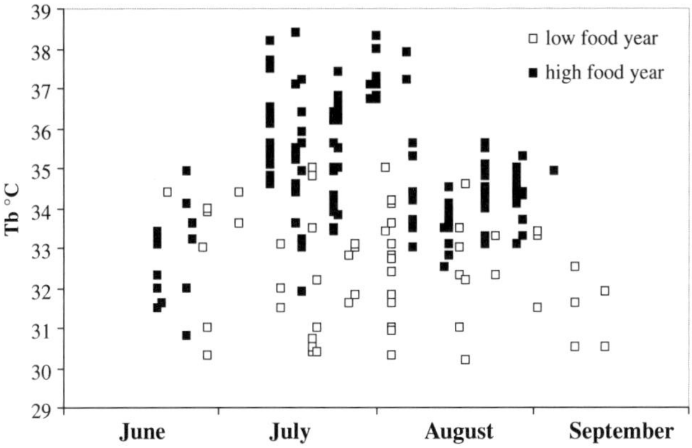

Fig. 43.2 Seasonal changes of T_b in male edible dormice measured during a high food and a low food year

Table 43.3 Tests and estimates of fixed effects of the linear mixed effect model for T_b values measured during 1999

Parameter	Num df	Den df	F	Significance	Estimate	Standard error	df	T	p
Intercept	1	48	4,798	<0.001	29.3	0.5	40	59	–
Month	2	38	71	<0.001	–	–	–	–	–
June	–	–	–	–	−1.15	0.17	20	−6.6	<0.001
July	–	–	–	–	1.6	0.18	64	8.8	<0.001
August	–	–	–	–	0[a]	–	–	–	–
Study area	1	30	14	0.001	–	–	–	–	–
HE	–	–	–	–	0.5	0.14	30	3.7	0.001
HL	–	–	–	–	0[a]	–	–	–	–
T_{nest}	1	57	86	<0.001	–	–	–	–	–
Time	6	19	13	<0.001	–	–	–	–	–

Formula: $T_b \sim$ month + time + Study area + T_{nest} + (1 | individual); Number of observations: 107, individuals: 61, AIC = 230. Note that we had only one data point for September, which was omitted from the analysis

[a] This parameter is set to zero because it is redundant

energetically costly organs, like the gut, that were inactive and atrophied during hibernation (Hume et al. 2002). While RMR values measured during June were within the range of RMR values measured in captive dormice (Heldmaier and Elvert 2004) and also of values expected for a similar-sized rodent (Hayssen and Lacy 1985), RMRs measured in late summer were markedly higher. A substantial part of this increase could possibly be explained by seasonal variations in food intake and the protein and energy content in the natural forage of edible dormice (Fietz et al. 2005; Sailer and Fietz 2009). MR is known to be substantially affected by food ingestion, particularly under protein-rich diet (Arnold et al. 2004).

Table 43.4 Tests and estimates of fixed effects of the linear mixed effect model

Parameter	Num df	Den df	*F*	Significance	Estimate	Standard error	df	*T*	*p*
Intercept	1	3.7	4,202	<0.001	29.6	0.45	4.3	66	–
T_{nest}	1	3	135	0.002	0.26	0.02	2.9	11.6	–
Study year	1	46	120	<0.001	–	–	–	–	–
1996	–	–	–	–	−1.93	0.18	46	−11.0	<0.001
1999	–	–	–	–	0[a]	–	–	–	–

T_b values were measured between July and August 1996 and 1999. Formula: $T_b \sim$ study year + T_{nest} + (1 | individual); Number of observations: 141, individuals: 93; AIC = 506
[a] This parameter is set to zero because it is redundant

Table 43.5 Tests and estimates of fixed effects of the linear mixed effect model for T_b values measured during 1996

Parameter	Num df	Den df	*F*	Significance	Estimate	Standard error	df	*T*	*p*
Intercept	1	12	79,974	<0.001	31.24	0.35	4.9	88.9	–
Month	3	11	12	0.001	–	–	–	–	–
June	–	–	–	–	1.76	0.35	4.9	5.0	0.004
July	–	–	–	–	1.08	0.40	8.7	2.6	0.028
August	–	–	–	–	1.42	0.40	8.2	3.5	0.007
September	–	–	–	–	0[a]	–	–	–	–

Formula: $T_b \sim$ month + (1 | individual); Number of observations: 61, individuals: 48; AIC = 224
[a] This parameter is set to zero because it is redundant

In accordance to seasonal variations observed in RMR, DEEs measured in early summer were markedly lower than that measured in late summer. We cannot completely exclude the possibility that low DEE values were caused by torpid individuals. However, we did not detect any torpid individuals during the nest box controls in 2003 and the metabolic scope (DEE/RMR) at 1.9 is not indicative of torpor use. Furthermore, most individuals in which we measured DEE during early summer were reproductively active and from former studies it is known that reproductively active edible dormice do not generally enter torpor (males: Jallageas and Assenmacher 1983; Fietz et al. 2004; females: Fietz personal communication). Taken this information together with the fact that DEE values measured in early summer were within the range of expected values for a mammal of the same body size (Nagy 2005), it seems extremely unlikely that comparatively low DEE values in early summer are caused by the use of torpor, but rather represent DEE of euthermic dormice.

In late summer the measured DEE was extremely and unexpectedly high. Although sample size of one is not really representative, a possible explanation for the high DEE value measured in late summer could be that locomotor activity of dormice was markedly higher in late summer than in early summer. During late summer 2002 food availability was comparatively low, with only 11–30% of the beeches producing seeds. Edible dormice are obligate fat-storing hibernators that have to accumulate sufficient body fat before the onset of hibernation, even if food availability is low. Thus it is possible that dormice were forced to a high level of locomotor activity in search for food during late summer 2002.

Mammals are known to lower their endogenous heat production together with the set-point of their T_bs as an energy-saving strategy. Accordingly, a decrease of T_b is a known response to food deprivation in small mammals like rodents (Severinsen and Munch 1999; Hambly and Speakman 2005; Speakman and Mitchell 2011). In captivity euthermic edible dormice had a mean core T_b of 35.6 ± 0.68°C (Wilz and Heldmaier 2000). This value corresponds well to our T_b measurements taken during high food availability in 1999. However, T_b measured during periods of low food availability in early summer, and the year of mast failure in 1996, were considerably lower. This strongly suggests that edible dormice reduce the set-point of their T_b to save energy during periods of food restriction. Unfortunately, we do not have information about corresponding energy consumption during a low food year.

43.5 Conclusions

In this study we demonstrated that energy turnover and T_b in edible dormice varied markedly over their active season and among years. Variations of physical activity, food intake and the heat increment of feeding are likely to cause a substantial part of these variations together with the reactivation of organs that are atrophied during hibernation. Our study suggests that edible dormice reduce their T_b to save energy during periods of limited food supply.

Acknowledgments W. Schlund, F. Scharfe, M. Regelmann, T. Kager and S. Schauer helped in many ways with this field project. Financial support provided by the Margarete von Wrangell Programme, the German Research Foundation (DFG, FI 831/3-1; 831/3-2), and the German Wildlife Foundation, all to JF, made this study possible. Our experiments were conducted under licence from the nature conservancy and the animal experiment department of the Regierungspräsidium Tübingen. Paula Redman and Peter Thomson provided technical assistance for analysis of the isotopes for DLW measurements.

References

Arnold W, Ruf T, Reimoser S, Tataruch F, Onderscheka K, Schober F (2004) Nocturnal hypometabolism as an overwintering strategy of red deer (*Cervus elaphus*). Am J Physiol-Reg I 286:R174–R181

Arnold W, Ruf T, Kuntz R (2006) Seasonal adjustment of energy budget in a large wild mammal, the Przewalski horse (*Equus ferus przewalskii*) II. Energy expenditure. J Exp Biol 209: 4566–4573

Bieber C (1998) Population dynamics, sexual activity, and reproduction failure in the fat dormouse (*Myoxus glis*). J Zool Lond 244(2):223–229

Bieber C, Ruf T (2004) Seasonal timing of hibernation and reproduction in the edible dormouse (*Glis glis*). In: Barnes B, Carey H (eds) Life in the cold: evolution, mechanisms, adaptation, and application. Twelfth hibernation symposium. Fairbanks, Alaska, pp 113–125

Blaxter K (1962) The fasting metabolism of adult wether sheep. Brit J Nutr 16:615–626

Butler PJ, Green JA, Boyd IL, Speakman JR (2004) Measuring metabolic rate in the field: the pros and cons of the doubly labelled water and heart rate methods. Funct Ecol 18(2):168–183

Eichhorn G, Groscolas R, Le Glaunec G, Parisel C, Arnold L, Medina P, Handrich Y (2011) Heterothermy in growing king penguins. Nature 2:435

Fietz J, Pflug M, Schlund W, Tataruch F (2005) Influences of the feeding ecology on body mass and possible implications for reproduction in the edible dormouse (*Glis glis*). J Comp Physiol B 175(1):45–55

Fietz J, Kager T, Schauer S (2009) Is energy supply the trigger for reproductive activity in male edible dormice (*Glis glis*)? J Comp Physiol B 179(7):829–837

Fietz J, Schlund W, Dausmann KH, Regelmann M, Heldmaier G (2004) Energetic constraints on sexual activity in the male edible dormouse (*Glis glis*). Oecologia 138(2):202–209

Geiser F, Ruf T (1995) Hibernation versus daily torpor in mammals and birds: physiological variables and classification of torpor patterns. Physiol Zool 68:935–966

Hambly C, Speakman JR (2005) Contribution of different mechanisms to compensation for energy restriction in the mouse. Obes Res 13(9):1548–1557

Hayssen V, Lacy RC (1985) Basal metabolic rates in mammals: taxonomic differences in the allometry of BMR and body mass. Comp Biochem Physiol A 81(4):741–754

Heldmaier G, Elvert R (2004) How to enter torpor: thermodynamic and physiological mechanisms of metabolic depression. In: Life in the cold: evolution, mechanisms, adaptation, and application, 12th international hibernation symposium, vol 27, pp 185–198

Heldmaier G (2004) Energiehaushalt von Tieren. In: Heldmaier G, Neuweiler G (eds) Vergleichende Tierphysiologie, vol 2. Springer, Berlin Heidelberg, pp 53–89

Heldmaier G, Ruf T (1992) Body temperature and metabolic rate during natural hypothermia in endotherms. J Comp Physiol B 162(8):696–706

Hume I, Biebach H (1996) Digestive tract function in the long-distance migratory garden warbler, *Sylvia borin*. J Comp Physiol B 166(6):388–395

Hume ID, Beiglböck C, Ruf T, Frey-Roos F, Bruns U, Arnold W (2002) Seasonal changes in morphology and function of the gastrointestinal tract of free-living alpine marmots (*Marmota marmota*). J Comp Physiol B 172:197–207

Humphries MM, Boutin S, Thomas DW, Ryan JD, Selman C, McAdam AG, Berteaux D, Speakman JR (2005) Expenditure freeze: the metabolic response of small mammals to cold environments. Ecol Lett 8(12):1326–1333

Jallageas M, Assenmacher I (1983) Annual plasma testosterone and thyroxine cycles in relation to hibernation in the edible dormouse *Glis glis*. Gen Comp End 50(3):452–462

Kager T, Fietz J (2009) Food availability in spring influences reproductive output in the seed-preying edible dormouse (*Glis glis*). Can J Zool 87(7):555–565

Lifson N, McClintock R (1966) Theory of use of the turnover rates of body water for measuring energy and material balance. J Theor Biol 12:46–74

McBride BW, Kelly JM (1990) Energy cost of absorption and metabolism in the ruminant gastrointestinal tract and liver: a review. J Anim Sci 68:2997–3010

Nagy KA (1983) Doubly-labelled water: a guide to its use. UCLA publications, Los Angeles

Nagy KA (2005) Field metabolic rate and body size. J Exp Biol 208(9):1621–1625. doi:10.1242/jeb.01553

Nagy KA, Gruchacz MJ (1994) Seasonal water an energy metabolism of the desert dwelling kangaroo rat (*Dipodomys merriami*). Physiol Zool 67(6):1461–1478

Piersma T, Drent J (2003) Phenotypic flexibility and the evolution of organismal design. Trendy Ecol Evol 18(5):228–233

Piersma T, Gill RE Jr (1998) Guts don't fly: small digestive organs in obese bar-tailed godwits. Auk 115(1):196–203

Sailer M, Fietz J (2009) Seasonal differences in the feeding ecology and behaviour of male edible dormice (*Glis glis*). Mamm Biol 74:114–124

Schlund W (1997) Length of tibia as a measure of body size and determining the age of edible dormice (*Myoxus glis* L.). Z Säugetierkd 62(3):187–190

Schlund W (2005) Siebenschläfer *Glis glis* (Linnaeus, 1766). In: Braun M, Dieterlen F (eds) Die Säugetiere Baden-Württembergs, vol 2. Ulmer GmbH & Co, Stuttgart, pp 199–210

Schlund W, Stauss MJ, Burkhardt JF (1993) Fat dormice in nest boxes: a long-term study on habitat choice. Carolinea 51:93–100

Schlund W, Scharfe F, Stauss MJ, Burkhardt JF (1997) Habitat fidelity and habitat utilization of an arboreal mammal (*Myoxus glis*) in two different forests. Z Säugetierkd 62(3):158–171

Schmid J, Ruf T, Heldmaier G (2000) Metabolism and temperature regulation during daily torpor in the smallest primate, the pygmy mouse lemur (*Microcebus myoxinus*) in Madagascar. J Comp Physiol B 170 (1):59–68

Schmidt-Nielsen K (1997) Animal physiology: adaptation and environment. Cambridge University Press, Cambridge

Severinsen T, Munch IC (1999) Body core temperature during food restriction in rats. Acta Physiol Scand 165:299–305

Speakman JR (1997) Doubly labelled water: theory and practice. Chapman and Hall, London

Speakman JR, Lemen C (1999) Doubly labelled water analyses program. http://www.natureware.com/double.htm

Speakman J, Krol E (2005) Limits to sustained energy intake IX: a review of hypotheses. J Comp Physiol B 175(6):375–394

Speakman JR, Mitchell SE (2011) Caloric restriction. Mol Aspects Med 32(3):159–221

Speakman JR, Racey PA (1988) Consequences of non steady-state CO_2 production for accuracy of the doubly labelled water technique—the importance of recapture interval. Comp Biochem Physiol A 90:337–340

Speakman JR, Nagy KA, Masman D, Mook WG, Poppitt SD, Strathearn GE, Racey PA (1990) Interlaboratory comparison of different analytical techniques for the determination of O-18 abundance. Anal Chem 62:703–708

SPSS (2004) SPSS base 13.0 for Windows user's guide. SPSS Inc., Chicago

Stevens CE, Hume ID (1995) Comparative physiology of the vertebrate digestive system. Cambridge University Press, Cambridge

Storch G (1978) Familie Gliridae Thomas, 1897—Schläfer. In: Niethammer J, Krapp F (eds) Handbuch der Säugetiere Europas. Band 1. Nagetiere. Akademische Verlagsgesellschaft, Wiesbaden, pp 201–280

Turbill C, Ruf T, Mang T, Arnold W (2011) Regulation of heart rate and rumen temperature in red deer: effects of season and food intake. J Exp Biol 214:963–970

Visser GH, Schekkerman H (1999) Validation of the doubly labeled water method in growing precocial birds: the importance of assumptions concerning evaporative water loss. Physiol Biochem Zool 72:740–750

Wilz M, Heldmaier G (2000) Comparison of hibernation, estivation and daily torpor in the edible dormouse, *Glis glis*. J Comp Physiol B 170(7):511–521

Chapter 44
The Effect of Ambient Temperature on Body Mass, Torpor, Food Intake, and Leptin Levels: Implications on the Regulation of Food Intake in Mammalian Hibernators

Gregory L. Florant, Melanie M. Richter and Susan K. Fried

Abstract The effect of ambient temperature (T_a) on the body mass, food intake, and leptin concentrations in the Golden-Mantled Ground Squirrels (*Callospermophilus lateralis*) (GMGS) was investigated. We hypothesized GMGS would begin feeding in winter due to declining plasma leptin levels, if forced to undergo torpor bouts at high T_a (22°C). Our rationale was that GMGS would have an elevated metabolism at a high T_a, and therefore would utilize their fat stores faster causing a decrease in serum leptin. The torpor bout results demonstrated that animals kept at high T_a spent significantly ($P < 0.01$) less time at low body temperature compared to cold room animals (52 vs. 142 h). Furthermore, the body mass of warm room GMGS was lower compared to cold room (5°C) animals during the winter months, suggesting a higher metabolism. Warm room animals that had stopped eating, started eating again when body mass declined to near their trapped body mass. Leptin levels were significantly ($P < 0.05$) lower in warm room animals compared to cold room animals (4.5 vs. 7.3 ng/ml), when they commenced eating again. We conclude that as fat mass decreased due to the higher metabolism in the warm room, this decreased leptin levels, which in turn stimulated food intake in warm room animals. The cold room animals had food at the same time as the warm room animals but did not feed and their leptin concentrations remained higher. We conclude that leptin is an important part of the feedback signal involved in the cessation of food intake during hibernation.

G. L. Florant (✉)
Department of Biology, Colorado State University, Fort Collins, CO 80523, USA
e-mail: Florant@Colostate.edu

M. M. Richter
Department of Biology, University of Alaska, Anchorage, AK 99508, USA

S. K. Fried
Department of Medicine, Boston University School of Medicine, Boston, MA 02118, USA

T. Ruf et al. (eds.), *Living in a Seasonal World*,
DOI: 10.1007/978-3-642-28678-0_44,

Keywords Sliding set-point · Body fat · Temperature regulation · Leptin

44.1 Introduction

The regulation of food intake and body mass in mammals that hibernate (hibernators) has received considerable study over the past several decades (for a review, Dark 2005). Many hibernators increase their food intake throughout the summer and early autumn in preparation for winter, when food will not be available. For example, the golden-mantled ground squirrel (*Callospermophilus lateralis*) will nearly double its body mass by the end of summer and its food intake will decrease to virtually zero by mid autumn (Pengelley 1968; Mrosovsky and Faust 1985; Heller and Poulson 1970). The increase in body mass is primarily due to the dramatic increase in body fat deposition. Throughout winter, animals voluntarily cease to feed relying on their fat depots to meet endogenous energy demands, and remarkably little is known about the regulation of this dramatic change in food intake.

Ambient temperature (T_a) is known to alter the yearly rhythm of food intake and body mass in GMGS and can change characteristics of the torpor bouts; at higher T_as torpor bouts are shorter (Pengelley and Fisher 1963; Mrosovsky 1980; Geiser and Kenagy 1988; Russell et al. 2010). The T_a in which GMGS hibernate plays a role in determining their metabolic rate and consequently, how fast they will lose body mass, particularly endogenous fat. At a high T_a, their metabolism will be higher and thus their rate of mass loss will be faster. Previous studies of the arctic ground squirrel (*Urocitellus. parryi*) have shown that there is an optimum T_a in which these animals undergo torpor: optimum is defined as the minimum metabolic rate possible. At this T_a, body mass lost was slower over time and arctic ground squirrels used the least amount of energy (Buck and Barnes 2000).

In hibernators, food intake and body mass increase concurrently throughout the summer and are thought to be under the control of the hormone, leptin, which has been shown to regulate food intake and total body energy expenditure in mammals (for a review, Ahima and Flier 2000; Schwartz et al. 2000; Leibel 2002). In several species that hibernate, it has been shown that leptin levels increase with fat mass (Dark 2005) and the injection of mouse recombinant leptin into pre-hibernatory arctic ground squirrels leads to a reduction in food intake and body mass (Ormseth et al. 1996; Boyer et al. 1997). However, studies in the little brown bat (Myotis lucifugus) suggest that high leptin levels may be incompatible with torpor (Kronfeld-Schor et al. 2000). In non-photoperiodic and photoperiodic hibernators, serum leptin levels rise with increasing adiposity in preparation for winter hibernation (Klingenspor et al. 2000; Concannon et al. 2001; Florant et al. 2004).

In this chapter, we report torpor bout length, food intake, body mass, and serum leptin levels in GMGS maintained at a high (22°C) and a low (5°C) T_a. We hypothesized that GMGS maintained at a high T_a would lose fat mass faster, and therefore leptin levels would drop sooner during the winter. This decrease in serum leptin would initiate food intake in these animals much earlier than in animals maintained in the cold T_a. Our results strongly suggest that leptin levels do drop with decreasing fat mass and we found that animals at the high T_a ate food sooner than animals kept in the cold T_a. These results support the hypothesis that circulating leptin levels may be a part of the food intake regulatory system in hibernators.

44.2 Materials and Methods

44.2.1 Animals

A total of 18 adult GMGS of both genders were collected in Larimer County and Summit County, Colorado, during the summer of 2005. Animals were transported to Colorado State University and maintained under an approved animal care and use protocol. From June to September the animals were maintained on a natural light–dark cycle at a T_a of 22°C. During the months of October, 2005 to end of March, 2006, they were assigned to either the cold room ($T_a = 5$°C) or warm room ($T_a = 22$°C). In both rooms, animals were maintained in constant darkness and under a red light (25 W). Both groups were provided with food (Teklad 8650) ad libitum until food intake ceased in October and then food was only available as per the experiment. Water was provided at all times throughout the experiment. Body mass (grams) was measured weekly in the summer, and at specific dates during the winter hibernation period. The average food intake for all animals from June/July to September 14th, 2005 was roughly 26 g per day. As animals prepared for hibernation and food intake declined, food intake was measured on average about every 5 days from September until April, 2006. Food intakes below 1 g per day were considered zero due to spillage (one brick of food weighed more than 3 g). Four animals from each experimental T_a were sacrificed in February to obtain heart and brain tissues for future analysis. Thus, the sample sizes for the cold and warm room decreased from nine in both groups to four in each group after February 4, 2006. However, one animal from each group died due to natural causes during the experimental period.

In October, at time point A (10/17/05; Figs. 44.1, 44.2 and 44.3), we anesthetized animals with a cocktail containing ketamine (150 mg/kg), acepromazine (1.0 mg/ml), and xylazine (5 mg/kg) and a sterile temperature sensitive data logger (I Button (DS1921-F5), Maxem Dallas Semiconductor Inc.TX) was implanted into the abdominal cavity to record body temperature (T_b) every 2 h. At the same time, we collected blood samples and measured body mass. All animals were provided with food following surgery, and allowed to recover for a week prior to being put into different T_as. In December, at time point B (12/09/05; Figs. 44.1, 44.2 and 44.3), we anesthetized animals as described above and collected blood and measured body mass. Due to the declining body mass of warm room animals, we gave a measured amount of food (50 g) to each animal in both groups. In January, at time point C (1/05/06; Figs. 44.1, 44.2 and 44.3), we recorded the amount of food eaten over the time between point B and point C. In addition, serum was collected and body mass was measured at point C.

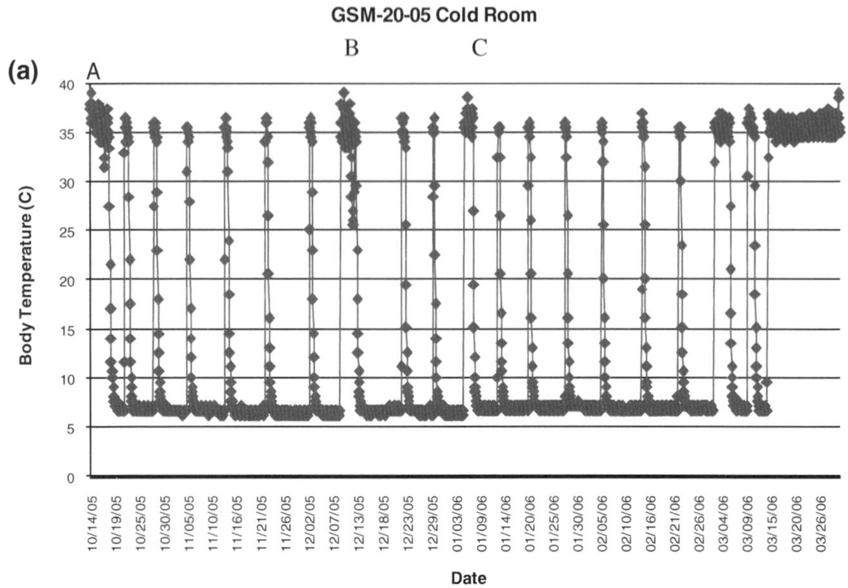

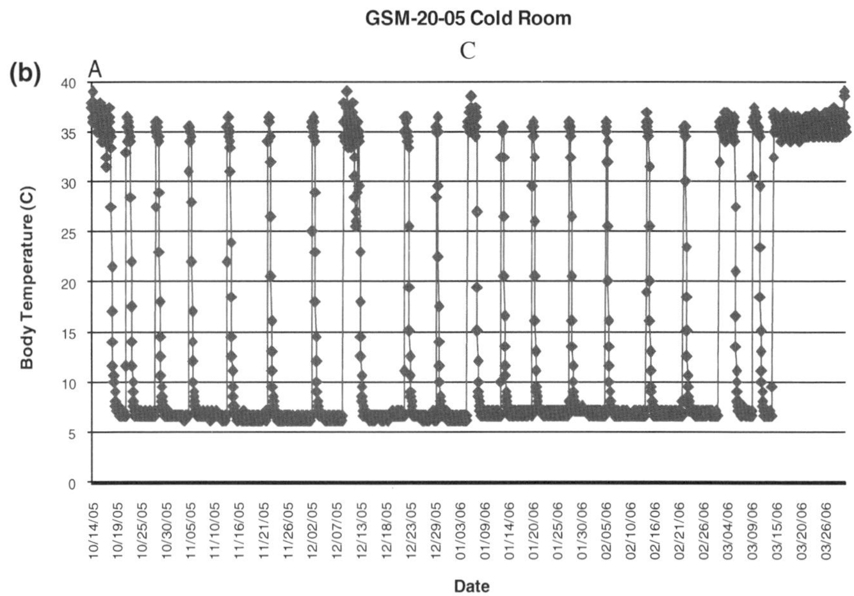

Fig. 44.1 **a** Illustration of torpor bouts in a typical cold room animal ($T_a = 5°C$). Point A (10/17/05) was when animals were placed in either a cold or warm room and food was removed from the cages. Point B (12/09/05), animals were aroused, weighed, serum samples collected, and food was returned to their cage. Point C (01/05/06), animals were aroused, weighed, serum collected, and food removed from cages after intake recorded. **b** A typical illustration of the torpor recorded in a warm room animal ($T_a = 22°C$). All letters refer to the same manipulations described in (**a**)

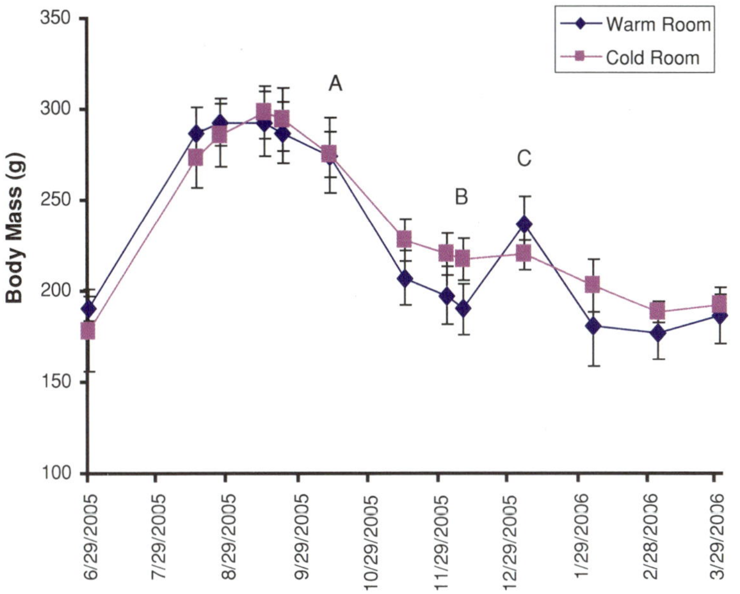

Fig. 44.2 The mean body mass of warm and cold room animals from August, 2005 to March, 2006. Letters refer to the same manipulation events described in Fig. 44.1a. Values are the mean ± SEM. Significance between groups ($P < 0.05$), as denoted by an asterisk (*) was determined using ANOVA. Samples size was nine until February and four through March

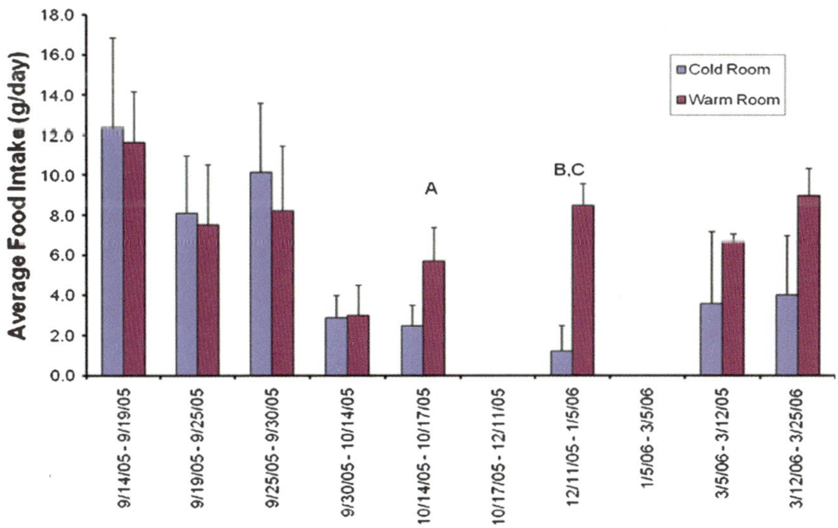

Fig. 44.3 A comparison of the mean food intake + SEM (g/day) in warm and cold room animals. Letters represent the same time and manipulation as described in Fig. 44.1a. Significance between groups ($P < 0.05$), as denoted by an asterisk (*) was determined using ANOVA. Sample size was nine until February and four through March

44.2.2 *Blood Collection and Leptin Analysis*

Serum samples (<1 ml) were collected via cardiac puncture from deeply anesthetized animals at the time points A, B, C, and in early February, as indicated on the figures and Table 44.2. The serum was immediately spun and frozen at −80°C for further analysis. Serum leptin levels were determined using Linco's multi-species 125I-labeled leptin RIA kit (Linco Research, cat. # XL-85K) and the lowest detectable limit was (<1 ng/ml). Ground squirrel leptin concentrations were validated by using a standard curve created from increasing concentrations of unlabeled leptin standards provided in the kit.

44.2.3 *Torpor Bout Measurement*

Body temperature measurements and torpor bout lengths were measured using I button data from October to March for all animals except those sacrificed in early February. An animal was considered to be in torpor when the $T_b < 30°C$ and aroused when the $T_b > 30°C$ for both warm and cold room groups.

44.2.4 *Data Analysis*

All values presented are the means ± SEM. Student T-tests and ANOVA were used to test mean differences in body mass, food intake, and torpor bout lengths. Results were considered significant if $P < 0.05$ or less.

44.3 Results

44.3.1 *Analysis of Torpor Patterns*

Animals maintained in the cold room from October to February or March had torpor bouts that were significantly longer (142 vs. 52 h; $P < 0.01$) than animals kept in the warm room (Table 44.1). Cold room animals spent significantly more time (3,166 vs. 1,806 h; $P < 0.01$) at lower Tb compared to warm room animals. Consequently, the number of torpor bouts exhibited by cold room animals was significantly less than that found for warm room animals (23 vs. 35, Table 44.1). The torpor bout results from animals that were sacrificed in February were not significantly different from the results reported for animals that were sacrificed in March (Table 44.1). A typical T_b record for a cold and warm room animal is illustrated in Fig. 44.1. Torpor bouts for cold room animals showed the normal

Table 44.1 Analysis of torpor bouts recorded from I buttons in warm and cold room animals

	Cold room	Warm room
T_b data from October to March ($N = 4$)		
$T_b < 30°C$ (h)	3,166	1,806
Number of torpor bouts[a]	23	35
Time in torpor[b] (h)	142	52
T_b data from October to February (animals sacrificed) ($N = 3$)		
$T_b < 30°C$ (h)	2,250	1,300
Number of torpor bouts[a]	19	23
Time in torpor[b] (h)	122	56

[a] A torpor bout was defined as the time between when T_b dropped below 30°C to to when T_b returned to 30°C
[b] The mean time spent below 30°C during a single torpor bout

Table 44.2 Serum leptin concentrations in cold and warm room animals during the body mass cycle

Point	Cold room (5°C)		Warm room (22°C)	
	Body mass (gms)	Leptin (ng/ml)	Body mass (gms)	Leptin (ng/ml)
A (10/17/05) N = 9, 9	275 ± 12.6	12.7 ± 1.91	280 ± 20.7	8.4 ± 2.9
B (12/9/05) N = 9, 9	217 ± 11.6	7.3 ± 2.5	185 ± 13.8	[a]4.6 ± 1.6
C (1/5/06) N = 8, 9	220 ± 8.3	5.2 ± 1.9	237 ± 15.5	5.7 ± 2.0
(2/04/06) N = 4, 4	201 ± 14.7	2.7 ± 1.3	180 ± 22.6	2.9 ± 1.4

[a] Significantly different from cold room animals ($P < 0.05$)
Samples sizes differ due to sacrificing animals in February. Values are mean ± SEM

pattern of test drops at the onset of hibernation, long torpor bouts in the middle of winter, and shorter bouts prior to the cessation of hibernation. Warm room animals also showed the typical torpor pattern for hibernators, but the bout lengths during the middle of winter were more variable and shorter in duration.

44.3.2 Food Intake Manipulations and Body Mass Changes

As was stated above, manipulations of both groups occurred at points A, B, and C (Fig. 44.1). At point A (October 17th, 2005), animals were separated into the two experimental groups and food was removed from their cages, as animals had stopped eating. On December 9th (point B), the body mass of warm room animals had decreased to or below their trapped weight recorded in June/July (Fig. 44.2). We felt that these animals were in need of food to survive, therefore, both groups of animals were provided with food. Prior to providing them with food at this point, they were aroused, anesthetized, body mass was recorded, and serum collected. Clearly, this disrupted the torpor bout pattern of the warm room animals more than of the cold room animals. Warm room animals took on average 2 weeks to return to torpor, while cold room animals returned to torpor within 1–2 days. As shown in Fig. 44.3, animals in the warm room that stayed euthermic after manipulation, ate a significant amount of food between points B (December 11th, 2005) and C (January 5th). Cold room animals ate little (~1–2 g) or nothing even though food was in their cage (Fig. 44.3) and thus, they began to undergo torpor bouts sooner after the manipulation at point B. During the nearly month period between points B and C, the warm room animals ate enough food to increase their body mass up to the mass recorded for the cold room animals (Fig. 44.2). At point C, the body mass of the cold and warm room animals was not significantly different. At this time, we removed food from both groups since the cold room animals were not eating and the warm room animals had stopped eating. Four animals from each experimental group were sacrificed at point C. The remaining animals in both groups continued to hibernate until sacrifice in March, however, the warm room animals again lost body mass faster than the cold room animals (Fig. 44.2). The body mass of the warm room group was significantly ($P < 0.05$) below that of the cold room group on March 8th, 2006. In order to prevent animals from dying (since they were below their summer trap weight), we gave back food and the warm room animals ate significantly more ($P < 0.05$) than the cold room animals (Fig. 44.3). As shown in Figs. 44.2 and 44.3, the warm room animals ate enough food to bring their body mass back up so that their mass was not significantly different from the cold room animals by the end of March.

44.3.3 Serum Leptin Levels

At time points A, B, and C, blood samples were collected in order to measure serum leptin levels. At the beginning of the experiment, when animals were initially put into two groups, body mass and serum leptin levels were not significantly different between the cold and warm room groups (Table 44.2). However, after nearly 2 months at their respective experimental temperature, body masses had clearly separated with the mean warm room mass lower than the cold room body mass (Fig. 44.2, Table 44.2) and the mean serum leptin level was significantly lower in the warm room animals compared to cold room animals. Between points B and C, animals in both groups had been given food, but only the warm room animals ate, mean serum leptin levels rose to 5.7 ng/ml in the warm room animals which was not significantly different from the value recorded for the cold room animals (5.2 ng/ml). The body mass of the two groups was not significantly different at that time. Serum leptin levels were recorded in the animals during February and interestingly, although warm room animals had a lower body mass compared to cold room animals, their mean serum leptin levels were not significantly different at this time.

44.4 Discussion

This is the first study to illustrate and document changes in torpor bout length, body mass, and food intake in GMGS at a low and high T_a while measuring serum leptin levels. Our study illustrates the characteristics of torpor bouts at a high

T_a and it is clear that torpor bout lengths are shorter and animals can be easily disturbed, as has been previously shown (Geiser and Kenagy 1988; Buck and Barnes 2000). The fact that GMGS hibernated at a higher T_b may support the hypothesis of Kondo et al. (2006) that the mechanisms governing torpor bout generation are independent of T_a. It is unknown whether the frequent arousals at a high T_a are due to changes in the hibernation-specific protein (HP) complex within the CSF and blood of GMGS. Previous studies have shown that torpor bout lengths are shorter at high T_a however, these studies did not include changes in body mass and blood leptin concentrations. In addition, some ground squirrels maintained at a high T_a will eat more when given the opportunity and therefore enter torpor less than animals maintained at a low T_a (Jameson 1965). Interestingly, Jameson (1965) did not determine if the animals were actually going into torpor; he called them "sluggish".

The effect of temperature on metabolism is so profound that it was not surprising leptin levels fell once animals were placed in the warm room. Serum leptin levels in GMGS from both warm and cold room groups were highest at peak body mass (point A) and declined with body mass (point B) as previously reported for hibernators (Concannon et al. 2001; Florant et al. 2004). Interestingly, when warm and cold room animals were presented with food (point B), only warm room animals fed. After staying awake and feeding for a short period of time, warm room GMGS leptin levels and body mass increased to values not significantly different from values found in cold room animals. The increased level of serum leptin coincided with the voluntary cessation of food intake in the warm room animals. The cold room animals did not feed during that time, even though food was in their cage. This suggests that leptin may be regulating food intake during the early portion of the body mass cycle, as previously suggested (Boyer et al.1997). We do not know whether the increase in food intake at point B was initiated due to falling leptin levels, body mass falling, and/or to some low set-point concentration of leptin, however, it appears that leptin levels are regulated since leptin concentration did increase to the value found in cold room animals. As the winter hibernation season progressed, leptin levels fell to about a quarter of the values recorded in both groups at peak body mass in autumn. Studies on marmots suggest that serum leptin levels decline in the late winter/early spring regardless of whether fat mass is significantly decreased (Florant et al. 2004). It would be interesting to perform a leptin injection study in order to artificially manipulated endogenous leptin levels and test whether increasing serum leptin levels actually prevented food intake in animals with a lower body mass in winter.

The hormone leptin was not known to exist when Mrovosky and Fisher (1970) put forth the "sliding-set" point hypothesis. Our results in ground squirrels and results found in hamsters (Klingenspor et al. 2000) suggest that leptin may play an important role in producing the "sliding set point" observed in body mass by changing food intake in hibernators. This conclusion is supported by the finding that as GMGS lose significant mass in the warm room (point B), leptin levels are about half the concentration found in the cold room animals at the same time. As has been previously reported (Forger et al. 1986; Dark et al. 1989), GMGS lose mostly fat during the weight loss phase of the body mass cycle and although we did not measure fat mass directly, data from our previous studies (Healy et al. 2008) indicated that animals kept at low body mass or who lost fat during this period of time have low plasma leptin concentrations. We hypothesize that if serum leptin concentrations are above some "sliding set-point" for hypothalamic serum leptin, then food intake will be decreased. Whether hibernators are above or below the set-point at which leptin initiates food intake may in part be due to the T_a in which they are hibernating: a high T_a increases overall metabolism and decreases body mass (fat) faster, putting them below the set-point, thus stimulating food intake. Furthermore, other factors including leptin resistance (Scherer and Buettner 2011) may alter the animal's food intake response.

We hypothesize that the hypothalamus changes its sensitivity to leptin at the onset of the reproductive season due to increasing gonadal steroid hormone production (Barnes et al. 1988). Reproductive status and lipid mass appear to play an important role in the termination of torpor in rodent hibernators (Kunz et al. 1999; Dark 2005). Regardless of body mass, animals in both groups had similarly low leptin levels in February. In March, warm room animals had a low body mass and began to eat prior to cold room animals which had a greater mass at the time. It would be interesting to measure serum leptin levels in the spring when fat mass and reproductive status are changing.

In summary, these results suggest that rodent hibernators can undergo torpor bouts at high T_a, and T_b duration at low tissue temperature is significantly shorter. Animals kept at a high T_a, lost body mass faster than animals maintained in a cold T_a and we assume that the body mass lost was predominately fat. We believe that the changes in serum leptin levels during the winter correspond to changes in body fat and that leptin concentration is involved in producing the sliding set-point for body mass in hibernators. It is possible that leptin concentrations may be under the control of a circannual clock producing an increase in sensitivity in the fall as animals prepare for hibernation and a decrease in sensitivity in the spring, when reproductive activity commences. Future experiments manipulating endogenous leptin concentrations and other hormones (e.g., insulin, ghrelin, neuropeptide Y) involved in food intake and body mass regulation are necessary to determine why food intake is so dramatically reduced during the winter hibernation period.

Acknowledgments Support for this grant was provided by NIH grant # (DK-067017) and the CSU lipid Foundation.

References

Ahima RS, Flier JS (2000) Leptin. Annu Rev Physiol 62:413–437

Barnes BM et al (1988) Plasma androgen and gonadotropin levels during hibernation and testicular maturation in golden-mantled ground squirrels. Biol Reprod 38(3):616–622

Boyer BB et al (1997) Leptin prevents posthibernation weight gain but does not reduce energy expenditure in arctic ground squirrels. Comp Biochem Physiol C Pharmacol Toxicol Endocrinol 118(3):405–412

Buck CL, Barnes BM (2000) Effects of ambient temperature on metabolic rate, respiratory quotient, and torpor in an arctic hibernator. Am J Physiol Regul Integr Comp Physiol 279(1): R255–R262

Concannon P et al (2001) Seasonal changes in serum leptin, food intake, and body weight in photoentrained woodchucks. Am J Physiol Regul Integr Comp Physiol 281(3):R951–R959

Dark J (2005) Annual lipid cycles in hibernators: integration of physiology and behavior. Annu Rev Nutr 25:469–497

Dark J, Stern JS, Zucker I (1989) Adipose tissue dynamics during cyclic weight loss and weight gain of ground squirrels. Am J Physiol 256(6 Pt 2):R1286–R1292

Florant GL et al (2004) Fat-cell mass, serum leptin and adiponectin changes during weight gain and loss in yellow-bellied marmots (*Marmota flaviventris*). J Comp Physiol B 174(8):633–639

Forger NG et al (1986) Fat ablation and food restriction influence reproductive development and hibernation in ground squirrels. Biol Reprod 34(5):831–840

Geiser F, Kenagy GJ (1988) Torpor duration in relation to temperature and metabolism in hibernating ground squirrels. Physiol Zool 61(5):442–449

Healy JE, Richter MM, Suu L, Fried SK, Florant GL (2008) Changes in serum leptin concentrations with fat mass in golden-mantled ground sfuirrels (*spermophilus lateralis*). In: Lovegrove BG, McKechnie AE (eds) Hypometabolism in animals: torpor, hibernation and cryobiology. University of KwaZulu-Natal, Pietermaritzburg

Heller HC, Poulson TL (1970) Circannian rhythms—II endogenous and exogenous factors controlling reproduction and hibernation in chipmunks (*Eutamias*) and ground squirrels (*Spermophilus*). Comp Biochem Physiol 33:357–383

Jameson EW Jr (1965) Food consumption of hibernating and nonhibernating *Citellus lateralis*. J Mammal 46(4):634–640

Klingenspor M, Niggemann H, Heldmaier G (2000) Modulation of leptin sensitivity by short photoperiod acclimation in the Djungarian hamster, *Phodopus sungorus*. J Comp Physiol B 170(1):37–43

Kondo N et al (2006) Circannual control of hibernation by HP complex in the brain. Cell 125(1):161–172

Kronfeld-Schor N et al (2000) Dissociation of leptin secretion and adiposity during prehibernatory fattening in little brown bats. Am J Physiol Regul Integr Comp Physiol 279(4): R1277–R1281

Kunz TH et al (1999) Plasma leptin decreases during lactation in insectivorous bats. J Comp Physiol B 169(1):61–66

Leibel RL (2002) The role of leptin in the control of body weight. Nutr Rev 60(10 Pt 2):S15–S19; discussion S68–S84, 85–87

Mrosovsky N, Fisher KC (1970) Sliding set points for body weight in ground squirrels during the hibernation season. Can J Zool 48(2):241–247

Mrosovsky N (1980) Circannual cycles in golden-mantled ground squirrels: experiments with food deprivation and effects of temperature on periodicity. J Comp Physiol 136:355–360

Mrosovsky N, Faust IM (1985) Cycles of body fat in hibernators. Int J Obesity 9(Suppl 1):93–98

Ormseth OA et al (1996) Leptin inhibits prehibernation hyperphagia and reduces body weight in arctic ground squirrels. Am J Physiol 271(6 Pt 2):R1775–R1779

Pengelley ET (1968) Interrelationships of circannian rhythms in the ground squirrel, *Citellus lateralis*. Comp Biochem Physiol 24:915–919

Pengelley ET, Fisher KC (1963) The effect of temperature and photoperiod on the yearly hibernating behavior of captive golden-mantled ground squirrels (*C. L. tescorum*). Can J Zool 41:1103–1120

Russell RL, O'Neill PH, Epperson LE, Martin SL (2010) Extensive use of torpor in 13-lined ground squirrels in the fall prior to cold exposure. J Comp Physiol B Nov 180(8):1165–1172

Scherer T, Buettner C (2011) Yin and Yan of hypothalamic insulin and leptin signaling in regulating white adipose tissue. Rev Endocr Metabol Disord Sept 12(3):235–243

Schwartz MW, Woods SC, Seeley RJ, Baskin DG (2000) Central nervous system control of food intake. Nature 404:661–671

Chapter 45
Ghrelin, Leptin, and Fatty Acids in Free-Living *Callospermophilus lateralis* (Golden-Mantled Ground Squirrels)

Jessica E. Healy and Gregory L. Florant

Abstract Although golden-mantled ground squirrels (GMGS, *Callospermophilus lateralis*) have long been used as laboratory animals, little is known about their physiology as wild-living animals under natural conditions. In examining hormone profiles of serum taken from wild-living GMGS during the euthermic (May–August) season, we found that post-hibernation levels of the orexigenic hormone ghrelin were low, and gradually increased through July and August (pre-hibernation). Females generally had higher concentrations of circulating ghrelin than did males, possibly reflecting the greater energy requirements imposed on them by pregnancy and lactation. Leptin concentrations were extremely low throughout the summer months, possibly reflecting the lean condition of field GMGS, particularly on their exit from hibernation in May. These data provide preliminary evidence that seasonal changes in energy-modulating hormones may be informed by different energy requirements inherent between the sexes.

Keywords Field hormones · Ghrelin · Leptin · NEFA · Sex differences · GMGS

This manuscript is derived from a thesis submitted to the Academic Faculty of Colorado State University in partial fulfillment of the requirements for the degree of Doctor of Philosophy.

J. E. Healy (✉)
Department of Basic Medical Sciences, University of Arizona College of Medicine—Phoenix, Phoenix, AZ 85004, USA
e-mail: jehealy@email.arizona.edu

G. L. Florant
Department of Biology, Colorado State University, Fort Collins, CO 85024, USA

T. Ruf et al. (eds.), *Living in a Seasonal World*,
DOI: 10.1007/978-3-642-28678-0_45,

45.1 Introduction

Hibernation is an adaptation characterized by significant physiological and behavioral changes, including metabolic and cardiovascular suppression and the reduction or cessation of food intake (aphagia). The family Sciuridae includes many species that hibernate (e.g., *Spermophilus, Urocitellus, Callospermophilus, Marmota*); in the field, individuals of the previously named genera enter winter hibernacula prior to the first torpor bout (Michener 1992; Wang 1979) and typically remain below-ground throughout the heterothermic season, emerging in the spring to reproduce and recommence food intake (Kenagy et al. 1989). Most male ground squirrels end heterothermy prior to females, undergo testicular recrudescence (Barnes et al. 1986), and establish territories (Buck and Barnes 2003) prior to female emergence. Females end hibernation when food resources are more abundant and are impregnated within days of emergence from their burrows (Kenagy et al. 1989). Young-of-year (juveniles) are weaned in mid-summer and spend the rest of the summer and autumn building up fat stores for the next hibernation season (Yousef and Bradley 1971). As such, hibernators exhibit extreme seasonal fluctuations in their desire to eat which make them interesting animal models for studying the controls of food intake that contributes to obesity and anorexia.

Golden-mantled ground squirrels (GMGS, *Callospermophilus lateralis*) have been used in physiological laboratory studies for many years (Jameson 1964; Pengelley and Fisher 1961; Twente and Twente 1964), but their physiology in the field under natural conditions is mostly unknown. In the lab, circulating levels of hormones involved with food intake [such as the orexigenic (appetite-stimulating) hormone ghrelin (produced by the stomach) and the anorexigenic (appetite-suppressing) hormone leptin (produced by white adipose tissue)] fluctuate with energy status and season in GMGS (Healy et al. 2008, 2010), and appear to work in opposition to help drive the seasonal changes in food intake. Wild-living animals, which are more susceptible to the vagaries of climate and food availability than ad libitum-fed laboratory hibernators, are likely to exhibit more pronounced differences in endogenous hormone production to compensate for these environmental changes. In conjunction with an examination of torpor patterns of wild-living GMGS, we set out to determine if age, sex, or season played a role in determining the concentrations of various hormones involved with food intake at different stages of GMGS euthermic season. Based on previous studies conducted in the lab and field (Concannon et al. 2001; Healy et al. 2010) we hypothesized that the orexigenic hormone ghrelin would be highest during periods of increased energy demand, such as during lactation for females, and during the late summer hyperphagic period. We hypothesized that serum leptin concentrations would generally fluctuate with body mass, being low in the spring and increasing as the euthermic season progresses. Finally, we hypothesized that circulating concentrations of non-esterified fatty acids (NEFA) would fluctuate based on endogenous energy stores and demands, and as such would be high when animals were attempting to increase their energy stores.

45.2 Methods and Materials

45.2.1 Animals and Surgical Procedure

All animal handling and surgeries were performed under a CSU-approved IACUC protocol in Larimer County, Red Feather Lakes, Colorado from July 2008 to June 2010. Body temperature (T_b) data loggers (iButtons; Maxim #DS1921G, Dallas, TX) were coated in paraffin wax (total mass $\sim$3.8 g) and programmed to record T_b every 3 h from September 1 to May 15 each year of the study. The recording time interval chosen (3 h) was the shortest interval that allowed the iButtons to record from September to May. The iButtons used were accurate to ±1°C and had a resolution of 0.5°C and a range of −40 to +85°C. In the pre-hibernation season (on multiple dates from July to August of 2008 and 2009), GMGS were live-trapped (Tomahawk Live Traps, Hazelhurst, WI) and anesthetized using isoflurane gas. Food intake at time of trapping was confirmed by behavioral observation, by examination of cheek pouches, or by visual examination of remaining bait in trap. Sex (determined by anal-genital distance) and approximate age (adult or juvenile, determined by body mass and length, and also by morphological characteristics where possible, such as scrotal testes for adult males, and pregnancy or lactation for adult females) were determined for each animal at initial trapping, and animals were ear tagged and marked with a colored pipe cleaner for identification. In addition, each animal was injected with a passive integrated transponder (PIT tags, AVID Identification Systems, Norco, CA) subcutaneously between the scapulae, was weighed (±1 g), measured for snout-anus length (±0.5 cm) and surgically implanted intraperitoneally (IP) with an iButton as previously described (Wilson et al. 2010). While animals were anesthetized, blood was drawn via cardiac puncture. A total of 32 individuals (9 adult males, 6 juvenile males, 11 adult females, and 6 juvenile females) were trapped and implanted with iButtons and blood samples taken. In the post-hibernation season (on multiple dates in May of 2009 and 2010), 21 surviving animals were re-trapped, anesthetized, weighed, measured, iButtons were removed, and a second blood sample was taken (3 adult males, 4 juvenile males, 8 adult females, and 6 juvenile females). The T_b data from iButtons were downloaded to a computer using 1-Wire software (1-Wire Driver, Maxim, Sunnyvale, CA). Date of first entry into torpor was determined as the date on which T_b first dropped below 30°C. Date of final arousal from hibernation was determined as the date on which T_b returned to euthermic regulation (T_b consistently 35–39°C).

45.2.2 Serum Hormone and Metabolite Analysis

Blood was collected via cardiac puncture, centrifuged, serum removed, and stored at −80°C until assayed. In May, blood was collected and handled as previously described and animals were allowed to recover from anesthesia before being released at their original trap sites. Serum removed from animals trapped in spring and summers of 2009 and 2012 was analyzed for various circulating hormones and fatty acids; due to limitations in amount of serum and assay space available, not all samples were tested for all hormones. Total ghrelin concentrations were measured with an enzyme immunoassay from Phoenix Pharmaceuticals (EK031-31) as previously described (Healy et al. 2010). Serum samples were measured from blood drawn from animals trapped in May (post hibernation), June, July, and August (prehibernation). Serum NEFA concentrations were determined by a colorimetric assay from Wako Chemicals [HR Series NEFA-HR(2)], using a BioTek Synergy HT microplate reader. NEFA concentrations were used as a measure of circulating endogenous energy availability and measured in blood samples from animals trapped in May, July, and August ($n = 3$–7 per sex per month). Serum leptin was measured with a multi-species radioimmunoassay kit from Linco-Millipore (XL-85K) as previously described ($n = 3$–11 per sex per month, Florant et al. 2004).

45.2.3 Statistical Analysis

Differences between groups were analyzed using the statistical software GraphPad Prism 5. Comparisons between sexes were made by Student's t-test. Correlations between torpor characteristics and hormone or morphological measurements were determined by linear regression analysis. Differences were considered significant at $p \leq 0.05$.

45.3 Results

45.3.1 Sex Differences in Circulating Hormones During the Euthermic Season

Circulating serum ghrelin concentrations gradually increased throughout the active season—May ghrelin concentrations were fairly low for both sexes, and tended to increase through July and August. There were no statistically significant differences in serum ghrelin concentrations between sexes within each month (Fig. 45.1a). Post hibernation (May) adult female GMGS had slightly elevated ghrelin concentrations compared to other ages and sexes, but again the differences were not statistically significant (Figs. 45.1a, b). Mean serum NEFA concentrations fluctuated widely between months, with no clear monthly pattern evident, but NEFA concentrations were generally (but not significantly) elevated in July; females had significantly ($p = 0.05$) higher NEFA concentrations than males in May, but there was no sex difference in NEFA concentrations in other months

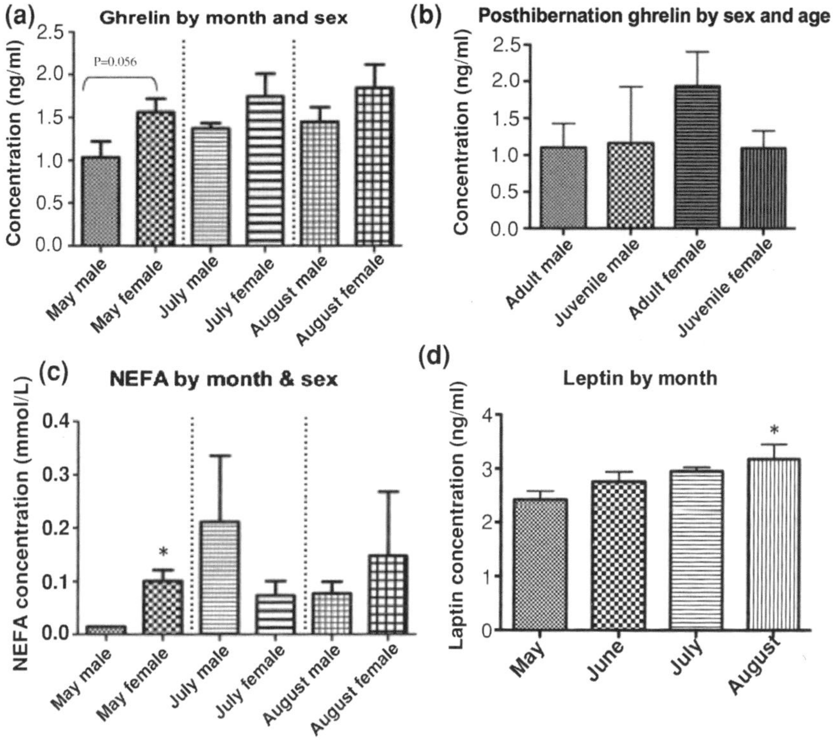

Fig. 45.1 **a** Sex differences in concentrations of circulating ghrelin from field GMGS in May (post-hibernation), July, and August (pre-hibernation) ($n = 4$–6 per month). **b** Sex and age differences in circulating ghrelin in May (post-hibernation). **c** Sex differences in circulating NEFA in May (post-hibernation), July, and August (pre-hibernation) [$n = 3$–10 per month, * = female different from male ($p < 0.05$)]. **d** Serum leptin concentration by month [$n = 3$–7 per month, * = August significantly different from May ($p < 0.05$)]

(Fig. 45.1c). Serum leptin concentrations (including all ages and sexes) increased slightly but significantly from May to August (Fig. 45.1d).

45.3.2 Correlations Between Torpor Characteristics and Physiological Parameters

There was no correlation between body mass at time of iButton implantation and date of entry into torpor. Nor was there a correlation between these characters when the amount of time between date of iButton implant and date of first torpor bout is taken into effect ($R^2 = 0.07$, $p = 0.2520$, Fig. 45.2a). However, there was a positive correlation between body mass at time of iButton removal and date of exit from heterothermy ($R^2 = 0.3992$, $p = 0.0021$, Fig. 45.2b). There was no correlation between ghrelin concentration at time of iButton removal and date of arousal from heterothermy when all age and sex cohorts were considered together, but if the groups are considered separately, there is a significant positive correlation between these characters in juvenile females (Fig. 45.2c). There was a positive relationship between body mass and leptin concentration when samples from all months were pooled ($R^2 = 0.6259, p = 0.0037$, Fig. 45.3a). There was no correlation between leptin concentration at time of iButton removal and date of arousal from heterothermy when all sexes are included ($R^2 = 0.005, p = 0.7933$, Fig. 45.3b); this lack of a relationship persists when leptin concentration is considered against the number of days between final arousal and iButton explant ($R^2 = 0.007, p = 0.7634$, Fig. 45.3c).

45.4 Discussion

Hibernators have distinct physiological seasons that follow environmental seasons. During spring and summer, hibernators are normophagic and homeothermic. In autumn, hibernators become hyperphagic, and then become aphagic and heterothermic in winter. Each of these physiological states is associated with changes in concentrations of circulating hormones, expression of enzymes in various tissues, and endogenous energy stores.

There is a trend toward sex differences in circulating concentrations of the orexigenic hormone ghrelin and in NEFA that changes by month. We have previously shown seasonal changes in circulating ghrelin concentrations in lab GMGS, but lab animals did not exhibit a sex difference (Healy et al. 2010, 2011). In the field, female GMGS are typically impregnated within days of emergence from hibernacula (Kenagy et al. 1989). In a laboratory setting, animals are not typically allowed to reproduce, and as such the usual pattern of energy expenditure is attenuated. In the field, adult male GMGS end hibernation earlier than other cohorts, and set up and defend a territory (Kenagy et al. 1989; Florant lab unpublished data). Males in the lab still arouse from hibernation and commence food intake earlier than females (Barnes 1996; Florant lab unpublished data), but because neither sex undergoes the energetic costs associated with reproduction, differences in circulating appetite hormones between sexes are minimal. In the field, female GMGS have slightly higher concentrations of circulating ghrelin than males in spring. In the same animals, NEFA were slightly lower in spring than during the summer, with a significant difference between sexes (males had lower circulating NEFA than did females). This would seem to indicate that females were mobilizing remaining fat stores to support a higher metabolic cost than males, presumably related to the costs of pregnancy. In this current study, field females tend to maintain higher concentrations of ghrelin than field males throughout the summer. This may be explained by the high metabolic costs of reproduction (from pregnancy to lactation) in females during the months of May–August. Higher metabolic costs require increased food intake, and increased ghrelin concentrations may be facilitating this increase in

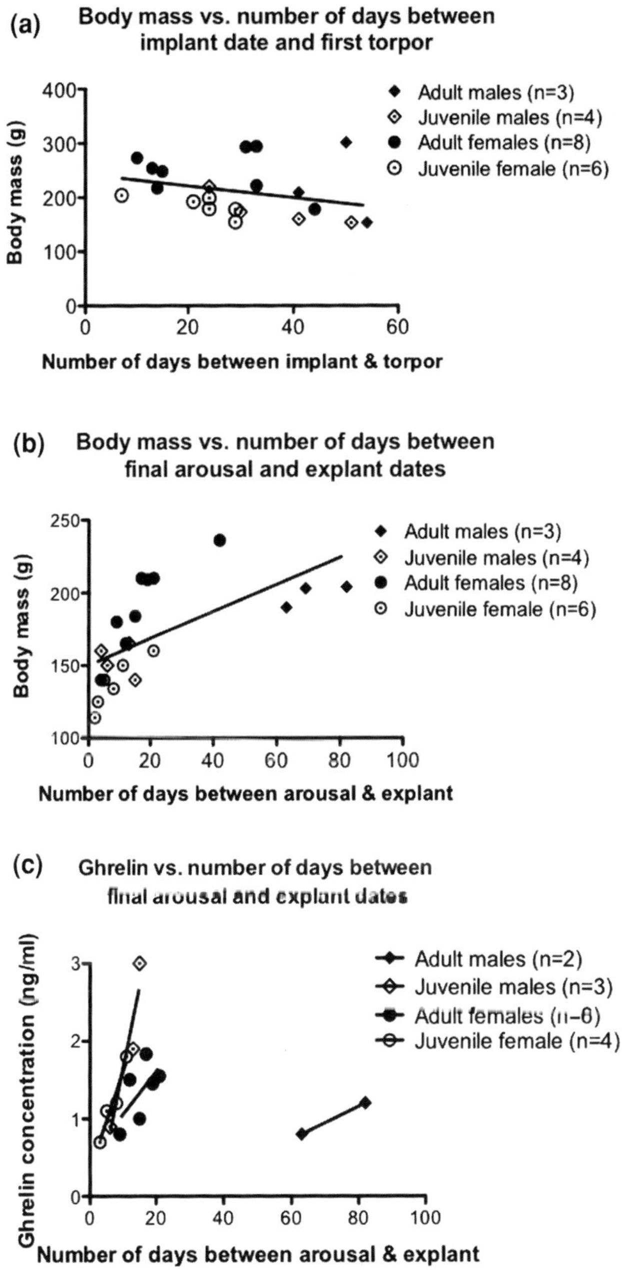

Fig. 45.2 Physiological characteristics plotted against time between iButton implant or explant dates versus dates of entry into or final arousal from heterothermy. **a** Body mass on date of iButton implantation is not correlated with timing of first torpor bout ($R^2 = 0.07, p = 0.2520$). **b** Body mass on date of iButton removal is positively correlated with timing of final arousal from heterothermy ($R^2 = 0.3992, p = 0.0021$). **c** Ghrelin concentration on date of iButton removal plotted against timing of arousal from heterothermy (adult males: $R^2 = 1, p =$ n.a.; juvenile males: $R^2 = 0.89$, $p = 0.2151$; adult females: $R^2 = 0.40, p = 0.1764$; juvenile females: $R^2 = 0.93, p = 0.0363$)

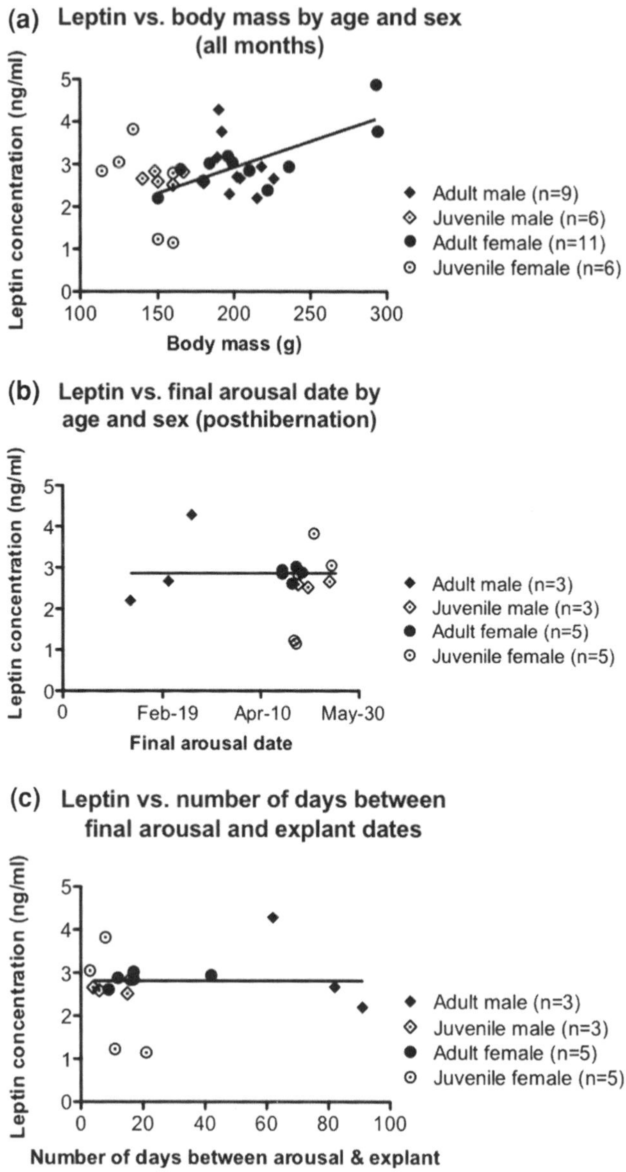

Fig. 45.3 Sex differences in serum leptin compared with body mass and arousal date. **a** Leptin concentration (ng/ml) is positively correlated with body mass (g) at time of capture ($R^2 = 0.6259$, $p = 0.0037$). **b** Leptin concentration has no correlation with date of final arousal from heterothermy ($R^2 = 0.005, p = 0.7933$). **c** Correcting for time between final arousal from heterothermy and time of recapture has no effect on the correlation between leptin and arousal date ($R^2 = 0.007$, $p = 0.7634$)

female food intake, which appears to be reflected in circulating NEFA concentrations in May.

In autumn (August–October), GMGS in the lab are hyperphagic for the first part of the season (late July–August), before gradually decreasing food intake to zero in preparation for hibernation. During this time, hibernating scuirils decrease metabolic costs in combination with increasing food intake to support dramatic lipogenesis (reviewed in Dark 2005). Ghrelin circulates at higher concentrations during the hyperphagic period, increased over spring and summer concentrations, theoretically driving the increase in food intake (Healy et al. 2010). In field animals, there was no correlation between body mass at time of iButton implantation and date of entry into heterothermy, but there was a positive correlation between body mass at time of iButton removal and date of arousal from the final torpor bout when the varying lengths of time between iButton implant/explant and the beginning/end of the heterothermic season were taken into account. Many animals were trapped and implanted with iButtons in July or August, and did not enter heterothermy until September–November. In the lab, the hyperphagic stage of the GMGS when most body mass is added occurs from late July-early September, and as such, field animals likely gained significant mass following their release after iButton implantation. In spring, however, recapture dates (mostly in May and early June) more closely corresponded with date of arousal from heterothermy (except in the case of reproductive males, who aroused in February and March). Therefore, later arousing animals had less time before their recapture to rebuild body mass than those animals that aroused earlier in the spring.

The hormone leptin is produced by white adipose tissue, and as such generally fluctuates in tandem with body fat. We found a weakly positive correlation between leptin concentration and body mass at time of iButton removal, which demonstrates the inadvisability of using body mass as a predictor of fat mass, as a small-bodied individual with extensive fat stores may weigh the same as a large-bodied individual with no fat. At time of post-hibernation recapture, most GMGS had very little to no body fat (observation at time of iButton removal), and as such leptin concentrations were extremely low. If serum leptin is accurately reflecting body fat stores, this may suggest that low serum leptin is part of the signal for earlier arousal in some hibernators, as has been suggested previously (Florant et al. 2004). Alternatively, it is possible that leptin is not directly correlated with body or fat mass in free-living GMGS, as has been suggested for certain other hibernators (see Dark 2005 for a review). However, our sample size is extremely small, and data from more animals are needed to draw confident conclusions about leptin and fat mass in free-living hibernators.

In conclusion, this study provides preliminary evidence that differences in sex and age may be reflected in circulating hormone concentrations in a group of naturally hibernating GMGS. These preliminary data provide intriguing possibilities for further study, including the correlation of profiles of circulating hormones and metabolic factors (including ghrelin, leptin, testosterone, GnRH, and circulating fatty acids) with torpor characteristics (e.g., torpor bout length and depth, dates of entry into and arousal from heterothermy) and ecological parameters (such as overwinter survival, snow fall, burrow site selection, and local flora). A long-term study of this population of GMGS will give us a far better understanding of the ecological physiology of the species, allowing us to track possible effects of climate change at the population level as well as adding to our knowledge of this species in its natural habitat.

Acknowledgments Thanks to Cara Ostrom, Kendra Burdett, Cassandra Gearhart, Jenna Bateman, Ashley Fenn, Thomas Barnett, Litto Suu, and Melanie Richter for help with trapping and data analysis, to Melanie Hoobler and Danielle Wantuck for help with ghrelin assays, and Susan Fried for performing leptin RIA. We also thank Denise Pearson of Fox Acres Country Club for allowing us to trap ground squirrels on their property. This work was supported by NIH R25DK067017 to GLF.

References

Barnes BM (1996) Relationships between hibernation and reproduction in male ground squirrels. In: Geiser F, Hulbert AJ, Nicol SC (eds) Adaptations to the cold: tenth international hibernation symposium. University of New England Press, Armidale, pp 71–80

Barnes BM, Kretzmann M, Licht P, Zucker I (1986) The influence of hibernation on testis growth and spermatogenesis in the golden-mantled ground squirrel, *Spermophilus lateralis*. Biol Reprod 35:1289–1297

Buck CL, Barnes BM (2003) Androgen in free-living arctic ground squirrels: seasonal changes and influence of staged male–male aggressive encounters. Horm Behav 43:318–326

Concannon P, Levac K, Rawson R, Tennant B, Bensadoun A (2001) Seasonal changes in serum leptin, food intake, and body weight in photoentrained woodchucks. Am J Physiol Regul Integr Comp Physiol 281:R951–R959

Dark J (2005) Annual lipid cycles in hibernators: integration of physiology and behavior. Annu Rev Nutr 25:469–497

Florant GL, Porst H, Peiffer A, Hudachek SF, Pittman C, Summers SA, Rajala MW, Scherer PE (2004) Fat-cell mass, serum leptin and adiponectin changes during weight gain and loss in yellow-bellied marmots (*Marmota flaviventris*). J Comp Physiol B 174:633–639

Healy JE, Richter MM, Suu L, Fried SK, Florant GL (2008) Changes in serum leptin concentrations with fat mass in golden-mantled ground squirrels (*Spermophilus lateralis*). In: Lovegrove BG, McKechnie AE (eds) Hypometabolism in animals: torpor, hibernation and cryobiology. University of KwaZulu-Natal, Pietermaritzburg

Healy JE, Ostrom CE, Wilkerson GK, Florant GL (2010) Plasma ghrelin concentrations change with physiological state in a sciurid hibernator (*Spermophilus lateralis*). Gen Comp Endocrinol 166:372–378

Healy JE, Bateman JL, Ostrom CE, Florant GL (2011) Peripheral ghrelin stimulates feeding behavior and positive energy balance in a sciurid hibernator. Horm Behav 59:512–519

Jameson EW Jr (1964) Patterns of hibernation of captive *Citellus lateralis* and *Eutamias speciosus*. J Mammal 45:455–460

Kenagy GJ, Sharbaugh SM, Nagy KA (1989) Annual cycle of energy and time expenditure in a golden-mantled ground squirrel population. Oecologia 78:269–282

Michener GR (1992) Sexual differences in over-winter torpor patterns of Richardson's ground squirrels in natural hibernacula. Oecologia 89:397–406

Pengelley ET, Fisher KC (1961) Rhythmical arousal from hibernation in the golden-mantled ground squirrel, *Citellus lateralis tescorum*. Can J Zool 39:105–120

Twente JW, Twente JA (1964) The duration of hibernation cycles as determined by body temperature. Am Zool 4:295

Wang LCH (1979) Time patterns and metabolic rates of natural torpor in the Richardson's ground squirrel. Can J Zool 57:149–155

Wilson WA, O'Riain MJ, Hetem RS, Fuller A, Fick LG (2010) Winter body temperature patterns in free-ranging Cape ground squirrel, Xerus inauris: no evidence for torpor. J Comp Physiol B 180:1099–1110

Yousef MK, Bradley WG (1971) Physiological and ecological studies on *Citellus lateralis*. Comp Biochem Physiol 39A:671–682

Chapter 46
Seasonal Variation in Brain Prostaglandin D_2 and E_2 of Marmots and *n*-6 Fatty Acid Availability

Walter Arnold, Paul Y. Kim, Kenneth G. D. Allen and Gregory L. Florant

Abstract Prostaglandins (PG) are involved in thermoregulation and reproduction. We investigated seasonal variation of PG in brain regions of two hibernators, the Alpine marmot (*Marmota marmota*) and the yellow-bellied marmot (*Marmota flaviventris*). The cerebrum, cerebellum, and basal forebrain homogenates of 24 animals were assayed for PGD_2 by competitive enzyme immunoassay, and of 18 *M. marmota* also for PGE_2 by radioimmunoassay. PGD_2 concentration was always higher than that of PGE_2. PGD_2 was already high at emergence from hibernation, rapidly increased during the following 2 weeks when marmots mate, and declined thereafter to late summer nadir. During winter, PGD_2 concentrations apparently increased, and, in contrast, PGE_2 concentrations seemed to be lower. These observations support the putative role of PGD_2 in lowering body temperature (T_b) during hibernation and of PGE_2 in elevating T_b during the summer-active season. This suggests that, with the exception of the spring mating season, a reciprocal production of PGD_2 and PGE_2 in brain regions of hibernators during different seasons. Results indicate further that the availability of polyunsaturated fatty acids (PUFA) of the *n*-6 series is important for PGD_2 synthesis and reproduction. During winter and spring, when fat is the major metabolic fuel, adult Alpine marmots metabolized less *n*-6 PUFA compared to other fatty acids; in contrast to younger, still immature individuals. Thus, high availability of *n*-6 PUFA may not only lower

W. Arnold (✉)
Department of Integrative Biology and Evolution,
University of Veterinary Medicine, Savoyenstr 1, 1160 Vienna, Austria
e-mail: walter.arnold@vetmeduni.ac.at

P. Y. Kim · K. G. D. Allen
Department of Food Science and Human Nutrition, Colorado State University, Fort Collins, CO 80523, USA

G. L. Florant
Department of Biology, Colorado State University, Fort Collins, CO 80523, USA

T. Ruf et al. (eds.), *Living in a Seasonal World*,
DOI: 10.1007/978-3-642-28678-0_46,

energy expenditure during hibernation, but also improve reproductive success in the subsequent year.

Keywords Hibernation · Thermoregulation · Reproduction · Body temperature · Prostaglandin D_2 · Prostaglandin E_2 · n-6 PUFA · Brain · Marmots

46.1 Introduction

During winter, many rodent hibernators depend on body fat stores that they have accumulated during summer to meet energy needs (Geiser 1993). Mobilization of fatty acids (FA) does not occur randomly: during lipolysis longer chain FA and unsaturated FA are released first (Connor et al. 1996; Raclot 2003). Nevertheless, polyunsaturated fatty acids (PUFA) of the n-6 series accumulate in white adipose tissue (WAT) in many hibernators (Falkenstein et al. 2001; Geiser et al. 1994; Hill and Florant 1999), suggesting selective retention of these PUFA instead of metabolization. The preferential use of other FA to fuel metabolism indicates alternative physiological functions of n-6 PUFA. One is apparently associated with cold tolerance. Consuming a diet rich in n-6 PUFA during summer enables hibernators to tolerate lower body temperatures (T_b) (Bruns et al. 2000; Frank 1992; Geiser 1993; Geiser and Kenagy 1987, 1993; Thorp et al. 1994) and to remain in deep torpor for longer uninterruptedly periods (French 1982; Geiser and Kenagy 1987). As a result, over-winter mass loss is lower the more n-6 PUFA an animal has in its WAT at begin of hibernation (Ruf and Arnold 2008). A potential physiological mechanism enabling tolerance of low T_b could be an increased activity of the sarcoplasmatic reticulum Ca^{2+}-Mg^{2+} pump (SERCA 2a) in a membrane containing a high phospholipid n-6 to n-3 ratio that mitigates temperature (Arrhenius) effects (Ruf and Arnold 2008). Improved activity of SERCA 2a would be most important in cardiac myocytes of hibernating animals in order to maintain proper heart function at temperatures close to the freezing point. Indeed, animals seem to prepare tissues for a life at low T_b as expected, independent of the current dietary intake of PUFA: Deer mice have been found to increase the amount of n-6 PUFA in leg muscle when exposed to short photoperiod (Geiser et al. 2007), and Alpine marmots transfer n-6 PUFA from WAT to heart and liver phospholipids at high rate shortly before hibernation (Arnold et al. 2011).

Another physiological reason for sparing n-6 PUFA from β-oxidation might be that arachidonic acid (AA, 20:4 n-6), a FA synthesized through chain prolongation and desaturation from linoleic acid (18:2 n-6) is the preferred substrate of cyclooxygenase (COX) and therefore the most important precursor of prostaglandins (PG). PG are well known for their function in reproduction and thermoregulation (Prendergast et al. 2002; Ruan et al. 2008; Saito et al. 2002a; Ueno et al. 1982), whereby PGD_2 and PGE_2 appear to act reciprocally in thermoregulation. In hibernators PGD_2 elicits hypothermia (Ueno et al. 1982), whereas PGE_2

infusion has been shown to cause arousal from hibernation concomitant with fever in Golden-mantled ground squirrels (*Spermophilus lateralis*) (Prendergast et al. 2002). The only previous study on brain PGD_2 and PGE_2 concentrations in a hibernator suggested that PGD_2, but not PGE_2, may change seasonally (Takahata et al. 1996). In that study, measurement of PG concentrations required pooling the brains of three animals to form a single sample and some brain regions were not assayed. The hibernating chipmunks were sacrificed during a torpor bout at low body temperatures (4–10°C) and compared with euthermic summer animals.

We investigated PGD_2 and PGE_2 concentrations in individual free-living Alpine marmots in the cerebrum, cerebellum, and basal forebrain to obtain a more detailed picture about seasonal changes. We supplemented this information with PGD_2 data from closely related Yellow-bellied marmots kept in the lab, to include direct information about the difference between summer and winter values of euthermic marmots. We hypothesize that brain PGD_2 concentration is higher, and PGE_2 concentration is lower in the hibernation season than in the summer. Furthermore, we hypothesize that sparing *n*-6 PUFA during winter may be required to enable peak production of PG necessary for reproductive function after hibernation,

46.2 Materials and Methods

46.2.1 Animals

We studied Alpine marmots (*Marmota marmota*) in the Swiss Canton Grison. Animals used for tissue sampling were shot in a population control program from populations in Bergala, Nufenen, and Avers by game wardens from the Wildlife and Fishery Department of the Swiss Canton Grisons. Thirteen males and five females shot between April 25th and September 16th with a single neck shot were used for determining PG concentrations in various parts of the brain. Cerebrum, cerebellum, and rostral basal forebrain were immediately excised and frozen in liquid nitrogen, then stored at −80°C until assayed.

Sex of the animals was determined by visual inspection, and age class (juvenile, yearling, 2-year old, adult) from body mass (Arnold 1990), measured to the nearest 50 g. For a subsample of males, we further measured testes width and length with a sliding caliper. A rough measure of testis volume was obtained by multiplying width and length. Alpine marmots live and hibernate in groups. From regular observations in spring we knew when the first marmot had emerged from a group's hibernaculum and assumed this date as end hibernation for all animals belonging to that group.

Samples of approximately 100 mg inguinal WAT were obtained from shot marmots, and, surgically under anesthesia, from animals trapped in a field study in the Bergalga valley. For further details on the study site, sampling procedure and FA analyses, see (Arnold et al. 2011). Treatment of Alpine marmots in this study was approved by the cantonal veterinary office of Grisons, Chur, Switzerland, no. 5/1997.

Four adult male and two adult female Yellow-bellied marmots (*Marmota flaviventris*) were trapped in Gunnison County, Colorado, during June and brought into the laboratory. The marmots were caged individually at 25°C under a long-day light cycle (14L:10D). Food and water were provided ad libitum. In July, three randomly selected marmots were sacrificed and the cerebrum, cerebellum, and rostral basal forebrain were excised, immediately frozen in liquid nitrogen and stored at −80°C until assayed for PGD_2. In September, the remaining three marmots were moved to a 5 ± 1°C cold room and complete darkness. The hibernation state was determined by the cessation of food consumption and body temperature of 8°C, monitored using thermocouples. The three hibernating animals were aroused on the same day between 9:00 a.m. and 12:00 p.m. and were euthermic ($T_b \geq 35$°C) at the time the animals were sacrificed and brains were collected. The cerebrum, cerebellum, and basal forebrain were excised, immediately frozen in liquid nitrogen and stored at –80°C until assayed. All procedures were approved by the Colorado State University Animal Care and Use Committee.

46.2.2 Prostaglandin Analyses

Each frozen brain region tissue sample was homogenized in a 10:90 w/v 50 mM (K^+) PO_4, 5 mM EDTA, pH 7.4 buffer containing 43 mM acetylsalicylic acid to prevent further synthesis of PG. Homogenates were centrifuged at 13,600 g for 2 min and supernatants were stored at −80°C. The next day, homogenates were thawed at room temperature and centrifuged at 13,600 g for 2 min. PGD_2 concentration was measured by a PGD_2 methoxime enzyme immunoassay kit from Cayman Chemical (Ann Arbor, MI). Samples were assayed in duplicate or triplicate. PGE_2 concentration was determined by double antibody radioimmunoassay (Iwakiri et al. 2002), but only for samples from Alpine marmots, and also in duplicate.

46.2.3 Data Analyses

All statistical analyses were carried out using the statistical package R (R Development Core Team 2011). Violation of assumptions to be met for parametric testing was identified by inspecting residuals from statistical models and eliminated, if necessary, by Box-Cox transformation of the response variable. Seasonal changes of PGD_2 and PGE_2 concentrations were analyzed with the procedure 'gam' of package 'mgcv' (generalized additive models with integrated smoothness estimation).

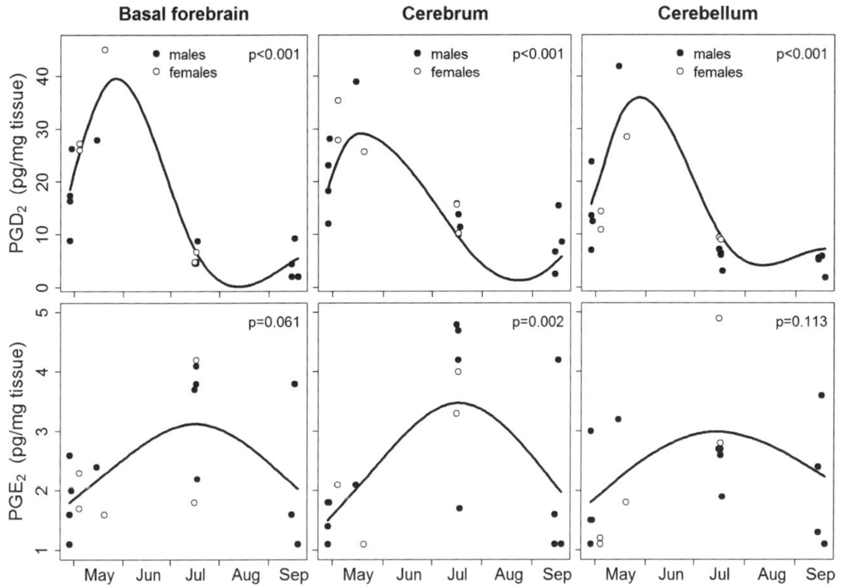

Fig. 46.1 The time course of PGD_2 (*upper panel*) and PGE_2 (*lower panel*) concentrations in the basal forebrain, cerebrum, and cerebellum of Alpine marmots during the summer-active season. Prostaglandin data are plotted as means ± SEM, averaged for each individual over assay repeats. The *p*-value in the *upper left corner* of each plot is the approximate significance of the smooth term. For each PG identical smooth parameters were forced for the three brain regions

46.3 Results

46.3.1 Seasonal Changes of PG and Thermoregulation

Both PGD_2 and PGE_2 concentrations varied with season and similarly in each brain region. At any time of the year, the concentrations of PGD_2 were in both sexes considerably higher than those of PGE_2 (Fig. 46.1). At the end of hibernation PGD_2 was more than two times higher than during July or September, suggesting higher levels during winter (Fig. 46.1, upper panel). In the closely related Yellow-bellied marmot, we measured high levels of PGD_2 during winter in two of the three brain regions [mean ± standard error of the mean (SEM), basal forebrain 30.0 ± 8.6; cerebrum 21.8 ± 11.4; cerebellum 36.4 ± 15.4; pg/mg tissue] compared with summer levels (basal forebrain 19.1 ± 8.9; cerebrum 23.6 ± 7.8; cerebellum 9.1 ± 5.4 pg/mg tissue), but these differences reached statistical significance only in the cerebellum ($p < 0.05$). PGD_2 in Alpine marmots continued to rapidly increase in spring and reached its peak in late May (Fig. 46.1, upper panel). This was about 2 weeks after termination of hibernation, thus in the midst of the mating season (Arnold 1990). In contrast to PGD_2, seasonal variation of PGE_2 was less pronounced, and, peaked in July, if at all.

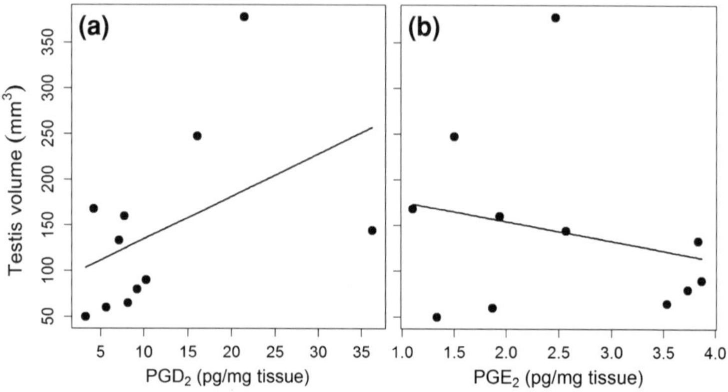

Fig. 46.2 The relation of PGD_2 (**a**) and PGE_2 (**b**) concentrations in the brain of Alpine marmot males to estimated testis volume. Each *point* represents the average over the three brain regions of an individual. See text for statistics

46.3.2 PG and Reproductive Function

The peak of PGD_2 during the mating season (Fig. 46.1, upper panel) and the large testis size found in adult Alpine marmots at that time of the year (Arnold and Dittami 1997) suggests that PGD_2 concentrations and reproductive function may be correlated. Male Alpine marmots reach adult testis size and androgen levels after their third hibernation (Arnold and Dittami 1997). Interestingly, testis volume showed a tendency to increase with average PGD_2 concentration in the brain (Fig. 46.2a; partial effect of PGD_2: $F_{(1,8)} = 4.50$, $p = 0.067$), independent of the increase with age (partial effect of age class: $F_{(1,8)} = 6.10$, $p = 0.034$). In contrast, there was no evidence for such a relation with PGE_2 (Fig. 46.2b; partial effect of PGE_2: $F_{(1,8)} = 0.16$, $p = 0.700$), but still a positive correlation with age (partial effect of age class: $F_{(1,8)} = 10.04$, $p = 0.013$), presumably reflecting the reproductive quiescence of young marmots (Arnold and Dittami 1997; Clasadonte et al. 2011). For females, we have no measure of reproductive function and thus cannot relate it to PG concentrations in the brain.

46.3.3 Age-Dependent Handling of Precursors

The differences in PG concentrations among age-classes reflected not only reproductive activity but also availability of the most important precursor, *n*-6 PUFA. Sparing of *n*-6 PUFA during winter in WAT did not occur in yearlings that never reproduce, to a higher degree in 2-year-old animals, and was highest in adults (>2-year-old animals, Fig. 46.3). These age-specific changes of the concentration of *n*-6 PUFA in the WAT continued after hibernation (Fig. 46.4).

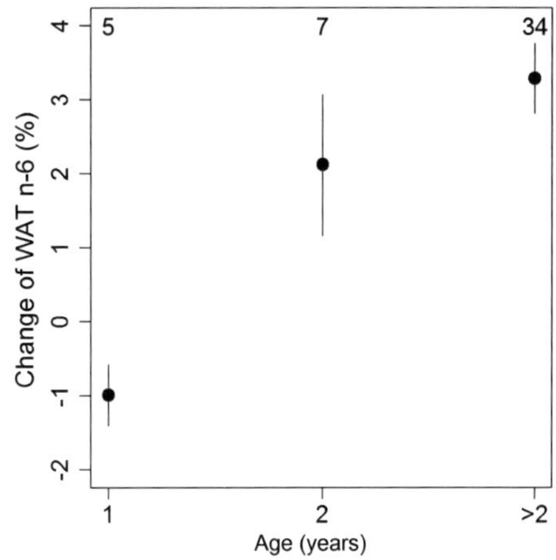

Fig. 46.3 Change of concentrations of *n*-6 PUFA in inguinal WAT of Alpine marmots during winter in different age-classes ($F_{(1,44)} = 11.32$, $p = 0.002$). Values (means ± SEM) represent the difference in concentrations before and after hibernation determined from WAT samples of repeatedly trapped individuals. Sample sizes are given on *top* of the graph

46.4 Discussion

This is the first study to report PGD_2 and PGE_2 concentrations in the cerebrum, cerebellum, and basal forebrain regions of hibernators. PG concentrations were determined in individual brain region homogenates in the presence of acetylsalicylic acid, which inhibits COX-dependent PGH_2 production, and therefore represent in vivo PG concentrations. The PGD_2 values found, 5–40 pg/mg tissue, are similar to those reported for rat brain (Abdel-Halim et al. 1977; Ujihara et al. 1988) and postmortem human brain (Ogorochi et al. 1984). The PGE_2 values reported here, 1–4 pg/mg tissue, are consistent with previously published observations in the rat brain (Ujihara et al. 1988). In the study by Takahata et al. (1996) on the Asian chipmunk (*Tamias asiaticus*), they stored frozen chipmunk heads from which brains were subsequently excised. This storage procedure may have contributed to PG degradation in situ and may account for their reported low PGD_2 and PGE_2 values (pg/g wet tissue). This may also explain the need to pool brains from three animals to provide a single sample in that study.

Since PGD_2 is known to decrease and PGE_2 to increase T_b, we expected that marmots would have higher levels of PGD_2 in the brain during winter, and hence on emergence from hibernation when euthermic T_b is still low and relapsing into torpor possible (Fig. 46.5, Arnold et al. 2011). Furthermore, we expected brain PGD_2 concentrations to be low during summer, and the seasonal changes of PGE_2 to mirror that of PGD_2. With the exception of the spring peak of PGD_2, which is most likely related to reproduction and not to thermoregulation, our results are in line with these expectations (cf. Fig. 46.1 and Fig. 46.5, PGD_2 data of Yellow-bellied marmots in the text). We suggest that the changes in PGD_2 concentration in the marmot brain during winter may be involved in the regulation of hibernation. Higher PGD_2 concentrations in the brain during winter may be due to seasonal changes in the expression of one or more enzymes involved in the production of PGD_2, either COX-derived PGH_2 or subsequent PGD synthase. A concomitant

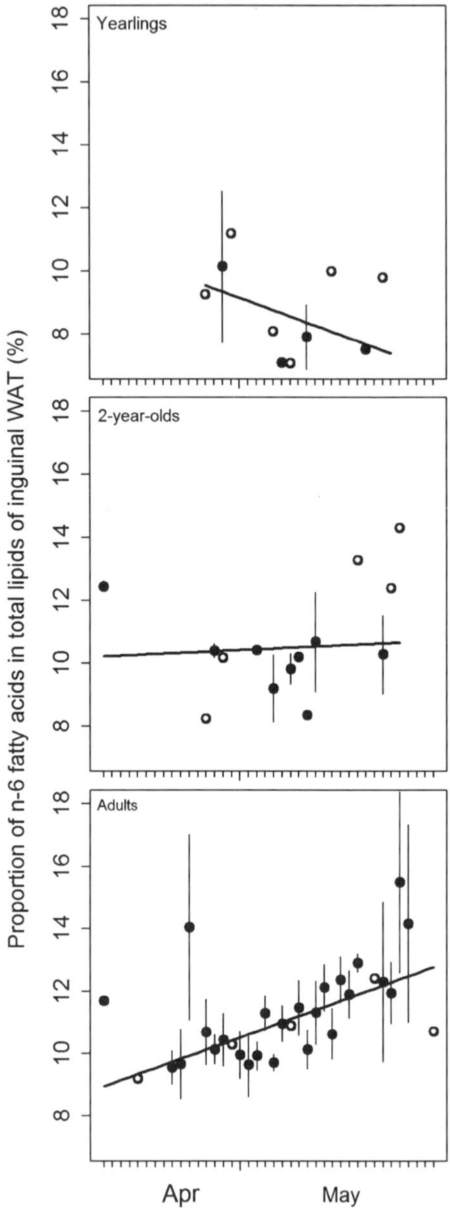

Fig. 46.4 Change of concentrations of *n*-6 PUFA in inguinal WAT of Alpine marmots during spring in different age-classes. (means ± SEM; SEM is not shown if smaller than symbol size, or if sample size equals one as indicated by symbols with a *dot*; Yearlings: $F_{(1,13)} = 2.49$, $p = 0.139$; 2-year-olds: $F_{(1,24)} = 0.11$, $p = 0.742$; Adults: $F_{(1,145)} = 18.29$, $p < 0.001$; differences between slopes of the three age-classes: $F_{(1,183)} = 7.56$, $p = 0.007$)

decrease in brain PGE_2 in winter would suggest that the increase in PGD_2 is due to changes in enzyme expression downstream of COX. In the golden-mantled ground squirrel (*Spermophilus lateralis*), PGD synthase mRNA declines significantly in the hypothalamus during late torpor and returns to its former level during arousal, whereas both the cortex and basal forebrain show a non-significant decline

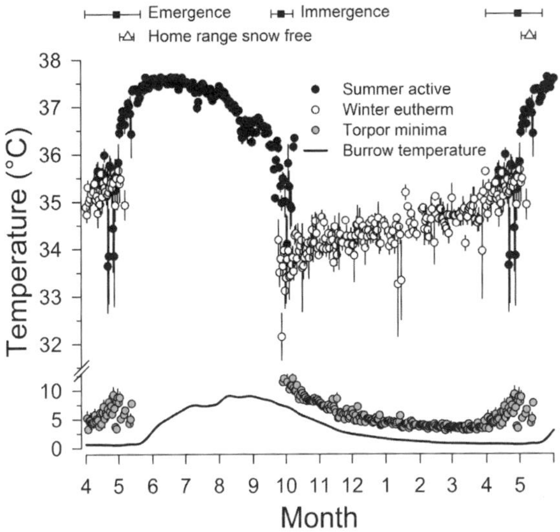

Fig. 46.5 Phenology of activity, hibernation and T_b of Alpine marmots at the field study site Bergalga, equipped with temperature-sensitive transmitters (modified from Arnold et al. 2011). Data are *double plotted* for months April and May to ease visualization of the seasonal pattern. *Top line* Means and ranges of dates of emergence from and immergence into hibernacula (*black squares* and *horizontal lines*). *Line below top line* Mean and range of dates when snow cover had disappeared from approximately ¾ of a marmots' home ranges as an approximation of the onset of vegetation growth (*open triangles* and *horizontal lines*). *Plot* T_b of summer-active (*black circles*), and winter euthermic (*white circles*) marmots, minimum T_b reached during bouts of torpor (*gray circles*) and burrow temperature (*line*). Symbols represent daily means. Error bars (± SEM) reflect variation between individuals, missing error bars indicate SEM smaller than symbol size. Burrow temperature is a *line plot* of daily means smoothed with a cubic spline

(O'Hara et al. 1999). On the other hand, Pinzar et al. (2000) found no differences in sleep-wake patterns between transgenic mice that overexpress human PGD synthase and wild-type mice, supporting the view that the production of PGH_2 is the rate-limiting step in PGD_2 synthesis.

The most important substrate for COX and hence PG production is AA. During winter it is impossible and during spring at least unlikely that *n*-6 PUFA necessary for PG syntheses are directly derived from food intake. Hibernating marmots do not feed at all (Arnold 1993), and spring emergence from hibernation is typically long before snow melt has progressed so far that food becomes accessible (Fig. 46.5). Therefore, sufficient availability of *n*-6 PUFA may not only be necessary for improved cold tolerance minimizing energy expenditure during hibernation, but also for success in the subsequent reproductive season.

Indeed, the accumulation of *n*-6 PUFA in WAT during winter and spring occurred only in reproductively active individuals (Figs. 46.3 and 46.4), and the tendency of a positive correlation between PGD_2 concentration and testis size (Fig. 46.2) suggests a causal link between the availability of *n*-6 PUFA and reproductive function. A potential mechanism responsible for sparing *n*-6 PUFA for this purpose is the selective reacylation of PUFA by monoacylglycerol acyltransferase (MGAT) after lipolysis (Xia et al. 1993). High activity of hepatic MGAT can prevent PUFA from metabolization despite high ongoing rates of β-oxidation as it is the case during hibernation. MGAT appears to prefer monoacylglycerols that contain certain PUFA, and is therefore a candidate enzyme for the selective retention of *n*-6 PUFA (Mostafa et al. 1993).

The importance of PG for reproduction is well known, however, most studies have focused on PG concentrations in reproductive tissue and serum (Allen and Harris 2001). There are only a few studies of PG concentrations in rodent brain, and it is assumed that brain concentrations reflect PG concentrations in the reproductive tissues. For instance, lipocalin-type PGD_2 synthase mRNA is expressed not only in the brain, but also in testis, prostate, and epididymis, and is necessary to maintain spermatogenesis (Saito et al. 2002b). Thus, during the spring mating season a high demand of PGD_2 would be expected, although its role for female reproduction is unclear. In contrast, the importance of PGE_2 for maintaining pregnancy is well known. Inhibition of PG synthesis before implantation causes pregnancy failure, most likely because PGE_2 is involved in the maternal recognition of pregnancy and acts as a luteoprotective/antiluteolytic factor (reviewed in Waclawik et al. 2009). Therefore, we would expect PGE_2 to increase during early pregnancy and to remain high through parturition (Allen and Harris 2001). However, PGE_2 is concomitantly high in the brain of males, too, suggesting that PGE_2 concentrations in the brain have a role in thermoregulation rather than for reproduction.

To summarize, our results suggest that PGD_2 and PGE_2 concentrations in the marmot brain change periodically with season and age. The availability of sufficient *n*-6 PUFA precursors for PG synthesis is apparently important in spring when the animals become reproductively active. Whether there is a definitive role for PGs in the regulation of hibernation is still unclear and the selective inhibition of PG synthases remains to be studied further. Furthermore, the selective retention of *n*-6 PUFA in reproductively active animals suggests a critical role of AA availability to COX and hence PG production. Measurement of COX and other PG synthases would help to elucidate the seasonal changes in brain PGs that may be involved in regulating hibernation.

Acknowledgments This study was supported by the Colorado Agricultural Experiment Station (Allen), NIDDK grant 60676-01, Fulbright Research Fellowship (Florant). The field study was financed by the Austrian Science Fund (FWF project P12430-BIO granted to W. Arnold). We thank P. Ratti, G. Brosi, and the team of the Grisons Game and Fishery Inspectorate, Switzerland, for enduring and substantial support of the study.

References

Abdel-Halim MS, Hamberg M, Sjöquist B, Ångård E (1977) Identification of prostaglandin D_2 as a major prostaglandin in homogenates of rat brain. Prostaglandins 14:633–643

Allen KGD, Harris MA (2001) The role of *n*-3 fatty acids in gestation and parturition. Exp Biol Med 226:498–506

Arnold W (1990) The evolution of marmot sociality: I. Why disperse late? Behav Ecol Sociobiol 27:229–237

Arnold W (1993) Energetics of social hibernation. In: Carey C, Florant GL, Wunder BA, Horwitz B (eds) Life in the cold: ecological, physiological, and molecular mechanisms. Westview Press, Boulder, pp 65–80

Arnold W, Dittami J (1997) Reproductive suppression in male alpine marmots. Anim Behav 53:53–66

Arnold W, Ruf T, Frey-Roos F, Bruns U (2011) Diet-independent remodeling of cellular membranes precedes seasonally changing body temperature in a hibernator. PLoS ONE 6:e18641

Bruns U, Frey-Roos F, Pudritz S, Tataruch F, Ruf T, Arnold W (2000) Essential fatty acids: their impact on free-living alpine marmots (*Marmota marmota*). In: Heldmaier G, Klingenspor M (eds) Life in the cold IV. Springer, Berlin, pp 215–220

Clasadonte J, Poulain P, Hanchate NK, Corfas G, Ojeda SR, Prevot V (2011) Prostaglandin E_2 release from astrocytes triggers gonadotropin-releasing hormone (GnRH) neuron firing via EP2 receptor activation. Proc Natl Acad Sci 108:16104–16109

Connor WE, Lin DS, Colvis C (1996) Differential mobilization of fatty acids from adipose tissue. J Lipid Res 37:290–298

Falkenstein F, Koertner G, Watson K, Geiser F (2001) Dietary fats and body lipid composition in relation to hibernation in free-ranging echidnas. J Comp Physiol B 171:189–194

Frank CL (1992) The influence of dietary fatty acids on hibernation by golden mantled ground squirrels (*Spermophilus lateralis*). Physiol Zool 65:906–920

French AR (1982) Effects of temperature on the duration of arousal episodes during hibernation. J Appl Physiol Respir Environ Exerc Physiol 52:216–220

Geiser F (1993) Dietary lipids and thermal physiology. In: Carey C, Florant GL, Wunder BA, Horwitz B (eds) Life in the cold: ecological, physiological, and molecular mechanisms. Westview Press, Boulder, pp 141–153

Geiser F, Kenagy GJ (1987) Polyunsaturated lipid diet lengthens torpor and reduces body temperature in a hibernator. Am J Physiol 252:R897–R901

Geiser F, Kenagy GJ (1993) Dietary fats and torpor patterns in hibernatin ground squirrels. Can J Zool Rev 71:1182–1185

Geiser F, McAllan B, Kenagy G, Hiebert S (2007) Photoperiod affects daily torpor and tissue fatty acid composition in deer mice. Naturwissenschaften 94:319–325

Geiser F, McAllan BM, Kenagy GJ (1994) The degree of dietary fatty acid unsaturation affects torpor patterns and lipid composition of a hibernator. J Comp Physiol B 164:299–305

Hill VL, Florant GL (1999) Patterns of fatty acid composition in free-ranging yellow-bellied marmos (*Marmota flaviventris*) and their diet. Can J Zool 77:1494–1503

Iwakiri Y, Sampson DS, Allen KGD (2002) Suppression of cyclooxygenase-2 and inducible nitric oxide synthase expression by conjugated linoleic acid in murine macrophages. Prostaglandins Leukot Essent Fatty Acids 67:435–443

Mostafa N, Everett DC, Chou SC, Kong PA, Florant GL, Coleman RA (1993) Seasonal changes in critical enzymes of lipogenesis and triacylglycerol synthesis in the marmot (*Marmota flaviventris*). J Comp Physiol B 163:463–469

O'Hara BF, Watson FL, Srere HK, Kumar H, Wiler SW, Welch SK, Bitting L, Heller HC, Kilduff TS (1999) Gene expression in the brain across the hibernation cycle. J Neurosci 19:3781–3790

Ogorochi T, Narumiya S, Mizuno N, Yamashita K, Miyazaki H, Hayaishi O (1984) Regional distribution of prostaglandins D_2, E_2, and $F_{2\alpha}$ and related enzymes in postmortem human brain. J Neurochem 43:71–82

Pinzar E, Kanaoka Y, Inui T, Eguchi N, Urade Y, Hayaishi O (2000) Prostaglandin D synthase gene is involved in the regulation of non-rapid eye movement sleep. Proc Natl Acad Sci 97:4903–4907

Prendergast BJ, Freeman DA, Zucker I, Nelson RJ (2002) Periodic arousal from hibernation is necessary for initiation of immune responses in ground squirrels. Am J Physiol Regul Integr Comp Physiol 282:R1054–R1082

R Development Core Team (2011) R: a language and environment for statistical computing. R Foundation for Statistical Computing, Vienna

Raclot T (2003) Selective mobilization of fatty acids from adipose tissue triacylglycerols. Prog Lipid Res 42:257–288

Ruan YC, Wang Z, Du JY, Zuo WL, Guo JH, Zhang J, Wu ZL, Wong HY, Chung YW, Chan HC, Zhou WL (2008) Regulation of smooth muscle contractility by the epithelium in rat vas deferens: role of ATP-induced release of PGE2. J Physiol 586:4843–4857

Ruf T, Arnold W (2008) Effects of polyunsaturated fatty acids on hibernation and torpor: a review and hypothesis. Am J Physiol Regul Integr Comp Physiol 294:R1044–R1052

Saito S, Tsuda H, Michimata T (2002a) Prostaglandin D2 and reproduction. Am J Reprod Immunol 47:295–302

Saito S, Tsuda H, Michimata T (2002b) Prostaglandin D_2 and reproduction. Am J Reprod Immunol 47:295–302

Takahata R, Matsumura H, Eguchi N, Sri Kantha S, Satoh S, Sakai T, Kondo N, Hayaishi O (1996) Seasonal variation in levels of prostaglandins D_2, E_2 and $F_{2\alpha}$ in the brain of a mammalian hibernator, the Asian chipmank. Prostaglandins Leukot Essent Fatty Acids 54:77–81

Thorp CR, Ram PK, Florant GL (1994) Diet alters metabolic rate in the yellow-bellied marmot (*Marmota flaviventris*) during hibernation. Physiol Zool 67:1213–1229

Ueno R, Narumiya S, Ogorochi T, Nakayama T, Ishikawa Y, Hayaishi O (1982) Role of prostaglandin D_2 in the hypothermia of rats caused by bacterial lipopolysaccharide. Proc Natl Acad Sci 79:6093–6097

Ujihara M, Urade Y, Eguchi N, Hayashi H, Ikai K, Hayaishi O (1988) Prostaglandin D_2 formation and characterization of its synthetases in various tissues of adult rats. Arch Biochem Biophys 260:521–531

Waclawik A, Jabbour HN, Blitek A, Ziecik AJ (2009) Estradiol-17β, prostaglandin E_2 (PGE_2), and the PGE_2 receptor are involved in PGE_2 positive feedback loop in the porcine endometrium. Endocrinology 150:3823–3832

Xia T, Mostafa BG, Bhat GI, Florant GL, Coleman RA (1993) Selective retention of essential fatty acids: the role of hepatic monoacylglycerol acyltransferase. Am J Physiol 265: R414–R419

Chapter 47
Expression of Orexigenic and Anorexigenic Neuropeptides Before and During Hibernation in the Daurian Ground Squirrel (*Spermophilus dauricus*)

Xin Xing, Ming-Yue Sun, Xia Peng, Shi-Yi Song and Ming Yang

Abstract Hibernation is one of the most important energy-saving strategies used by some mammals during periods of relative food shortage. Fat-storing hibernators need to store enough body fat to survive the winter. To understand the roles leptin played before and during hibernation, we measured changes in body weight, food intake, serum leptin level, and the expression of genes related to food intake regulation in the neurons of the hypothalamic arcuate nucleus (ARC), in Daurian ground squirrels (*Spermophilus dauricus*). During the fattening period, the expression of neuropeptide Y (NPY) significantly decreased ($P < 0.01$, $df = 8.348$, $t = -4.649$), and the expression of cocaine- and amphetamine-regulated transcript (CART) peptide ($P < 0.01$, $df = 12$, $t = 3.324$) and serum leptin level ($P < 0.05$, $df = 10$, $t = 3.070$) significantly increased compared to the levels before fattening started. However, changes in neuropeptide expression did not inhibit food intake during the fattening period. The expression of pro-opiomelanocortin reached the peak value before hibernation that might be involved in reducing food intake in ground squirrels. There was no significant change in the expression of peptides during hibernation compared with the euthermic stage before fattening had started. These results suggest that leptin can regulate the expressions of Ob-Rb, CART, and NPY during the fattening period but food intake and body weight continued to rise despite the changes of serum leptin level and expressions of neuropeptides.

X. Xing · M.-Y. Sun · S.-Y. Song · M. Yang (✉)
College of Chemistry and Life Science, Shenyang Normal University, 110034 Shenyang, China
e-mail: yangming@synu.edu.cn

X. Peng · S.-Y. Song
Liaoning Key Laboratory of Biological Evolution and Agricultural Ecology, 110034 Shenyang, China

T. Ruf et al. (eds.), *Living in a Seasonal World*,
DOI: 10.1007/978-3-642-28678-0_47,

Keywords Hibernation · Leptin · Fattening · Neuropeptides · Daurian ground squirrels (*Spermophilus dauricus*)

47.1 Introduction

Hibernation is an important energy-saving strategy used by some mammals to survive food shortages during the winter months (Humphries et al. 2003; Boyer and Barnes 1999; Heldmaier et al. 2004). To prepare for hibernation, hibernators need to store enough body fat during the pre-hibernation fattening period. Species such as ground squirrels (*Callospermophilus*, *Urocitellus*, *Spermophilus*) (Buck and Barnes 1999; Michener 1992; Healy et al. 2010; Yuan 2008), woodchucks (*Marmota*) (Concannon et al. 2001), marmots (*Marmota*), bats (*Rhinolophus*), hedgehogs (*Erinaceus*, *Setifer*, *Tenrec*), and bears (*Ursus*) (Humphries et al. 2003) increase their body mass sharply before hibernation. During deep hibernation, hibernators do not eat, use body fat stores as fuel, and can reach the nadir of body mass before vernal emergence (Andrews 2007; Humphries et al. 2003; Carey et al. 2003).

Body weight increase in fat-storing hibernators is mostly due to the accumulation of adipose tissue, especially white adipose tissue (WAT). WAT is not only an energy storage tissue, but also an endocrine organ that can synthesize and secrete an important protein hormone, leptin. Leptin is crucial for the regulation of energy intake and expenditure (Zhang et al. 1994). Circulating leptin is positively related to body mass and can cross the blood–brain barrier (BBB) by a saturable system independent of insulin (Banks et al. 1996). Leptin acts on leptin receptors, especially the long form of the receptor (Ob-Rb), in the hypothalamus to modulate food intake and energy balance (Flier 2004). High levels of leptin in the hypothalamus can increase the expression of anorectic peptides pro-opiomelanocortin (POMC) and cocaine- and amphetamine-regulated transcript (CART), and decrease the expression of orexigenic peptides neuropeptide Y (NPY) and agouti-related peptide (AgRP) (Flier 2004; Morton et al. 2006). Thus, high serum leptin levels can reduce energy intake and stimulate energy expenditure, whereas low serum leptin levels have the opposite effect. Consequently, leptin acts on orexigenic and anorexic neurons to keep body weight in the range of dynamic balance (Schwartz et al. 2000). It has been reported that suppressor of cytokine signaling 3 (SOCS-3), a leptin-inducible inhibitor of leptin signaling and a potential mediator of leptin resistance (Bjorbaek et al. 1998, 1999), may be a mediator of hypothalamic leptin resistance in the aged Wistar rat (Peralta et al. 2002) and pregnant rodents (Tang et al. 2008; Ladyman 2008). Leptin has been shown to lose its function during the fattening period at times of seasonal physiological "obesity" in fat-storing hibernators (Florant et al. 2004). Data on the relationship between leptin and hibernation is scarce (Ormseth et al. 1996; Boyer et al. 1997). Furthermore, neuropeptide changes in the hypothalamus during fattening and hibernation periods are still unclear. We hypothesized that SOCS-3 and other hypothalamic orexigenic and anorexic neuropeptides such as NPY, AgRP, CART, and POMC may be involved in the regulation of energy intake and energy expenditure in fat-storing hibernators during the fattening period, in a way that is different from non-hibernators.

The Daurian ground squirrel (*Spermophilus dauricus*) is a typical seasonal fat-storing hibernator, which can double its body weight during the fattening period (Yuan 2008; Chen et al. 2008). This is mainly due to the accumulation of body fat. We hypothesized that there might be a leptin resistance mechanism during the prehibernation fattening period.

47.2 Material and Method

47.2.1 Animals

Male Daurian ground squirrels (*Spermophilus dauricus*) were captured at Tongliao (122.5°N, 43.9°E), Inner Mongolia, China, in May 2009 and 2010, respectively, and were brought to the laboratory. The animals were kept at room temperature (25 ± 1°C) and under a natural light cycle. Animals were housed individually in plastic cages (48 × 35 × 20 cm^3) with wood shavings as bedding, and were fed with standard rodent pellets (maintenance diet) supplemented with sunflower seeds, apple pieces, and carrot pieces within the first 2 weeks of captivity. Water and food were available ad libitum. After animal acclimation, we monitored the changes in body weight once per week. All animals experienced three different stages before hibernation: (1) the "pre-fattening stage" where body weights remained stable, (2) "fattening stage" where body weights increased significantly compared to the "pre-fattening period", and (3) "post-fattening stage" or "pre-hibernation stage" where body weights decreased. When body weight began to decrease and the animals started to show torpor behavior, characterized by a curving body, motor retardation, and hypothermia, they were transferred to a dark room with constant temperature (5 ± 1°C).

47.2.2 Experimental Protocols

The animals were sacrificed before fattening (June, $n = 7$), during fattening (July, $n = 7$), before hibernation (August, $n = 7$), and at torpid state during hibernation (December, $n = 7$), respectively. The animals were anaesthetized by CO_2, and blood samples were collected immediately after they were sacrificed by decapitation. Trunk blood was collected and centrifuged at 2,500 r/min for 30 min at 4°C. Serum was collected and stored at −75°C before leptin detection. The hypothalami of ground squirrels were removed immediately according to the method described by Tang et al. (2008). After the collection of trunk blood, the hypothalami were immediately frozen in 0.6 ml TRIzol® reagent (Cat. No. 15596-026, Invitrogen, USA) and liquid nitrogen and stored at −75°C.

47.2.3 Measurement of Body Weight and Food Intake

Body weight was measured every week before hibernation and in 2-week intervals after hibernation. Food intake was measured in 3-day intervals before hibernation and at each interbout arousal.

47.2.4 Serum Leptin Assays

Serum leptin levels were measured by radioimmunoassay (RIA) with a ^{125}I multi-species kit (Cat. No. XL-85K, Linco Research Inc., St Charles, MO, USA), a method that has been validated previously in Daurian ground squirrels (Chen et al. 2008).

47.2.5 Primer Design

Using the recorded gene sequences of the Norway rat (*Rattus norvegicus*), and the house mouse (*Mus musculus*) in NCBI, and the thirteen-lined ground squirrel (*Ictidomys tridecemlineatus*) in Ensembl Genome Browser, we designed primers for Ob-Rb, SOCS-3, NPY, AgRP, POMC, and CART as the target genes, and β-actin as the reference gene. The nucleotide fragments were confirmed as the target genes using homology analysis. Species-specific primers were designed based on the cloned sequences.

47.2.6 Isolation of Total RNA and RT-QPCR Protocols

Total RNA was isolated from hypothalami of Daurian ground squirrels using TRIzol Reagent (Cat. No. 15596-026, Invitrogen, USA). To remove the contaminating gDNA, the samples were treated with DNase I (Cat. No. D2215, TaKaRa, Japan) for 30 min at 37°C. We used phenol/chloroform/isoamyl alcohol extraction to eliminate residual DNase I. The A260/280 ratio of total RNA was measured using a UV-spectrophotometer, and found to be approximately 1.8. An equal amount (400 ng) of total RNA was transcribed into first strand cDNA for each sample using a reverse transcription kit (Cat. No. DRR037A, TaKaRa, Japan) according to the manufacturer's instructions. The cDNA was subjected to real-time PCR amplification using the SYBR Green I qPCR kit (Cat. NO. DRR041A, TaKaRa, Japan) and the Rotor Gene 3000 quantitative PCR system (Corbett Research, Australian). Real-time RT-PCR reaction volume used was 25 μl and contained 0.5 μl 10 mM forward and reverse primers (0.2 μmol/l), 2 μl cDNA templates, 12.5 μl 2 × SYBR® Premix Ex Taq™, and 9.5 μl 0.1% DEPC treated water. The cycling parameters consisted of a 5 min pre-denaturation at 95°C followed by 45 cycles of 10 s denaturing at 95°C, 15 s annealing at 58°C, and 20 s extension at 72°C. After thermal cycling, melting was performed at 58–95°C. A non-template control (NTC) was run with every assay. The specific PCR products were separated on a 2% agarose gel (Biowest) and detected using ethidium bromide (EB) under UV illumination.

The standard curves for each gene were constructed using serial dilutions of cDNA ($1–2^{10}$-fold dilutions for NPY and CART, and $1–2^{4}$-fold dilutions for all other genes, which can cover the range of unknowns). Analysis of the standard curves of target genes and β-actin showed that they had similar amplification efficiency, which ensures the validity of the comparative C_q method (Bustin et al. 2009; Nolan et al. 2006).

47.2.7 *Statistical Analysis*

Differences between groups in body fat mass, serum leptin level, and mRNA levels of hypothalamic Ob-Rb, NPY, AgRP, POMC, and CART peptides were assessed by one-way ANOVA followed by the Tukey post hoc test. The data for AgRP gene expression did not follow a normal distribution. We transformed the data by SQRT and used Pearson correlation analysis to examine the relationships among serum leptin level, body fat mass, and expression level of the various genes. The statistics software SPSS 13.0 (SPSS, Chicago, IL) was used for all data analysis. Data were expressed as mean ± the standard error of the mean (SE). The level of statistical significance was set at $P < 0.05$.

47.3 Results

47.3.1 *Body Weight and Food Intake Changes*

The body weights of the Daurian ground squirrels increased sharply during the fattening period. The body weights in July (296.62 ± 13.80 g) were significantly higher than those in June (182.9 ± 7.98 g, $P < 0.01$, Fig. 47.1a). After the fattening period, body weights decreased but not significantly compared with those

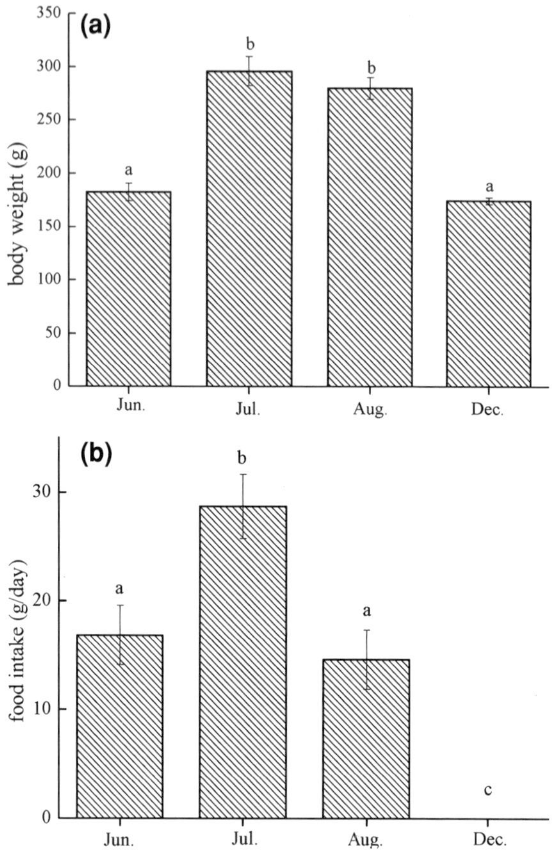

Fig. 47.1 Changes in body weight (**a**) and food intake (**b**) of *S. dauricus* in different fattening stages and during hibernation. Significant differences are indicated by differing letters

in July ($P < 0.05$). During hibernation, body weights decreased to levels seen before fattening (Fig. 47.1a). Animals increased their food intake during the fattening period and reached peak body weight in July. After the fatting period, food intake dropped to levels seen before fattening. Animals did not eat during deep hibernation (Fig. 47.1b).

47.3.2 Serum Leptin Levels

Serum leptin concentration increased during the fattening period coincident with body weight and food intake increases. Serum leptin concentration reached its peak in August (Fig. 47.2), whereas peak body weight and food intake was reached in July (Figs. 47.1 and 47.3a). Serum leptin levels were positively correlated with body weight ($r = 0.643$, $P = 0.001 < 0.01$) (Fig. 47.3b).

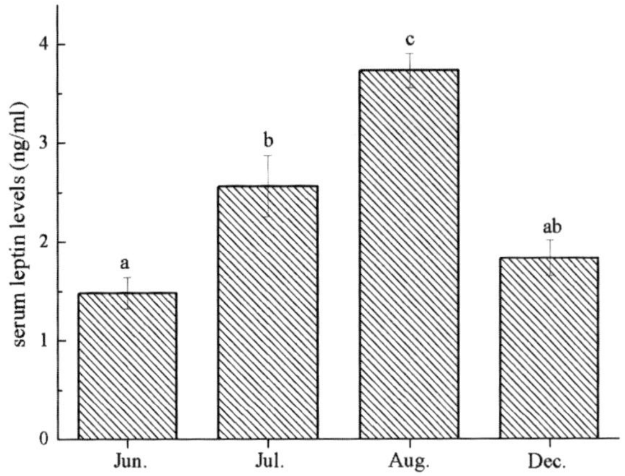

Fig. 47.2 Serum leptin levels of *S. dauricus* in different stages of fattening period and during hibernation

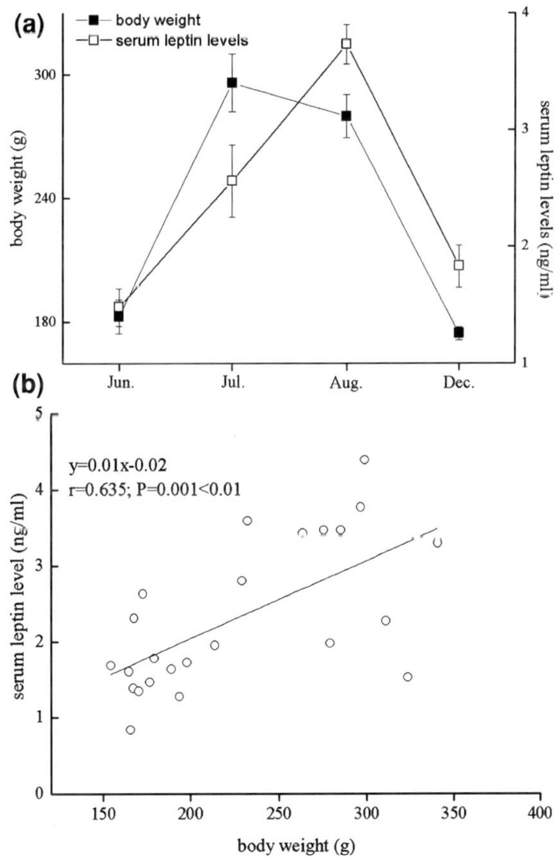

Fig. 47.3 Relationships between serum leptin levels and body weight of *S. dauricus* in different fattening stages and during hibernation

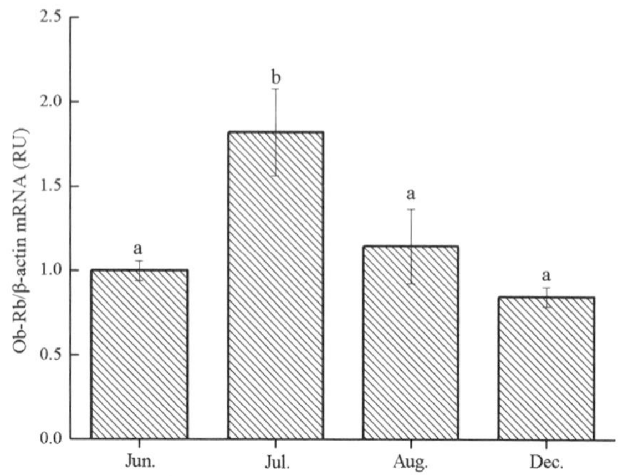

Fig. 47.4 Ob-Rb gene expressions in the hypothalamus in different stages of fattening period and during hibernation in *S. dauricus*

47.3.3 Gene Expression Level of Ob-Rb and Peptides Related to Energy Balance and Food Intake in the Hypothalamus

To investigate changes of gene expression related to energy balance and food intake in the hypothalamus during the fattening period and hibernation, we collected the hypothalami of ground squirrels in June, July, August, and December. This represents the different stages of the fattening period and hibernation. We found that the mRNA of Ob-Rb, which plays an important role in leptin regulation, increased significantly in July ($P < 0.01$, $F_{(3,22)} = 6.057$). However, the mRNA levels of Ob-Rb in August and December did not change significantly compare with June (Fig. 47.4).

The mRNA levels of NPY decreased during the fattening period. There was an increase in NPY after fattening and in the hibernation stage, but this was not significant (Fig. 47.5a). AgRP mRNA levels did not change significantly during the fattening period. However, AgRP mRNA levels significantly increased after the fattening period and significantly decreased during hibernation compared to the post-fattening period ($P < 0.01$, $F_{(3,\ 21)} = 10.449$) (Fig. 47.5b). The mRNA level of CART was elevated during the fattening period and gradually decreased thereafter (Fig. 47.5c). The mRNA level of POMC continued to rise before hibernation. There was a non-significant trend toward an increase in POMC level between June and July ($P = 0.311$, df $= 10$, $t = 1.067$). The peak expression of POMC was reached in August (Fig. 47.5d).

Figure 47.6 shows the mRNA level of SOCS-3, the leptin-inducible inhibitor of leptin signaling, which was approximately three times higher in August than in other months.

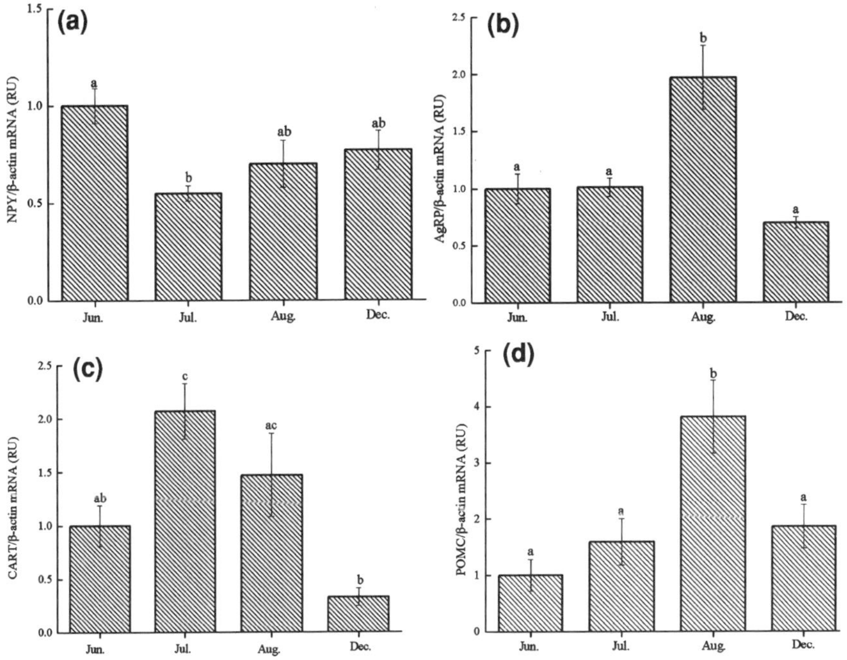

Fig. 47.5 Gene expressions of NPY (**a**), AgRP (**b**), POMC (**c**), and CART (**d**) in the hypothalamus in different stages of fattening period and during hibernation in *S. dauricus*

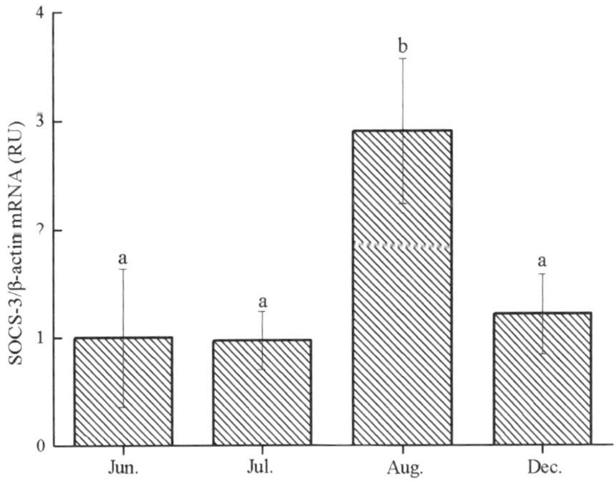

Fig. 47.6 SOCS-3 gene expressions in the hypothalamus in different stages of fattening period and during hibernation in *S. dauricus*

47.4 Discussion

In this study, we found that during the pre-hibernation fattening period the expression of Ob-Rb and, anorectic peptide, CART increased, while the expression of orexigenic NPY decreased with increasing serum leptin concentration. However, these orexigenic changes in leptin concentration and peptide expression did not reduce food intake or body weight in ground squirrels during the fattening period. Consequently, food intake and body weight all reached their yearly peak values during the fattening period.

Pre-hibernation fattening is important for fat-storing hibernators because fat is the main source of energy throughout the hibernation season (Boyer and Barnes 1999; Carey et al. 2003). During the fattening period, animals accumulate energy by increasing food intake (Concannon et al. 2001). As a typical fat-storing hibernator the Daurian ground squirrel begins a fattening period before hibernation. In this study, the food intake of Daurian ground squirrels was significantly increased, and body weights almost doubled during the fattening period. The increase in serum leptin level was followed by increases in body weight and food intake. This has been shown to occur in other hibernators in previous reports of seasonal changes of leptin and body weight (Concannon et al. 2001). Although serum leptin levels in the fattening period increased sharply, food intake did not decrease. The change in gene expression of Ob-Rb, NPY, and CART indicates that leptin can regulate the expression of these genes during fattening. However, these changes in gene expression did not regulate food intake and body weight at the level of the individual organism. This paradoxical phenomenon does not support the reported effects of leptin on body weight and food intake (Münzberg et al. 2005; Schwartz et al. 2000). However, it may be the mechanism responsible for prehibernation fattening. There may also be other signals used to regulate food intake and energy balance during the fattening stage, such as insulin and ghrelin. Both hormones have been shown to be involved in the regulation of food intake in hibernators (Healy et al. 2010; Gluck et al. 2006; Martin 2008). An alternative hypothesis is that food intake was not regulated by the hypothalamus to ensure energy accumulation during the fattening stage.

After the fattening period, animals prepared to enter hibernation. Leptin levels reached peak value, and food intake and body weight decreased. The high levels of serum leptin may play an important role in the decrease of food intake and body weight. Exogenous leptin injection can inhibit pre-hibernation hyperphagia in arctic ground squirrels (*Urocitellus parryii*) (Ormseth et al. 1996). As a signal of energy storage, a high serum leptin level might be a trigger of hibernation because ample energy storage is essential for survival over winter (Boyer and Barnes 1999; Carey et al. 2003; Humphries et al. 2003). In Siberian hamsters (*Phodopus sungorus*), short light can increase leptin sensitivity (Klingenspor et al. 2000) and gene expression of AgRP, but decrease the gene expressions of POMC and Ob-Rb (Mercer et al. 2000). In our study, Ob-Rb expression was decreased and AgRP gene expression was increased after the fattening period in autumn, compared to

July. This suggests that Ob-Rb and AgRP peptides may be regulated by short light in autumn. At the same time, POMC expression increased sharply. The relatively high expression of POMC before hibernation might be one factor leading to reduced food intake. To be effective in reducing food intake, POMC has to be cleaved into different peptide products. A key product is α-melanocyte-stimulating hormone (α-MSH), which can bind to MC4-receptors to reduce food intake and increase energy expenditure (Schwartz et al. 2000; Flier 2004). Chronic CNS administration of AgRP decreases oxygen consumption and decreases the capacity of brown adipose tissue (BAT) to expend energy (Small et al. 2003). Hibernators have to reduce their metabolic rate to prepare to enter hibernation (Boyer and Barnes 1999). Therefore, the high expression level of AgRP might play an important role in the depression of metabolic rate before hibernation. There was no difference in the gene expression of SOCS-3 between the fattening period and hibernation stage. It was highest after fattening or before hibernation in August coinciding with the peak in serum leptin level. This is in line with previous data showing that SOCS-3 is a leptin-inducible neuropeptide (Bjorbaek et al. 1998, 1999). The high expression level of SOCS-3 did not reduce signaling by leptin to promote food intake and body weight increases before hibernation.

Gene expression levels during the hibernation stage were all decreased to prefattening levels. Although animal physical activity indexes were at nadir during hibernation, the expression level of genes in the hypothalamus was that of an active stage. However, protein translation was inhibited in hibernation and mRNA, therefore, may not be effective during hibernation.

Limited data on the relationship between leptin and hibernation are available at present, especially for the fat-storing hibernators. Our data suggest that neuropeptides in the hypothalamus might participate in the regulation of energy intake and energy expenditure before and after the fattening period. We investigated the mechanisms of regulation on energy balance and food intake in CNS in mRNA level; the precise changes in protein level are presently unclear but need to be determined to get more insight in these mechanisms.

Acknowledgments Thanks to Professor Dehua Wang and Dr. Gangbin Tang, Institute of Zoology of the Chinese Academy of Sciences, for their helpful suggestions. This study was supported by grants from the National Natural Science Foundation of China (No.30770337), (No.31170380).

Appendix

The qRT-PCR transcripts of neuropeptides in hypothalamus of Daurian ground squirrel

1: Ob-Rb

AAACGTTACAGCTGAGATCTTGAGCAGGATTCTGCCTGTATTAGTGA
CCAGGGCAACAGCCCTCACATCTCTGAGGCTGGGAGCATCCCGGCTGT

CTGTGAGGACGAAAGCCAGAGACAACCCTCAGTTAAATACGCCACAC
TGGTCAGTAATTCCAAGTCAAGCGAAACGGATGACC

2: AgRP
TACTCGTCCTAATGTGCGCCTCTTCTGAGCGATCTACAGGCCCTTCTT
TAGCAGAGCTTGCAGGACTAATCCTGGGGCAGACATGTGAGGACAGGA
CCCCCATGCACAATCCCAGGGCTGGGATTGGGAAGGGGGAACGGAGG
GATTATGCCAGA

3: NPY
CCGAAGCATACGAGCTACTCTCCAGCCGGACACCCGGGAGAGGACG
CTCCAGCGGAGGACATGGCCAGATACTACTCGGCGCTGCGACACTACA
TCAACCTCATCACCAGGCAGAGATACGGCAAGAGAA

4: POMC
CCAAGCGTTACGTGTAGACTCACCACGGAAGCACCTGCTGGTATGT
GGGCCTTGGGTGCCACCTCTGGGCATCAGATGGGA

5: CART
CAGTCGCGGGCTGGCAGGACAGTCCTATAATCTGTGTCGGAGGAGT
ATTAACTGAAGATGAATTCTTCACCTAGAGAAGAATTCCCTGAGTTGG
AAGAGACCTTGAATCTCGA

6: SOCS-3
TGCACCTTCTCTACGTGGCCACCCTTCAGCATCTCTGTCGGAGACGG
TCAATGGACACCTGGACA

7: *β*-actin
AGTGTCAGTAGGCGTGATAGTGAGGCCAGGATGGAGCCACCGATCC
ACACAGAGTACTTGCGCTCAGGGGGAGCGATAATCTTGATCTTCATGG
TGCTGGGTGCCAGGGCTGTGATCTCCTTCTGCATCCTGTCAGCGATGCC
TGGGTACATGGTGGTGCCGCCCGACAGCACGGTGTTGGATAGAGGTCT
TAGA

References

Andrews MT (2007) Advances in molecular biology of hibernation in mammals. BioEssays 29:431–440

Banks WA, Kastin AB, Huang W, Jaspan JB, Maness LM (1996) Leptin enters the brain by a saturable system independent of insulin. Peptides 17(2):305–311

Bjorbaek C, Elmquist J, Frantz D, Shoelson SE, Flier JS (1998) Identification of SOCS-3 as a potential mediator of central leptin resistance. Mol Cell 1:619–625

Bjorbaek C, El-Haschim K, Frantz JD, Flier JS (1999) The role of SOCS-3 in leptin signaling and leptin resistance. J Biol Chem 274(42):30059–30065

Boyer BB, Barnes BM (1999) Molecular and metabolic aspects of mammalian hibernation. Bioscience 49(9):713–723

Boyer BB, Ormseth OA, Buck L, Nicolson M, Pelleymounter MA, Barnes BM (1997) Leptin prevents post hibernation weight gain but dose not reduce energy expenditure in arctic ground squirrels. Comp Biochem Physiol 118(3):405–412

Buck CL, Barnes BM (1999) Annual cycle of body composition and hibernation in free-living arctic ground squirrels. J Mammal 80:1264–1276

Bustin SA, Benes V, Garson JA et al (2009) The MIQE guildelines: minimun information for publication of quantitative real-time PCR experiments. Clin Chem 55(4):611–622

Carey HV, Andrews MT, Martin SL (2003) Mammalian hibernation: cellular and molecular responses to depressed metabolism and low temperature. Physiol Rev 83:1153–1181

Chen JF, Yang M, Zhong WQ et al (2008) Seasonal changes in body mass, serum leptin concentration and UCP1 content in Daurian Ground Squirrels (*Spermophilus dauricus*). In: Lovegrove BG, McKechnie AE (eds) Hypometabolism in animals: torpor, hibernation and cryobiology. University of KwaZulu-Natal, Pietermaritzburg

Concannon P, Levac K, Rawson R, Tennant B, Bensadoun A (2001) Seasonal changes in serum leptin, food intake, and body weight in photoentrained woodchucks. Am J Physiol Regulatory Integrative Comp Physiol 281:951–959

Flier JS (2004) Obesity war: molecular progress confronts an expanding epidemic. Cell 116:337–350

Florant GL, Porst H, Peiffer A, Hudachek SF, Pittman C, Summers SA, Rajala MW, Scherer PE (2004) Fat-cell mass, serum leptin and adiponectin changes during weight gain and loss in yellow-bellied marmots (*Marmota flaviventris*). J Comp Physiol B 174:633–639

Gluck EF, Stephens N, Swoap SJ (2006) Peripheral ghrelin deepens torpor bouts in mice through the arcuate nucleus neuropeptide Y signaling pathway. Am J Physiol Regul Integr Comp Physiol 291:1303–1309

Healy JE, Ostrom CE, Wilkerson GK, Florant GL (2010) Plasma ghrelin concentrations change with physiological state in a sciurid hibernator (*Spermophilus lateralis*). Gen Comp Endocrinol 166(2):372–378

Heldmaier G, Ortmann S, Elvert R (2004) Natural hypometabolism during hibernation and daily torpor in mammals. Respir Physiol Neurobiol 141:317–329

Humphries MM, Thomas DW, Kramer DL (2003) The role of energy availability in mammalian hibernation: a cost-benefit approach. Physiol Biochem Zool 76(2):165–179

Klingenspor M, Niggemann H, Heldmaier G (2000) Modulation of leptin sensitivity by short photoperiod acclimation in the Djungarian hamster, *Phodopus sungorus*. J Comp Physiol B 170:37–43

Ladyman SR (2008) Leptin resistance during pregnancy in the rat. J Neuroendocrin 20:269–277

Martin SL (2008) Mammalian hibernation: a naturally reversible model for insulin resistance in man? Diab Vasc Dis Res 5(2):76–81

Mercer JG, Moar KM, Ross AW, Morgan PJ (2000) Regulation of leptin receptor, POMC and AGRP gene expression by photoperiod and food deprivation in the hypothalamic arcuate nucleus of the male Siberian hamster (*Phodopus sungorus*). Appetite 34:109–111

Michener GR (1992) Sexual differences in over-winter torpor patterns of Richardson's ground squirrels in natural hibernation. Oecologia 89:397–406

Morton GJ, Cummings DE, Baskin DG, Barsh GS, Schwartz MW (2006) Central nervous system control of food intake and body weight. Nature 443:289–295

Münzberg H, Björnholm M, Bates SH, Myers MG Jr (2005) Leptin receptor action and mechanisms of leptin resistance. Cell Mol Life Sci 62:642–652

Nolan T, Hands RE, Bustin SA (2006) Quantification of mRNA using real-time RT-PCR. Nat Protoc 1(3):1559–1582

Ormseth OA, Nicolson M, Pelleymounter MA, Boyer BB (1996) Leptin inhibits prehibernation hyperphagia and reduces body weight in arctic ground squirrels. Am J Physiol Regul Integr Comp Physiol 271:1775–1779

Peralta S, Carrascosa JM, Gallardo N, Ros M, Arribas C (2002) Ageing increases SOCS-3 expression in rat hypothalamus: effects of food restriction. Biochem Biophys Res Commun 296:425–428

Schwartz MW, Woods SC, Porte D Jr, Seeley RJ, Baskin DG (2000) Central nervous system control of food intake. Nature 404:661–671

Small CJ, Liu YL, Stanley SA, Connoley IP, Kennedy A, Stock MJ, Bloom SR (2003) Chronic CNS administration of Agouti-related protein (Agrp) reduces energy expenditure. Int J Obes Relat Metab Disord 27:530–533

Tang GB, Cui JG, Wang DH (2008) Hypothalamic suppressor-of-cytokine -signalling 3 mRNA is elevated and pro-opiomelanocortin mRNA is reduced during pregnancy in Brandt's Voles (*Lasiopodomys brandtii*). J Neuroendocrinol 20:1038–1044

Yuan LL (2008) A study on energy metabolism characteristics and hibernation patterns in Daurian ground squirrel (*Spermophilus dauricus*). Dissertation, Shenyang Normal University, Shenyang (Liaoning)

Zhang Y, Proenca R, Maffei M et al (1994) Positional cloning of the mouse obese gene and it's human homologue. Nature 372:425–432

Index

DOI: 10.1007/978-3-642-28678-0, © Springer-Verlag Berlin Heidelberg 2012

Printed by Printforce, the Netherlands